Unconventional Starch Sources for Sustainable Bioplastics

With growing environmental concerns and rising consumer demands for sustainable alternatives, industries are increasingly shifting toward eco-friendly materials. Traditional petroleum-based plastics are being re-evaluated due to their harmful ecological footprint, leading to a surge of interest in biopolymers. Among these, starch-based bioplastics have emerged as a promising solution due to their biodegradability, renewability, and reduced environmental impact. As the urgency for greener materials intensifies, there is a pressing need for comprehensive resources that explore both conventional and novel approaches to bioplastic development.

Unconventional Starch Sources for Sustainable Bioplastics addresses this need by offering in-depth insights into the latest advancements in starch-based bioplastics, with a unique focus on underutilized and nontraditional starch sources. Going beyond common crops, the book explores innovative materials such as marine, seaweeds, underutilized seeds, stems, tubers, and industrial by-products. It examines starch extraction techniques, processing methods, challenges in scalability, and the environmental implications of these alternatives. This book illustrates how these novel materials are being integrated into sectors such as packaging, agriculture, and biomedicine. Bridging scientific principles with practical applications, this book serves as a vital guide for researchers and students, as well as professionals in biopolymers, materials science, environmental science, food science, and related disciplines contributing meaningfully to the global pursuit of sustainable solutions.

Unconventional Starch Sources for Sustainable Bioplastics

Edited by
Pinki Saini and Mazia Ahmed

CRC Press
Taylor & Francis Group
Boca Raton London New York

CRC Press is an imprint of the
Taylor & Francis Group, an **informa** business

Designed cover image: Shutterstock

First edition published 2026
by CRC Press
2385 NW Executive Center Drive, Suite 320, Boca Raton, FL 33431

and by CRC Press
4 Park Square, Milton Park, Abingdon, Oxon, OX14 4RN

CRC Press is an imprint of Taylor & Francis Group, LLC

ISBN: 9781032878720 (hbk)
ISBN: 9781032878737 (pbk)
ISBN: 9781003534914 (ebk)

DOI: 10.1201/9781003534914

Typeset in Times
by Newgen Publishing UK

Dedication

To our mothers – whose unconditional love, sacrifices, and unwavering support have been the foundation of all our achievements.

Contents

Preface

There is a growing awareness among consumers about the environmental impact of products. The demand for eco-friendly and sustainable alternatives has driven industries to explore biopolymers as a way to meet consumer expectations and address environmental concerns. As the demand for eco-conscious alternatives to traditional plastics intensifies, this book serves as a valuable resource for professionals seeking in-depth insights into the latest developments in starch-based bioplastics. Authored by experts in the field, the book covers a wide range of topics, from the fundamentals of conventional starch-based polymers, to novel sources along with their diverse applications in packaging, food, agriculture, and biomedical fields. Readers will gain a profound understanding of the sustainable solutions offered by unconventional starch-based bioplastics, including their biodegradability, renewability, and reduced environmental impact.

Unconventional Starch Sources for Sustainable Bioplastics will specifically delve into the unexplored frontiers of material science, redefining the landscape of bioplastics. The book ventures beyond the traditional boundaries of starch-rich materials, navigating through the alternatives of fruit waste, millets, aquatic plants, fungi, sea weeds, underutilized seeds, industrial byproducts, and more. With a focus on practicality, the book delves into extracting starch from these lesser-known materials. It provides insights into the processing techniques, challenges, and the feasibility of adopting these unconventional sources on a larger scale. Case studies illustrate how these innovative materials are being successfully integrated into various industries, showcasing the potential for widespread adoption. With a focus on both the scientific principles and practical implementations, the book bridges the gap between theory and application.

This book will serve as a crucial contribution to society by addressing the urgent need for environmentally friendly alternatives to traditional plastics. In a world grappling with plastic pollution and ecological degradation, this book provides insights and solutions that hold the potential to positively impact various facets of our society. The book is appropriate for researchers, advanced students, and industry professionals working and studying in the areas of biopolymers, bioplastics, polymer sciences, polymer chemistry, materials science, food science, environmental sciences, microbial prospecting, nanotechnology, and biochemical engineering. The book will serve as an indispensable guide for anyone looking for a comprehensive overview of the subject.

Pinki Saini
Mazia Ahmed

Acknowledgments

First and foremost, we extend our heartfelt thanks to our esteemed contributors, whose valuable insights, research, and scholarly inputs have enriched the content and academic merit of this volume. Their dedication and commitment to excellence have been instrumental in shaping this book.

We are thankful to our families and friends for their constant encouragement and understanding, which kept us motivated during this endeavor.

We also acknowledge the efforts of the publishing team for their professionalism, patience, and attention to detail during the production of this book.

This book is a product of collective effort and intellectual collaboration, and we sincerely hope it contributes meaningfully to the academic community.

Pinki Saini
Mazia Ahmed

About the Editors

Pinki Saini is presently working as Associate Professor at the Centre of Food Technology, University of Allahabad. She completed her MSc in Food Science and Technology in 2005 and PhD in 2008 from G. B. Pant University of Agriculture and Technology, Pantnagar, Uttarakhand. She has 16 years of teaching and 18 years of research experience. She has supervised six PhD theses and 45 MSc dissertations. She has published more than 75 research/review papers in peer-reviewed journals and conferences. She has a Council of Scientific & Industrial Research (CSIR)-funded ongoing project: "Folate fortification of selected millets through probiotic fermentation." She has completed two sponsored research projects funded by Uttar Pradesh Council of Agricultural Research and has been granted one patent (Patent no. 402984) by the Indian Patent Office. She has received an accession number MCC 4716 for depositing novel isolated strain as *Lactiplantibacillus plantarum* sp. strain ANPSCRLB at NCMR (Pune, Maharashtra). She has authored a book entitled *Bioprospecting of Microbial Resources for Agriculture, Environment and Bio-Chemical Industry* published by Springer. She has edited two books: *Food and Industrial Bioprocessing* published by New India Publishing Agency, India, and *Biosorption Processes for Heavy Removal* published by IGI Global, USA. She also has more than 20 book chapters to her credit in national and international published books. Her research interests include bioprocessing of foods, fermentations studies, probiotics, biosorption of heavy metals, biofilm formation, and sustainable polysaccharide-based edible films from plant and microbial sources.

Mazia Ahmed has successfully completed her PhD in Nutritional Sciences from the Centre of Food Technology, University of Allahabad, Prayagraj, India. Her doctoral research focused on microbial cellulose production and its innovative applications in food packaging systems. She holds a Master's degree in Nutritional Sciences from the Centre of Food Technology, University of Allahabad (2018), and gained valuable practical experience as a dietetics intern at Sir Ganga Ram Hospital, New Delhi. Her professional experience includes serving as an analyst at the NABL-accredited Food Analysis and Research Laboratory (FARL). Dr Ahmed has demonstrated her expertise through various publications, including research papers, review articles, and book chapters. Additionally, she is the author of her debut nonacademic book, *Eulogy for the Color Lavender*, published in 2020.

Contributors

E. I. O. Ajayi
Metabesity Complications and Infectious Diseases-Related Movement Disorders Research Group, Computation Membrane Biochemistry and Nanobiotechnology Unit, Department of Biochemistry, Faculty of Basic and Applied Sciences, College of Science, Engineering and Technology
Osun State University, Osogbo, Nigeria
The STREEAAMMIE Centre, Ebelola Bioenergetic Solutions
Kajola Ijesa, Osun State, Nigeria

Sk Asraful Ali
Division of Agronomy
ICAR-Indian Agricultural Research Institute
New Delhi, India

S. K. Ayiti
Biomedical Sciences Division
Osun State University Teaching Hospital
Osogbo, Nigeria

Sahith Battula
Department of Mechanical Engineering, VNR Vignana Jyothi Institute of Engineering and Technology, Hyderabad, India
and
Ruhr-Universität Bochum, Bochum, Germany

Edla Maria Bezerra Lima
EMBRAPA Food Technology
Rio de Janeiro, Brazil

Aashita Bhardwaj
Department of Food Science and Technology
National Institute of Food Technology Entrepreneurship and Management
Sonipat, India

Yograj Bist
Department of Biotechnology, Graphic Era (Deemed to be University),
Clement Town, Dehradun 248002, India

Eliseo Cárdenas-Hernández
Facultad de Ciencias Químicas
Universidad Autónoma de Coahuila
Saltillo, México

Ana Carolina Corrêa
Graduate Program in Materials Science and Engineering
Universidade Federal de São Carlos (UFSCar)
São Paulo, Brazil

Jyotsna Cherukuri
Department of Chemistry, VNR Vignana Jyothi Institute of Engineering and Technology
Hyderabad, India

Poonam Choudhary
ICAR-Central Institute of Post-Harvest Engineering and Technology
Ludhiana, India

Subhankar Das
Biotechnology Unit, Mangalore University
Mangalore, India

H. O. Faleke
Department of Chemistry and Biochemistry
Texas Tech University
Texas, USA

Luis Fernando Flores
Wikimedistas de Jujuy, San Salvador de Jujuy, Jujuy, Argentina

K. Haripriya
Department of Mechanical Engineering
VNR Vignana Jyothi Institute of Engineering and Technology
Hyderabad, India

Gitanjali Jothiprakash
School of Computing, Engineering & the Built Environment
Edinburgh Napier University
Edinburgh, UK

Manjula Ishwara Kalyani
Department of Microbiology, Mangalore University
Mangalore, India

Sumedha Arora Kapoor
Agricultural Structure and Environment Control Division
ICAR-Central Institute of Post-Harvest Engineering and Technology
Ludhiana, India

P. Karthik
Centre for Food Nanotechnology
and
Department of Food Technology
Karpagam Academy of Higher Education
Coimbatore, India

Subburamu Karthikeyan
Centre for Post-Harvest Technology
Agricultural Engineering College and Research Institute, Tamil Nadu Agricultural University
Coimbatore, India

Mine Kocyigit
Department of Pharmaceutical Botany
Faculty of Pharmacy, Istanbul University
Türkiye

Yogesh Kumar
Department of Food Science and Technology, Graphic Era (Deemed to be University)
Clement Town, Dehradun, India

Sapna Kumari
Department of Food Science and Technology
National Institute of Food Technology Entrepreneurship and Management
Sonipat, India

Sandeep Mann
Agricultural Structure and Environment Control Division
ICAR-Central Institute of Post-Harvest Engineering and Technology
Ludhiana, India

José Manoel Marconcini
National Nanotechnology Laboratory for Agriculture (LNNA)
EMBRAPA Instrumentation
São Carlos, Brazil

Antonieta Middea
Centre for Mineral Technology (CETEM)
Rio de Janeiro, Brazil

Mohit Nagar
Department of Food Engineering and Technology
Sant Longowal Institute of Engineering and Technology
Longowal, India

C. D. Naidu
Department of Electronics and Communications
VNR Vignana Jyothi Institute of Engineering and Technology
Hyderabad, India

Reiner Neumann
Centre for Mineral Technology (CETEM)
Rio de Janeiro, Brazil

A. Nkumah
University of Ibadan, Nigeria
University of Tübingen, Germany

B. O. Olatunbosun
Osun State University
Osogbo, Nigeria

Renata Nunes Oliveira
Chemical Engineering Department
Universidade Federal Rural do Rio de Janeiro (UFRRJ)
Rio de Janeiro, Brazil

Surabhi Pandey
Department of Food Technology
Harcourt Butler Technical University
Kanpur, India

Ravi Prakash
Agricultural Structure and Environment Control Division
ICAR-Central Institute of Post-Harvest Engineering and Technology
Ludhiana, India

R. Rajam
Department of Food Technology
Kalasalingam Academy of Research and Education
Krishnakoil, India

Rafiq A. Rather
Department of Clinical Biochemistry
Government Degree College Shopian
Jammu & Kashmir, India

R. Sai Nandhini
Department of Biotechnology
Vel Tech Rangarajan Dr. Sagunthala R&D Institute of Science and Technology
Chennai, India

Linganaboina Srinivasa Rao
Centre for Nanoscience and Technology,
VNR Vignana Jyothi Institute of Engineering and Technology
Hyderabad, India

Pinki Saini
Centre of Food Technology
University of Allahabad
Prayagraj, India

Chandrasekaran Sankaranarayanan
Assistant Professor, Department of Biochemistry and Biotechnology, Annamalai University Annamalainagar, Tamil Nadu, India

D. C. Saxena
Department of Food Engineering and Technology
Sant Longowal Institute of Engineering and Technology,
Longowal, India

Shahwar Siddiqui
University of Allahabad
Prayagraj, India

Alice Singh
Department of Food Science and Technology
National Institute of Food Technology Entrepreneurship and Management
Sonipat, India

Anurag Singh
Department of Food Technology
Harcourt Butler Technical University
Kanpur, India

Chetna Singh
Department of Food Science and Technology
National Institute of Food Technology Entrepreneurship and Management
Sonipat, India

Poornima Singh
Department of Food Technology
Harcourt Butler Technical University
Kanpur, India

Akansha Srivastava
Centre of Food Technology, University of Allahabad
Prayagraj, India

Shivangi Srivastava
Department of Food Technology, Harcourt Butler Technical University
Kanpur, India

P. Srujana
Department of Mechanical Engineering
VNR Vignana Jyothi Institute of Engineering and Technology
Hyderabad, India

Tanya L. Swer
Department of Food Science and Technology
National Institute of Food Technology Entrepreneurship and Management
Coimbatore, India

Rossana Mara da Silva Moreira Thiré
Program of Metallurgical and Materials Engineering/COPPE
Universidade Federal do Rio de Janeiro (UFRJ)
Rio de Janeiro, Brazil

T. Venkateshwari
Department of Food Technology
Karpagam Academy of Higher Education
Coimbatore, India

Neelam Yadav
Centre of Food Technology
University of Allahabad
Prayagraj, India

1 Introduction to Starch-Based Bioplastics

Subhankar Das and Manjula Ishwara Kalyani

1.1 INTRODUCTION

With the discovery of Bakelite, a thermosetting phenol-formaldehyde resin, it holds the distinction of being the first entirely synthetic plastic. Developed in 1907 by chemist Leo Baekeland, Bakelite was initially used as an electrical insulator. The use of synthetic plastic has seen an increasing demand in various products, including components of cars, mobiles, telephones, toys, radios, etc. Furthermore, the discovery has paved the way for the development of a wide range of plastics. This boom in production has been accompanied by a staggering rise in plastic waste globally, highlighting the environmental challenges associated with these innovations. Synthetic petroleum-based plastics have several drawbacks, including the release of harmful toxins that can trigger health issues in humans, the insidious formation of micro- and nanoplastics, and their slow decomposition (Danie Kingsley et al., 2020; Klun et al., 2022; Rusdi et al., 2020). Moreover, with the increasing daily use of petroleum-based synthetic plastics, they have not only encircled the surrounding ecosystems but also posed significant threats to aquatic marine life, disrupting delicate aquatic ecosystems and endangering countless species (Figure 1.1). It has been estimated that each year, around 380 million tonnes of plastic are produced globally, with a significant portion ultimately discarded. Therefore, in the recent years, growing environmental awareness has spurred the search for eco-friendly and sustainable substitutes for traditional plastics produced from petroleum (Krishnamurthy & Amritkumar et al., 2019; Flury & Narayan et al., 2021).

Starch, a naturally occurring polysaccharide derived from various renewable resources like pea, corn, yam, potato, tapioca, etc., offers an excellent material for the production of eco-friendly, sustainable bioplastics. Starch-based bioplastics offer a unique blend of functionality and sustainability, making them adaptable for various applications, including packaging materials, disposable cutlery, agricultural films, and biodegradable bags. Starch consists of glucose monomers interconnected by glycosidic linkages. It has two unique components: amylose, an unbranched chain of glucose molecules, and amylopectin, which possesses a branched structure (Figure 1.2). The combination of structures significantly influences the physical properties and functionality of starch across many applications. Utilising waste-derived sources of starch—such as fruits, vegetables, roots, and tubers—not only promotes the valorisation of plant-based food waste but also contributes to reducing global food waste, which amounts to approximately 1.3 billion tonnes each year (Oluwasina et al., 2021; Perotto et al., 2018; Das & Kalyani, 2023). The applications of starch in material science are limited, as its granular structure must be fully broken down through gelatinisation to produce thermoplastic starch (TPS). Starch displays thermoplastic properties when combined with

DOI: 10.1201/9781003534914-1

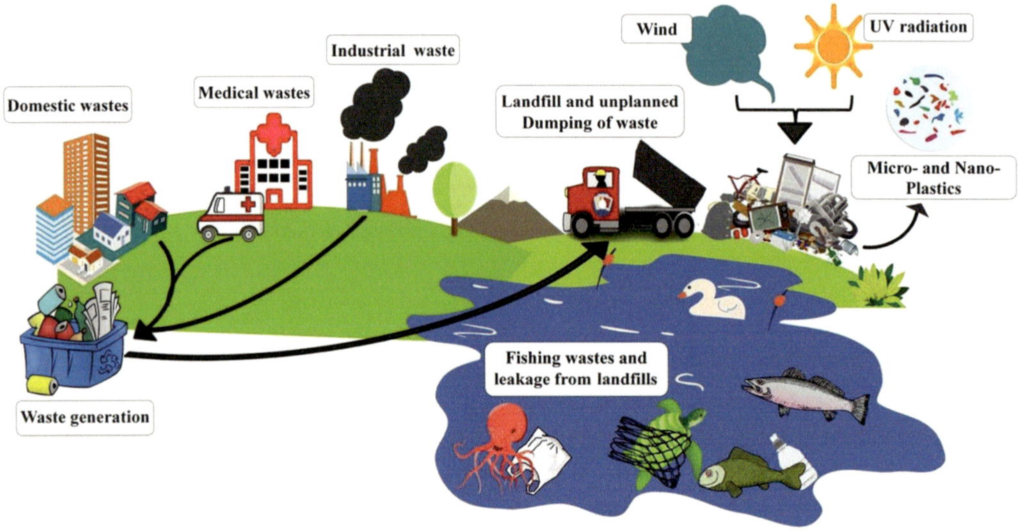

FIGURE 1.1 A diagrammatic illustration showing the effects of fossil-based plastic waste on the environment.

plasticisers, subjected to elevated temperatures, and processed with shear (Alonso-Gonzalez et al., 2021; Zhang et al., 2014). However, starch-based bioplastics exhibit certain limitations, particularly in their mechanical and barrier properties, notably high permeability to gases and water vapour. Furthermore, owing to the hydrophilic characteristics of starch, its moisture content may vary with alterations in relative humidity. Thus, there has been much research whereby the blending of the starch-based bioplastics with other materials can enhance their physiochemical properties (Oluba et al., 2021; Judawisastra et al., 2017; Oluwasina et al., 2019).

Starch-based bioplastics are easily biodegradable, offering advantages over synthetic plastics by reducing landfill needs and eliminating the release of toxic gases during combustion. Biodegradable plastics are often referred to as hydro-biodegradable plastics and are able to break down into water, carbon dioxide (CO_2), methane, and biomass. This decomposition process enhances their environmental appeal, making them a viable alternative to conventional plastics. Despite methane and CO_2 being released in the degradation process, both recognised as greenhouse gases, the majority of researchers deem their cumulative impact on the total concentration of greenhouse gases in the atmosphere to be insignificant, as the source of the starch is from photosynthetic plants that intake CO_2during their lifetime (Maheshwari et al., 2013; Özdamar & Ateş, 2018).

Collectively, this chapter offers a comprehensive insight into starch and its utilisation in the production of bioplastics, highlighting both the progress and the obstacles in this domain. Furthermore, the chapter will delve into the degradation process of the starch-based bioplastics and its role in achieving sustainable goals. Thus, this chapter underscores the significance of starch as a renewable resource in the ongoing quest for environmentally friendly alternatives to traditional plastics.

1.2 BRIEF OVERVIEW ON STARCH-BASED BIOPLASTICS AND ITS COMPONENTS

Starch-based bioplastics are an environmentally friendly substitute for traditional petroleum-derived plastics, providing a viable approach for mitigating environmental effects. Starch-based bioplastics mostly consist of the following components:

1.2.1 Starch

Starch is regarded as the second most abundant biomass generated on Earth. Starch is a primary carbohydrate reserve synthesised by plants during photosynthesis. In plants, starch serves an essential physiological function by aiding in the mitigation of phosphate limitations during photosynthesis. In conditions of low phosphate, starch can be broken down to supply the necessary energy and carbon for metabolic functions. Starch further functions as an energy and carbon store in storage organs, enabling survival during adverse developmental circumstances. Due to its physiochemical properties, starch has a wide range of applications across various industries, including food and pharmaceuticals such as drug delivery systems, hydrogels, bioplastics, bakery, snacks, dairy, alcohol production, adhesives, etc. Starch granules are commonly found in various plant parts like roots, seeds, tubers, leaves, stems, fruits, and pollen. For instance, maize kernels can contain as much as 75% starch, highlighting their role as a major energy reserve (Kossmann & Lloyd 2000; Masina et al., 2017; Zia-ud-Din et al., 2017). Starch is a polysaccharide primarily made up of two types of polymers of D-glucose, viz., amylose, which is linear, and amylopectin, which has a branching structure (as shown in Figure 1.2). One of the main components in the majority of starches is amylopectin.

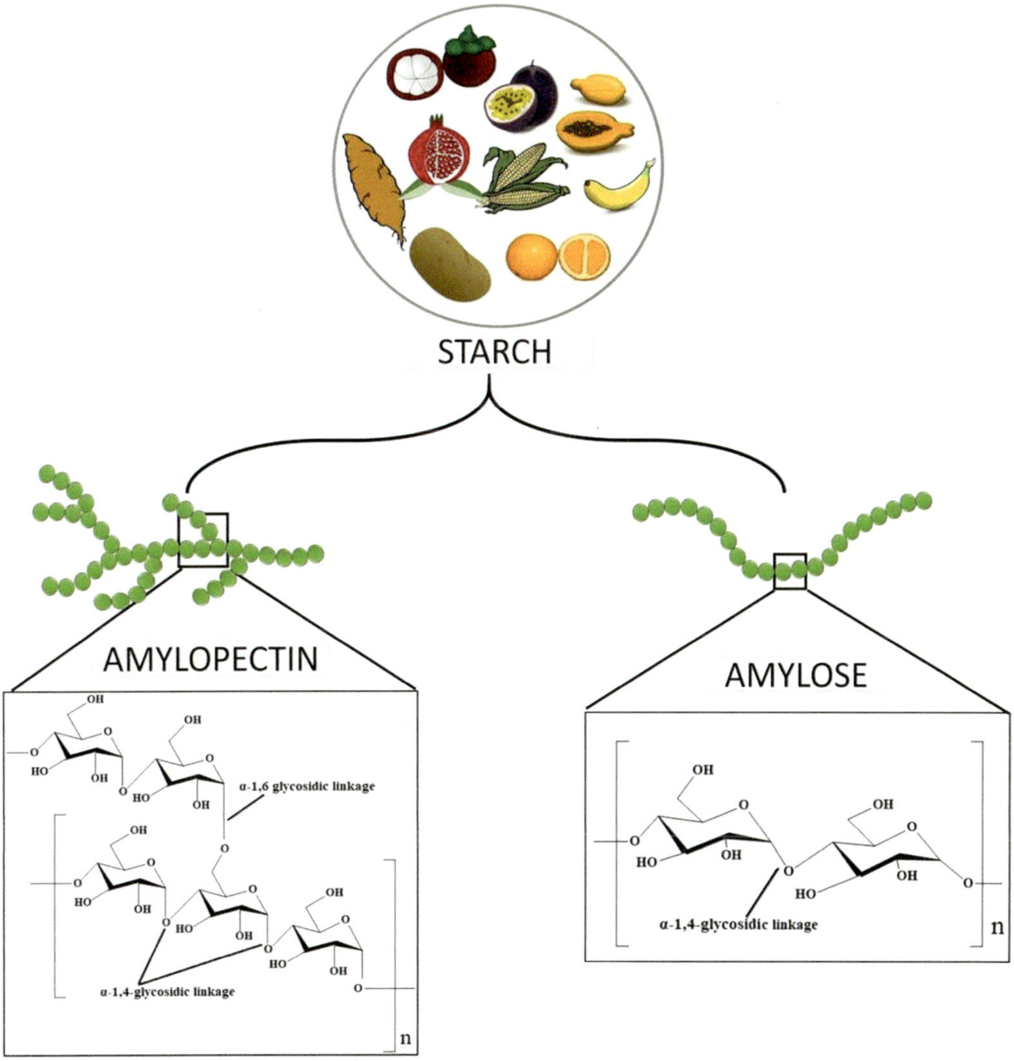

FIGURE 1.2 Schematic representation of starch components and their molecular structure.

Amylopectin constitutes around 70–80% of starch and is organised in a highly branching structure with short chains of D-glucosyl units linked together by α- (1,4)-linkages and α- (1,6)-linkages. On the other hand, amylose typically makes up 20–30% of the starch, depending on the type of plant species, and has a linear structure longer than amylopectin. Starch granules also include traces of proteins, fatty acids, and minerals, which influence their physiochemical characteristics (Vamadevan & Bertoft, 2015; Kong, 2020; Wang et al., 2015; Hu et al., 2016). According to its crystallinity, starch can be further categorised into A-type, B-type, and C-type. The A- and B-type starches consist of A-type and B-type crystallinity, while the C-type starch contains both A- and B-type crystallinity, making it more complex than the former types. The A-type starch contains shorter average branch chains in its amylopectin compared to the B-type starch, which consists of amylopectin with longer branch chains. In contrast, the C-type starch comprises amylopectin, which features both long and short branch chains. Furthermore, amylopectin functions as the structural framework for the crystalline domains within the starch granule (Rolland-Sabaté et al., 2012; He & Wei, 2017).

The amylose to amylopectin ratio in the starch of a certain plant species, combined with the structural makeup of the starch granules, influences qualities such as solubility, gelatinisation temperature, viscosity, gelation, and retrogradation. The starch-based bioplastics were further significantly influenced by these factors. Therefore, there are studies that blend different starches for the production of starch-based bioplastics. There have been approaches to blend two different starches to get a starch-based bioplastics with better physicochemical properties. A study conducted by Sultan and Johari in 2017 examined the combination of maize as a co-biopolymer with banana peels serving as the primary biopolymer (Schirmer et al., 2013; Sultan & Johari, 2017). In another instance, a mixture of 55% corn starch and 45% cassava starch has been combined to create a starch-based bioplastic, incorporating passion fruit waste as a reinforcing additive (Moro et al., 2017). The global annual starch production has reached an impressive 3 billion tonnes, with refined starch alone exceeding 60 million tonnes. Overall, the world starch market has been estimated to be around 48 million tonnes, valued at €15 billion annually, as this surge is fuelled by industrialists and researchers exploring innovative properties and high-value applications, driving a vibrant and evolving market (Dejene & Geletaw, 2024).

Starch is utilised in the manufacture of starch-based bioplastics owing to its cost-effectiveness, widespread availability, and thermoplastic characteristics. Almost 50% of commercially used starch-based bioplastics are derived from starch. In 2018, the global bioplastics market reached a remarkable $3.02 billion, with projections indicating an increase to $12.4 billion by 2027, driven by an escalating need for sustainable materials and innovative approaches (Enwere et al., 2024). When heated, starch becomes soluble in water, causing the granules to enlarge and rupture. Consequently, the semi-crystalline structure of the starch is disrupted, leading to the percolation of small amylose particles from the granule, which forms an interconnected structure. The resulting structure compresses water and enhances the mixture's viscosity. This process is referred to as starch gelatinisation. However, without any reinforcements or additives, bioplastics made of native starch have poor mechanical qualities and cannot be characterised as basic bioplastics, even though the final product has better degradability (Marichelvam et al., 2019; Abe et al., 2021; Gawande et al., 2024). Bananas, a highly demanded fruit worldwide, generate significant waste in the form of peels, pseudostems, etc. The peel of the banana constitutes 35–40% of the fruit. The banana peel serves as a source for starch extraction, which can subsequently be utilised in the production of starch-based bioplastics (Bello et al., 2020; Alcivar-Gavilanes et al., 2022). The study conducted by Azieyanti et al. in 2020 demonstrated the preparation of bioplastics utilising starch derived from banana peel, incorporating additives such as glycerol as a plasticiser and hydrochloric acid. The findings on the bioplastics derived from banana starch further indicated a significant modulus of elasticity (Azieyanti et al., 2020).

In comparison to its natural state, starch may be chemically modified or have its structure changed to improve its physical and mechanical properties. Among the frequent chemical derivatisation processes of starch are oxidation, etherification (hydroxypropylation), and esterification (acetylation). In contrast to oxidised starch-based bioplastics, starch acetylation may produce a more robust and hydrophobic bioplastic. The enhancement in acetylated starch arises from the substitution of hydrophilic groups with acetyl, leading to the development of less flexible and hygroscopic bioplastic characteristics. Furthermore, acetyl has more hydrophobicity than carbonyl and carboxyl groups, perhaps due to the chemical composition of the hydroxyl substituents. Consequently, studies have observed that bioplastics derived from acetylated starch have reduced water retention, resulting in a denser matrix with enhanced polymer adhesion (Abe et al., 2021; Kim et al., 2017; Bourtoom, 2008).

1.2.2 Plasticiser

Plasticisers, which are generally non-volatile compounds that have a high boiling point, are added to improve flexibility as well as permeability toward gas and water vapour, and enhance mechanical properties. As starch-based bioplastics have a propensity to absorb water, that in turn affects their mechanical properties, due to which additives are added. Plasticisers further reduce the intermolecular force accumulation in long-chain molecules. Glycerol, phthalate ester, beeswax, urea, sebacic acid, polyvinyl alcohol, and sorbitol are a few examples of plasticisers. The bioplastic synthesis method involves the use of plasticisers, which enhance the flexibility and tensile strength of the bioplastics while reducing the glass transition temperature. For instance, glycerol is used in various starch-based bioplastics research as an ideal plasticiser, with a high boiling point of 290 °C, enabling it to resist evaporation during heating. Furthermore, the plasticiser is essential for ensuring interoperability within the polymer, significantly enhancing factors like polarity, hydrogen bonding, dielectric constant, and solubility properties (Chapain et al., 2021; Al Fath et al., 2024; Irmayanti & Anwar, 2024; Azsarinka et al., 2020; Paluch et al., 2022).

For instance, Santana et al. (2018) reported the use of glycerol in the synthesis of starch-based bioplastics derived from jackfruit seeds, which demonstrated enhanced mechanical stability of the bioplastics. Moreover, the study revealed that glycerol concentration has no correlation with the solubility of starch bioplastics. This is attributed to the fact that glycerol molecule possesses three free hydroxyl groups, whereas starch molecule contains multiple free hydroxyl groups, facilitating greater interaction with water molecules. Polyols like glycerol and sorbitol are the most frequently utilised plasticisers. Apart from this, polyols like xylitol, mannitol, ethylene glycol, butane diol, etc., are also used as plasticisers. Research studies have demonstrated that the use of both glycerol and sorbitol as plasticisers resulted in the synthesised biodegradable polymer exhibiting reduced fragility and an improved smooth surface. Moreover, research has demonstrated that the use of glycerol as a plasticiser results in greater flexibility, stickiness to the film, a reduction in gaps in the polymer, and low crystallinity. The addition of glycerol reduces transparency in bioplastic as compared to sorbitol. The sorbitol plasticiser film exhibited a reduced water vapour transmission rate compared to glycerol, which is linked to its hydrophilic properties. Moreover, a higher glycerol content leads to reduced tensile strength due to increased formation of hydrogen bonds within the polymer, resulting in enhanced free movement and flexibility (Santana et al., 2018; Ng et al., 2022; Inayati et al., 2019; Imoisili & Jen, 2023). In addition, various oils function as efficient, naturally obtained plasticisers, providing additional advantages, such as antioxidant characteristics for starch-based bioplastics. For example, castor oil was utilised as a natural plasticiser, resulting in an increased elongation of the bioplastics (García et al., 2020; Hamid et al., 2022).

1.2.3 Water

Water is often considered a plasticiser due to its ability to hydrolyse the molecular bond structure of starch when subjected to heat. The development of starch-based bioplastics is associated with starch gelatinisation, a process whereby starch thickens in the presence of water due to the breakdown of intermolecular bonds within the starch molecules (Ashok et al., 2016). Water plays a crucial role in gelatinisation, where it facilitates the swelling of starch granules, the melting of crystalline structures (double helices), and the leaching of amylose in the colloidal system. Furthermore, hot water dismantles starch's crystalline structures by disrupting intra- and inter-chain hydrogen bonds, converting it from a hard to a more pliable state and preventing recrystallisation. However, a secondary plasticiser is further required in the process of preparation of starch-based bioplastics (Mallakpour, 2018).

1.2.4 Acid

The main primary function of acid in the production of starch-based bioplastics is to assist in the breaking down of the branched structure of the amylopectin into a straight polymeric chain. Acids also play a crucial role in facilitating the hydrolysis of starch, helping to break it down into simpler components (Hamid et al., 2022; Muhammad et al., 2019; Ismail et al., 2016). Research studies have also demonstrated the use of acetic acid and acetic acid anhydride in the presence of sulphuric acid as a catalyst (Syarif et al., 2022). For instance, it has been observed that the addition of citric acid enhances the solubility of starch while also functioning as a plasticiser. Research indicates that citric acid can modify the characteristics of cationic starch-pectin mixtures, thereby facilitating plasticisation and yielding uniform polymer blends. A study by Marichelvam et al. (2022) demonstrated the application of the *Prosopis juliflora* plant in the formulation of starch-based bioplastics, including citric acid along with additional components (Moro et al., 2017; Marichelvam et al., 2022; Zuraida et al., 2012).

1.2.5 Crosslinkers

The use of crosslinking agents enhances the characteristics of starch-based bioplastics for various applications. The most popular crosslinking agents are sodium hexametaphosphate, glutaraldehyde, adipic acid, boric acid, and epichlorohydrin. Citric acid, characterised by its multi-carboxylic structure, is a non-toxic chemical compound that serves as a viable crosslinking agent to improve compatibility between starch and fibres. Citric acid diminishes the molecular interactions of starch, leading to the formation of intermolecular ester linkages between the hydroxyl groups of starch and the carboxylic groups of citric acid, while excess citric acid acts as a plasticiser in starch-based bioplastics. Furthermore, it has been stated that crosslinking agents enhance the properties of the polymer, such as interfacial compatibility, as well as mechanical, thermal, and water barrier characteristics (Yang et al., 2021; Olivato et al.,2012; Sun et al., 2018; Wang et al., 2009; Widiarto, 2005; Chakraborty et al., 2022). Tartaric acid serves as a crosslinker, and studies have demonstrated that citric acid and tartaric acid can lead to crosslinked TPS films in an acidic environment. Moreover, in the presence of acidic conditions along with excess water at elevated temperatures, the thermoplastic starch chain interacted with a carboxylic acid, leading to concurrent crosslinking and acid hydrolysis reactions (Lai et al., 2024). Boric acid, as an additional crosslinker, can be used in small amounts in the starch–glycerol mix to further improve the elongation and transparency of the starch-based film owing to the formation of an interconnections network (Widdiarto, 2005). In a study by Tongdeesoontorn et al. (2011), it has been shown that carboxymethylase as a crosslinker interacts with its carboxylic acid groups to form a stable structure with the composite cassava. Additionally, malic acid, another poly (carboxylic acid), is utilised as a crosslinker in the preparation of bioplastics derived from thermoplastic starch. To improve the compatibility of the polymeric phase, starch/poly (butyene adipate co-terephthalate) composites were cross-linked using citric, malic, and tartaric acids. These organic acids aid in the hydrolysis of starch, which breaks it down into low molecular weight fragments and dextrin (Castaño et al., 2014).

1.2.6 Additives/Fillers

Starch-based bioplastics possess inherent disadvantages, resulting in low mechanical properties and high water absorption. Therefore, it has been demonstrated that the mechanical characteristics of starch-based bioplastics can further be improved by adding several organic and inorganic reinforcement agents, such as nanoclay, lignocellulose, rice husk, passion fruit peel, cellulose, polylactic acid, chicken feather, coconut shell, palm fruit filler, cocoa pod husk, etc. (Abdullah et al., 2020; Oluba et al., 2021; Gowman et al., 2019; Shafqat et al., 2021). A study conducted by Hamin et al. in (2023) demonstrated that bioplastics derived from potato starch, when combined with sugarcane bagasse, exhibited enhanced thermal stability and mechanical properties. In another study conducted by Kasmuri and Zait (2018), potato starch-based bioplastics have been enforced using eggshells or chitosan as fillers. It has been noted that the incorporation of eggshells into potato starch-based bioplastics enhances tensile strength by 4.94%, while simultaneously decreasing water absorption by 10.95%. In contrast, chitosan exhibited a tensile strength increase of 1.28% and a more significant reduction in water absorption of 27.59%. In biodegradability assessments conducted over a span of 20 days, eggshells exhibited a weight reduction of 21.06%, whereas chitosan demonstrated a mere 7.9% loss in weight (Kasmuri & Zait 2018). Graphene oxide is a single layer of oxidised graphite, acquired through chemical synthesis or various methods of graphite oxidation, and serves as one of the fillers utilised in starch-based bioplastics. Studies have shown that their incorporation to the biocomposite film results in improved mechanical characteristics like tensile strength, elongation, and Young's modulus (Amri et al., 2018). In an alternative investigation carried out by Oluba et al. in (2021), starch obtained from ginger was reinforced with keratin derived from chicken feather waste. The resulting biocomposite exhibited a smooth surface, and the tensile strength increased with a higher amount of keratin. However, with an increased level of keratin, the solubility and transparency of the starch-based biocomposite film decrease (Oluba et al., 2021; Oladayo et al., 2016).

Furthermore, recent research has implemented micro- and nanofillers into starch-based bioplastic biocomposites, significantly enhancing their physio-mechanical properties. The nanofillers can be categorised into organic types, such as chitosan and cellulose, as well as inorganic types, including metallic and metal oxide nanoparticles (Othman & Fadzil 2021; Siddiqui et al., 2024). According to research conducted by Abdullah et al. (2020), the incorporation of zinc oxide has improved the tensile strength of starch-based bioplastic films. Furthermore, the incorporation of metallic nanoparticles or metal oxide nanoparticles enhances the properties of bioplastics by inhibiting microbial growth (Abdullah et al., 2020). In another study, clay nanoparticles were incorporated into cassava-derived starch-based bioplastics, resulting in enhanced the tensile strength of cassava-based bioplastics. The further addition of clay nanoparticles did not affect the transparency of the bioplastics, and the resulting mixture remained homogeneous (Harunsyah et al., 2018).

1.3 DEGRADATION MECHANISMS OF STARCH-BASED BIOPLASTICS

Starch-based bioplastics, recognised for their biodegradable properties, are compostable polymers that can be effectively transformed into harmless waste after a specified duration in a composting environment. In contrast to synthetic plastics, which require years to decompose owing to their slow breakdown rate and eventually breaking down into toxic compounds in addition to the formation of harmful micro- and nanoplastics (Figure 1.1), the breakdown process of starch-based bioplastics allows them to readily decompose in water or soil. Moreover, the degradation of starch-based bioplastics occurs when subjected to various stressors, such as sunlight, heat, etc. The soil microorganisms further facilitate the degradation process of starch-based bioplastics by disintegrating the strong covalent bonds. The degradation of starch-based bioplastics results in the generation of methane, CO_2, and water as end products (Reshmy et al., 2021; Atiwesh et al., 2021; Degli-Innocenti et al., 2022).

Bioplastics, derived from starch, are biodegradable in aquatic environments and easily decompose in the presence of water. Since starch-based bioplastics are hydrophilic in nature, it assists them to break down easily in water (Rizwan & Amal 2022; Syuhada et al., 2020). Starch-based bioplastics degrade significantly in the presence of UV light, causing photo-oxidative processes that break polymer chains, produce free radicals, and decrease molar mass (Quispe et al., 2019). A variety of research has investigated the integration of starch with synthetic polymers to produce films for various applications and to enhance their biodegradability. This approach has been linked to the release of microplastic particles, rising environmental and health concerns. The resultant composite consists of synthetic polymer and starch, experiences microbial breakdown of the starch component, resulting in non-biodegradable synthetic plastic as microplastic remnants (Hurley et al., 2024; Jayarathna et al., 2022).

It has been shown in a study that taro starch combined with bentonite strengthens it to produce a biodegradable film. Investigations have demonstrated that starch-based polymeric films with bentonite display a slower breakdown rate compared to those without bentonite. As the concentration of bentonite increased, the degradation rate decreased. Consequently, it can be asserted that starch-based composites with additives exhibit differing degradation rates contingent upon the specific type of additives used (Shanmathy et al., 2021). Microbes play an essential function in the breakdown of starch-based bioplastics in soil environments. Microorganisms such as bacteria, fungi, and algae can readily decompose the polymeric structure of starch-based bioplastics (Rao et al., 2022). Research indicates the presence of microorganisms, including *Pseudomonas* spp., *Streptococcus* spp., *Staphylococcus* spp., *Micrococcus* spp., and *Moraxella* spp., along with fungi such as *Aspergillus niger* and *Aspergillus glaucus*, as well as actinomycetes, on the surface of polyethylene/starch films (Prabhat et al., 2013). Plastic-degrading microbes deteriorate plastics, which serve as a source of nutrition for the growth of microorganisms. Furthermore, the degradation of the microbes leads to discolouration and further deterioration of the texture, like disintegration, crack formation, phase separation, etc., of the plastic polymer (Ghosh et al., 2013; Trivedi et al., 2016). Microorganisms biodegrade bioplastics via two methods: aerobically (in the presence of oxygen) and anaerobically (in the absence of oxygen). However, the aerobic degradation of bioplastics has been the subject of extensive studies. Aerobic degradation, which characterises composting, occurs in an oxygen-rich environment where microorganisms utilise biodegradable polymers as carbon and energy sources, resulting in the production of CO_2 and water as by-products. In anaerobic degradation, microorganisms facilitate the breakdown of polymers in the absence of oxygen, resulting in the production of methane gas, CO_2, water, hydrogen sulphide, ammonia, and hydrogen. The activity entails a sequence of metabolic interactions between diverse types of microorganisms (Bátori et al., 2018). The breakdown process of bioplastics can be influenced favourably or adversely by factors such as weather, humidity, soil moisture, the kind of microbial population in the soil, the presence of plant bioactive components in the biodegradable film, the nature of the starch, etc. (Oluwasina et al., 2019; Nasution et al., 2021).

When biodegradable plastic is broken down by microorganisms, CO_2 and water are produced as by-products. However, the CO_2 produced as a by-product causes the soil to become acidified by producing carbonic acid, H_2CO_3 (aqueous), which might hinder the biodegradation of bioplastics by lowering microbial activity. Moreover, elevated temperatures further impact soil microbes, diminishing their ability to break down organic matter. While it has been found that low moisture content impacts microbial population in soil, high moisture content fosters microbial development (Zoungranan et al., 2020). Also the storage of starch-based bioplastics adversely affects their strength and contributes to ageing, which gradually disintegrates the polymer and negatively impacts its tensile strength. Additionally, storage temperature might expedite the migration of additives in bioplastics (Hurley, 2024; Wahyuningtiyas & Suryanto, 2017).

1.4 APPLICATIONS OF STARCH-BASED BIOPLASTICS

Despite ongoing exploration and development, starch-based bioplastics hold significant promise in various industries due to their biodegradable nature, renewable resources, and versatile properties. Many applications are already using starch-based bioplastics, fostering a more sustainable future, despite existing challenges such as the enhancement of mechanical strength and water resistance (Jayarathna et al., 2022; Saiful et al., 2019; Ibrahim et al., 2021). Moreover, significant domains employ starch-based bioplastics, demonstrating their potential to replace traditional plastics as research continues to mitigate their shortcomings.

1.4.1 Packaging Materials

Given the escalating demand for synthetic single-use plastics, starch-based bioplastics emerge as a main alternative to mitigate their detrimental effects. Recently, the primary use of starch-based bioplastics is in the packaging sector, offering an environmentally friendly and non-toxic alternative to synthetic plastics (Zołek-Tryznowska & Kałuza, 2021). Food packaging represents a significant sector where starch-based bioplastics have found extensive application. Despite the challenges associated with inherent starch packaging with increased food shelf life, various studies have successfully addressed these issues through the use of additives, such as silica nanocrystals, chitosan, silica, titanium dioxide nanoparticles, etc., enhancing its physio-mechanical properties for food packaging applications (Oluwasina et al., 2021; Abdullah et al., 2024; Fathanah et al., 2015; Amin et al., 2019). A study suggests that films made by blending PLA and cassava starch are appropriate for packaging purposes. Biodegradable packaging can preserve the quality of capsicum for up to 12 and 24 days at 25 °C and 8 °C, respectively, in contrast to the shelf life of unpackaged capsicum, which lasts between 4 and 9 days at the same temperatures (Mangaraj et al., 2022).

The food industry has increasingly concentrated on research to create starch-based antibacterial food packaging. Citric acid is a weak organic acid with antibacterial characteristics that serves as a preservative to prevent food degradation. The study included citric acid in starch film to enhance food preservation and extend shelf life. Reports indicate that the starch citrate packaging film is effective in suppressing microbial development, extending the shelf life of cake by up to 10 days and bread by up to 40 days (Romainor et al., 2022). Subsequently, research studies have integrated starch with antimicrobial agents such as nisin, lysozymes, and ethylenediaminetetraacetate, assessing their synergistic effects as antimicrobial packaging to prolong shelf life. Similarly, in another study, starch blended with poly (3-hydroxybutyrate) and poly (lactic-co-glycolide) has also been evaluated for its antimicrobial activity (Mlalila et al., 2018; Bhatia & Bharti, 2015). Pomegranate peel extract, which functions as an antibacterial agent, has also been integrated into starch-based films, demonstrating inhibitory effects on the development of both gram-positive *S. aureus* and gram-negative *Salmonella* sp. bacteria. Moreover, the incorporation of pomegranate peel extract significantly enhances Young's modulus, tensile strength, and stiffness of the starch-based films (Ali et al., 2019). Essential oils, including clove, basil, and cinnamon, have been integrated into starch-based films to enhance the antimicrobial properties of the packaging materials (Mallick et al., 2020). On the other hand, it is essential to consider the possibility that antimicrobial agents might migrate to the packaged food (Kuorwel et al., 2013).

1.4.2 Agricultural Sector

Mulching is a long-established agricultural technique involving the application of materials directly onto the soil surface for purposes such as insulating seedlings and young shoots, reducing evaporation, controlling weed growth, and preventing soil erosion. The worldwide mulch market was valued at $3.5 billion in 2020 and is anticipated to reach $5.1 billion by 2027. The most popular and extensively used agricultural mulch film is produced from synthetic polymers; thus, a biodegradable film derived from starch presents a superior, sustainable, and eco-friendly alternative (Mansoor et al., 2022; Kasirajan & Ngouajio, 2012). Studies have described the use of empty fruit bunch fibres as reinforcement in polybutylene succinate combined with modified tapioca starch, produced using the hot press process, to produce agricultural mulch film. Another work explored the phosphorylation of starch, followed by surface functionalisation with chitosan, in starch-based films produced from maize (*Zea mays*) starch for agricultural mulch applications (Ayu et al., 2020; Merino et al., 2019). Another investigation discovered that oxidised starch used in the production of thermoplastic films demonstrated water resistance. The research developed starch-based films for agricultural mulch, using native and oxidised corn starch nanocomposites produced by extrusion with natural and chitosan-modified bentonite fillers, which also possess antibacterial properties (Merino et al., 2018).

1.4.3 Medical Applications

Starch as a polymer has been reported to enhance compatibility between materials and shows potential for use in medicine. Owing to its plant origin, starch has been promoted for use as a capsule for medicine delivery and holds potential to replace gelatin capsules. Starch capsules are also appropriate for drug delivery in the treatment of conditions such as ulcerative colitis, irritable bowel syndrome, etc. (Nasir & Othman, 2020). Starch has therefore been an excellent option in wound healing applications as starch has been used for the preparation of surgical suture. Furthermore, in tissue engineering, biological scaffolds are constructs that replicate the shape of the natural extracellular matrix. They facilitate the regeneration, reconstruction, or repair of damaged tissues by supporting cells. Studies have reported the fabrication of starch-based nanofibrous scaffolds by electrospinning for wound healing applications. As demonstrated by using L929 mouse fibroblast cells, scaffolds have the capacity to stimulate cellular proliferation without having any detrimental effects on the cells (Herrera-Ibarra et al., 2023; Waghmare et al., 2018).

1.4.4 Other Applications

Starch-based bioplastics are an emerging domain in bioplastics production, with potential to be used in the automotive industry. In addition, thermoplastic starch is utilised in the creation of several items, including cardboard (Rusu et al., 2011; Unger et al., 2017). Furthermore, recent investigations have employed starch to develop starch-based foams for industrial use (Amaraweera et al., 2022).

1.5 ADVANTAGES AND CHALLENGES OF STARCH-BASED BIOPLASTICS

Starch-based bioplastics are making significant strides in transforming the plastic industry by offering alternatives to synthetic plastics; however, there are still numerous challenges and limitations that require attention. Additional investigation and advancements are necessary in this area. Starch is cost-effective, readily accessible, sustainable, environmentally friendly, and easily degradable without emitting any harmful substances into the ecosystem. The primary drawbacks of starch-based bioplastics lie in their brittleness, hydrophobicity, low water and gas barrier properties, as well as reduced mechanical characteristics over time, primarily due to the material's recrystallisation (Sidek et al., 2019; Mose & Maranga, 2011). However, chemical modifications, surface functionalisation, blends, and composites with other biopolymers can address these limitations, potentially transforming starch into a more viable alternative to synthetic plastics (Gamage et al., 2022).

The ongoing challenges regarding scalability, cost competitiveness, and final disposal also pose a limitation to starch-based bioplastics. To address this challenge, collaboration among academia, industry, and policymakers is essential (Jabeen et al., 2024). Fossil fuel-based synthetic polymers produce pollutants and toxicity throughout every stage of their life cycle. Starch-based bioplastics, which are sourced from plants, contribute in the reduction of greenhouse gas emissions. The production method for bioplastics utilises significantly less energy in comparison to conventional plastics. Nonetheless, utilising plant materials for bioplastic production may still raise environmental issues, including acidification and eutrophication, attributed to the significant application of fertilisers and chemicals in farming practices (Sarkingobir & Lawal 2021).

Starch-based bioplastics are favourable in terms of their carbon footprint, although it is not zero. Starch-based bioplastics exhibit a substantially lower carbon footprint relative to petroleum-based synthetic polymers. As starch originates from biological sources, specifically plants, it sequesters CO_2 during the process of photosynthesis. Despite undergoing many manufacturing stages, the CO_2 sequestered by the plants stays confined within the plastic substance. Moreover, the carbon footprint of starch-based bioplastics may be diminished by methodical initiatives, including the use of renewable energy and plant-derived waste in the manufacturing process (Shafqat et al., 2020; Chauhan et al., 2024).

1.6 CONCLUSION AND FUTURE OPPORTUNITIES

Starch-based bioplastics represent a substantial advancement in our endeavour to mitigate the environmental consequences of conventional plastic manufacturing. Starch-based bioplastics, sourced from renewable plant materials, provide a sustainable alternative to petroleum-derived plastics, which substantially contribute to pollution, greenhouse gas emissions, and environmental toxicity. Starch, being a natural polymer, is biodegradable and decomposes more readily in the environment, offering a reduced long-term waste issue in comparison to traditional plastics. A primary benefit of starch-based bioplastics is their comparatively negligible carbon footprint. However, we must also address issues such as the scalability and performance of starch-based bioplastics in various applications. Additionally, by utilising a variety of starch resources including starch derived from waste and implementing biotechnological innovations, these starch-based bioplastics will not only contribute to a circular economy but also pave the way for a more sustainable future in the field of materials science.

REFERENCES

Abdullah, A.H.D., Putri, O.D., Fikriyyah, A.K., Nissa, R.C. and Intadiana, S., 2020. Effect of microcrystalline cellulose on characteristics of cassava starch-based bioplastic. *Polymer-Plastics Technology and Materials, 59* (12), pp. 1250–1258. https://doi.org/10.1080/25740881.2020.1738465

Abdullah, A.H.D., Putri, O.D., Fikriyyah, A.K., Nissa, R.C., Hidayat, S., Septiyanto, R.F., Karina, M. and Satoto, R., 2020. Harnessing the excellent mechanical, barrier and antimicrobial properties of zinc oxide (ZnO) to improve the performance of starch-based bioplastic. *Polymer-Plastics Technology and Materials, 59* (12), pp. 1259–1267. https://doi.org/10.1080/25740881.2020.1738466

Abdullah, S.H.Y.S., Diah, S.B.M., Yew, B.S., Fatah, I.Y.A., Rahman, N.I.A., Hafizulhaq, F. and Alias, N.N., 2024. Impact of starch-based bioplastic film reinforced with alkaline treated lemongrass fibre and lemongrass essential oil on the barrier and mechanical properties. *Journal of Advanced Research in Applied Mechanics, 123* (1), pp. 120–134. https://doi.org/10.37934/aram.123.1.120134

Abe, M.M., Martins, J.R., Sanvezzo, P.B., Macedo, J.V., Branciforti, M.C., Halley, P., Botaro, V.R. and Brienzo, M., 2021. Advantages and disadvantages of bioplastics production from starch and lignocellulosic components. *Polymers, 13* (15), p. 2484. https://doi.org/10.3390/polym13152484

Al Fath, M.T., Ayu, G.E., Hasibuan, G.C.R., Dalimunthe, N.F. and Alexander, V., 2024. The effect of glycerol and sago starch addition on the characteristics of bioplastics based on orange peel pectin: Original scientific paper. *Chemical Industry & Chemical Engineering Quarterly, 30* (4), pp. 359–365. https://doi.org/10.2298/CICEQ231214007A

Alcivar-Gavilanes, M.G., Carrillo-Anchundia, K.L. and Rieral, M.A., 2022. Development of a bioplastic from banana peel. *Ingeniería e Investigación, 42* (3). https://doi.org/10.15446/ing.investig.92768

Ali, A., Chen, Y., Liu, H., Yu, L., Baloch, Z., Khalid, S., Zhu, J. and Chen, L., 2019. Starch-based antimicrobial films functionalized by pomegranate peel. *International Journal of Biological Macromolecules, 129*, pp. 1120–1126. https://doi.org/10.1016/j.ijbiomac.2018.09.068

Alonso-González, M., Felix, M., Guerrero, A. and Romero, A., 2021. Rice bran-based bioplastics: Effects of the mixing temperature on starch plastification and final properties. *International Journal of Biological Macromolecules, 188*, pp. 932–940. https://doi.org/10.1016/j.ijbiomac.2021.08.043

Amaraweera, S.M., Gunathilake, C., Gunawardene, O.H., Dassanayake, R.S., Fernando, N.M., Wanninayaka, D.B., Rajapaksha, S.M., Manamperi, A., Gangoda, M., Manchanda, A. and Fernando, C.A., 2022. Preparation and characterization of dual-modified cassava starch-based biodegradable foams for sustainable packaging applications. *ACS Omega, 7* (23), pp. 19579–19590. https://doi.org/10.1021/acsomega.2c01292

Amin, M.R., Chowdhury, M.A. and Kowser, M.A., 2019. Characterization and performance analysis of composite bioplastics synthesized using titanium dioxide nanoparticles with corn starch. *Heliyon, 5* (8). https://doi.org/10.1016/j.heliyon.2019.e02009

Amri, A., Ekawati, L., Herman, S., Yenti, S.R., Aziz, Y. and Utami, S.P., 2018, April. Properties enhancement of cassava starch based bioplastics with addition of graphene oxide. In IOP Conference Series: *Materials Science and Engineering* (Vol. 345, No. 1, p. 012025). IOP Publishing. Doi: 10.1088/1757-899X/345/1/012025

Ashok, A., Rejeesh, C. and Renjith, R., 2016. Biodegradable polymers for sustainable packaging applications: A review. *International Journal of Bionics and Biomaterials, 2* (2), 1 (11).

Atiwesh, G., Mikhael, A., Parrish, C.C., Banoub, J. and Le, T.A.T., 2021. Environmental impact of bioplastic use: A review. *Heliyon, 7* (9). https://doi.org/10.1016/j.heliyon.2021.e07918

Ayu, R.S., Khalina, A., Harmaen, A.S., Zaman, K., Mohd Nurrazi, N., Isma, T. and Lee, C.H., 2020. Effect of empty fruit brunch reinforcement in polybutylene-succinate/modified tapioca starch blend for agricultural mulch films. *Scientific Reports, 10* (1), p. 1166. https://doi.org/10.1038/s41598-020-58278-y

Azieyanti, N.A., Amirul, A., Othman, S.Z. and Misran, H., 2020, April. Mechanical and morphology studies of bioplastic-based banana peels. In *Journal of Physics: Conference Series* (Vol. 1529, No. 3, p. 032091). IOP Publishing. doi:10.1088/1742-6596/1529/3/032091

Azsarinka, R., Saleh, D. and Djonaedi, E., 2020, July. Synthesis of biodegradable plastic from corn starch and corn husk filler with addition of glycerol as plasticizer and variation of chitosan composition. In IOP Conference Series: *Materials Science and Engineering* (Vol. 902, No. 1, p. 012037). IOP Publishing. Doi: 10.1088/1757-899X/902/1/012037

Bátori, V., Åkesson, D., Zamani, A., Taherzadeh, M.J. and Horváth, I.S., 2018. Anaerobic degradation of bioplastics: A review. *Waste Management, 80*, pp. 406–413. https://doi.org/10.1016/j.wasman.2018.09.040

Bello, A., Morales, K., Sánchez, L., Lidueñez, V., Leal, A. and Gelves, G., 2020, October. Computational implementation of required industrial unit operations for bio-plastic production from starch extracted from banana peels by aerobic fermentation using *Rizophus oryzae*. In *Journal of Physics: Conference Series* (Vol. 1655, No. 1, p. 012078). IOP Publishing. Doi: 10.1088/1742-6596/1655/1/012078

Bhatia, S. and Bharti, A., 2015. Evaluating the antimicrobial activity of nisin, lysozyme and ethylenediaminetetraacetate incorporated in starch based active food packaging film. *Journal of Food Science and Technology, 52*, pp. 3504–3512. Doi: 10.1007/s13197-014-1414-7

Bourtoom, T., 2008. Plasticizer effect on the properties of biodegradable blend film from rice starch-chitosan. *Songklanakarin Journal of Science & Technology, 30*, 149–165.

Castaño, J., Rodríguez-Llamazares, S., Contreras, K., Carrasco, C., Pozo, C., Bouza, R., Franco, C.M.L. and Giraldo, D., 2014. Horse chestnut (*Aesculus hippocastanum* L.) starch: Basic physico-chemical characteristics and use as thermoplastic material. *Carbohydrate Polymers, 112*, pp. 677–685. http://dx.doi.org/10.1016/j.carbpol.2014.06.046

Chakraborty, I., Pooja, N., Govindaraju, I., Managuli, V., Banik, S., Mahato, K.K. and Mazumder, N., 2022. Preparation and characterization of citric acid crosslinked starch based bioplastic. *Materials Today: Proceedings, 55*, pp. 26–30. https://doi.org/10.1016/j.matpr.2021.11.649

Chapain, K., Shah, S., Shrestha, B., Joshi, R., Raut, N. and Pandit, R., 2021. Effect of plasticizers on the physicochemical properties of bioplastic extracted from banana peels. *Journal of Institute of Science and Technology, 26* (2), pp. 61–66. https://doi.org/10.3126/jist.v26i2.41423

Chauhan, K., Kaur, R. and Chauhan, I., 2024. Sustainable bioplastic: A comprehensive review on sources, methods, advantages, and applications of bioplastics. *Polymer-Plastics Technology and Materials, 63* (8), pp. 913–938. https://doi.org/10.1080/25740881.2024.2307369

Danie Kingsley, J., Ghosh, K., Bhattacharya, T., Biswas, A. and Mandal, R., 2020. Eco-friendly bioplastics from natural raw materials. *International Research Journal of Engineering & Technology*, 1680–1691.

Das, S. and Kalyani, M.I., 2023. From trash to treasure: review on upcycling of fruit and vegetable wastes into starch based bioplastics. *Preparative Biochemistry & Biotechnology, 53* (7), pp. 713–727. Doi: 10.1080/10826068.2022.2158470

Degli-Innocenti, F., Barbale, M., Chinaglia, S., Esposito, E., Pecchiari, M., Razza, F. and Tosin, M., 2022. Analysis of the microplastic emission potential of a starch-based biodegradable plastic material. *Polymer Degradation and Stability, 199,* 109934. https://doi.org/10.1016/j.polymdegradstab.2022.109934

Dejene, B.K. and Geletaw, T.M., 2024. Development of fully green composites utilizing thermoplastic starch and cellulosic fibers from agro-waste: A critical review. *Polymer-Plastics Technology and Materials, 63* (5), pp. 540–569. https://doi.org/10.1080/25740881.2023.2296649

Enwere, C.F., Okafor, I.S., Adeleke, A.A., Petrus, N., Jakada, K., Olosho, A.I., Ikubanni, P.P., Paramasivam, P. and Ayuba, S., 2024. Production of bioplastic films from wild cocoyam (Caladium bicolor) starch. *Results in Engineering, 24,* 103132. https://doi.org/10.1016/j.rineng.2024.103132

Fathanah, U., Lubis, M.R. and Moulana, R., 2015, November. Biopolymer from starch and chitosan as bioplastic material for food packaging. In *Proceedings of the Annual International Conference, Syiah Kuala University-Life Sciences & Engineering Chapter* (Vol. 5, No. 1).

Flury, M. and Narayan, R., 2021. Biodegradable plastic as an integral part of the solution to plastic waste pollution of the environment. *Current Opinion in Green and Sustainable Chemistry, 30,* 100490. https://doi.org/10.1016/j.cogsc.2021.100490

Gamage, A., Liyanapathiranage, A., Manamperi, A., Gunathilake, C., Mani, S., Merah, O. and Madhujith, T., 2022. Applications of starch biopolymers for a sustainable modern agriculture. *Sustainability, 14* (10), p. 6085. https://doi.org/10.3390/su14106085

García, A.V., Álvarez-Pérez, O.B., Rojas, R., Aguilar, C.N. and Garrigós, M.C., 2020. Impact of olive extract addition on corn starch-based active edible films properties for food packaging applications. *Foods, 9* (9), p. 1339. Doi: 10.3390/foods9091339

Gawande, G., Khiratkar, T., Urkude, Y., Bombarde, S., Sonwane, U., Nikhade, S. and Sanap, J., 2024. Bioplastic production from corn and potato starch and its industrial applications. https://doi.org/10.21203/rs.3.rs-3865690/v1

Ghosh, S.K., Pal, S. and Ray, S., 2013. Study of microbes having potentiality for biodegradation of plastics. *Environmental Science and Pollution Research, 20,* pp. 4339–4355. Doi: 10.1007/s11356-013-1706-x

Gowman, A.C., Picard, M.C., Lim, L.T., Misra, M. and Mohanty, A.K., 2019. Fruit waste valorization for biodegradable biocomposite applications: A review. *BioResources, 14* (4), pp. 10047–10092.

Hamid, L., Elhady, S., Abdelkareem, A. and Fahim, I., 2022. Fabricating starch-based bioplastic reinforced with bagasse for food packaging. *Circular Economy and Sustainability, 2* (3), pp. 1065–1076. https://doi.org/10.1007/s43615-021-00139-5

Hamin, S.H., Sayid Abdullah, S.H.Y., Lananan, F., Abdul Hamid, S.H., Kasan, N.A., Mohamed, N.N. and Endut, A., 2023. Effect of chemical treatment on the structural, thermal, and mechanical properties of sugarcane bagasse as filler for starch-based bioplastic. *Journal of Chemical Technology & Biotechnology, 98* (3), pp. 625–632. Doi: 10.1002/jctb.7218

Harunsyah, Sariadi and Raudah, 2018, January. The effect of clay nanoparticles as reinforcement on mechanical properties of bioplastic base on cassava starch. *Journal of Physics: Conference Series*, *953*, p. 012021. https://doi.org/10.1088/1742-6596/953/1/012021

He Wei, H.W. and Wei CunXu, W.C., 2017. Progress in C-type starches from different plant sources. https://doi.org/10.1016/j.foodhyd.2017.07.003

Herrera-Ibarra, E., Salazar-Hernández, M., Talavera-López, A., Solis-Marcial, O.J., Hernandez-Soto, R., Ruelas-Leyva, J.P. and Hernández, J.A., 2023. Preparation of surgical thread from a bioplastic based on nopal mucilage. *Polymers, 15* (9), p. 2112.https://doi.org/10.3390/polym15092112

Hu, L., Zheng, Y., Peng, Y., Yao, C. and Zhang, H., 2016. The optimization of isoamylase processing conditions for the preparation of high-amylose ginkgo starch. *International Journal of Biological Macromolecules, 86*, pp. 105–111. http://dx.doi.org/10.1016/j.ijbiomac.2016.01.045

Hurley, R., Binda, G., Briassoulis, D., Carroccio, S.C., Cerruti, P., Convertino, F., Dvořáková, D., Kernchen, S., Laforsch, C., Löder, M.G. and Pulkrabova, J., 2024. Production and characterisation of environmentally relevant microplastic test materials derived from agricultural plastics. *Science of the Total Environment, 946*, 174325. https://doi.org/10.1016/j.scitotenv.2024.174325

Ibrahim, N.I., Shahar, F.S., Sultan, M.T.H., Shah, A.U.M., Safri, S.N.A. and Mat Yazik, M.H., 2021. Overview of bioplastic introduction and its applications in product packaging. *Coatings, 11* (11), p. 1423. https://doi.org/10.3390/coatings11111423

Imoisili, P.E. and Jen, T.C., 2023. Synthesis and characterization of bioplastic films from potato peel starch; effect of glycerol as plasticizer. *Materials Today: Proceedings*. https://doi.org/10.1016/j.matpr.2023.05.565

Inayati, I., Pamungkas, D.J. and Matovanni, M.P.N., 2019, April. Effect of glycerol concentration on mechanical characteristics of biodegradable plastic from rice straw cellulose. In *AIP Conference Proceedings* (Vol. 2097, No. 1). AIP Publishing. https://doi.org/10.1063/1.5098285

Irmayanti and Anwar, C., 2024. Synthesis of biodegradable plastic from Wak banana peel starch with glycerol as plasticizer. *Техника и технологияпищевыхпроизводств, 54* (3), pp. 546–557. https://doi.org/10.21603/2074-9414-2024-3-2527

Ismail, N.A., Mohd Tahir, S., Yahya, N., Abdul Wahid, M.F., Khairuddin, N.E., Hashim, I., Rosli, N. and Abdullah, M.A., 2016, April. Synthesis and characterization of biodegradable starch-based bioplastics. In *Materials Science Forum* (Vol. 846, pp. 673–678). Trans Tech Publications.

Jabeen, M., Tarıq, K. and Hussain, S.U., 2024. Bioplastic an alternative to plastic in modern world: A systemized review. *Environmental Research and Technology, 7* (4), pp. 614–625. Doi: https://10.35208/ert.1467590

Jayarathna, S., Andersson, M. and Andersson, R., 2022. Recent advances in starch-based blends and composites for bioplastics applications. *Polymers, 14* (21), p. 4557. https://doi.org/10.3390/polym14214557

Judawisastra, H., Sitohang, R.D.R., Marta, L. and Mardiyati 2017, July. Water absorption and its effect on the tensile properties of tapioca starch/polyvinyl alcohol bioplastics. In *IOP Conference Series: Materials Science and Engineering* (Vol. 223, No. 1, p. 012066). IOP Publishing. Doi: 10.1088/1757-899X/223/1/012066

Kasirajan, S. and Ngouajio, M., 2012. Polyethylene and biodegradable mulches for agricultural applications: A review. *Agronomy for Sustainable Development, 32,* pp. 501–529. Doi: 10.1007/s13593-011-0068-3

Kasmuri, N. and Zait, M.S.A., 2018. Enhancement of bio-plastic using eggshells and chitosan on potato starch based. *International Journal of Engineering and Technology*, 7, pp. 110–115.

Kim, H.Y., Jane, J.L. and Lamsal, B., 2017. Hydroxypropylation improves film properties of high amylose corn starch. *Industrial Crops and Products, 95*, pp. 175–183. https://doi.org/10.1016/j.indcrop.2016.10.025

Klun, B., Rozman, U., Ogrizek, M. and Kalčíková, G., 2022. The first plastic produced, but the latest studied in microplastics research: The assessment of leaching, ecotoxicity and bioadhesion of Bakelite microplastics. *Environmental Pollution, 307,* 119454. https://doi.org/10.1016/j.envpol.2022.119454

Kong, X., 2020. Fine structure of amylose and amylopectin. In S. Wang (Ed.), *Starch Structure, Functionality and Application in Foods* (pp. 29–39). Elsevier. https://doi.org/10.1007/978-981-15-0622-2_3

Kossmann, J. and Lloyd, J., 2000. Understanding and influencing starch biochemistry. *Critical Reviews in Plant Sciences, 19* (3), pp. 171–226.

Krishnamurthy, A. and Amritkumar, P., 2019. Synthesis and characterization of eco-friendly bioplastic from low-cost plant resources. *SN Applied Sciences, 1* (11), p. 1432. https://doi.org/10.1007/s42452-019-1460-x

Kuorwel, K.K., Cran, M.J., Sonneveld, K., Miltz, J. and Bigger, S.W., 2013. Migration of antimicrobial agents from starch-based films into a food simulant. *LWT-Food Science and Technology, 50* (2), pp. 432–438. http://dx.doi.org/10.1016/j.lwt.2012.08.023

Lai, D.S., Osman, A.F., Adnan, S.A., Ibrahim, I., Ahmad Salimi, M.N. and Mustapha, M.J., 2024. Thermoplastic starch hybrid biocomposite films with improved strength and flexibility produced through crosslinking via carboxylic acid. *Journal of Thermoplastic Composite Materials, 37* (3), pp. 1150–1186. Doi: 10.1177/08927057231193372

Maheshwari, R., Rani, B., Sangeeta, P. and Sharma, A., 2013. Eco-friendly bioplastic for uncontaminated environment. *Research Journal of Chemical and Environmental Sciences, 1* (1), pp. 44–49.

Mallakpour, S., 2018. Fructose functionalized MWCNT as a filler for starch nanocomposites: Fabrication and characterizations. *Progress in Organic Coatings, 114,* pp. 244–249. http://dx.doi.org/10.1016/j.porgcoat.2017.08.013

Mallick, N., Pal, D., Soni, A.B., Jhariya, D.C. and Singh, D., 2020, December. Starch based antimicrobial food packaging film towards a sustainable environment. In *IOP Conference Series: Earth and Environmental Science* (Vol. 597, No. 1, p. 012019). IOP Publishing. Doi: 10.1088/1755-1315/597/1/012019

Mangaraj, S., Thakur, R.R. and Yadav, A., 2022. Development and characterization of PLA and Cassava starch-based novel biodegradable film used for food packaging application. *Journal of Food Processing and Preservation, 46* (9), e16314. Doi: 10.1111/jfpp.16314

Mansoor, Z., Tchuenbou-Magaia, F., Kowalczuk, M., Adamus, G., Manning, G., Parati, M., Radecka, I. and Khan, H., 2022. Polymers use as mulch films in agriculture—A review of history, problems and current trends. *Polymers, 14* (23), p. 5062. https://doi.org/10.3390/polym14235062

Marichelvam, M.K., Jawaid, M. and Asim, M., 2019. Corn and rice starch-based bio-plastics as alternative packaging materials. *Fibers, 7* (4), p. 32. Doi: 10.3390/fib7040032

Marichelvam, M.K., Manimaran, P., Sanjay, M.R., Siengchin, S., Geetha, M., Kandakodeeswaran, K., Boonyasopon, P. and Gorbatyuk, S., 2022. Extraction and development of starch-based bioplastics from *Prosopis juliflora* plant: Eco-friendly and sustainability aspects. *Current Research in Green and Sustainable Chemistry, 5,* 100296. https://doi.org/10.1016/j.crgsc.2022.100296

Masina, N., Choonara, Y.E., Kumar, P., du Toit, L.C., Govender, M., Indermun, S. and Pillay, V., 2017. A review of the chemical modification techniques of starch. *Carbohydrate Polymers, 157,* pp. 1226–1236. http://dx.doi.org/10.1016/j.carbpol.2016.09.094

Merino, D., Gutiérrez, T.J. and Alvarez, V.A., 2019. Potential agricultural mulch films based on native and phosphorylated corn starch with and without surface functionalization with chitosan. *Journal of Polymers and the Environment, 27,* pp. 97–105. https://doi.org/10.1007/s10924-018-1325-1

Merino, D., Gutiérrez, T.J., Mansilla, A.Y., Casalongué, C.A. and Alvarez, V.A., 2018. Critical evaluation of starch-based antibacterial nanocomposites as agricultural mulch films: Study on their interactions with water and light. *ACS Sustainable Chemistry & Engineering, 6* (11), pp. 15662–15672. Doi: 10.1021/acssuschemeng.8b04162

Mlalila, N., Hilonga, A., Swai, H., Devlieghere, F. and Ragaert, P., 2018. Antimicrobial packaging based on starch, poly (3-hydroxybutyrate) and poly (lactic-co-glycolide) materials and application challenges. *Trends in Food Science & Technology, 74*, pp. 1–11. https://doi.org/10.1016/j.tifs.2018.01.015

Moro, T.M., Ascheri, J.L., Ortiz, J.A., Carvalho, C.W. and Meléndez-Arévalo, A., 2017. Bioplastics of native starches reinforced with passion fruit peel. *Food and Bioprocess Technology, 10*, pp. 1798–1808. Doi: 10.1007/s11947-017-1944-x

Mose, B.R. and Maranga, S.M., 2011. A review on starch based nanocomposites for bioplastic materials. *Journal of Materials Science and Engineering B, 1* (2B), p. 239.

Muhammad, A., Roslan, A., Sanusi, S.N.A., Shahimi, M.Q. and Nazari, N.Z., 2019, November. Mechanical properties of bioplastic form cellulose nanocrystal (CNC) mangosteen peel using glycerol as plasticizer. In *Journal of Physics: Conference Series* (Vol. 1349, No. 1, 012099). IOP Publishing. Doi: 10.1088/1742-6596/1349/1/012099

Nasir, N.N. and Othman, S.A., 2020. Application of bioplastic packaging in industry. *Journal of Advanced Research in Materials Science, 74* (1), pp. 19–28. https://doi.org/10.37934/arms.74.1.1928

Nasution, H., Julianti, E., Dalimunthe, N.F. and Wulandari, G., 2021, November. The role of betel (Piper betle) leaf extract and glycerol on physical properties of bioplastic based on sago starch. In *IOP Conference Series: Earth and Environmental Science* (Vol. 912, No. 1, 012042). IOP Publishing. Doi: 10.1088/1755-1315/912/1/012042

Ng, J.S., Kiew, P.L., Lam, M.K., Yeoh, W.M. and Ho, M.Y., 2022. Preliminary evaluation of the properties and biodegradability of glycerol-and sorbitol-plasticized potato-based bioplastics. *International Journal of Environmental Science and Technology, 19* (3), pp. 1–10. https://doi.org/10.1007/s13762-021-03213-5

Oladayo, O.O., Joseph, O.S. and Oluwasegun, W., 2016. Physicochemical properties of cassava starch and starch-keratin prepared biofilm. *Songklanakarin Journal of Science & Technology, 38* (4), pp. 349–355.

Olivato, J.B., Grossmann, M.V.E., Bilck, A.P. and Yamashita, F., 2012. Effect of organic acids as additives on the performance of thermoplastic starch/polyester blown films. *Carbohydrate Polymers, 90* (1), pp. 159–164. http://dx.doi.org/10.1016/j.carbpol.2012.05.009

Oluba, O.M., Obi, C.F., Akpor, O.B., Ojeaburu, S.I., Ogunrotimi, F.D., Adediran, A.A. and Oki, M., 2021. Fabrication and characterization of keratin starch biocomposite film from chicken feather waste and ginger starch. *Scientific Reports, 11* (1), p. 8768. https://doi.org/10.1038/s41598-021-88002-3

Oluba, O.M., Osayame, E. and Shoyombo, A.O., 2021. Production and characterization of keratin-starch bio-composite film from chicken feather waste and turmeric starch. *Biocatalysis and Agricultural Biotechnology, 33*, p. 101996. https://doi.org/10.1016/j.bcab.2021.101996

Oluwasina, O.O., Akinyele, B.P., Olusegun, S.J., Oluwasina, O.O. and Mohallem, N.D., 2021. Evaluation of the effects of additives on the properties of starch-based bioplastic film. *SN Applied Sciences, 3*, pp. 1–12. https://doi.org/10.1007/s42452-021-04433-7

Oluwasina, O.O., Olaleye, F.K., Olusegun, S.J., Oluwasina, O.O. and Mohallem, N.D., 2019. Influence of oxidized starch on physicomechanical, thermal properties, and atomic force micrographs of cassava starch bioplastic film. *International Journal of Biological Macromolecules, 135,* pp. 282–293. https://doi.org/10.1016/j.ijbiomac.2019.05.150

Othman, S.A. and Fadzil, N.F., 2021. Implementation of nanofiller for bioplastic reinforcement: A review. *International Journal of Advanced Research in Technology and Innovation, 3* (2), pp. 19–27.

Özdamar, E.G. and Ateş, M., 2018. Rethinking sustainability: A research on starch based bioplastic. *Journal of Sustainable Construction Materials and Technologies, 3* (3), pp. 249–260. https://doi.org/10.29187/jscmt.2018.28

Paluch, M., Ostrowska, J., Tyński, P., Sadurski, W. and Konkol, M., 2022. Structural and thermal properties of starch plasticized with glycerol/urea mixture. *Journal of Polymers and the Environment,* pp. 1–13. https://doi.org/10.1007/s10924-021-02235-x

Perotto, G., Ceseracciu, L., Simonutti, R., Paul, U.C., Guzman-Puyol, S., Tran, T.N., Bayer, I.S. and Athanassiou, A., 2018. Bioplastics from vegetable waste via an eco-friendly water-based process. *Green Chemistry, 20* (4), pp. 894–902. Doi: 10.1039/x0xx00000x

Prabhat, S., Bhattacharyya, S., Vishal, V., Kalyan, R.K., Vijai, K., Pandey, K.N. and Singh, M., 2013. Studies on isolation and identification of active microorganisms during degradation of polyethylene/starch film. *International Research Journal of Environmental Sciences, 2* (9), pp. 83–85.

Quispe, M.M., López, O.V. and Villar, M.A., 2019. Oxidative degradation of thermoplastic starch induced by UV radiation. *Journal of Renewable Materials, 7* (4), pp. 383–391. Doi: 10.32604/jrm.2019.04276

Rao, L.S., Naidu, C.D. and Tiwari, S., 2022. Investigation on synthesis, structure and degradability of starch based bioplastics. *Materials Today: Proceedings, 49*, pp. 257–261. https://doi.org/10.1016/j.matpr.2021.01.917

Reshmy, R., Philip, E., Vaisakh, P.H., Raj, S., Paul, S.A., Madhavan, A., Sindhu, R., Binod, P., Sirohi, R., Pugazhendhi, A. and Pandey, A., 2021. Development of an eco-friendly biodegradable plastic from jack fruit peel cellulose with different plasticizers and Boswellia serrata as filler. *Science of the Total Environment, 767,* 144285. https://doi.org/10.1016/j.scitotenv.2020.144285

Rizwan, M., and Jamal T., 2022. Degradation of bioplastics under the influence of several environmental conditions. *International Journal of Innovations in Science & Technology, 74* (4). Doi: 10.33411/IJIST/2021030302

Rolland-Sabaté, A., Sanchez, T., Buléon, A., Colonna, P., Jaillais, B., Ceballos, H. and Dufour, D., 2012. Structural characterization of novel cassava starches with low and high amylose contents in comparison with other commercial sources. *Food Hydrocolloids, 27* (1), 161–174. https://doi.org/10.1016/j.foodhyd.2011.07.008

Romainor, A.N., Chin, S.F. and Lihan, S., 2022. Antimicrobial starch-based film for food packaging application. *Starch-Stärke, 74* (3–4), p. 2100207. Doi: 10.1002/star.202100207

Rusdi, S., Destian, R.A., Rahman, F. and Chafidz, A., 2020, April. Preparation and characterization of biodegradable plastic from banana Kepok peel waste. In *Materials Science Forum* (Vol. 981, pp. 132–137). Trans Tech Publications. Doi: 10.4028/www.scientific.net/MSF.981.132

Rusu, D., Boyer, S.A., Lacrampe, M.F. and Krawczak, P., 2011. Bioplastics and vegetal fiber reinforced bioplastics for automotive applications. In S. Pilla (Ed.), *Handbook of Bioplastics and Biocomposites Engineering Applications* (pp. 397–449). John Wiley & Sons.

Saiful, Helwati, H.I.R.A., Saleha, S.I.T.T.I. and Iqbalsyah, T.M., 2019, May. Development of bioplastic from wheat Janeng starch for food packaging. In *IOP Conference Series: Materials Science and Engineering* (Vol. 523, No. 1, p. 012015). IOP Publishing.

Santana, R.F., Bonomo, R.C.F., Gandolfi, O.R.R., Rodrigues, L.B., Santos, L.S., dos Santos Pires, A.C., de Oliveira, C.P., da Costa Ilhéu Fontan, R. and Veloso, C.M., 2018. Characterization of starch-based bioplastics from jackfruit seed plasticized with glycerol. *Journal of Food Science and Technology, 55,* pp. 278–286. https://doi.org/10.1007/s13197-017-2936-6

Sarkingobir, Y. and Lawal, A.A., 2021. Bioplastics: Their advantages and concerns. *Journal of Materials and Metallurgical Engineering, 11* (1), pp. 13–18.

Schirmer, M., Höchstötter, A., Jekle, M., Arendt, E. and Becker, T., 2013. Physicochemical and morphological characterization of different starches with variable amylose/amylopectin ratio. *Food Hydrocolloids, 32* (1), pp. 52–63. http://dx.doi.org/10.1016/j.foodhyd.2012.11.032

Shafqat, A., Tahir, A., Mahmood, A., Tabinda, A.B., Yasar, A. and Pugazhendhi, A., 2020. A review on environmental significance carbon foot prints of starch based bio-plastic: A substitute of conventional plastics. *Biocatalysis and Agricultural Biotechnology, 27,* 101540. https://doi.org/10.1016/j.bcab.2020.101540

Shafqat, A.R.I.F.A., Tahir, A., Khan, W.U., Mahmood, A.D.E.E.L. and Abbasi, G.H., 2021. Production and characterization of rice starch and corn starch based biodegradable bioplastic using various plasticizers and natural reinforcing fillers. *Cellulose Chemistry and Technology, 55,* pp. 867–881.

Shanmathy, M., Mohanta, M. and Thirugnanam, A., 2021. Development of biodegradable bioplastic films from Taro starch reinforced with bentonite. *Carbohydrate Polymer Technologies and Applications, 2,* 100173. https://doi.org/10.1016/j.carpta.2021.100173

Siddiqui, S.A., Yang, X., Deshmukh, R.K., Gaikwad, K.K., Bahmid, N.A. and Castro-Muñoz, R., 2024. Recent advances in reinforced bioplastics for food packaging–A critical review. *International Journal of Biological Macromolecules, 263* (2), 130399. https://doi.org/10.1016/j.ijbiomac.2024.130399

Sidek, I.S., Draman, S.F.S., Abdullah, S.R.S. and Anuar, N., 2019. Current development on bioplastics and its future prospects: an introductory review. *INWASCON Technology Magazine, 1,* pp. 3–8. http://doi.org/10.26480/itechmag.01.2019.03.08

Sultan, N.F.K. and Johari, W.L.W., 2017. The development of banana peel/corn starch bioplastic film: A preliminary study. *Bioremediation Science and Technology Research, 5* (1), pp. 12–17.

Sun, S., Liu, P., Ji, N., Hou, H. and Dong, H., 2018. Effects of various cross-linking agents on the physicochemical properties of starch/PHA composite films produced by extrusion blowing. *Food Hydrocolloids, 77,* pp. 964–975. https://doi.org/10.1016/j.foodhyd.2017.11.046

Syarif, M.A., Fahma, F. and Sailah, I., 2022, July. Bioplastic beads composite production based on cellulose acetate-starch blend: A literature study. In IOP Conference Series: *Earth and Environmental Science* (Vol. 1063, No. 1, 012015). IOP Publishing. Doi: 10.1088/1755-1315/1063/1/012015

Syuhada, M., Sofa, S.A. and Sedyadi, E., 2020. The effect of cassava peel starch addition to bioplastic biodegradation based on chitosan on soil and river water media. *Biology, Medicine, & Natural Product Chemistry, 9* (1), pp. 7–13.

Tongdeesoontorn, W., Mauer, L.J., Wongruong, S., Sriburi, P. and Rachtanapun, P., 2011. Effect of carboxymethyl cellulose concentration on physical properties of biodegradable cassava starch-based films. *Chemistry Central Journal, 5,* pp. 1–8.

Trivedi, P., Hasan, A., Akhtar, S., Siddiqui, M.H., Sayeed, U. and Khan, M.K.A., 2016. Role of microbes in degradation of synthetic plastics and manufacture of bioplastics. *Journal of Chemical and Pharmaceutical Research, 8* (3), pp. 211–216.

Unger, S.R., Hottle, T.A., Hobbs, S.R., Thiel, C.L., Campion, N., Bilec, M.M. and Landis, A.E., 2017. Do single-use medical devices containing biopolymers reduce the environmental impacts of surgical procedures compared with their plastic equivalents?. *Journal of Health Services Research & Policy, 22* (4), pp. 218–225.

Vamadevan, V. and Bertoft, E., 2015. Structure-function relationships of starch components. *Starch-Stärke, 67* (1–2), pp. 55–68. Doi: 10.1002/star.201400188

Waghmare, V.S., Wadke, P.R., Dyawanapelly, S., Deshpande, A., Jain, R. and Dandekar, P., 2018. Starch based nanofibrous scaffolds for wound healing applications. *Bioactive Materials, 3* (3), pp. 255–266. https://doi.org/10.1016/j.bioactmat.2017.11.006

Wahyuningtiyas, N.E. and Suryanto, H., 2017. Analysis of biodegradation of bioplastics made of cassava starch. *Journal of Mechanical Engineering Science and Technology, 1* (1), pp. 24–31.

Wang, N., Zhang, X., Han, N. and Bai, S., 2009. Effect of citric acid and processing on the performance of thermoplastic starch/montmorillonite nanocomposites. *Carbohydrate Polymers, 76* (1), pp. 68–73. Doi: 10.1016/j.carbpol.2008.09.021

Wang, S., Li, C., Copeland, L., Niu, Q. and Wang, S., 2015. Starch retrogradation: A comprehensive review. *Comprehensive Reviews in Food Science and Food Safety, 14* (5), pp. 568–585. Doi: 10.1111/1541-4337.12143

Widiarto, S., 2005. Effect of borax on mechanical properties and biodegradability of sago starch—poly (vinylalcohol) blend films. *Jurnal Sains and Teknologi, 2*, pp. 151–157.

Yang, J., Ching, Y.C., Chuah, C.H., Hai, N.D., Singh, R. and Nor, A.R.M., 2021. Preparation and characterization of starch-based bioplastic composites with treated oil palm empty fruit bunch fibers and citric acid. *Cellulose, 28,* pp. 4191–4210. https://doi.org/10.1007/s10570-021-03816-8

Zhang, Y., Rempel, C. and Liu, Q., 2014. Thermoplastic starch processing and characteristics—A review. *Critical Reviews in Food Science and Nutrition, 54* (10), pp. 1353–1370. Doi: 10.1080/10408398.2011.636156

Zia-ud-Din, Xiong, H. and Fei, P., 2017. Physical and chemical modification of starches: A review. *Critical Reviews in Food Science and Nutrition, 57* (12), pp. 2691–2705. http://dx.doi.org/10.1080/10408398.2015.1087379

Zołek-Tryznowska, Z. and Kałuza, A., 2021. The influence of starch origin on the properties of starch films: Packaging performance. *Materials, 14,* p. 1146. https://doi.org/10.3390/ma14051146

Zoungranan, Y., Lynda, E., Dobi-Brice, K.K., Tchirioua, E., Bakary, C. and Yannick, D.D., 2020. Influence of natural factors on the biodegradation of simple and composite bioplastics based on cassava starch and corn starch. *Journal of Environmental Chemical Engineering, 8* (5), 104396. https://doi.org/10.1016/j.jece.2020.104396

Zuraida, A., Yusliza, Y., Anuar, H. and Mohd Khairul Muhaimin, R., 2012. The effect of water and citric acid on sago starch bio-plastics. *International Food Research Journal, 19* (2), pp. 715–719.

2 Unconventional Sources of Starch in Bioplastics

Rafiq A. Rather

2.1 INTRODUCTION

Starch is a polysaccharide (a type of carbohydrate) and occurs as a natural component of nearly all plants, including fruits, vegetables, seeds, tubers, palms, roots, grains, and agricultural wastes (Kringel et al., 2020). Starch plays a significant role in our daily diet and serves as a highly useful and versatile ingredient in both food and nonfood applications (Yazid et al., 2018). In addition to its basic nutritional benefits, starch is utilized in brewing and as a thickening agent in baked goods and confections (Mason, 2009). Starch is also used in paper production to enhance the strength of the paper and in surface sizing applications (Li et al., 2019; Wei et al., 2021). Many plant products are rich in starch, and the properties of starch play a crucial role in determining the quality of these products (dos Silveira Junior & de Francisco, 2020). Major conventional sources of starch, which hold substantial economic value, include maize, potatoes, cassava, and wheat (Pfister et al., 2016). The starches extracted from these sources have been the subject of extensive research in industrial, pharmaceutical, and academic fields. These commercial starches are prominent in today's markets and find applications across various industries. However, the composition, structure, and properties of starch can differ widely among plant sources, suggesting a variety of potential applications based on these differences (Ai & Jane, 2024). Recently, there has been a surge in research exploring underutilized, unconventional, and novel starches from a range of different fruit and vegetable wastes (Tagliapietra et al., 2021). Unconventional starches, that is, those not yet commercialized starches, are derived from nontraditional or less common sources compared to the conventional ones like maize, potatoes, or wheat (Dini et al., 2012). Unconventional starches offer unique properties like improved flexibility, distinct thickening abilities, textures, lower cost, enhanced biodegradability, and specific nutritional benefits, setting them apart from more commonly used starches. The adoption of unconventional starches as substitutes for commercial starches has generated considerable interest (Zhu, 2020). Unconventional starches can provide technological benefits that are advantageous for consumers. Additionally, growing social concerns about sustainable development drive the exploration of new starch sources. Unconventional starches may have properties similar to or different from those of commercial starches. However, there is a lack of standardization of methodologies for studying unconventional starches. Therefore, a thorough analysis is necessary to determine their suitable technological applications, such as their use in bioplastic production. This chapter examines the significance of various unconventional starch sources that could be valuable for biomaterial production, highlighting specific examples and emphasizing their increasing technological applications in this field.

The application of unconventional starches in bioplastic production has gained considerable attention over the last several decades (Pérez-Pacheco et al., 2024; Porras-Saavedra et al., 2024).

DOI: 10.1201/9781003534914-2

Petroleum-based plastics, although effective in performance, contribute to issues like greenhouse gas emissions and persistent pollution (Coppola et al., 2021). In comparison, bioplastics or biodegradable plastics offer a more sustainable alternative, meeting the urgent demand for eco-friendly materials (Ancy et al., 2024). The term "bioplastics" refers to biobased polymers produced from biomass sources (vegetable oils, fats, starch, etc.) and recycled food waste, which can be divided into biodegradable, bio-based, or both (Arikan & Ozsoy, 2015). Thus, bioplastics are produced from renewable resources including carbohydrates, vegetable oils, etc. (Lackner, 2000). As reported by the European Bioplastics Organisation (2024), bioplastics currently account for approximately 0.5% of the more than 400 million tons of plastics produced each year. However, as demand increases and more advanced materials, applications, and products are developed, the market is already experiencing significant growth. Bioplastics are an alternative plastic source with similar physical properties to synthetic plastics (Thomas et al., 2023). Bioplastics can be appropriately managed using a variety of end-of-life options, including reuse, mechanical recycling, organic recycling, and energy recovery.

Bioplastics come in various forms, including polylactic acid (PLA), polyhydroxyalkanoate (PHA) and polyglycolic acid (Adetunji et al., 2023). Furthermore, there are also bioplastics that are cellulose-based, protein-based polymers, and starch-based. Among these, PLA and PHA are the major types of bioplastics produced in large quantities. Starch-based bioplastics, made from plant-derived starches, are versatile biomaterials used in a wide variety of applications, such as food packaging, protective coatings, and film coatings (Gadhave et al., 2018). In the packaging sector, the future of starch-based bioplastics looks increasingly promising as the demand for sustainable plastics fuels ongoing innovation and development. Bioplastics differ from conventional plastics in several key ways (Gironi et al., 2011; Muneer et al., 2021), as summarized briefly in Table 2.1.

Bioplastics offer several environmental benefits. They contribute to a cleaner environment by lowering greenhouse gas emissions and generally require less energy for manufacturing compared to traditional petroleum-based plastics, thereby promoting energy efficiency (Reddy et al., 2013). By utilizing renewable biomass resources, bioplastics reduce dependence on nonrenewable fossil fuels and support carbon neutrality (Muneer et al., 2021). Additionally, biodegradable bioplastics provide an eco-friendly end-of-life solution, helping to reduce landfill waste and environmental pollution (Emadian et al., 2017; Rosenboom et al., 2022). Unlike conventional plastics, bioplastics do not contain harmful additives, such as bisphenol A or phthalates. The versatility of bioplastics allows them to be used in various applications, from packaging to medical devices, offering a broad

TABLE 2.1
Characteristic differences between bioplastics and conventional plastics

Bioplastics vs. Conventional Plastics		
Characteristics	**Bioplastics**	**Conventional Plastics**
Source	Derived from natural, renewable resources	Derived from fossil fuels and petroleum-based sources
Resource Availability	Utilize renewable resources	Reliant on finite fossil fuel resources
Biodegradability	Many bioplastics are biodegradable	Most conventional plastics are non-biodegradable
Environmental Impact	Often considered more eco-friendly	Contributes to environmental pollution
Common Applications	Applied in biodegradable food packaging, biomedical tools, automotive parts and more	Used in packaging, bags, bottles, construction, textiles, electronics, and more
Examples of Materials	Polyhydroxyalkanoate, polylactic acid, polyhydroxyurethanes	Polyethylene terephthalate, polyvinyl chloride, polystyrene

range of sustainable solutions. An important biomaterial used in the production of bioplastics is starch, derived from different parts of agricultural by-products and waste materials.

2.2 UNCONVENTIONAL SOURCES OF STARCH

The growth of bioplastics has been supported by new biotechnologies, yet high manufacturing costs and poor performance restrict their use (Bhagwat et al., 2020; Li et al., 2022). Simultaneously, as global food production rises, the disposal of food waste, especially from fruits and vegetables, raises environmental concerns. One potential solution is to develop a platform to refine fruit and vegetable waste and integrate it into the production of eco-friendly biodegradable bio-based plastics (Chan et al., 2021; George et al., 2021). Although bioplastics frequently face high manufacturing costs, incorporating fruit and vegetable waste into the creation of green biomaterials such as bioplastics presents a promising new approach (Li et al., 2022). Starch from agricultural wastes is regarded as one of the strongest raw material candidates for manufacturing bioplastics because of its high capacity to form good films, in addition to its low cost, abundance, eco-friendliness, and renewable features. Unconventional starch sources hold significant potential for bioplastic production (Henning et al., 2022). These unconventional starch sources are often by-products or wastes from agricultural and food industries, providing the dual benefit of waste valorization and bioplastic manufacturing. Some important sources of unconventional starch with potential for bioplastic production are discussed next.

2.2.1 Banana Peels

Banana (*Musa* sp., Musaceae family) is a widely grown fruit crop that thrives in tropical and subtropical regions of the world. Green banana fruit is a significant source of starch, comprising both flesh and peel. While the physicochemical properties of the flesh starch have been thoroughly studied, the properties of the peel starch have not been adequately investigated, leading to the waste of this part of the fruit (Zhang et al., 2005). Due to their high starch content, bananas are considered a promising source of starch, with about 22.6% in the peels and 69.5% in the dry flesh (Li et al., 2018). Peel starch is one of the major unconventional sources in the development of bioplastics (Azieyanti et al., 2020; Chandrasekar et al., 2023). In banana, the starch occurs as irregular, oval-shaped granules with off-centered hila. Both the flesh- and peel-starch show B-type crystallinity with comparable lamellar structures and similar levels of relative crystallinity (Li et al., 2018). In India, several industries produce banana-based products like banana cakes, banana chips, and banana fritters. These industries utilize the banana flesh as their primary material and typically discard the peels as waste. The disposal of this significant amount of wet organic waste negatively impacts the environment and poses health risks. However, due to their abundance, low cost, renewability, and biodegradability, banana peels could be effectively utilized in the production of bioplastics (Chandrasekar et al., 2023; Sharma et al., 2024).

2.2.2 Mango Seeds

Mango (*Mangifera indica*) is a popular tropical fruit extensively cultivated in India, Myanmar, Malaysia, and Thailand (Bally, 2006). The edible portion of mangoes, which varies by variety, constitutes approximately 33–85% of the fresh weight and is used for processing and consumption. The peels and seeds represent the processing waste, comprising about 7–24% and 9–40% of the fruit, respectively (Punia Bangar et al., 2021). In India, mango seeds are often discarded as waste after the fruit is consumed. Thus, due to their availability and their potential as a reliable source of starch, mango seeds could be utilized to isolate starch for the synthesis of bioplastics and other value-added products (Maulida et al., 2018b). The extracted mango seed starch granules showed oval and irregular or cuboidal-shaped with smooth surface despite different extraction methods. X-ray diffraction studies revealed that the starch, isolated from mango seeds, is an A-type starch (Shahrim et al., 2018). However, the yield of starch isolated from mango seeds varies depending on the extraction method employed. For example, one study found that starch isolated using the alkali method had the highest yield at 91.72%, compared to yields from distillation (81.87%), sedimentation (74.82%), and centrifugation (74.56%) (Shahrim et al., 2018). This suggests that mango seeds could be a significant source of unconventional starch for the preparation of bioplastics (Joseph et al., 2024; Maulida et al., 2018a). In addition to their application in bioplastics, mango seed starch provides a sustainable and eco-friendly alternative to meet growing industrial needs. Its digestibility profile and glycemic index indicate a higher presence of resistant starch compared to conventional starches, making it a suitable ingredient for diabetes management. Further, the structure of mango seed starch can be effectively modified using biological, chemical, and physical methods to enhance its applications in bioplastics (Amaraweera et al., 2021; Maulida et al., 2018a).

2.2.3 Jackfruit Seeds

Jackfruit seeds are predominantly composed of starch (70–85% when dry), which is utilized in food processing and various other applications as thickeners, stabilizers, microencapsulation agents, coagulants, and bioplastics (Brahma et al., 2023). Additionally, they offer notable health benefits. Jackfruit seeds represent about 18–25% of the total weight of a jackfruit and each fruit contains approximately 100–500 seeds, making jackfruit seeds (JFS) an important source of unconventional starch (Brahma et al., 2023). There has been a growing interest in the development of starch-based bioplastics from jackfruit seeds. While these bioplastics may have lower mechanical properties compared to petroleum-based plastics, their environmental friendliness and high potential added value suggest that they could be a key material for the future (Nguyen et al., 2022). However, the mechanical properties of these bioplastics can be greatly enhanced by incorporating glycerol as a plasticizer and by blending starch with varying ratios of polypropylene.

2.2.4 Cassava Tubers

Cassava (*Manihot esculenta*) is the third-largest source of starch in tropical and subtropical regions, making it a crucial staple food for over 500 million people. Cassava is widely grown as an annual crop primarily for its edible starchy root tubers. The tuberous roots of this plant are rich in starch (average 84.5%, dry basis) (Hsieh et al., 2019). Cassava starch (also known as 'tapioca') is a valuable raw material used to produce various native and modified starches for cooking. It also serves multiple industrial purposes, including as a staple food for humans, animal feed component, food processing ingredient, edible coatings, locally made alcoholic beverages, and ethanol production. Cassava contains a large amount of starch, which can suitably be used for the production of bioplastics (Wahyuningtiyas et al., 2017). In one study, bioplastics made from cassava peel starch were plasticized with sorbitol, and the addition of microcrystalline cellulose led to enhanced tensile strength compared to non-reinforced bioplastics (Maulida & Tarigan, 2016). Due to the low product costs and oversupply of cassava production, cassava starch is a potential feedstock for bioplastic production.

2.2.5 Litchi Seeds

Litchi (*Litchi chinensis*), a member of Sapindaceae family (soapberry family), is cultivated in tropical and warmer subtropical regions worldwide and is commercially valued for its delicious and nutritious fruit. The peel and seed of litchi, which account for about 30–40% (w/w) of the fruit, are often discarded during processing or consumption, leading to waste management and environmental issues (Bangar et al., 2021). Rich in complex carbohydrates, litchi seeds contain around 40% starch, which has unique functional and structural properties due to its low amylose and high amylopectin content (Bangar et al., 2021). Litchi seed starch displays C-type starch, which is a blend of A- and B-type allomorphs, with a predominant A-type crystalline structure. The starch granules exhibit a bimodal granular size distribution, consisting of both large and small granules. As the primary by-products of litchi fruit processing for beverages, jellies, wine, and squash, litchi seeds contribute to the over 540,000 metric tons of waste generated annually by the litchi processing industry. Utilizing litchi seed starch to produce bioplastics could improve the economic feasibility of this industry and help reduce waste and environmental impact (Esua et al., 2024). Given their high starch content, litchi seeds offer a potential nontraditional starch source with diverse applications, including in bioplastics and formulation of biodegradable packaging (Fatima et al., 2024).

2.2.6 Longan Seeds

The longan (*Dimocarpus longan*), also called the dragon's eye, is a tropical tree species that bears edible fruit. It is a prominent member of the soapberry family, Sapindaceae, which also includes the litchi. Each year, a significant portion of longan fruit is processed into various products, including black longan. The dried, dark-brown flesh of black longan is used in traditional medicine, while the starch-rich seeds are typically discarded. Hu et al. (2018) found that longan seeds contained up to 49.5% w/w starch, with amylose levels ranging from 26.6% to 30.1%. Likewise, Guo et al.reported that the starch content was 59.4% w/w, with an amylose content of 25.1% (Kittipongpatana et al., 2021). These starch-rich seeds are well-suited for starch extraction (Hu et al., 2018; Manzoor & Ahmad, 2024). Given that petroleum-based plastics pose serious environmental, wildlife, and health risks, and that longan fruit consumption results in a substantial amount of seed waste, utilizing starch from these seeds to create biodegradable bioplastics could tackle both problems. However, research on this topic remains limited. For example, enhancing the tensile strength of longan starch-based bioplastics needs to be investigated before evaluating their potential as a replacement for existing petroleum-based plastics. The significant quantity of longan seeds generated from both fresh fruit and the dried longan industry, along with prior reports on starch extraction, indicates that this biowaste could serve as a novel source of unconventional starch for bioplastic production.

2.2.7 Loquat Seeds

The loquat (*Eriobotrya japonica*), a member of Rosaceae family, is a large evergreen shrub or tree that is cultivated for its orange fruit and also used as an ornamental plant. This plant is also known as yellow plum tree or Japanese plum tree. The large kernel in loquat fruit comprises approximately 17–30% of the fruit's fresh weight and starch content varied from 59% to 71% (25% amylose and 75% amylopectin), with typical type-C crystallinity (Tosif et al., 2023). Loquat starches appear as spherical granules that are similar in appearance to those found in other starches and exhibit a blend of irregular and techno-functional properties when compared to commercial starches (Bangar et al., 2024; Tosif et al., 2023). Loquat seed starch is emerging as a novel nontraditional starch source, distinguished by its high amylose content, small granule size, and high viscosity. Loquat seed starch is a cost-effective alternative to traditional starch sources. Due to its biodegradability, loquat starch is employed in making bioplastics, which have applications in packaging, agricultural films, and other disposable products (Costa et al., 2023). The use of loquat starch offers an eco-friendly alternative to petroleum-based plastics, leveraging renewable resources and reducing environmental impact.

2.2.8 Avocado Seeds

Avocado, also known as alligator pear or avocado pear, is a medium-sized, evergreen tree belonging to the laurel family (Lauraceae). The fruit of this plant is highly nutritious and offers multiple health benefits (Bhore et al., 2021). Increased consumption of avocado fruit can result in agricultural waste, including avocado seeds, which, if not managed properly, may contribute to environmental pollution and negatively impact human health. One way to address this issue is to explore value-added uses, such as the production of avocado seed starch. Avocado seed starch is environmentally friendly, exhibits a crystalline structure with a high yield rate (over 40%), and has an oval shape (Kahn, 2006). It is rich in amylose (42–48%), demonstrates high gelatinization and low retrogradation rates, and has superior functional properties compared to other plant seed-based starches (Wang et al., 2022). Consequently, avocado seed starch holds potential for use in the production of bioplastics. Furthermore, avocado seed starch has applications in both food and nonfood industries.

2.2.9 Turmeric Rhizomes

Turmeric (*Curcuma longa*), a flowering plant in the ginger family Zingiberaceae, is a perennial, rhizomatous, herbaceous plant native to the Indian subcontinent and Southeast Asia. The rhizomes of this plant are either used fresh or boiled, dried, and ground into a deep orange-yellow powder which is used as a coloring and flavoring agent. The turmeric rhizome contains a major fraction of starch (47–56% w/w) on a dry basis. X-ray diffractogram indicates that turmeric starch is of B-type crystallinity (Kuttigounder et al., 2011). The isolation and characterization of starch from an unconventional source such as turmeric rhizome suggest its potential use as a functional ingredient in both the food and bioplastic industries, including applications in natural bioactive starch films (Hornung et al., 2018; Rasmi et al., 2024).

2.2.10 Sweet Potatoes

Sweet potato (*Ipomoea batatas* Lam) is an annual herb belonging to the Convolvulaceae family. Sweet potato is cultivated as a root crop in tropical and subtropical regions. Sweet potato is notably rich in starch, especially in certain varieties, with starch making up about 50–80% of its dry weight (Wang et al., 2020). The starch in sweet potatoes consists of approximately one-third amylose and two-thirds amylopectin, making them an excellent source of starch and energy (Adams, 2004). In Asia, sweet potatoes are widely consumed due to their high energy value and are cultivated extensively because of their relatively high yield. Additionally, they have a lower environmental impact compared to some other crops, which makes sweet potato starch a promising material for bioplastic production.

2.2.11 Yams

The yam (*Dioscorea alata* L.) is a subsoil herbaceous plant with an underground rhizome belonging to the Dioscoreaceae family. Yam is cultivated globally and presents a viable alternative to more traditional starch sources. Yam is a rich source of starch, containing between 60% and 85% of starch relative to its other nutritional components, which make it suitable for extraction of starch (Oliveira et al., 2021). Starch is typically isolated from fresh root tubers of yam. A study by Oliveira et al. revealed that yam starch comprises 96.58% of total starch, with 25.77% amylose and 74.23% amylopectin. Yam starch shows type-B crystallinity pattern (Oliveira et al., 2021). Although yam starch has low breakdown viscosity when compared to cassava starch, its technological characteristics make it a promising material for bioplastic production and biomaterials (Behera et al., 2022; Kumar et al., 2021). The yam starch-based bioplastic can be utilized as a food packaging material (Behera et al., 2022).

2.2.12 Beetroot

Beetroot (*Beta vulgaris* L.) is a horticultural product that generates large quantities of agro-industrial by-products. Although beetroot is not generally recognized for its high starch content compared to potatoes or cassava, certain varieties and processing methods can produce appreciable amounts of starch (Punia Bangar et al., 2022). In addition, incorporating beetroot powder into cassava starch films has been shown to reduce water vapor permeability and enhance the films' hydrophobic properties (Otálora González et al., 2020). Beetroot is widely cultivated and can be a good candidate for bioplastic production if starch yield is optimized. Sugar beet pulp, fibrous by-product left behind after sugar extraction, contains a variety of components that make it a promising option for bioplastic production (Marzo-Gago et al., 2023).

2.2.13 Arrowroot

Arrowroot (*Maranta arundinacea*) is a large, perennial herb native to tropical rainforests, typically growing along riverbanks and clearings, but it is also cultivated in tropical regions around the world. Arrowroot starch is obtained from the roots of this plant and other related species. Known for its high digestibility, arrowroot starch is used as a home remedy for gastrointestinal disorders, especially in England. The roots of this plant can contain over 20% starch, of which 17–18% is extractable (Corbishley & Miller, 1984). Arrowroot starch granules are large (15–70 μm), round or polygonal in shape, and possess A-type X-ray diffraction pattern. Arrowroot starch has a high content of amylose (~35.20%) which makes it ideal for superior film production (Tarique et al., 2022). Furthermore, arrowroot starch can be plasticized, blended with other polymers, or reinforced with fibers to improve its properties (Jamal et al., 2022). Arrowroot starch is known for its high purity and fine texture. The starch from arrowroot has excellent thickening properties and could be useful in specialized bioplastic applications.

2.2.14 Taro

Taro (*Colocasia esculenta*) is a root vegetable from the Araceae family, valued for its corms, leaves, stems, and petioles. Taro is grown in tropical regions and can be a sustainable source of starch. It is ranked as the fifth major root crop due to its roles in medicine, decoration, and culinary uses. Taro is rich in starch, with its underground corms containing 70–80% starch, surpassing even potatoes, sweet potatoes, and cassava in starch content (Ferdaus et al., 2023). This high starch content makes taro an economical ingredient for the food industry, where it serves a variety of functions including as a stabilizer, emulsifier, fat replacer, and filling agent. However, taro root is a potential source of starch with applications beyond food, including bioplastics (Shanmathy et al., 2021). The amylopectin in taro starch is relatively sparse but maintains long chain lengths, which aid in creating firm, elastic gels. Taro starch exhibits A-type crystallinity, characterized by a tightly packed crystalline matrix formed by double helices (Tattiyakul et al., 2006). This structure allows it to endure higher temperatures and shear forces, making it well-suited for bioplastic production.

2.2.15 Lotus Rhizome

Lotus (*Nelumbo nucifera*) is a perennial aquatic plant from the Nelumbonaceae family, a small family that contains just one genus with two species. The rhizomes of the lotus are a sustainable yet underutilized source of starch, with up to 20% starch content (Dhull et al., 2023). The starch granules in lotus rhizomes are oval and elongated, measuring 10–50 μm along the long axis and 10–35 μm along the short axis. The rhizome starch is predominantly reported as C-type polymorph, with some reports also indicating B-type and A-type forms (Yu et al., 2013). The chemical composition of lotus rhizome starch, along with its techno-functional properties such as swelling power, solubility, in vitro digestibility, pasting characteristics, and gelatinization, make it an ideal candidate for bioplastic production (Dhull et al., 2023).

2.2.16 Jerusalem Artichoke

The Jerusalem artichoke (*Helianthus tuberosus* L.) is a North American herbaceous plant from the Asteraceae family. Unlike other plants that store carbohydrates as starch, Jerusalem artichoke tubers store carbohydrates mainly as fructans, including inulin and fructooligosaccharides (Krochmal-Marczak et al., 2022). These tubers contain about 10–22% carbohydrates, with 80–90% of this being fructans, 7–14% sucrose, and 3–6% reducing sugars. The tubers of this plant are high yielding and can grow in various climates, making them a versatile option for bioplastic production.

2.2.17 Kudzu

Kudzu (*Pueraria montana*), a species in the Fabaceae family, is a perennial vine with tuberous roots and long, dark brown, rope-like stems. Kudzu root, which contains up to 51.6% starch (dry basis), is a promising alternative source of commercial starch (Zhao et al., 2021). Kudzu root starch has been thoroughly studied for various properties, such as gelatinization, retrogradation, digestion, viscosity, and interfacial activity. These characteristics have enabled its use in producing edible films, stabilizing emulsions, encapsulating oxidizable substances, altering food textures, and developing functional foods (Zhao et al., 2021). While native kudzu starch is widely used in both food- and non-food applications, the use of modified kudzu root starch is still limited, despite improvements in its functional properties through various modification techniques.

2.2.18 Sago palm

Sago palm (*Metroxylon* spp.) is a tropical plant that thrives in harsh, swampy peat conditions. Sago starch accumulates in the pith of the palm's stem, starting from the base and rising upward. When mature, the trunk is completely filled with starch, reaching nearly to the crown (Pei-Lang et al., 2006). In this plant, most of the starch is stored in the pith of the palm which contains approximately 250 kg of starch (Pei-Lang et al., 2006). It is estimated that about 60 million tonnes of sago starch are produced annually in South East Asia. According to Cecil et al. (1982), the pith's chemical composition includes 6–12% of soluble solids (dry substance), 1–3% of ash, and 79–88% starch plus sugar. The amylose content ranges between 24% and 31% (Pei-Lang et al., 2006). The starch granules are oval, averaging 30 μm in diameter, and display a C-type X-ray diffraction pattern (Okazaki, 2018). With growing concerns about environmental pollution from non-biodegradable materials, starch-based biomaterials, particularly sago starch, are recognized as effective biodegradable sources (Cheong et al., 2010; Lusiana et al., 2019).

2.2.19 Inhambu Tuber

The inhambu (*Dioscorea trifida* L.), the largest and most significant genus in the Dioscoreaceae family with around 600 species, is notable for its high carbohydrate content and medicinal properties. Despite this diversity, only a few species are extensively cultivated in tropical and subtropical regions. The extraction yield of inhambu starch is notably high, with percentages of 84.56% for total starch, 36.82% for amylose content, and 22.76% for amylopectin contents (Silva et al., 2020). The starch displays a type C diffraction pattern, which is an intermediate crystalline form comprising a mixture of type A and type B starches.

Before utilizing these starch sources for bioplastic production, it is important to consider several factors. The feasibility of using these sources depends on their starch yield, the simplicity of processing, and their economic viability. Evaluating the environmental impact, such as land use and sustainability, is essential to identify the most suitable starch source for bioplastics. Each source offers distinct opportunities and challenges, and continued research is necessary to enhance their application in bioplastic production and to address concerns related to processing, cost, and performance.

2.3 STARCH EXTRACTION METHODS

The extraction of starch from unconventional sources has been a vital process across various industries, from food production to pharmaceuticals. This involves isolating starch granules from plant tissues using a wide range of methods, from traditional techniques to cutting-edge mechanical strategies (Dorantes-Fuertes et al., 2024). The specific method chosen depends on the plant material and the desired purity of the extracted starch. Each method has its own benefits and drawbacks, emphasizing the importance of understanding the various approaches and their effects on starch yield, sustainability, environmental impact, and quality. Some common techniques are presented here.

2.3.1 Traditional Methods

Various methods are used to extract starch from different plant sources, including roots, tubers, and seeds. These methods are broadly categorized into primary methods, which typically involve washing and grinding, and other secondary techniques that incorporate filtration and sedimentation and operate typically through wet and/or dry pathways.

2.3.1.1 Wet Pathway

Starch extraction methods from tubers often vary depending on the specific plant. Taro, for example, is typically processed by cleaning, peeling, and cutting, followed by filtration of the resulting puree through a muslin bag to obtain starch slurry. This starch slurry is allowed to rest so that the supernatant can settle. Once settled, the supernatant is removed to produce a concentrated wet starch filtrate. This wet starch filtrate is dried in the sun, then ground into a fine powder, and stored for further use or used in the production of bioplastics (Ahmed & Khan, 2013; Moorthy et al., 1993; Singla et al., 2020).

Low-quality potatoes, often discarded or used as animal feed, are excellent source of starch. The extraction of starch from these potatoes involves preparing the raw material, soaking the potatoes in water, separating the starch from the aqueous solution, washing to remove impurities, and then drying the starch (Ali et al., 2023). The starch may be modified to improve its techno-functional properties that make it suitable for bioplastic production (Sonawane et al., 2021). Chinese yucca, a starch-rich tuber, is processed for starch extraction through a series of steps: cutting, homogenizing with water, filtering through a 200-mesh sieve, allowing the filtered product to settle, and decanting the supernatant. The remaining starch precipitate is then washed, resuspended in deionized water, and dried (Shao et al., 2023). Taro peel, an important nonconventional source of starch, is processed for starch extraction using the following method employed by a researcher: 8 kg of taro peel is cleaned, cut, and crushed to make a paste. This paste is mixed with distilled water at a 1:2 weight/volume (w/v) ratio. After several oil washes to remove impurities, the mixture is left overnight to allow starch to settle. The sedimented starch is collected, washed, dried at 30 °C for 48 hours, and ground into a fine powder. The powder is passed through 0.804 μm and 0.104 μm sieves for uniformity (Bidari et al., 2023). The final starch is stored and used for bioplastic production.

The extraction of starch from palmyra tubers involves several steps. Initially, the tubers are washed to eliminate impurities, then peeled and cut into small cubes. These cubes are milled with distilled water in a 1:4 w/v ratio. The resulting mixture is filtered through double cotton fabric multiple times. After filtering, the mixture is allowed to rest for 24 hours, enabling the starch to settle. The top liquid layer is decanted and discarded. Distilled water is added to the sediment, which is then shaken, decanted, and the top liquid is removed. This washing process is repeated until the starch is pure. Finally, the starch is dried overnight at 55 °C, milled, sifted, and stored (Sobini et al., 2022). The methods for extracting starch from potato, yucca, and sweet potato follow a similar wet extraction process. To extract starch from Indian Yam (*D. trifida*), the roots are first cleaned, rinsed with water, cut, and crushed in a blender for 2 minutes with a sufficient volume of distilled water. The resulting suspension is passed through a 200-mesh sieve several times. The slurry is then centrifuged at 1,500 rpm for 15 minutes. The sediment is washed with distilled water and dried in an oven at 45°C. Once dried, the starch is sieved using a 60-mesh sieve and stored for future use (Bou Rached et al., 2006; Pérez et al., 2011). Similarly, to extract starch from the tubers of *Colocasia esculenta* (L.), the tubers are first peeled, washed with water, cut into small pieces and subsequently soaked in water to release the starch. The mixture was then stirred and filtered through a fine mesh to obtain starch slurry. The starch suspension is centrifuged to separate starch from residue and washed several times with water to remove residue and impurities. The starch is then dried in a low temperature oven and sieved to obtain a fine starch powder (Simsek & El, 2012). Finally, the obtained starch is stored for future use. In the process of starch extraction from roots of *Canna edulis* starch, the roots are disintegrated in a blender to form a suspension which is passed through a nylon filter medium (200 μm). The starch-rich filtrate is transferred to a container for allowing starch to sediment. Once starch is settled, the supernatant is decanted, and water is added. This step is repeated at least twice. Finally, the starch is dried in an oven at 40 °C with a moisture content of approximately 11% (weight) (Cisneros et al., 2009). Starch is stored in a closed container for future use.

2.3.1.2 Dry Pathway

Starch extraction via the dry pathway technique involves drying, milling, and sifting the starch source. Each processing step varies depending on the type of raw material used. Dry pathway methods employed in starch isolation from different sources of unconventional starch can change the structural and functional properties of the extracted starch. For example, starch extraction from yucca using the dry pathway involves two techniques. The first technique includes washing and peeling the yucca, then grating it to crumble the pulp and break cell walls for easier starch release. Next, the yucca undergoes pre-dehydration at 45 °C with residual wetness levels of 20%, 30%, or 45%. It is then pre-milled under intermediate humidity to separate fiber, using a ball mill for 10 minutes. Final dehydration occurs at 60 °C until the humidity reaches 12%. The sifting process uses screens with mesh sizes of 100, 90, 71, and 45 μm (Cobana & Antezana, 2007).

The starch extraction process for tannia (*Xanthosoma sagittifolium*) and taro (*Colocasia esculenta*) involves several steps. First, clean tubers from both plants are prepared, with commercial starch from cassava (*M. esculenta*) used as a control. The preparation begins by peeling, weighing, cutting, and grinding the tubers in a blender with distilled water at high speed for 2 minutes. The homogenate (mixture) is then passed through an 80-mesh sieve, with the grinding and sieving repeated four times. The resulting suspension is filtered through 200- and 270-mesh sieves and centrifuged for 20 minutes at 1500 rpm. After removing the mucilaginous layer, the pellet is washed several times with water and the centrifuged until it is free of non-starch materials. Finally, the sediment is dried in an oven at 45 °C (Pérez et al., 2005). Two methods exist for extracting starch from the legume tuberous roots of *Pachyrhizus ahipa*: the traditional method and the proposed method. In the traditional method, water is used as a solvent for starch extraction. In this method, the chip roots are crushed, mixed with water, filtered, and allowed to rest for 24 hours. This mixing and filtration process is repeated six times to obtain the starch. This method serves as a benchmark for comparison with the proposed method. The proposed method involves using a buffer solvent instead of water during the initial extraction stages. Here, the crushed chip roots are mixed with a $(PO_4)^{-3}$/NaCl buffer (2 liters of solvent per kilogram of root), filtered, and allowed to rest for 24 hours. Following this, the mixing and filtration are repeated six times with water to extract the starch (López et al., 2010). This approach has shown a significant improvement in protein extraction while maintaining the starch yield.

2.3.2 Mechanical Methods

Mechanical methods encompass traditional techniques such as wet and dry processing, which involve crushing, pressing the raw material, and other modern techniques such as centrifugation to enhance starch yield.

2.3.2.1 Pressing and Crushing

One mechanical method for extracting starch from taro husk and modified palmyrah tuber for biodegradable bioplastics involves milling, a process that mechanically breaks down raw materials into smaller particles. In this process, the dried starch sample is ground into a powder using a commercial blender, and the powder is then sifted through a sieve (fine mesh strainer or sift) to obtain a uniform particle size (Shao et al., 2023; Sobini et al., 2022). To extract starch from purple yam (*D. trifida*), the sample is subjected to grinding using two methods: alkaline and aqueous extraction. Briefly, the sample is soaked in a 0.1% NaOH solution for 18 hours. After soaking, the yam is blended for 2 minutes at high speed. The resulting mixture is then passed through a 63-μm sieve to separate the starch from other components (Santos et al., 2022). The starch may be stored for use in different applications at this stage.

When extracting starch from potatoes (*Solanum tuberosum*), the potatoes are sliced and blended into a pulp using a commercial blender. This process makes it easier for the mixture to pass through the filter because of the pulp's consistency. The pulp is then washed multiple times with distilled water to separate the starch from other potato components (Vera Bravo & Chavarría Chavarría, 2020). The process of extracting starch from inhambu begins with removing impurities, followed by peeling and grinding the tubers using a food processor. The resulting pulp is passed through 0.149 mm and 0.079 mm meshes to remove undesired fibers and other debris. The suspended pulp is washed several times with cold water and then purified with absolute alcohol to further eliminate impurities. After filtration, the pulp is dried at 35 °C for 12 hours in an air-circulation oven. Subsequently, the dried starch is ground through a 48-mesh screen to produce a fine powder (Silva et al., 2020). This extraction method has been shown to yield starch with high levels of both starch and amylose, making it suitable for a bioplastic production.

2.3.2.2 Centrifugation

Centrifugation is essential in the starch extraction process. In one study focusing on isolating starch from sweet potatoes, cleaned roots are peeled, sliced, and blended with a 1:2 (w/v) water ratio for 3.5 minutes. The resulting slurry is then filtered through a 100-mesh sieve and washed with water (1:2 w/v) to eliminate fibrous residues. The starch is then separated by centrifugation at 3000g for 10 minutes, after which the supernatant water is discarded. The starch is washed twice more with water (1:2 w/v), with centrifugation performed after each wash to settle the starch. Finally, the starch is dried at 45 °C for 6–7 hours and then stored (Prabha et al., 2020).

The extraction of oxalate-free starch from taro flour involves several steps. It starts with treating the taro flour with oxalate oxidase, followed by dispersing various concentrations of the flour in a buffer and incubating the mixture with continuous shaking for different durations. The suspension is then homogenized, sieved, and centrifuged. Finally, the sediment is washed with distilled water and isopropanol before being dried to yield the starch (Jacob Kizhakedathil et al., 2021). To extract starch from *Dioscorea opposita* Thunb, the Chinese yam tubers are first washed, peeled, and cut into small pieces. These pieces are then dried in an aluminum tray at 45 °C for 24 hours. Once dried, the tubers are ground and sifted through a 100-mesh sieve to produce Chinese yam flour. This flour is soaked in an ethanol-sodium hydroxide solution (pH 8.5) for 12 hours at room temperature. The supernatant is carefully poured off, and the precipitated starch is washed five times with deionized water. The starch is then collected by centrifugation at 4000g for 15 minutes and dried at 45 °C for 24 hours. Now the starch is ground again and passed through a 100-mesh sieve to obtain native Chinese yam starch (Zou et al., 2020).

2.3.3 Chemical Methods

The use of chemicals (acids, bases, and solvents) can enhance the yield of starch extracted from nonconventional sources such as tubers and rhizomes. When combined with traditional and mechanical methods, these techniques can create optimal conditions for starch extraction. There are several chemical methods used for starch extraction. One method is alkaline extraction, which treats plant material with a dilute alkaline solution to dissolve non-starch components like proteins and lipids, allowing for separation of starch through centrifugation or filtration. Another method, acid hydrolysis, involves using acid to break down cell walls and release starch, although it may compromise the starch structure. To extract starch from Chinese cassava, tubers are chopped and homogenized with water in the blender to create slurry. The slurry is then filtered through a 200-mesh sieve, with the residue being washed twice with distilled water. The filtrate is allowed to stand for 2 hours, after which the supernatant is decanted, and the precipitated starch layer is resuspended with water. This process is repeated eight times. Finally, the starch is resuspended in ethyl alcohol and dried at 45 °C for 24 h (about 10% moisture), and collected by filtration through a 100-mesh sieve (Shao et al., 2023).

For starch extraction from wild yam, the tubers are washed with distilled water, peeled, sliced, and blended at room temperature using a blender. The resulting slurry is mixed with a 4% $Na_2S_2O_3$ solution and filtered through a muslin bag. The filtrate is allowed to settle for several hours, enabling the wet starch cake to form at the bottom. The supernatant is decanted. The wet cake is washed multiple times with distilled water to achieve a pure wet starch cake. This cake is subsequently dried in aluminum foil at 50 °C for 24 hours and stored in transparent polythene film at 4 °C prior to analysis. This process is crucial for obtaining pure starch for the subsequent evaluation of its physicochemical properties (Oladebeye et al., 2023). Starch was extracted from purple yam (*D. trifida*) using the alkaline extraction method; the purple yam tubers are soaked in a 0.1% NaOH solution for 18 hours, then blended at high speed for 2 minutes. The resulting mixture is passed through a 63-mesh sieve to separate the starch from other components. The mixture is then centrifuged to separate the layers and recover the starch. The recovered starch is washed with NaOH solution and deionized water to eliminate impurities. It is then neutralized with 1.0 mol/L hydrochloric acid until a pH of 6.5 is achieved. Finally, the starch is dried at 40 °C for 24 hours and stored at room temperature for further analysis (Santos et al., 2022). This alkaline extraction method effectively yields purple yam starch for subsequent characterization and evaluation of its properties. To extract oxalate-free starch from taro flour using an oxalate-oxidase-assisted process, a 4% suspension of taro flour is prepared in a succinate/NaOH buffer (pH 3.8) containing 1 mM $MnSO_4$, and then supplemented with 100 U/mg of microbial oxalate oxidase. The mixture is incubated at 55 °C for 150 minutes (Jacob Kizhakedathil et al., 2021). This innovative method enables the production of taro starch with unique properties while eliminating oxalates, thereby broadening its potential uses in various industrial applications.

2.3.4 Enzymatic Methods

Enzymatic method involves using enzymes (such as cellulases, pectinases, xylanases, and amylases) to break down the nonstarch components of the plant material, such as proteins and fiber. Starch can then be separated by filtration, centrifugation, or by any other appropriate approach. The enzymatic extraction of starch may influence the physiochemical properties of extracted starch (Liu et al., 2024). In the extraction of starch from cassava and sweet potato, enzymes like cellulases and pectinase play a crucial role. They break down the pectin-cellulosic matrix of cell membranes, which helps release starch granules more efficiently (Hameed et al., 2020). This approach has been shown to significantly enhance starch extraction yields and produce high-quality starch. In the study conducted by Sit et al. (2015), starch is extracted from taro tubers through multiple steps. Initially, the taro tubers are washed, peeled, and cut into approximately 1 cm cubes. Next, 100 g of these cubes is milled in a laboratory blender for 1 minute and 30 seconds. For the enzymatic treatment, a combination of cellulase and xylanase is used, varying the concentrations of cellulase (ranging from 0 to 100 U per 100 g of tubers) and xylanase (0 or 100 U per 100 g of tubers), along with different incubation temperatures (30 to 50 °C) and times (1–5 h). A rotative central composite design is used to optimize the starch extraction process by examining the effects of these various factors on starch yield and determining the optimal conditions for maximizing yield.

2.3.5 Combined Methods

A combination of these methods can be used to optimize starch extraction efficiency and purity, depending on the specific plant source and desired application (Sit et al., 2015). Several factors influence the choice of the combined method, including plant source as different plants have different starch structures and associated components. The purity, viscosity, and other properties of the extracted starch may influence the choice of method. The cost and efficiency of the extraction process are also important considerations. By carefully selecting the appropriate combination of extraction methods, it is possible to obtain high-quality starch with specific properties for various applications, including food, pharmaceuticals, and industrial products.

2.4 CHALLENGES AND OPPORTUNITIES

Starch-based bioplastics are a promising alternative to conventional petroleum-based plastics, but they come with their own set of challenges and limitations (Abe et al., 2021). Here are some key points to consider.

2.4.1 Mechanical Properties

Starch-based bioplastics often lack the strength and durability of conventional plastics. They can be more brittle and less resistant to impact and stress. They can also have lower flexibility and elongation properties, making them less suitable for applications requiring flexibility or high tensile strength. Efforts have been made to enhance their mechanical properties and capabilities by integrating reinforcement materials, including inorganic and lignocellulosic fibers (Boey et al., 2022).

2.4.2 Moisture Sensitivity

Because of the hydrophilic nature of starch, starch-based bioplastics are sensitive to moisture which can lead to degradation or loss of mechanical properties of starch-based bioplastics when exposed to water or high humidity (Abral & Hartono, 2017). This sensitivity limits their use in environments where moisture resistance is crucial.

2.4.3 Degradation and Stability

While biodegradability is a positive feature of starch-based bioplastics, it can also be a limitation for certain applications. Starch-based bioplastics may start to degrade under certain conditions before the end of their intended use (Nandakumar et al., 2021). Their shorter shelf life compared to conventional plastics can be problematic for long-term storage or use.

2.4.4 Processing Challenges

The processing of starch-based bioplastics requires specific conditions, such as precise temperature and humidity control. This can complicate manufacturing and increase production costs. Therefore, to improve the properties of starch-based bioplastics, various additives or modifiers are often required, which can add to the complexity and cost of production (Ahmed et al., 2024).

2.4.5 Cost

The cost of producing starch-based bioplastics can be higher than conventional plastics, particularly due to the need for specialized processing equipment and the addition of other materials to improve performance (Wellenreuther et al., 2022).

2.4.6 Resource Competition

Agricultural Resources: The production of starch-based bioplastics may compete with food production for agricultural resources. This raises concerns about the impact on food supply and land use.

2.4.7 Temperature Resistance

Starch-based bioplastics often have lower thermal resistance compared to conventional plastics, limiting their use in high-temperature applications.

2.4.8 End-of-Life Issues

Even though they are biodegradable, the conditions required for complete degradation (disposal) are not always met in natural environments or conventional waste management systems (Van Roijen & Miller, 2022). This can lead to partial degradation and potential environmental impact.

2.4.9 Consumer Perception

There may be a lack of consumer confidence in the performance and reliability of starch-based bioplastics compared to established conventional plastics. Hence, bioplastics may not be acceptable for the general masses (Filho et al., 2022; Uehara et al., 2023). Consumer acceptance of bioplastics is essential for the shift toward a bioeconomy. This requires a clear understanding of consumer preferences and perceptions regarding bioplastics, along with effective education to enable informed decision-making.

Addressing these challenges requires ongoing research and development to enhance the properties of starch-based bioplastics, improve processing techniques, and find balanced solutions that consider both environmental benefits and practical applications.

2.5 CONCLUSION

In conclusion, starch, a versatile polysaccharide found in various plant sources, plays a crucial role in both dietary and industrial applications. Its significant economic value is underscored by its use in food production, brewing, and paper manufacturing, with conventional sources like maize, potatoes, cassava, and wheat being extensively studied. The rise of unconventional starches from underutilized agricultural by-products presents exciting opportunities for innovation, particularly in the development of sustainable bioplastics. These bioplastics, derived from renewable resources, offer a more environmentally friendly alternative to traditional petroleum-based plastics, contributing to reduced greenhouse gas emissions and promoting biodegradability. As research continues to explore the unique properties of unconventional starches, their potential in creating eco-friendly materials becomes increasingly apparent. Thus, the integration of starch in bioplastic production not only supports sustainable development but also provides technological advancements, which is beneficial for both consumers and the environment. Furthermore, focused research is needed to enhance the value of starch-based bioplastics by making them simpler yet smarter, which would enable consumers to evaluate the quality, safety, and shelf-life of these eco-friendly materials while keeping costs manageable.

ACKNOWLEDGMENTS

The author is grateful to SERB-National Postdoctoral Fellowship program and Science and Engineering Research Board, Department of Science and Technology, New Delhi, India (Grant Number: PDF/2016/002730) for financial support.

REFERENCES

Abe, M. M., Martins, J. R., Sanvezzo, P. B., Macedo, J. V., Branciforti, M. C., Halley, P., Botaro, V. R., & Brienzo, M. (2021). Advantages and disadvantages of bioplastics production from starch and lignocellulosic components. *Polymers (Basel), 13* (15). https://doi.org/10.3390/polym13152484

Abral, H., & Hartono, J. (2017). Moisture absorption of starch based biocomposites reinforced with water hyacinth fibers. *IOP Conference Series: Materials Science and Engineering.*

Adams, J. B. (2004). Raw materials quality and the texture of processed vegetables. In D. Kilcast (Ed.), *Texture in Food* (pp. 342–363). Woodhead Publishing. https://doi.org/https://doi.org/10.1533/978185538362.3.342

Adetunji, A. I., Oberholster, P. J., & Erasmus, M. (2023). From garbage to treasure: A review on biorefinery of organic solid wastes into valuable biobased products. *Bioresource Technology, 24,* 101610. https://doi.org/10.1016/j.biteb.2023.101610

Ahmed, A., & Khan, F. J. I. R. N. E. (2013). Extraction of starch from taro (Colocasia esculenta) and evaluating it and further using taro starch as disintegrating agent in tablet formulation with over all evaluation. *Inventi Rapid: Novel Excipients, 2,* 1–5.

Ahmed, S., Rehman, H., & Ahmed, N. (2024). Potato starch extraction: Techniques, challenges, and future opportunities. *Journal of Pharmacognosy and Phytochemistry, 13,* 512–524. https://doi.org/10.22271/phyto.2024.v13.i4f.15046

Ai, Y., & Jane, J.-l. (2024). Understanding starch structure and functionality. In Nilsson, L. (Ed.), *Starch in Food* (pp. 55–77). Elsevier.

Ali, S. M., Siddique, Y., Mehnaz, S., & Sadiq, M. B. (2023). Extraction and characterization of starch from low-grade potatoes and formulation of gluten-free cookies containing modified potato starch. *Heliyon, 9*(9), e19581. https://doi.org/10.1016/j.heliyon.2023.e19581

Amaraweera, S. M., Gunathilake, C., Gunawardene, O. H., Fernando, N. M., Wanninayaka, D. B., Dassanayake, R. S., Rajapaksha, S. M., Manamperi, A., Fernando, C. A., & Kulatunga, A. K. J. M. (2021). Development of starch-based materials using current modification techniques and their applications: A review. *Molecules, 26* (22), 6880.

Ancy, A., Lazar, M., Chandran, A. S., Ushamani, M. J. S. C., & Pharmacy. (2024). Development of ecofriendly and sustainable bioplastics from cassava starch: Tailoring the properties using nanoparticles. *Sustainable Chemistry and Pharmacy, 37,* 101377.

Arikan, E. B., & Ozsoy, H. D. (2015). A review: Investigation of bioplastics. *Journal of Civil Engineering and Architecture, 9* (1), 188–192.

Azieyanti, N., Amirul, A., Othman, S., & Misran, H. (2020). Mechanical and morphology studies of bioplastic-based banana peels. *Journal of Physics: Conference Series, 1529* (3):032091. https://doi.org/10.1088/1742-6596/1529/3/032091

Bally, I. S. E. (2006). Mangifera indica (mango). *Species Profiles for Pacific Island Agroforestry,* 4, 1–25.

Bangar, S. P., Dhull, S. B., Manzoor, M., Chandak, A., & Esua, O. J. J. S. S. (2024). Functionality and applications of nonels. *Journal of Physics: Conference Series*, *76* (1–2), 2300073.

Bangar, S. P., Kumar, M., Whiteside, W. S., Tomar, M., Kennedy, J. F. J. C. P. T., & Applications. (2021). Litchi (*Litchi chinensis*) seed starch: Structure, properties, and applications-A review. *Carbohydrate Polymer Technologies and Applications*, *2*, 100080. https://doi.org/10.1016/j.carpta.2021.100080

Behera, L., Mohanta, M., Thirugnanam, A. (2022). Intensification of yam-starch based biodegradable bioplastic film with bentonite for food packaging application. *Environmental Technology and Innovation*, *25*, 102180. https://doi.org/10.1016/j.eti.2021.102180

Bhagwat, G., Gray, K., Wilson, S. P., Muniyasamy, S., Vincent, S. G. T., Bush, R., & Palanisami, T. (2020). Benchmarking bioplastics: A natural step towards a sustainable future. *Journal of Polymers and the Environment, 28* (12), 3055–3075. https://doi.org/10.1007/s10924-020-01830-8

Bhore, S., Salgado, D., Houssari, A., Zelaya, A., Yang, R., Chen, Z., Deeya, S., Sens, S., Schumann, M., Zhang, Z., & Eltantawy, a. (2021). *The Avocado (Persea americana Mill.): A Review and Sustainability Perspectives.* https://doi.org/10.20944/preprints202112.0523.v1

Bidari, R., Abdillah, A. A., Ponce, R. A. B., & Charles, A. L. (2023). Characterization of biodegradable films made from taro peel (*Colocasia esculenta*) starch. *Polymers*, *15* (2), 338. https://doi.org/10.3390/polym15020338

Boey, J. Y., Lee, C. K., & Tay, G. S. (2022). Factors affecting mechanical properties of reinforced bioplastics: A review. *Polymers (Basel), 14* (18). https://doi.org/10.3390/polym14183737

Bou Rached, L., de Vizcarrondo, C. A., Rincón, A. M., & Padilla, F. (2006). Evaluación de harinas y almidones de mapuey (*Dioscorea trifida*), variedades blanco y morado. *Archivos Latinoamericanos de Nutrición*, *56* (4), 375–383.

Brahma, R., Ray, S. (2023). Finding out various potentials and possibilities of jackfruit seed and its usage in the industry: A review. *Food Production Processing and Nutrition*, *5* (1), 55. https://doi.org/10.1186/s43014-023-00170-z

Cecil, J. E., Lau, G., Heng, S. H., & Ku, C. K. (1982). *The Sago Starch Industry: A Technical Profile Based on a Preliminary Study Made in Sarawak*. London, UK: Tropical Products Inst.

Chan, J. X., Wong, J. F., Hassan, A., & Zakaria, Z. (2021). Bioplastics from agricultural waste. In *Biopolymers and Biocomposites from Agro-Waste for Packaging Applications* (pp. 141–169). Elsevier.

Chandrasekar, C. M., Krishnamachari, H., Farris, S., & Romano, D. (2023). Development and characterization of starch-based bioactive thermoplastic packaging films derived from banana peels. *Carbohydrate Polymer Technologies and Applications*, *5*, 100328. https://doi.org/https://doi.org/10.1016/j.carpta.2023.100328

Cheong, K. S., Balasubramaniam, J.-R., Hung, Y. P., Chuong, W. S., & Amartalingam, R. (2010). Development of biodegradable plastic composite blends based on sago derived starch and natural rubber. *Pertanika Journal of Science and Technology*, *18* (2), 411–420.

Cisneros, F. H., Zevillanos, R., Cisneros-Zevallos, L. J. J. o. A., & Chemistry, F. (2009). Characterization of starch from two ecotypes of andean achira roots (*Canna edulis*). *Journal of Agricultural and Food Chemistry*, *57* (16), 7363–7368.

Cobana, M., & Antezana, R. (2007). Proceso de extracción de almidón de yuca por vía seca. *Revista Boliviana de Química, 24* (1), 77–83.

Coppola, G., Gaudio, M. T., Lopresto, C. G., Calabro, V., Curcio, S., Chakraborty, S. (2021). Bioplastic from renewable biomass: A facile solution for a greener environment. *Earth Systems and Environment*, *5*, 231–251.

Corbishley, D. A., & Miller, W. (1984). Chapter XIII – tapioca, arrowroot, and sago starches: Production. In R. L. Whistler, J. N. Bemiller, & E. F. Paschall (Eds.), *Starch: Chemistry and Technology (Second Edition)* (pp. 469–478). Academic Press. https://doi.org/https://doi.org/10.1016/B978-0-12-746270-7.50019-7

Costa, B. P., Carpiné, D., Ikeda, M., Pazzini, I. A. E., Alves, F. E. d. S. B., de Melo, A. M., & Ribani, R. (2023). Bioactive coatings from non-conventional loquat (*Eriobotrya japonica* Lindl.) seed starch to extend strawberries shelf-life: An antioxidant packaging. *Progress in Organic Coatings*, *175*, 107320. https://doi.org/10.1016/j.porgcoat.2022.107320

Dhull, S. B., Chandak, A., Chawla, P., Goksen, G., Rose, P. K., & Rani, J. (2023). Modifications of native lotus (*Nelumbo nucifera* G.) rhizome starch and its overall characterization: A review. *International Journal of Biological Macromolecules, 253*, 127543. https://doi.org/https://doi.org/10.1016/j.ijbiomac.2023.127543

Dini, C., García, M. A., Viña, S. (2012). Non-traditional flours: frontiers between ancestral heritage and innovation. *Food & Function, 3* (6), 606–620.

Dorantes-Fuertes, M.-G., López-Méndez, M. C., Martínez-Castellanos, G., Meléndez-Armenta, R. Á., & Jiménez-Martínez, H.-E. (2024). Starch extraction methods in tubers and roots: A systematic review. *Agronomy, 14* (4), 865. https://doi.org/10.3390/agronomy14040865

Emadian, S. M., Onay, T. T., & Demirel, B. (2017). Biodegradation of bioplastics in natural environments. *Waste management, 59*, 526–536.

Esua, J., Okonkwo, C., Ajani, C., Bassey, E., Ali, M., Hussain, N., Bassey, A., Adeyanyu, A., Chin, N., Zeng, X.-A., & Liu, H. (2024). Litchi seed starch-isolation, modification, and characterization. In J. M. Lorenzo & S. Punia Bangar (Eds.), *Non-Conventional Starch Sources* (pp. 21–59). Academic Press. https://doi.org/10.1016/B978-0-443-18981-4.00002-1

European Bioplastics Organisation. (2024). *Bioplastics market development update*. www.european-bioplastics.org/bioplastics-market-development-update-2024/

Fatima, S., Khan, M. R., Ahmad, I., & Sadiq, M. B. (2024). Recent advances in modified starch based biodegradable food packaging: A review. *Heliyon, 10* (6), e27453. https://doi.org/https://doi.org/10.1016/j.heliyon.2024.e27453

Ferdaus, M. J., Chukwu-Munsen, E., Foguel, A., & da Silva, R. C. (2023). Taro roots: An underexploited root crop. *Nutrients, 15* (15). https://doi.org/10.3390/nu15153337

Filho, W. L., Barbir, J., Abubakar, I. R., Paço, A., Stasiskiene, Z., Hornbogen, M., Christin Fendt, M. T., Voronova, V., & Klõga, M. (2022). Consumer attitudes and concerns with bioplastics use: An international study. *PloS One, 17* (4), e0266918. https://doi.org/10.1371/journal.pone.0266918

Gadhave, R. V., Das, A., Mahanwar, P. A., & Gadekar, P. T. (2018). Starch based bio-plastics: The future of sustainable packaging. *Open Journal of Polymer Chemistry, 8*, 21–33. https://doi.org/10.4236/ojpchem.2018.82003

George, N., Debroy, A., Bhat, S., Bindal, S., & Singh, S. (2021). Biowaste to bioplastics: An ecofriendly approach for a sustainable future. *Journal of Applied Biotechnology Reports, 8* (3), 221–233.

Gironi, F., Piemonte, V. (2011). Bioplastics and petroleum-based plastics: strengths and weaknesses. *Energy Sources, Part A: Recovery, Utilization, and Environmental Effects, 33* (21), 1949–1959.

Hameed, M., Ahmad, S. W., Ahmad, S., Qutab, H. G., Dasih, M., & Imran, M. (2020). Enzymatic extraction of potato starch: A parametric optimization study using response surface methodology. *Polish Journal of Chemical Technology, 22* (3), 48–54.

Henning, F. G., Ito, V. C., Demiate, I. M., & Lacerda, L. G. (2022). Non-conventional starches for biodegradable films: A review focussing on characterisation and recent applications in food packaging. *Carbohydrate Polymer Technologies and Applications, 4*, 100157. https://doi.org/https://doi.org/10.1016/j.carpta.2021.100157

Hornung, P. S., Masisi, K., Malunga, L. N., Beta, T., & Ribani, R. H. (2018). Natural bioactive starch film from Amazon turmeric (*Curcuma longa* L.). *Polymer Bulletin, 75* (10), 4735–4752. https://doi.org/10.1007/s00289-018-2285-2

Hsieh, C.-F., Liu, W., Whaley, J. K., Shi, Y.-C.. (2019). Structure, properties, and potential applications of waxy tapioca starches–A review. *Trends in Food Science & Technology, 83*, 225–234.

Hu, Z., Zhao, L., Hu, Z., & Wang, K. (2018). Hierarchical structure, gelatinization, and digestion characteristics of starch from Longan (*Dimocarpus longan* Lour.) seeds. *Molecules, 23* (12), 3262. https://doi.org/10.3390/molecules23123262

Jacob Kizhakedathil, M. P., Suvarna, S., Belur, P. D., Wongsagonsup, R., Agoo, E. M. G., Janairo, J. I. B. (2021). Optimization of oxalate-free starch production from Taro flour by oxalate oxidase assisted process. *Preparative Biochemistry & Biotechnology, 51* (2), 105–111. https://doi.org/10.1080/10826068.2020.1795672

Jamal, T., Sapuan, S., Abdan, K., R. A, I., Zainudin, E. S., Sherwani, S., M.M, H., & Kamaruddin, Z. H. (2022). *Mechanical Performance of Arrowroot (Maranta Arundinacea) Starch-Based Biopolymer Composites: A Review*. In *Proceedings of the 8th Postgraduate Seminar on Natural Fibre Composites*

2022 and Undergraduate Seminar on Fabrication and Characterization of Composite Materials (FCCM 2022). Universiti Putra Malaysia.

Joseph, S., Hegde, A. R., Gopalakrishnan, V., Yallappa, S., Nadzri, N., Joseph, K., & Meenakshi, K. (2024). Biodegradable plastics from mango seed starch for sustainable food packaging-effect of citric acid and fillers. *ChemistrySelect, 9* (22), e202401312.

Kahn, V. (2006). Characterization of starch isolated from avocado seeds. *Journal of Food Science*, *52*, 1646–1648. https://doi.org/10.1111/j.1365-2621.1987.tb05896.x

Kittipongpatana, N., Wiriyacharee, P., Phongphisutthinant, R., Chaipoot, S., Somjai, C., & Kittipongpatana, O. S. (2021). Resistant starch contents of starch isolated from black Longan seeds. *Molecules*, *26* (11), 3405. https://doi.org/10.3390/molecules26113405

Kringel, D. H., Dias, A. R. G., Zavareze, E. d. R., & Gandra, E. (2020). Fruit wastes as promising sources of starch: Extraction, properties, and applications. *Starch – Stärke*, *72* (3–4), 1900200. https://doi.org/10.1002/star.201900200

Krochmal-Marczak, B., Sawicka, B., & Barbaś, P. (2022). Utility Meaning of Jerusalem Artichoke. In B. Sawicka & B. Krochmal-Marczak (Eds.), *Jerusalem Artichoke Food Science and Technology: Helianthus Tuberosus* (pp. 91–138). Springer Nature Singapore. https://doi.org/10.1007/978-981-19-0805-7_4

Kumar, S., Pattanayak, S., & Pandey, K. M. (2021). Bio-plastic from Yam—An ecofriendly prospective. Recent *Advances in Mechanical Engineering: Select Proceedings of* ICRAME 2020.

Kuttigounder, D., Rao, L., & Bhattacharya, S. (2011). Turmeric powder and starch: Selected physical, physicochemical, and microstructural properties. *Journal of Food Science*, *76*, C1284–C1291. https://doi.org/10.1111/j.1750-3841.2011.02403.x

Lackner, M. (2000). Bioplastics. In Kirk-Othmer Encyclopedia of Chemical Technology. John Wiley & Sons. https://doi.org/10.1002/0471238961.koe00006

Li, H., Qi, Y., Zhao, Y., Chi, J., & Cheng, S. (2019). Starch and its derivatives for paper coatings: A review. *Progress in Organic Coatings*, *135*, 213–227.

Li, H., Zhou, M., Mohammed, A. E. A. Y., Chen, L., Zhou, C. (2022). From fruit and vegetable waste to degradable bioplastic films and advanced materials: A review. *Sustainable Chemistry and Pharmacy*, *30*, 100859. https://doi.org/10.1016/j.scp.2022.100859

Li, Z., Guo, K., Lin, L., He, W., Zhang, L., & Wei, C. (2018). Comparison of physicochemical properties of starches from flesh and peel of green banana fruit. *Molecules*, *23* (9). https://doi.org/10.3390/molecules23092312

Liu, Z., Zhao, Y., Zheng, J., Wang, Z., Yan, X., & Zhang, T. (2024). Influence of enzymatic extraction on the properties of corn starch. *Food Bioscience*, *58*, 103775. https://doi.org/https://doi.org/10.1016/j.fbio.2024.103775

López, O. V., Vina, S. Z., Pachas, A. N., Sisterna, M. N., Rohatsch, P. H., Mugridge, A., Fassola, H. E., García, M. (2010). Composition and food properties of Pachyrhizus ahipa roots and starch. *International Journal of Food Science & Technology*, *45* (2), 223–233. https://doi.org/10.1111/j.1365-2621.2009.02125.x

Lusiana, S., Putri, D., & Nurazizah, I. Z. (2019). Bioplastic properties of sago-PVA starch with glycerol and sorbitol plasticizers. *Journal of Physics: Conference Series*, *1351* (1), 012102. https://doi.org/10.1088/1742-6596/1351/1/012102 .

Manzoor, A., & Ahmad, S. (2024). Chapter 4 – Longan seed starch: structure, functionality and applications. In J. M. Lorenzo & S. P. Bangar (Eds.), *Non-Conventional Starch Sources* (pp. 103–121). Academic Press. https://doi.org/https://doi.org/10.1016/B978-0-443-18981-4.00004-5

Marzo-Gago, C., Díaz, A. B., & Blandino, A. (2023). Sugar beet pulp as raw material for the production of bioplastics. *Fermentation*, *9* (7), 655. https://doi.org/10.3390/fermentation9070655

Mason, W. (2009). Starch use in foods. In J. BeMiller & R. Whistler (Eds.), *Starch in Food: Structure, Function and Applications* (2nd ed.). Woodhead Publishing. https://doi.org/10.1016/B978-0-12-746275-2.00020-3

Maulida, Kartika, T., Harahap, M., & Ginting, M. H. S. (2018). Utilization of mango seed starch in manufacture of bioplastic reinforced with microparticle clay using glycerol as plasticizer. *IOP Conference Series: Materials Science and Engineering*, *309*, 012068. https://doi.org/10.1088/1757-899X/309/1/012068

Maulida, M., & Tarigan, P. (2016). Production of starch based bioplastic from cassava peel reinforced with microcrystalline celllulose avicel PH101 using sorbitol as plasticizer. *Journal of Physics: Conference Series*, *710* (1), 012012. https://doi.org/10.1088/1742-6596/710/1/012012 .

Moorthy, S., Pillai, P. K., & Unnikrishnan, M. (1993). Variability in starch extracted from taro. *Carbohydrate Polymers, 20* (3), 169–173.

Muneer, F., Hussain, S., Sidra-tul-Muntaha, M. R., & Nadeem, H. (2021). Plastics versus bioplastics. In *Sustainable Polymers and Composites* (pp. 123–142). Springer. https://doi.org/10.21741/9781644901335-9

Nandakumar, A., Chuah, J.-A., & Sudesh, K. (2021). Bioplastics: A boon or bane? *Renewable and Sustainable Energy Reviews, 147*, 111237. https://doi.org/https://doi.org/10.1016/j.rser.2021.111237

Nguyen, T. K., That, N. T. T., Nguyen, N. T., & Nguyen, H. T. J. A. i. P. T. (2022). Development of starch-based bioplastic from jackfruit seed. *Advances in Polymer Technology, 2022* (1), 6547461. https://doi.org/10.1155/2022/6547461

Okazaki, M. (2018). The structure and characteristics of sago starch. In Ehara, H., Toyoda, Y., Johnson, D. (Eds.), *Starch: Chemistry and Technology* (4th ed., pp. 377–392). Springer. https://doi.org/10.1007/978-981-10-5269-9_18

Oladebeye, A. O., Oladebeye, A. A., & Arawande, J. O. (2023). Physicochemical properties of wild yam (*Dioscorea villosa*) starch. *International Journal of Food Science, 2023* (1), 8868218. https://doi.org/10.1155/2023/8868218

Oliveira, A. R., Chaves Ribeiro, A. E., Gondim, Í. C., Alves dos Santos, E., Resende de Oliveira, É., Mendes Coutinho, G. S., Soares Júnior, M. S., & Caliari, M. (2021). Isolation and characterization of yam (*Dioscorea alata* L.) starch from Brazil. *LWT, 149*, 111843. https://doi.org/https://doi.org/10.1016/j.lwt.2021.111843

Otálora González, C., Flores, S., Basanta, M., & Gerschenson, L. (2020). Effect of beetroot (Beta vulgaris L. var conditiva) fiber filler and corona treatment on cassava starch films properties. *Food Packaging and Shelf Life, 25*, 100605. https://doi.org/10.1016/j.fpsl.2020.100605

Pei-Lang, A. T., Mohamed, A. M. D., & Karim, A. A. (2006). Sago starch and composition of associated components in palms of different growth stages. *Carbohydrate Polymers, 63* (2), 283–286. https://doi.org/https://doi.org/10.1016/j.carbpol.2005.08.061

Pérez-Pacheco, E., Ortiz-Fernández, A., Ríos-Soberanis, C., Estrada-León, R., Moo-Huchín, V., Pérez-Padilla, Y., Canto-Pinto, J. C., & Dzul-Cervantes, M. A. (2024). Characterization of unconventional sources of starch: Physicochemical and thermal properties [Preprint]. *Research Square.* https://doi.org/10.21203/rs.3.rs-4745824/v1

Pérez, E., Gibert, O., Rolland-Sabaté, A., Segovia, X., Sánchez, T., Reynes, M., & Dufour, D. J. I. j. o. c. c. (2011). Evaluation of the functional properties of promising *Dioscorea trifida* L. waxy starches for food innovation. *Carbohydrate Polymers, 2011* (1), 165638.

Pérez, E., Schultz, F. S., & de Delahaye, E. P. J. C. p. (2005). Characterization of some properties of starches isolated from *Xanthosoma sagittifolium* (tannia) and *Colocassia esculenta* (taro). *Carbohydrate Polymers, 60* (2), 139–145.

Pfister, B., Zeeman, S. C. (2016). Formation of starch in plant cells. *CMLS, 73* (14), 2781–2807. https://doi.org/10.1007/s00018-016-2250-x

Porras-Saavedra, J., Palacios-González, E., Tovar-Benítez, T., Contreras-Lavida, H., Houbron, E. P., & Canul-Chan, M. (2024). Bioplastics made from starch as conventional plastic substitutes: Evaluation of physical properties and biodegradation. *Evaluation of physical properties and biodegradation* [Preprint]. Research Square. https://doi.org/10.21203/rs.3.rs-3783931/v1

Prabha, V., Dash, S. K., Rayaguru, K., Panda, M., & Nedunchezhiyan, M. (2020). Optimization of starch isolation process for sweet potato and characterization of the prepared starch. *Journal of Food Measurement and Characterization, 14*. https://doi.org/10.1007/s11694-020-00401-8

Punia Bangar, S., Kumar, M., & Whiteside, W. S. (2021). Mango seed starch: A sustainable and eco-friendly alternative to increasing industrial requirements. *International Journal of Biological Macromolecules, 183*, 1807–1817. https://doi.org/https://doi.org/10.1016/j.ijbiomac.2021.05.157

Punia Bangar, S., Singh, A., Chaudhary, V., Sharma, N., & Lorenzo, J. M. (2022). Beetroot as a novel ingredient for its versatile food applications. *Critical Reviews in Food Science and Nutrition, 63*. https://doi.org/10.1080/10408398.2022.2055529

Rasmi, Y., Kırboğa, K. K., Tekin, B., & Demir, M. (2024). Chapter 13 – Turmeric starch: Structure, functionality, and applications. In J. M. Lorenzo & S. P. Bangar (Eds.), *Non-Conventional Starch Sources* (pp. 377–405). Academic Press. https://doi.org/https://doi.org/10.1016/B978-0-443-18981-4.00013-6

Reddy, R. L., Reddy, V. S., Gupta, G. A. J. I. J. o. E. T., & Engineering, A. (2013). Study of bio-plastics as green and sustainable alternative to plastics. *3* (5), 76–81.

Rosenboom, J.-G., Langer, R., & Traverso, G. (2022). Bioplastics for a circular economy. *Nature Reviews Materials*, *7* (2), 117–137. https://doi.org/10.1038/s41578-021-00407-8

Santos, S. J. L., Pires, M. B., Amante, E. R., da Cruz Rodrigues, A. M., da Silva, L. H. M. (2022). Isolation and characterization of starch from purple yam (*Dioscorea trifida*). *Journal of Food Science and Technology*, *59* (2), 715–723. https://doi.org/10.1007/s13197-021-05066-9

Shahrim, N., Sarifuddin, N., & Ismail, H. (2018). Extraction and characterization of starch from mango seeds. *Journal of Physics: Conference Series*. *1082* (1), 012019. https://doi.org/10.1088/1742-6596/1082/1/012019

Shanmathy, M., Mohanta, M., & Arunachalam, T. (2021). Development of biodegradable bioplastic films from taro starch reinforced with bentonite. *Carbohydrate Polymer Technologies and Applications*, *2*, 100173. https://doi.org/10.1016/j.carpta.2021.100173

Shao, Y., Jiao, R., Wu, Y., Xu, F., Li, Y., Jiang, Q., Zhang, L., Mao, L. (2023). Physicochemical and functional properties of the protein–starch interaction in Chinese yam. *Food Science & Nutrition*, *11* (3), 1499–1506. https://doi.org/10.1002/fsn3.3189

Sharma, K., Kalra, P., & Kaur, B. (2024). Production of bioplastics from banana peels. In R. K. Arya, G. D. Verros, O. P. Verma, & C. M. Hussain (Eds.), *From Waste to Wealth* (pp. 1471–1491). Springer Nature Singapore. https://doi.org/10.1007/978-981-99-7552-5_67

Silva, G. M. S., Veloso, C. M., Santos, L. S., Melo Neto, B. A., Fontan, R. D. C. I., Bonomo, R. C. F. (2020). Extraction and characterization of native starch obtained from the inhambu tuber. *Journal of Food Science and Technology*, *57*(5), 1830–1839. https://doi.org/10.1007/s13197-019-04216-4

dos Silveira Junior, J. F., & de Francisco, A. (2020). Unconventional food plants as an alternative in starch production. *Cereal Foods World*, *65*, 2.

Simsek, S., & El, S. N. (2012). Production of resistant starch from taro (*Colocasia esculenta* L. Schott) corm and determination of its effects on health by in vitro methods. *Carbohydrate Polymers*, *90* (3), 1204–1209. https://doi.org/10.1016/j.carbpol.2012.06.039

Singla, D., Singh, A., Dhull, S. B., Kumar, P., Malik, T., & Kumar, P. (2020). Taro starch: Isolation, morphology, modification and novel applications concern-A review. *International Journal of Biological Macromolecules*, *163*, 1283–1290. https://doi.org/10.1016/j.ijbiomac.2020.07.093

Sit, N., Deka, S. C., & Misra, S. (2015). Optimization of starch isolation from taro using combination of enzymes and comparison of properties of starches isolated by enzymatic and conventional methods. *Journal of Food Science and Technology*, *52* (7), 4324–4332. https://doi.org/10.1007/s13197-014-1462-z

Sobini, N., Darsiga, S., Kananke, T., & Srivijeindran, S. J. F. C. A. (2022). Characterization of modified palmyrah tuber starch by pre-gelatinization, acid and dextrinization processes and its applicability. *1*, 100143.

Sonawane, P., Panchal, A., Naik, S., Mundaye, B., & Padalia, U. (2021). Potato starch based bioplastic as alternative packaging materials. *African Journal of Biological Sciences*, *3* (3), 111–113. https://ssrn.com/abstract=3902389

Tagliapietra, B. L., Felisberto, M. H. F., Sanches, E. A., Campelo, P. H., & Clerici, M. T. P. S. (2021). Non-conventional starch sources. *Current Opinion in Food Science*, *39*, 93–102. https://doi.org/10.1016/j.cofs.2020.02.005

Tarique, J., Sapuan, S. M., Khalina, A., Sherwani, S. F. K., Yusuf, J., & Ilyas, R. A. (2022). Recent developments in sustainable arrowroot (*Maranta arundinacea* Linn) starch biopolymers, fibres, biopolymer composites and their potential industrial applications: A review. *Journal of Materials Research and Technology*, *13*, 1191–1219. https://doi.org/https://doi.org/10.1016/j.jmrt.2021.05.047

Tattiyakul, J., Asavasaksakul, S., & Pradipasena, P. (2006). Chemical and physical properties of flour extracted from taro *Colocasia esculenta* (L.) Schott grown in different regions of Thailand. *ScienceAsia*, *32* (3), 279–284. https://doi.org/10.2306/scienceasia1513-1874.2006.32.279

Thomas, A. P., Kasa, V. P., Dubey, B. K., Sen, R., & Sarmah, A. K. (2023). Synthesis and commercialization of bioplastics: Organic waste as a sustainable feedstock. *Science of The Total Environment*, *904*, 167243. https://doi.org/https://doi.org/10.1016/j.scitotenv.2023.167243

Tosif, M. M., Bains, A., Sadh, P. K., Sarangi, P. K., Kaushik, R., Burla, S. V. S., Chawla, P., & Sridhar, K. J. I. J. o. B. M. (2023). Loquat seed starch—Emerging source of non-conventional starch: Structure, properties, and novel applications. *International Journal of Biological Macromolecules*, *244*, 125230. https://doi.org/10.1016/j.ijbiomac.2023.125230

Uehara, T., Nakatani, J., Tsuge, T., & Asari, M. (2023). Consumer preferences and understanding of bio-based and biodegradable plastics. *Journal of Cleaner Production*, *417*, 137979. https://doi.org/https://doi.org/10.1016/j.jclepro.2023.137979

Van Roijen, E. C., & Miller, S. A. (2022). A review of bioplastics at end-of-life: Linking experimental biodegradation studies and life cycle impact assessments. *Resources, Conservation and Recycling*, *181*, 106236. https://doi.org/https://doi.org/10.1016/j.resconrec.2022.106236

Bravo, A., & Chavarría, M. (2020). Extracción y caracterización del almidón de papa (Solanum tuberosum) variedad leona blanca. *Revista Ciencia y Tecnología El Higo*, *10* (2), 26–34. https://doi.org/10.5377/elhigo.v10i2.10550

Wahyuningtiyas, N., & Suryanto, H. (2017). Analysis of biodegradation of bioplastics made of cassava starch. *Journal of Mechanical Engineering Science and Technology*, *1* (1), 41–54. https://doi.org/10.17977/um016v1i12017p024

Wang, H., Yang, Q., Ferdinand, U., Gong, X., Qu, Y., Gao, W., Ivanistau, A., Feng, B., & Liu, M. (2020). Isolation and characterization of starch from light yellow, orange, and purple sweet potatoes. *International Journal of Biological Macromolecules*, *160*, 660–668. https://doi.org/https://doi.org/10.1016/j.ijbiomac.2020.05.259

Wang, J., Li, Y., Jin, Z., & Cheng, Y. (2022). Physicochemical, morphological, and functional properties of starches isolated from avocado seeds, a potential source for resistant starch. *Biomolecules*, *12*, 1121. https://doi.org/10.3390/biom12081121

Wei, Q., Zheng, H., Zhu, M., Han, X., Li, Y., Zhou, J. J. P., & Biomaterials. (2021). Starch-based surface-sizing agents in paper industry: An overview. *Biomolecules*, *6* (4), 54–61.

Wellenreuther, C., Wolf, A., & Zander, N. (2022). Cost competitiveness of sustainable bioplastic feedstocks – A Monte Carlo analysis for polylactic acid. *Cleaner Engineering and Technology*, *6*, 100411. https://doi.org/https://doi.org/10.1016/j.clet.2022.100411

Yazid, N. S. M., Abdullah, N., Muhammad, N., & Matias-Peralta, H. M. (2018). Application of starch and starch-based products in food industry. *Journal of Science and Technology*, *10* (2). https://publisher.uthm.edu.my/ojs/index.php/JST/article/view/3022

Yu, H., Cheng, L., Yin, J., Yan, S., Liu, K., Zhang, F., Xu, B., Li, L. J. F. S., & Nutrition. (2013). Structure and physicochemical properties of starches in lotus (N elumbo nucifera Gaertn.) rhizome. *Food Science & Nutrition*, *1* (4), 273–283. https://doi.org/10.1002/fsn3.37

Zhang, P., Whistler, R. L., BeMiller, J. N., & Hamaker, B. R. J. C. p. (2005). Banana starch: Production, physicochemical properties, and digestibility—A review. *Carbohydrate Polymers*, *59* (4), 443–458.

Zhao, Y., Zhu, X., & Fang, Y. (2021). Structure, properties and applications of kudzu starch. *Food Hydrocolloids*, *119*, 106817. https://doi.org/https://doi.org/10.1016/j.foodhyd.2021.106817

Zhu, F. (2020). Underutilized and unconventional starches: Why should we care? *Trends in Food Science & Technology*, *100*. https://doi.org/10.1016/j.tifs.2020.04.018

Zou, J., Xu, M., Wen, L., & Yang, B. J. C. P. (2020). Structure and physicochemical properties of native starch and resistant starch in Chinese yam (*Dioscorea opposita* Thunb.). *237*, 116188.

3 Starch from Industrial and Agricultural Byproducts

Eliseo Cárdenas-Hernández, Sk Asraful Ali, and Luis Fernando Flores

3.1 INTRODUCTION

Synthetic polymers, or petrochemical-based plastics, are widely used in various applications that are integral to daily life. However, their extensive use has escalated global concerns for both the environment and human health, underscoring the urgent need for sustainable alternatives to conventional plastics (Rosenboom et al., 2022). Derived primarily from the nonrenewable petroleum resources, traditional plastics contribute to the saturation of landfill sites, exacerbate water and land pollution, and accelerate the depletion of fossil fuel reserves (Singh et al., 2024). Each year, more than 9 million tons of plastic waste enter the oceans, contributing to 165 million tons of plastic pollution already present in our waters. Shockingly, only 9% of plastic is recycled; the remaining waste either pollutes the environment or is sent to landfills, where it can take as long as 500 years to decompose, all the while leaching harmful chemicals into the soil (Anitha et al., 2024). These challenges, along with the harmful impacts of traditional plastics on human health, have prompted research into eco-friendly alternatives. Figure 3.1 illustrates the classification of plastics into petroleum-derived plastics and bioplastics, highlighting the diverse range of bioplastics available as sustainable alternatives to conventional petroleum-based plastics.

Starch, a naturally abundant carbohydrate found in plants, is not only renewable but also biodegradable, making it an attractive candidate for bioplastic production. Traditionally, starch is extracted from food crops such as corn, potatoes, and cassava (George et al., 2021). While these sources are renewable, using food crops for bioplastic production raises ethical concerns about food security and competition between food and industrial uses (Mhaske et al., 2024; Weligama Thuppahige et al., 2023). This has driven research toward identifying alternative, nonfood sources of starch and other polysaccharides, which could mitigate the environmental and societal impacts of bioplastic production. In this context, agricultural and industrial byproducts present an untapped resource with immense potential for the sustainable production of bioplastics (Nalini & Sathiyamurthi, 2023).

Industrial and agricultural byproducts, often referred to as renewable residues, are frequently discarded or underutilized despite containing valuable polysaccharides such as starch, cellulose, and hemicellulose, which can be harnessed for bioplastic production (Choudhary et al., 2024; Flores et al., 2024). Hence, utilizing these waste materials offers a dual benefit: reducing the environmental burden of waste disposal and providing a sustainable feedstock for the bioplastics industry. Moreover, this approach also supports broader goals of food security, soil health, and water conservation. Therefore, by shifting toward waste-derived bioplastics, there is a significant opportunity to reduce the ecological and carbon footprint of plastic production, contribute to a more sustainable economy, and promote the conservation of natural resources (Gong et al., 2024; Mujtaba et al., 2023; Rahardiyan et al., 2023; Sadh et al., 2023). Despite the promising outlook, several challenges must

 DOI: 10.1201/9781003534914-3

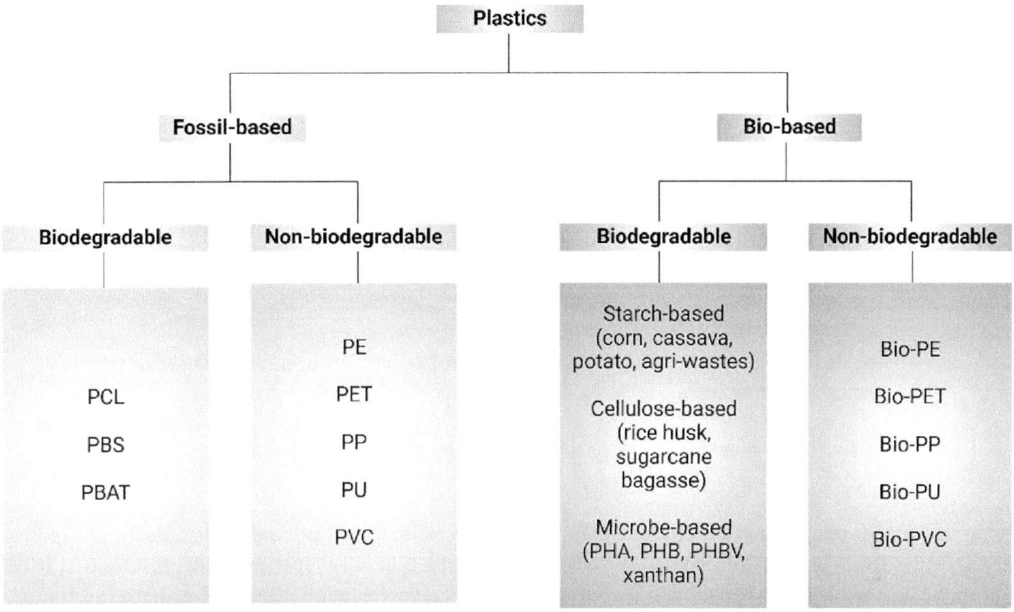

FIGURE 3.1 Different types of plastics.

be addressed to fully harness the potential of byproduct-derived starch for bioplastic applications. Extracting high-quality starch from renewable residues requires the development of optimized methods that ensure both efficiency and scalability, enabling starch recovery processes to be adapted for industrial-scale production (Nagarajan et al., 2024; Puri et al., 2024; Show et al., 2017). By fine-tuning these methods, starch-based bioplastics can be engineered to meet specific performance requirements, thereby providing a competitive alternative to petroleum-based plastics. This chapter examines the potential of agricultural and industrial byproducts as sustainable, alternative sources for advancing the starch-based bioplastics industry.

3.2 INDUSTRIAL AND AGRICULTURAL BYPRODUCTS AS FEEDSTOCK FOR STARCH-BASED BIOPLASTICS

Approximately 80% of total bioplastic production is contributed by starch-based bioplastics (Chakraborty et al., 2022) – sourced from corn (75%), cassava (12%), wheat (8%), and potatoes (5%) (Vilpoux & Santos Silveira Junior, 2023) – highlighting its significance in the industry. Despite its advantages, the reliance on primary agricultural commodities like corn and potatoes for starch production raises significant concerns about food security and sustainability. This has led to increased interest in utilizing agricultural byproducts as alternative sources of starch for bioplastics (Mhaske et al., 2024; Pham et al., 2024).

Agricultural byproducts, also referred to as agricultural residues, include a wide range of materials, such as straw, husks, peels, stalks, and leaves, which are left behind after the primary crop has been harvested. Examples of such byproducts include rice husks, wheat straw, corn stover, and banana peels, among others (Vigneswari et al., 2024). These byproducts are typically rich in polysaccharides, including starch, cellulose, and hemicellulose (Abrha et al., 2022; Maraveas, 2020), which can be extracted and utilized in bioplastics production within the principles of a circular economy, wherein waste materials are repurposed and given a new value, further enhancing the sustainability of the bioplastics industry. In addition to agricultural residues, various industrial

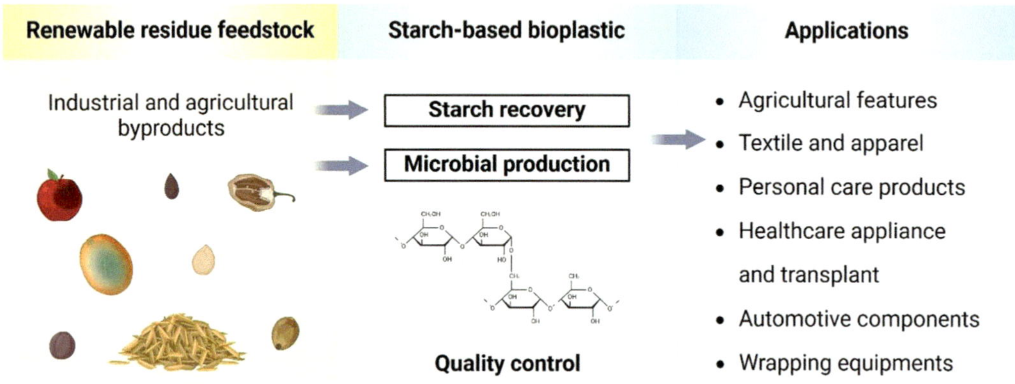

FIGURE 3.2 Industrial and agricultural byproducts as feedstock for starch-based bioplastics.

byproducts also hold potential for use in the bioplastics industry. Many industries, particularly those involved in food processing, paper manufacturing, and bioenergy production, generate substantial quantities of organic waste rich in polysaccharides. For instance, the food processing industry generates large amounts of potato peels, fruit pomace, and wheat bran, all of which are rich in starch and other polysaccharides that could serve as raw materials for bioplastic production (Kadam et al., 2024). Considering the growing food demand, food processing has been increasing, generating a large amount of industrial waste, which is considered a global problem (Roy et al., 2023). Fruit processing, for example, generates approximately 1.81 million tons of waste worldwide (Kowalski & Gumul, 2024). In addition, when fruit is processed, between 10% and 35% of the total weight is discarded (Saba et al., 2023).

Similarly, the pulp and paper industry generates lignocellulosic waste that is rich in cellulose and hemicellulose, which are suitable for bioplastic applications. Additionally, bioenergy production, particularly bioethanol plants, results in lignocellulosic residues such as bagasse and distillers' grains, which are abundant and underutilized (Shamim et al., 2023). By transforming these industrial byproducts into feedstocks for bioplastics, industries can enhance their resource efficiency and reduce their environmental footprint (Figure 3.2). Notably, industrial byproducts are different from agricultural byproducts, although in some cases they are often combined, which is known as agro-industrial waste (Yaashikaa et al., 2022). Industrial wastes are generated during food processing, while agricultural wastes are generated at harvesting stages. Industrial wastes can appear at different stages of the processes (Freitas et al., 2021).

3.3 INDUSTRIAL AND AGRICULTURAL STARCH-CONTAINING BYPRODUCTS

Byproducts from food processing rich in starch are varied, among them are those generated in the cereal industry, such as corn, for example, bran and low quality flour (R. Zhang et al., 2021); wheat, mainly wheat bran (Glaser et al., 2023); and rice, for example, bran, cooking water, and rice husk (Chen et al., 2023). The tuber industry also constitutes an important part as a source of starch, for example, in byproducts generated from potato and cassava processing, where the major byproduct are the peels (Almeida et al., 2023; Weligama Thuppahige et al., 2023). The starch present in industrial byproducts, i.e., non-conventional sources, is not different from that obtained from conventional sources. In fact, the main difference is a function of the plant (Makroo et al., 2021). Some parts of the plant can have a higher or lower concentration of starch. Moreover, it has been shown that, depending on the part of the plant where the starch is located, it can have a different amylose and amylopectin composition (Obadi & Xu, 2021); so even the same plant can have different starch compositions. This also depends on the maturity of the plant (Kringel et al., 2020). In addition, byproduct starch can form interaction with proteins and polyphenols, which can affect their properties (Raza et al., 2023).

Potato peels are a rich source of starch as they contain approximately 75% highly digestible starch (Vannini et al., 2021). The starch yield from fresh potatoes can range from 11% to 30%, depending on the cultivar and growth stage. This byproduct, generated during potato processing, is often wasted yet holds significant potential for starch extraction. Utilizing potato peels can contribute to waste reduction and added value in food processing, as starch can be further processed for various applications such as thickening agents, food additives, and biodegradable materials (Masurkar, 2024). Sweet potato peels contain between 54.72% and 95.09% starch on a dry matter basis, depending on the variety (Sasikala et al., 2022). The extraction of starch from sweet potato peels presents an excellent opportunity for utilizing agricultural residues, as these peels are often discarded. Sweet potato starch has unique properties that make it suitable for various applications, such as in food thickeners, bakery products, and health supplements. By converting sweet potato peels into starch, both economic and environmental benefits can be achieved, enhancing the sustainability of sweet potato farming (Peixoto et al., 2023; Taha et al., 2024).

Cassava byproducts, particularly the residue from processing, have an impressive starch content of about 45–55% in dry matter. Extracting starch from cassava residue not only maximizes the utility of the tubers but also provides valuable feed for livestock and raw materials for the food industry. The high starch yield allows for multiple applications, such as in the production of biofuels, biodegradable materials, and specialty food products (Fronza et al., 2024; Saputri et al., 2024; Weligama Thuppahige et al., 2023). In addition, some starches extracted from cassava byproducts such as husks and bagasse have been shown to have morphology characteristics, functional groups, crystalline phases, and thermal decomposition patterns similar to those obtained for commercial cassava starch (Weligama Thuppahige et al., 2023). Rice bran, a byproduct of rice milling, boasts a starch content of 34–62% depending on the variety and processing methods (Chen et al., 2023). This material is often discarded despite its functional compounds, including protein and vitamins. Extracting starch from rice bran offers a valuable resource for food applications such as gluten-free products, flavor enhancers, and nutritional supplements. The high bioavailability of nutrients in rice bran also aids dietary health. Utilizing rice bran for starch production helps reduce agricultural waste while maximizing the nutritional potential of rice processing (Alonso-González et al., 2025). Wheat bran is another significant agricultural byproduct that contains approximately 45% small granule starch. It comprises part of the outer layer of wheat grains left during milling. The starch from wheat bran is not only used in food products but also used in animal feed, enhancing their nutritional value (Aliotta et al., 2022).

Corn husks are notable agricultural byproducts with a starch content ranging from 9% to 23% (R. Zhang et al., 2021). They are typically discarded during corn processing but can be utilized for starch extraction, thus enhancing the economic value of corn production (Jiao et al., 2022). Starch obtained from corn husks undergoes various applications in food and nonfood industries, including the manufacture of biodegradable plastics, adhesives, and food thickening agents. Starch from pine nut (*Araucaria brasiliensis*) seeds exhibited lower retrogradation with higher consistency than corn starch, thus having higher swelling and solubility (Makroo et al., 2021). Starch extracted from avocado seeds has shown good water absorption properties, transmittance, high amylose content, and thermal stability up to 366 °C, in addition to oval granules with smooth surface, suggesting that it may have applications in biofilm development (Martins et al., 2022).

Currently, a lot of research has focused on byproducts from fruit processing, as they constitute one of the most important industries worldwide. Moreover, they contribute to a large generation of byproducts (Makroo et al., 2021). Mango peels represent a significant byproduct of mango processing and have been noted to yield about 20% starch. These peels are often overlooked yet offer a promising avenue for starch extraction. The starch obtained from mango peels can be applied in food processing, including thickening agents and bioplastics. Utilizing mango peels adds value to mango processing operations while reducing waste (Amaba et al., 2023). Pineapple processing generates considerable amounts of waste, particularly from peels and pomace, which are rich in starch. Pineapple peels are known for their unique starch properties, including high amylose content that influences gelatinization characteristics. The potential starch extraction from pineapple waste can lead to new applications in food products, bioplastics, and dietary supplements (Abdul Rahman et al., 2023). Banana peels, often discarded after the fruit is consumed, contain beneficial compounds and have been explored for starch extraction. While the starch content varies, it is useful for food applications such as probiotic supplements or nutrient-enriched products. By leveraging banana peels for starch-based bioplastic production, waste from banana processing can be significantly reduced, offering an eco-friendly solution to waste management in the banana industry (Choudhary et al., 2024; Ferreiro et al., 2023; Kringel et al., 2020). Most studies with fruit byproducts have focused on the extraction of bioactive compounds such as polyphenols, but they are also an important source of starch that can be exploited (Ochoa & Osorio-Tobón, 2024).

Jackfruit seeds contain a significant amount of starch, ranging from 70% to 85% on a dry weight basis. This starch is composed of approximately 25–45% amylose and 45–80% amylopectin, differing by variety. The high starch content highlights the seeds' potential as a valuable source for starch extraction, which can be utilized in various food and industrial applications (Nguyen et al., 2022). Byproducts from legume processing have also been investigated for some years; for example, shells and other parts that are not usually utilized are important sources of starch (X. Zhang et al., 2022). Finally, the bread industry has the potential to become a very important source of starch, as a large amount of byproducts and waste that are rich sources of starch are usually generated (Ben Rejeb et al., 2024). Thus, the growing demand for eco-friendly products, coupled with the emergence of renewable residue-based feedstocks, highlights the potential of starch-based materials to support a circular economy and reduce fossil fuel dependency in plastic production. This shift not only enhances sustainability but also adds value to agricultural byproducts by transforming waste into valuable resources for diverse industrial applications (Groff et ail., 2025).

3.4 STARCH EXTRACTION FROM BYPRODUCTS

The most commonly used starch extraction methods are traditional techniques, where the material is washed and milled and filtration and sedimentation steps are added (Rashwan et al., 2024). However, due to the interest in obtaining starch from non-conventional sources, new methodologies have been developed to recover a greater amount of starch (Makroo et al., 2021).

3.4.1 Maceration and Physical Separation

The common procedure includes milling, maceration, and separation; these processes can be carried out in two ways: wet and dry. The wet way is the most common (Ahmed et al., 2024). It is based on the soaking of the byproduct in a solution, generally water or a dilute solution of sodium metabisulfite, and milling and/or liquefaction processes are carried out to generate the separation of starch by sedimentation and filtration (Almeida et al., 2023; Rashwan et al., 2024). On the other hand, dry way extraction involves drying the byproduct, grinding and sieving. The purpose is to release the starch granules contained in the plant cell, and to separate them at a size that allows them to be separated without breaking the granules and without affecting their properties (Makroo et al., 2021). Physical processes also include pressing and crushing (Dorantes-Fuertes et al., 2024). For traditional methods, it has been reported that the dry way generates starches with poor functional characteristics compared to the wet way (Makroo et al., 2021). In addition, the wet way achieves starch purities up to 99.5% (Rashwan et al., 2024). Although the traditional wet milling method is efficient, it is not scalable due to the long process times and the high amount of water needed at each stage, so its use is limited to laboratory scale (Ahmed et al., 2024).

3.4.2 Acid and Alkali Extraction

Some of the acids and alkalis used include hydrochloric acid and sodium hydroxide (Dorantes-Fuertes et al., 2024). Acid extraction commonly includes the use of citric acid, ascorbic acid, and starch from banana peels and litchi seeds has been extracted from these acids (Kringel et al., 2020). Extractions by acid and alkali methods have also been compared in litchi seeds, where differences in amylose content were found, being higher in the acid method (Kringel et al., 2020; Rashwan et al., 2024). For alkaline extraction, sodium hydroxide or sodium metabisulfite solutions are used (Almeida et al., 2023). The use of alkalis has been employed to obtain bread waste starch, which facilitates its release, but requires pH neutralization steps (Ben Rejeb et al., 2024).

3.4.3 Enzymatic Extraction

This method involves the use of enzymes such as pectinase and cellulase to improve starch extraction yields. The enzymes alter the plant matrix which facilitates the release of starch granules (Dorantes-Fuertes et al., 2024). Only enzymatic treatments have been able to aid in the efficient extraction of starch from cassava bagasse (Weligama Thuppahige et al., 2023). Mainly because it breaks down the cell walls, in addition, it requires less mechanical effort and energy, compared to traditional methods. However, some disadvantages include cost, inactivation of enzymes due to pH and temperature, slow reaction times, complexity, and residual enzymes (Ahmed et al., 2024).

3.4.4 New Extraction Methods

Due to the increasing research on starch extraction, some emerging technologies have been incorporated as tools for this purpose, for example, the use of ultrasound and microwaves (Ahmed et al., 2024; Dorantes-Fuertes et al., 2024). The use of ultrasound and microwaves can affect the granular structure of starch, affecting its ability to gelatinize, unlike enzymatic methods that can help preserve the structure (Neeraj et al., 2021). Fermentation has also been used to improve extraction yields (Raza et al., 2023) and to improve starch properties such as swelling power, solubility and gelatinization temperature, although other properties such as viscosity appear to be compromised (Oyeyinka et al., 2020). Recently, pulsed electric field technology (Castro et al., 2023) and supercritical carbon dioxide-assisted extraction (Makroo et al., 2021) have been used as possible tools to improve starch extraction processes from industrial byproducts. Table 3.1 shows different byproducts from different non-conventional sources, as well as the techniques used and their yield. It can be seen that some byproducts are extremely high in starch. However, many of these have low purity, so they may require additional treatments to improve their characteristics (Almeida et al., 2023).

Extraction methods affect starch recovery, as it is found within a complex matrix where it interacts with fibers, proteins, and polyphenols, which can affect its recovery and ensure its purity (Raza et al., 2023). Extraction methods have a direct effect on the quality and properties of starch obtained from industrial byproducts. Factors affecting starch recovery are related to the structure of the byproducts, since some, such as cassava peel, can only be treated by enzymatic methods to obtain starch (Weligama Thuppahige et al., 2023), and many byproducts do not have high amounts of starch (Gumul et al., 2024). In addition, extraction conditions such as pH, time, and temperature must be considered (Makroo et al., 2021).

For example, starch from bread waste obtained by conventional water extraction, alkaline extraction, and enzymatic extraction showed that the enzymatic technique has higher yields, higher

TABLE 3.1
Industrial food byproducts as a source of starch: extraction methods and yields

Byproduct	Composition	Extraction Method	Yield	Reference
Mango seed (*Mangifera indica* L.)	Mango seed starch has a composition of 9.86% moisture, 3.63% protein, 4.63% lipid, and 0.03% ash.	Wet milling in 0.2% (v/v) sodium metabisulfite solution.	53.89%	(Souza et al., 2021)
Cassava (*Manihot esculenta*) peel	N.A.	Solution with 4% sample in 0.05 M sodium acetate buffer. Heating at 120 °C and physical treatment with sonication	30.17 ± 2.18%	(Weligama Thuppahige et al., 2023)
Cassava (*Manihot esculenta*) bagasse	N.A.	Solution with 4% sample in 0.05 M sodium acetate buffer. Heating at 120 °C and physical treatment with sonication	7.5 ± 1.18%	(Weligama Thuppahige et al., 2023)
Fiber flour from purple pulp sweet potato starch processing	The fiber flour has a soluble and insoluble fiber content between 1.20–1.63 and 13.53–21.91%, respectively.	1:3 ratio (w/v) with distilled water, 2,000 ppm sodium metabisulfite solution and 2,000 ppm citric acid	3.23–5.11%	(Julianti et al., 2021)
Bread waste	The bread waste contained 6.77% moisture, 13.61% protein, 1.83% ash and 70.79% carbohydrate.	Different techniques: distilled water, alkaline steeping, Enzyme-Assisted Extraction	81–88%	(Ben Rejeb et al., 2024)
Wheat bran	The composition of wheat bran is approximately 10% protein, 15% lignin, and less than 1% ash.	Hydrolysis by liquefaction with α-amylase and amyloglucosidase	48.9%	(Glaser et al., 2023)
Potato peel	Protein content 11.98–17.19%, lignin 22–35%, lipids 0.81–2.6% and ash 4.83–11.05%.	Enzymatic hydrolysis using glucoamylase	16.83%	(Almeida et al., 2023)
Litchi seed (*Litchi chinensis* Sonn.)	N.A.	Wet milling in 0.16% (v/v) sodium bisulfite solution.	212.4 g kg−1	(Kaur et al., 2024)
Avocado seed (*Persea americana* mill)	Starch had moisture (41.35%), ash (0.33%), lipids (1.68%), protein (1.60%) and carbohydrates (55.07%).	Wet milling with 0.2% (v/v) sodium metabisulphite solution	19.54%	(Martins et al., 2022)
Waste of anthocyanin extraction from purple yam (*Dioscorea alata*)	Starch had a composition of moisture (9.83–10.07%), protein (0.13–0.15%), lipids (0.11–0.14%) and ash (0.10–0.11%).	Different techniques: Wet milling in acid pH, alkaline pH, and water	41.23–50.86%	(Ochoa & Osorio-Tobón, 2024)
Banana peel	N.A.	Wet milling 1:2 w/v ratio in 0.2% sodium metabisulfite solution	13%	(Adi Sulianto et al., 2023)

N.A.: Not available.

swelling power and solubility than the other techniques, and the starch obtained had a structure similar to commercial quality starch. The alkaline and enzymatic methods revealed an A-type crystalline structure, indicating a starch with low lipid content compared to the conventional method. Enzymatic extraction results in high yields of pure starch and is more efficient (Ben Rejeb et al.,

2024; Rashwan et al., 2024). Therefore, each technique is chosen depending on the biowaste source and the final characteristics desired (Kringel et al., 2020; Rashwan et al., 2024). Although the best physicochemical and morphological characteristics have been obtained for combined extractions (physical, chemical, and enzymatic) (Almeida et al., 2023), this could reflect the need for novel studies evaluating the combination of these techniques to reduce the disadvantages and increase the recovery and quality of starch byproducts.

3.5 PHYSICOCHEMICAL PROPERTIES OF STARCH FROM BYPRODUCTS

Starch obtained from unconventional sources has different physicochemical properties that depend on its structural characteristics; these properties may differ from conventional sources, mainly because of the type of plant matrix and the extraction methods used (Makroo et al., 2021; Obadi & Xu, 2021). Some of the properties of starches from nonconventional sources include granule size and shape, amylose to amylopectin ratio, and crystallinity. On these characteristics will depend their functional properties, such as gelatinization temperature, swelling capacity and solubility, viscosity properties, and stability (Almeida et al., 2023).

3.5.1 Shape, Size, and Morphology

For industrial application, starch must have certain morphological properties. Therefore, particle size analysis helps to ensure uniformity of texture and shape (Ahmed et al., 2024). There is a wide variety of shapes and sizes in starch, which depend on the plant material and extraction technique used (Dereje, 2021). These differences have an effect on their functional properties. For example, starch with smaller, angular granules has a higher resistance to gelatinization than large, spherical starches (Makroo et al., 2021). Starch granules with spherical and irregular shapes with average sizes of 20 μm have been obtained from mango seed. Compared to cassava starch, which has rounded and smooth shapes, there may be differences, but parameters such as absolute density are maintained (Souza et al., 2021). Cassava byproducts show irregular, round, oval and polygonal shapes, as well as varied granule sizes. On the other hand, commercial starch has uniform particles because it undergoes advanced separation methods. The absence of these methods during laboratory extraction causes poor uniformity of extracted starches (Weligama Thuppahige et al., 2023). Sorghum starch granules are irregular in shape, ranging from oval to spherical. They may also have facets or irregular surfaces. The size of granules can vary, and some are relatively large compared to other starch sources (Rashwan et al., 2024).

3.5.2 Crystallinity

This property has been attributed to the ordered structure of amylopectin within starch granules, native starch is semi-crystalline, and this crystallinity is differentiated into 3 types: A, B or C (Dereje, 2021). Some studies have reported that type A starches contain low amylose contents and high amylopectin contents (Weligama Thuppahige et al., 2023).

Mango seed starch has type A crystalline, which is similar to that of some cereals. Amylose content, crystal size, and the number of crystalline regions can affect the degree of crystalline; as amylose content increases, crystallinity also increases (Souza et al., 2021). Another study found that the properties of wheat starch varied with the proportion of A and B granules. Granules A being larger in size showed higher gelatinization temperatures, where the completion temperature was 80.28 °C. In contrast, smaller B granules had higher peak temperatures of 66.00 °C (Rashwan et al., 2024).

3.5.3 Thermal Properties

Thermal properties are useful to study the gelatinization and retrogradation processes of starch. Gelatinization is related to starch components and granule size distribution, and gelatinization is determined by amylose, fat, and protein in starch (Dereje, 2021). Some non-conventional sources, such as mango seed starch, have shown better thermal properties than conventional sources such as corn, wheat, cassava, and potato starches (Souza et al., 2021).

3.5.4 Swelling Power and Solubility

The swelling power represents the amounts of water that a starch can absorb per gram of starch at an assigned temperature, while the solubilities at the assigned temperature correspond to the percentages of amylose and amylopectin leached (Obadi & Xu, 2021). The swelling of the granules and the solubilization of amylose and amylopectin induce the loss of granular integrity with the generation of a viscous paste. When these molecules begin to reassociate into a more ordered structure, it is called retrogradation process (Souza et al., 2021). The amylose content has a significant influence on the swelling behavior of starches. The solubility and swelling behavior of the extracted starch are affected by the processing conditions employed (Makroo et al., 2021).

3.5.5 Water Absorption Capacity

Water absorption capacity measures the ability of starch to absorb and retain water, which is essential for its application in food products (Ahmed et al., 2024). Water absorption capacity depends on the crystalline and amorphous regions within the starch, molecular structure, and granular size distribution, as it depends on its physical entrapment in the starch structure (Souza et al., 2021). It is also highly dependent on the source, modification, size, and structure of starch granules; furthermore, increasing temperature increases the water holding capacity due to the gelatinization property of starch (Makroo et al., 2021).

3.5.6 Other Properties

Other interesting properties of starch include pasting properties, i.e., the changes after heating following gelatinization, which is related to viscosity (Dereje, 2021). Viscosity evaluates the thickening capacity of starch, which is useful for industrial applications (Ahmed et al., 2024). Finally, amylose and amylopectin content is fundamental, as it affects all the properties mentioned above, and this content varies according to the type of matrix, demonstrating the great variability of starches that can be obtained from both conventional and nonconventional sources (Makroo et al., 2021).

3.6 UNCONVENTIONAL SOURCED-STARCH FOR THE BIOPLASTIC INDUSTRY

Starch, in its pure form, is primarily used as an ingredient in a wide range of food products and formulations, including candies, ice creams, sauces, ketchup, mayonnaise, jams, yogurts, puddings, prebiotics, fruit fillings for confectionery, canned foods such as meats and vegetables, as well as in frozen products, among others. However, beyond its importance in the food industry, starch is a highly versatile raw material with broad applications in the paper, chemical, textile, pharmaceutical, and biotechnology industries. Starch-based bioplastics also exhibit this versatility, enabling their use in food packaging, cutlery production, agriculture, horticulture, healthcare, diapers, and electronics, among others (Dejene & Geletaw, 2024; Makroo et al., 2021).

The production of bioplastic films from starch can be achieved through two primary methods: wet process and dry process. The wet process, which is typically conducted on a laboratory scale, involves casting techniques that can be categorized into two types: solvent casting and tape casting. In solvent casting, starch is gelatinized and heated with additives to create a filmogenic solution, which is then dried on Teflon sheets or petri plates to form thin plastic films. Conversely, tape casting employs controlled heat convection for drying the fibrogenic solution. On the other hand, the dry process encompasses advanced techniques such as extrusion and molding, making it suitable for industrial-scale production (Gupta et al., 2022). However, bioplastics are generally more brittle and less durable than petroleum-based plastics, often requiring blending with other polymers or additives to enhance their strength and flexibility. To produce high-quality plastic products, the modified thermoplastic starch can be subjected to advanced methods like thermoforming, compression molding, injection molding, and extrusion (Do Val Siqueira et al., 2021).

Bioplastics derived from unconventional starch sources are ideal for food packaging due to their low toxicity, biodegradability, and protection from spoilage. They offer a sustainable and eco-friendly alternative for single-use applications such as bags, wraps, and pouches. However, improvements are still needed in the properties of starch-based bioplastics, including hydrophilicity, moisture sensitivity, and barrier and mechanical properties (Weligama Thuppahige et al., 2023). Additionally, starch-based bioplastics derived from unconventional sources can be used to create edible plates or cutlery, providing green solutions for festivals, events, and restaurants (Dejene & Geletaw, 2024; Rahardiyan et al., 2023). One of the most promising applications of starch fibers is the encapsulation of bioactive compounds to promote their controlled and specific release. Their encapsulation is crucial for food products or food packaging, aiming to protect them from external elements such as temperature, light, moisture, and other environmental interactions (Pires et al., 2024). Starch-based bioplastics can also be designed to control the release of drugs in the body, creating matrices that dissolve at controlled rates. This facilitates the precise administration of medications and minimizes side effects. Biodegradable blister packs, capsules, and wraps made from starch-based bioplastics can help reduce medical waste (Makroo et al., 2021; Mhaske et al., 2024).

In horticulture and agriculture, mulch films are used to mitigate soil degradation and reduce plastic contamination. Starch-based biodegradable bioplastic pots represent an eco-friendly option, as they can be planted directly into the soil without harming the plant roots. These bioplastics can also be used to manufacture controlled-release systems for fertilizers and pesticides, ensuring the gradual release of these substances into the soil. These solutions offer advantages for the agricultural sector, as they are biodegradable, photodegradable, and compostable, breaking down into carbon dioxide and water through the action of microorganisms (Choudhary et al., 2024; Gong et al., 2024; Makroo et al., 2021). Other applications of starch-based bioplastics include their use as alternatives to synthetic fibers and coatings for textiles. In the electronics industry, starch-based bioplastics can be used to create flexible electronic devices, contributing to more manageable and eco-friendly electronic waste (Nagarajan et al., 2024; Xiang et al., 2022).

Each of these applications depends on the type of starch selected, in accordance with the specific industrial purpose, considering the type of feedstock byproduct and its physicochemical properties. It should also be noted that starch recovered from renewable waste can either remain native or be modified, which expands the number of applications for starch-derived bioplastics (Makroo et al., 2021). Figure 3.3 shows a schematic of the general starch extraction process and starch-based bioplastic production.

3.7 BENEFITS

Starch-based bioplastics derived from agricultural waste offer numerous environmental, economic, and functional benefits, making them an attractive alternative to conventional petroleum-based plastics. These bioplastics are primarily made from the starch found in agricultural byproducts such as corn, potatoes, or cassava, which makes them a renewable and sustainable resource. This not only reduces reliance on fossil fuels but also helps in managing agricultural waste effectively (Venkatachalam & Palaniswamy, 2020). One of the most significant benefits of starch-based bioplastics is their biodegradability. Unlike traditional petroleum-based plastics, which can take

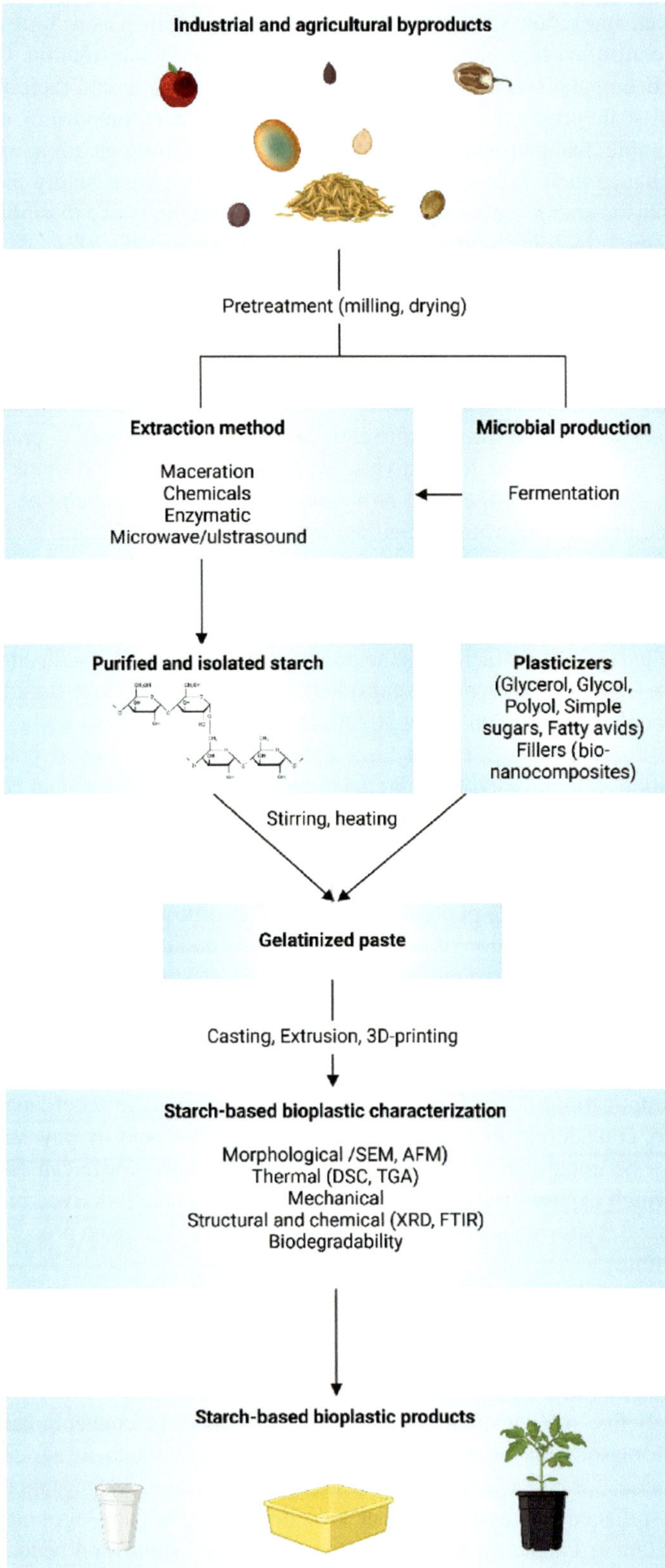

FIGURE 3.3 Starch-based bioplastic production from renewable residue-derived starch.

hundreds of years to decompose, starch-based bioplastics can break down naturally in the environment within a matter of months to a few years (Hubbe et al., 2020). The biodegradation process involves microorganisms such as bacteria and fungi that convert these materials back into natural substances, ensuring minimal environmental impact. This characteristic is crucial in mitigating plastic pollution, as it reduces the volume of waste that ends up in landfills and oceans, thereby protecting wildlife and ecosystems from harmful plastic debris.

In addition to being biodegradable, starch-based bioplastics have a lower carbon footprint compared to their petroleum counterparts (Korol et al., 2020). The production process for these bioplastics typically requires less energy and emits significantly fewer greenhouse gases. For instance, studies indicate that the production of starch-based materials can result in up to 80% less carbon dioxide emissions than traditional plastics. This reduction is vital in the fight against climate change, as it contributes to lowering overall greenhouse gas concentrations in the atmosphere (Polman et al., 2021). Moreover, in many agricultural practices, agri-residues are either burned, contributing to air pollution, or left to decompose, releasing greenhouse gases like methane and carbon dioxide. By diverting these byproducts for use in bioplastic production, both environmental harm and waste disposal issues can be mitigated (McNeill et al., 2024).

The economic advantages of starch-based bioplastics are also noteworthy. Utilizing agricultural waste not only provides a cost-effective raw material but also creates new revenue streams for farmers and manufacturers. By converting what would otherwise be discarded into valuable products, this approach promotes a circular economy where resources are reused and repurposed (Dejene & Geletaw, 2024; Pham et al., 2024). Furthermore, the growing market for sustainable materials opens up new job opportunities in production, research, and development sectors (Elkaliny et al., 2024). Versatility is another key benefit of starch-based bioplastics. Industrial byproducts containing high amounts of starch can be a sustainable alternative to conventional sources, as these sources can be destined for human food, contributing to food security and zero hunger challenges (Haque et al., 2023). The differences in the composition of the byproducts can help to obtain starches with different properties, even these can be blended to obtain high-quality starches, which extends their use for the development of novel biopolymers and starch-derived materials. They can be processed into various forms suitable for multiple applications, including packaging materials, disposable cutlery, agricultural films, and even medical devices. The properties of these bioplastics can be customized through blending with other natural or synthetic polymers to meet specific performance requirements, such as flexibility, strength, and moisture resistance, making them adaptable for diverse industries (Nagarajan et al., 2024; Taha et al., 2024).

Moreover, starch-based bioplastics are generally nontoxic and safe for food contact applications. This characteristic makes them an appealing choice for food packaging solutions, where consumer safety is paramount (Pham et al., 2024). Unlike some conventional plastics that may leach harmful chemicals into food products, starch-based options provide a safer alternative that aligns with increasing consumer demand for eco-friendly products. In this regard, the growing demand for fresher, higher-quality products with extended shelf life has driven innovations in packaging design as shown in Figure 3.4. These advancements focus on advanced technologies such as selective gas regulation, moisture permeability control, ultraviolet protection, antimicrobial systems, and the incorporation of active preservatives. Consequently, innovations in starch-based food packaging will continue to evolve, adapting to the needs of the food industry based on functionality.

Finally, the consumer appeal of starch-based bioplastics cannot be overlooked. As awareness of environmental issues rises, more consumers are seeking sustainable alternatives to traditional plastics. Starch-based bioplastics cater to this demand by offering a product that not only performs well but also contributes positively to environmental conservation efforts (Dorantes-Fuertes et al., 2024).

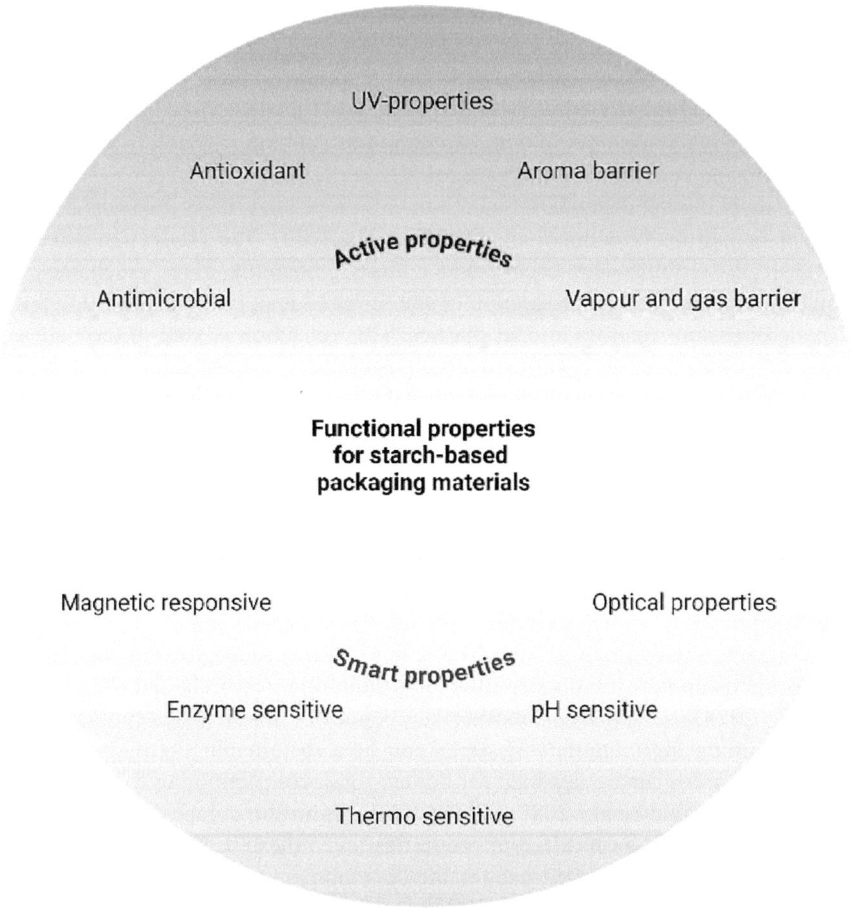

FIGURE 3.4 Smart and active properties for starch-based packaging materials.

3.8 LIMITATIONS

The availability of byproducts is influenced by both geographical and seasonal factors, as well as the production volume of each specific industry (Feyisa et al., 2024). While many of these byproducts benefit from being widely generated across various regions globally, suggesting a continuous and consistent production, there are notable limitations. A key challenge lies in the fact that numerous industrial byproducts do not contain high concentrations of starch, which could render the extraction of starch economically unfeasible if the yield is insufficient (Gumul et al., 2024). Moreover, another key point is that there are byproducts that are generated in large quantities, such as orange and apple peel. However, orange peel has in its composition pectin instead of starch and apple peels only contain starch in an immature stage, when it matures the starch is converted into simple sugars (Gumul et al., 2024). These aspects should be considered when searching for unconventional sources of starch.

Starch-based bioplastics offer substantial environmental benefits as renewable alternatives to fossil fuel-derived plastics. However, their widespread adoption is constrained by several significant limitations. One of the primary challenges lies in their manufacturing processes, particularly in achieving optimal miscibility between starch and other polymers, which can hinder scalability and increase production complexity. There are several factors to consider during the starch-based bioplastic production process, as shown in Figure 3.5. Additionally, these bioplastics often suffer

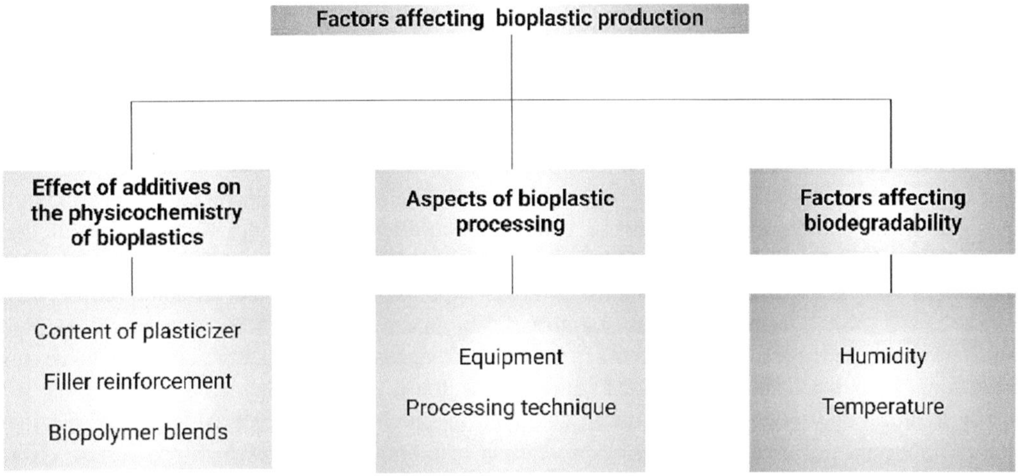

FIGURE 3.5 Factors affecting starch-based bioplastic production.

from mechanical property deterioration, especially at high starch contents, leading to increased brittleness and reduced durability compared to conventional plastics. As a result, blending with other polymers or additives is frequently required to enhance their strength and flexibility, which limits their applicability in demanding use cases (Abe et al., 2021; Pham et al., 2024; Taha et al., 2024).

Moisture sensitivity further complicates the usability of starch-based bioplastics. Without the incorporation of plasticizers, these materials are rigid and challenging to process, making them unsuitable for standard manufacturing techniques. Their tendency to absorb water results in swelling and subsequent loss of mechanical strength, thereby restricting their application in moisture-sensitive applications like liquid packaging. Moreover, although these bioplastics are biodegradable, the accelerated degradation induced by moisture can compromise structural integrity before disposal (Tecorralco-Bobadilla et al., 2024). Furthermore, the biodegradation of these materials is highly variable, depending on environmental conditions, with decomposition times ranging from months to years. During this period, there is potential for contributing to litter and pollution, particularly if biodegradation releases methane, a potent greenhouse gas (Samer, 2019).

The economic feasibility of starch-based bioplastics is another critical limitation. The production costs are typically higher than those of traditional plastics due to more expensive raw materials and the need for additional treatments to improve properties like strength and moisture resistance. This price disadvantage makes them less competitive in cost-sensitive markets (Choudhary et al., 2024; Padi & Chimphango, 2020). Finally, starch-based bioplastics present challenges in waste management and recycling. The inclusion of biodegradable additives can contaminate recycling streams, complicating efforts to integrate them into conventional recycling systems. Due to these limitations in strength, moisture resistance, and biodegradability control, starch-based bioplastics are primarily confined to niche applications, such as disposable cutlery, bags, and specific packaging solutions, and struggle to compete with traditional plastics in high-performance applications like automotive or electronics (Dejene & Geletaw, 2024; Elkaliny et al., 2024; Gong et al., 2024).

3.9 FUTURE PERSPECTIVES

The future of bioplastics derived from starch sourced from agricultural and industrial byproducts looks particularly promising, especially in response to the growing demand for sustainable substitutes for traditional plastics. As global awareness of environmental concerns increases, there is a coordinated effort to reduce plastic waste and the resulting pollution. In this context, starch-bioplastics are becoming an innovative solution.

Scholars and companies are investigating new materials and technologies to improve the performance, biodegradability, and environmental footprint of bioplastics, particularly those derived from agricultural and industrial waste (Elkaliny et al., 2024). The alignment of the United Nations Sustainable Development Goals with international policy initiatives will likely act as a primary catalyst for the expansion of this industry (Rahardiyan et al., 2023). Thus, it is crucial for governments, businesses, the public, and consumers to collaborate in alleviating the environmental challenges posed by food waste and to create methods for producing biopolymers from discarded and residual materials (Gong et al., 2024). For instance, government regulations aimed at reducing plastic waste influence the growth trajectory of the starch-based bioplastics market. Moreover, stricter policies on single-use plastics are driving manufacturers to seek sustainable alternatives, thereby increasing investment in research and development within this sector.

Agricultural and industrial byproducts, abundant in starch and other valuable polysaccharides, are an important resource for the future production of bioplastics. As research progresses, efforts will likely be directed toward optimizing the extraction techniques for these unconventional materials, ensuring high yields and purity while keeping costs manageable. Such advancements will enable more efficient integration of agricultural and industrial waste into bioplastic production, promoting a more sustainable future. Future methods of production will also lean toward more efficient and eco-friendly processes for producing bioplastics from starch obtained from renewable waste. As consumer awareness of environmental issues continues to grow, bioplastics are poised to become a significant alternative to petroleum-based plastics (Rahardiyan et al., 2023). Emerging methods, such as ultrasonic-assisted extraction or pulsed electric fields, offer promising solutions to enhance starch performance and quality while minimizing environmental impact. Technological advances in starch extraction and bioplastic production could revolutionize agricultural and industrial waste, transforming it into a valuable and sustainable resource, creating new opportunities for maximizing starch recovery and tailoring it to specific industrial applications. The starch extraction industry must evolve toward more sustainable, low-impact practices to ensure a transition to a circular economy (Dorantes-Fuertes et al., 2024). Looking ahead, the integration of agricultural and industrial byproducts into the production of starch-based bioplastics holds significant promise for enhancing sustainability within the industry.

The future also indicates an expansion in the variety of raw materials, reducing the reliance on food crops and strengthening the principles of a circular economy. Diversifying sources will allow agricultural and industrial waste to play a central role in bioplastic production, turning what was once considered waste into a valuable resource. Furthermore, fostering collaborations between agricultural producers and bioplastic manufacturers could facilitate a more sustainable supply chain that benefits both sectors. The European market, in particular, is witnessing a surge in demand for biodegradable packaging solutions due to stringent environmental regulations. In addition, research into the life cycle assessments (LCA) of bioplastics made from byproducts is also necessary to quantify their overall environmental impact and identify areas for improvement (Tecorralco-Bobadilla et al., 2024). Thus, LCA will be instrumental in measuring the environmental benefits of starch-based bioplastics, identifying the most sustainable production and disposal practices, and fostering a more efficient waste management approach (Rahardiyan et al., 2023).

The packaging sector, particularly in food packaging, is poised to integrate starch-based bioplastics in the near future. This shift will be driven by innovations aimed at optimizing renewable resources and minimizing the environmental impact of manufacturing processes. Biodegradable bioplastics derived from starch are set to play a key role in the development of sustainable packaging materials, offering a lower carbon footprint than conventional petroleum-based plastics (Mhaske et al., 2024). This shift mirrors the growing consumer demand for more eco-friendly packaging solutions (Rahardiyan et al., 2023). Starch-based bioplastics are especially well-suited for applications such as single-use packaging, disposable cutlery, and agricultural films, owing to their biodegradability and nontoxic nature. Future developments will likely focus on improving the barrier properties of these materials, safeguarding products from moisture, oxygen, and other contaminants, thus extending their shelf life. Additionally, the incorporation of smart packaging technologies, capable of adapting to environmental changes or monitoring food freshness, will further drive innovation in starch-based packaging solutions.

Starch-based bioplastics sourced from agricultural and industrial waste are also gaining attention in industries such as agriculture, horticulture, and healthcare. In these fields, they can be used to create biodegradable products like mulching films, seed coatings, and medical devices. Future research may lead to the development of biopolymers with antimicrobial properties, reducing the risk of infections in medical environments. Moreover, the biocompatibility of starch makes it an ideal candidate for tissue engineering applications, such as scaffolds for cell growth and wound healing.

As governments worldwide increasingly focus on reducing plastic waste, many countries are banning single-use plastics and promoting biodegradable alternatives. This regulatory shift is fostering a supportive environment for the development of starch-based bioplastics. The future of this sector will depend on the formulation of supportive policies, grants, and incentives that encourage the widespread adoption of bioplastic solutions. Additionally, the growing consumer demand for environmentally friendly products is driving investment in research and development, which is expected to accelerate the commercialization of new bioplastic technologies, contributing to a more sustainable future.

3.10 CONCLUSION

The growing concern over plastic waste and its harmful environmental impacts has driven extensive research into bioplastics as eco-friendly alternatives to conventional petrochemical plastics. Global demand for bioplastics is on the rise, with many standard polymers now having bioplastic counterparts that offer comparable material properties. This shift is fueled by global environmental challenges, pushing nations to speed up the transition toward a modern bioeconomy. Cost reductions in starch-based bioplastic production can be achieved by leveraging inexpensive and abundant raw materials, including fruit and vegetable waste, supported by biorefineries and circular economy principles. Exploring new, non-food starch sources or underutilized starch-rich crops could help avoid competition with current food resources. Improving starch extraction techniques is vital for maximizing efficiency, while fully eco-friendly composites made from thermoplastic starch and natural fibers provide durable substitutes for traditional materials across sectors such as packaging, automotive, and construction. However, starch-based bioplastics face challenges like limited structural integrity and hydrophilicity. To scale up production effectively, it is crucial to focus on improving biomass availability, streamlining preprocessing, and optimizing logistics and management, all the while advancing closed-loop systems that support sustainable living and reduce reliance on nonrenewable resources. Efficient starch extraction processes will be pivotal in driving both economic viability and environmental sustainability, ensuring the future success of this promising alternative to traditional plastics.

ACKNOWLEDGMENTS

Luis Fernando Flores acknowledges the Wikipedia Library of the Wikimedia Foundation, through his Wikimedia user account (Luis Fernando Flores LAB), for providing access to essential academic literature during the preparation of this chapter.

REFERENCES

Abdul Rahman, W., Kamaruddin, M. S., Samsudin, N., Azizan, N. A., & Nor Muhammed, N. I. (2023). Bioplastic from peel and rind of tropical fruits in Southeast Asia: A mini-review. *Scientific Research Journal*, 20, 55–74.

Abe, M. M., Martins, J. R., Sanvezzo, P. B., Macedo, J. V., Branciforti, M. C., Halley, P., Botaro, V. R., & Brienzo, M. (2021). Advantages and disadvantages of bioplastics production from starch and lignocellulosic components. *Polymers*, 13 (15), 2484. https://doi.org/10.3390/polym13152484

Abrha, H., Cabrera, J., Dai, Y., Irfan, M., Toma, A., Jiao, S., & Liu, X. (2022). Bio-based plastics production, impact and end of life: A literature review and content analysis. *Sustainability*, 14 (8), 4855. https://doi.org/10.3390/su14084855

Adi Sulianto, A., Adiyaksa, I. P., Wibisono, Y., Khan, E., Ivanov, A., Drannikov, A., Ozaltin, K., & Di Martino, A. (2023). From fruit waste to hydrogels for agricultural applications. *Clean Technologies*, 6 (1), 1–17. https://doi.org/10.3390/cleantechnol6010001

Ahmed, S., Rehman, H. U., & Ahmed, N. (2024). Potato starch extraction: Techniques, challenges, and future opportunities. *Journal of Pharmacognosy and Phytochemistry*, 13 (4), 512–524. https://doi.org/10.22271/phyto.2024.v13.i4f.15046

Aliotta, L., Vannozzi, A., Cinelli, P., Fiori, S., Coltelli, M.-B., & Lazzeri, A. (2022). Wheat bran addition as potential alternative to control the plasticizer migration into PLA/PBSA blends. *Journal of Materials Science*, 57 (30), 14511–14527. https://doi.org/10.1007/s10853-022-07534-9

Almeida, P. V., Gando-Ferreira, L. M., & Quina, M. J. (2023). Biorefinery perspective for industrial potato peel management: Technology readiness level and economic assessment. *Journal of Environmental Chemical Engineering*, 11 (3), 110049. https://doi.org/10.1016/j.jece.2023.110049

Alonso-González, M., Felix, M., Romero, A., Sergi, C., Bavasso, I., & Sarasini, F. (2025). Optimization of processing conditions for rice bran-based bioplastics through extrusion and injection molding. *Journal of Polymers and the Environment*, 33, 512–527. https://doi.org/10.1007/s10924-024-03377-4

Amaba, A. S., Villanueva, K. C., Tan, N. P., Siacor, F. D., & Paler, M. K. (2023). Preparation of bioplastic film from chitosan and mango (Mangifera Indica L. Anacardiaceae) kernel starch by casting method. *Applied Chemical Engineering*, 6 (3). https://doi.org/10.24294/ace.v6i3.2819

Anitha, R., Jayakumar, K., Samuel, G. V., Joice, M. E., Sneha, M., & Seeli, D. S. (2024). Synthesis and characterization of starch-based bioplastics: A promising alternative for a sustainable future. *The International Conference on Processing and Performance of Materials (ICPPM 2023),* 30. https://doi.org/10.3390/engproc2024061030

Ben Rejeb, I., Baraketi, S., Charfi, I., Khwaldia, K., & Gargouri, M. (2024). Valorization of bread waste, a nonconventional feedstock for starch extraction using different methods: A comparative study. *Euro-Mediterranean Journal for Environmental Integration*, 9 (3), 1485–1498. https://doi.org/10.1007/s41207-024-00572-6

Castro, L. M. G., Alexandre, E. M. C., Saraiva, J. A., & Pintado, M. (2023). Starch extraction and modification by pulsed electric fields. *Food Reviews International*, 39 (4), 2161–2182. https://doi.org/10.1080/87559129.2021.1945620

Chakraborty, I., N., P., Banik, S., Govindaraju, I., Das, K., Mal, S. S., Zhuo, G., Rather, M. A., Mandal, M., Neog, A., Biswas, R., Managuli, V., Datta, A., Mahato, K. K., & Mazumder, N. (2022). Synthesis and detailed characterization of sustainable starch-based bioplastic. *Journal of Applied Polymer Science*, 139 (39), e52924. https://doi.org/10.1002/app.52924

Chen, B., Qiao, Y., Wang, X., Zhang, Y., & Fu, L. (2023). Extraction, structural characterization, biological functions, and application of rice bran polysaccharides: A review. *Foods,* 12 (3), 639. https://doi.org/10.3390/foods12030639

Choudhary, P., Pathak, A., Kumar, P., S, C., & Sharma, N. (2024). Commercial production of bioplastic from organic waste–derived biopolymers viz-a-viz waste treatment: A minireview. *Biomass Conversion and Biorefinery*, 14 (10), 10817–10827. https://doi.org/10.1007/s13399-022-03145-1

Dejene, B. K., & Geletaw, T. M. (2024). Development of fully green composites utilizing thermoplastic starch and cellulosic fibers from agro-waste: A critical review. *Polymer-Plastics Technology and Materials*, 63 (5), 540–569. https://doi.org/10.1080/25740881.2023.2296649

Dereje, B. (2021). Composition, morphology and physicochemical properties of starches derived from indigenous Ethiopian tuber crops: A review. *International Journal of Biological Macromolecules*, 187, 911–921. https://doi.org/10.1016/j.ijbiomac.2021.07.188

Do Val Siqueira, L., Arias, C. I. L. F., Maniglia, B. C., & Tadini, C. C. (2021). Starch-based biodegradable plastics: Methods of production, challenges and future perspectives. *Current Opinion in Food Science*, 38, 122–130. https://doi.org/10.1016/j.cofs.2020.10.020

Dorantes-Fuertes, M.-G., López-Méndez, M. C., Martínez-Castellanos, G., Meléndez-Armenta, R. Á., & Jiménez-Martínez, H.-E. (2024). Starch extraction methods in tubers and roots: A systematic review. *Agronomy*, 14 (4), 865. https://doi.org/10.3390/agronomy14040865

Elkaliny, N. E., Alzamel, N. M., Moussa, S. H., Elodamy, N. I., Madkor, E. A., Ibrahim, E. M., Elshobary, M. E., & Ismail, G. A. (2024). Macroalgae bioplastics: A sustainable shift to mitigate the ecological impact of petroleum-based plastics. *Polymers*, 16 (9), 1246. https://doi.org/10.3390/polym16091246

Ferreiro, O., Martin, L. A., Chacon, W. D. C., Duarte, S., Perilla, J. E., & Valencia, G. A. (2023). Valorization of agro-industrial plantain (Musa × paradisiaca) by-products: Alternative sources of carbohydrates and bioactive compounds. Wiley Online Library. https://onlinelibrary.wiley.com/action/showCitFormats?doi=10.1002%2Fstar.202300210

Feyisa, T., Tolera, A., Nurfeta, A., Balehegn, M., & Adesogan, A. (2024). Availability, distribution and quality of agro-industrial byproducts and compound feeds in Ethiopia. *Frontiers in Animal Science*, 5, 1408050. https://doi.org/10.3389/fanim.2024.1408050

Flores, L. F., Osorio-Gonzalez, C. S., Saini, R., & Brar, S. K. (2024). Renewable residues as feedstock for drop-in biofuel production. In S. K. Brar, C. S. Osorio Gonzalez, C. R. Soccol, & R. Saini (Eds.), *The Microbiology of the Drop-in Biofuel Production* (pp. 41–74). Springer Nature Switzerland. https://doi.org/10.1007/978-3-031-61637-2_3

Freitas, L. C., Barbosa, J. R., Da Costa, A. L. C., Bezerra, F. W. F., Pinto, R. H. H., & Carvalho Junior, R. N. D. (2021). From waste to sustainable industry: How can agro-industrial wastes help in the development of new products? *Resources, Conservation and Recycling*, 169, 105466. https://doi.org/10.1016/j.resconrec.2021.105466

Fronza, P., Batista, M. J. P. A., Franca, A. S., & Oliveira, L. S. (2024). Bionanocomposite based on cassava waste starch, locust bean galactomannan, and cassava waste cellulose nanofibers. *Foods*, 13 (2), Article 2. https://doi.org/10.3390/foods13020202

George, N., Debroy, A., Bhat, S., Bindal, S., & Singh, S. (2021). Biowaste to bioplastics: An ecofriendly approach for a sustainable future. *Journal of Applied Biotechnology Reports*, 8 (3). https://doi.org/10.30491/jabr.2021.259403.1318

Glaser, S. J., Al-Rudainy, B., Hatti-Kaul, R., & Galbe, M. (2023). Wheat bran fractionation: Effect of steam explosion and hydrotropic extraction conditions on the recovery of sugars and lignin. *Industrial Crops and Products*, 195, 116405. https://doi.org/10.1016/j.indcrop.2023.116405

Gong, L., Passari, A. K., Yin, C., Kumar Thakur, V., Newbold, J., Clark, W., Jiang, Y., Kumar, S., & Gupta, V. K. (2024). Sustainable utilization of fruit and vegetable waste bioresources for bioplastics production. *Critical Reviews in Biotechnology*, 44 (2), 236–254. https://doi.org/10.1080/07388551.2022.2157241

Groff, M.C., Flores, L.F., Vaidya, K., & Priyanka, P.P. (2025). Agricultural renewable residues for a sustainable agro-industrial production. In Shaaban, M. (ed.), *Soils and Sustainable Agriculture. Frontier Studies in Soil Science*. Springer, Cham. https://doi.org/10.1007/978-3-031-91114-9_34

Gumul, D., Korus, J., Orczykowska, M., Rosicka-Kaczmarek, J., Oracz, J., & Areczuk, A. (2024). Starch from unripe apples (Malus domestica Borkh) as an alternative for application in the food industry. *Molecules*, 29 (8), 1707. https://doi.org/10.3390/molecules29081707

Gupta, P., Toksha, B., & Rahaman, M. (2022). A review on biodegradable packaging films from vegetative and food waste. *The Chemical Record*, 22 (7), e202100326. https://doi.org/10.1002/tcr.202100326

Haque, F., Fan, C., & Lee, Y.-Y. (2023). From waste to value: Addressing the relevance of waste recovery to agricultural sector in line with circular economy. *Journal of Cleaner Production*, 415, 137873. https://doi.org/10.1016/j.jclepro.2023.137873

Hubbe, M. A., Lavoine, N., Lucia, L. A., & Dou, C. (2020). Formulating bioplastic composites for biodegradability, recycling, and performance: A Review. *BioResources,* 16 (1), 2021–2083. https://doi.org/10.15376/biores.16.1.Hubbe

Jiao, Y., Chen, H.-D., Han, H., & Chang, Y. (2022). Development and utilization of corn processing by-products: A review. *Foods*, 11 (22), 3709. https://doi.org/10.3390/foods11223709

Julianti, E., Lubis, Z., Yusraini, E., & Ridwansyah. (2021). Physicochemical characteristics of fiber rich flour from solid waste of purple sweet potato starch processing. *IOP Conference Series: Earth and Environmental Science*, 924 (1), 012038. https://doi.org/10.1088/1755-1315/924/1/012038

Kadam, D., Arte, P., De, A., Tungare, K., Bhori, M., & Kadam, A. (2024). Phytochemicals from fruit wastes. In *Adding Value to Fruit W*astes (pp. 167–188). Elsevier. https://doi.org/10.1016/B978-0-443-13842-3.00006-X

Kaur, J., Borah, A., Chutia, H., & Gupta, P. (2024). Extraction, modification, and characterization of native litchi seed (LITCHI CHINESIS Sonn.) starch. *Journal of the Science of Food and Agriculture*, 104 (1), 215–224. https://doi.org/10.1002/jsfa.12913

Korol, J., Hejna, A., Burchart-Korol, D., & Wachowicz, J. (2020). Comparative analysis of carbon, ecological, and water footprints of polypropylene-based composites filled with cotton, jute and kenaf fibers. *Materials*, 13 (16), 3541. https://doi.org/10.3390/ma13163541

Kowalski, S., & Gumul, D. (2024). The use of waste products from the food industry to obtain high value-added products. *Foods*, 13 (6), 847. https://doi.org/10.3390/foods13060847

Kringel, D. H., Dias, A. R. G., Zavareze, E. D. R., & Gandra, E. A. (2020). Fruit wastes as promising sources of starch: Extraction, properties, and applications. *Starch – Stärke*, 72 (3–4), 1900200. https://doi.org/10.1002/star.201900200

Makroo, H. A., Naqash, S., Saxena, J., Sharma, S., Majid, D., & Dar, B. N. (2021). Recovery and characteristics of starches from unconventional sources and their potential applications: A review. *Applied Food Research*, 1 (1), 100001. https://doi.org/10.1016/j.afres.2021.100001

Maraveas, C. (2020). Production of sustainable and biodegradable polymers from agricultural waste. *Polymers*, 12 (5), 1127. https://doi.org/10.3390/polym12051127

Martins, S. H. F., Pontes, K. V., Fialho, R. L., & Fakhouri, F. M. (2022). Extraction and characterization of the starch present in the avocado seed (Persea americana mill) for future applications. *Journal of Agriculture and Food Research*, 8, 100303. https://doi.org/10.1016/j.jafr.2022.100303

Masurkar, S. (2024). Production and biodegradability studies of edible bioplastics from potato peel waste. *Bulletin of Pure and Applied Sciences-Zoology*, 13 (1), 110–121.

McNeill, D. C., Pal, A. K., Nath, D., Rodriguez-Uribe, A., Mohanty, A. K., Pilla, S., Gregori, S., Dick, P., & Misra, M. (2024). Upcycling of ligno-cellulosic nutshells waste biomass in biodegradable plastic-based biocomposites uses—A comprehensive review. *Composites Part C: Open Access*, 14, 100478. https://doi.org/10.1016/j.jcomc.2024.100478

Mhaske, S. T., Mohanty, J. D., & Borse, P. Y. (2024). Starch from agro-waste for food packaging applications. In Kumar, S., Mukherjee, A. & Katiyar, V. (Eds.), *Agro-Waste Derived Biopolymers and Biocomposites* (pp. 241–265). Wiley. https://doi.org/10.1002/9781394175161.ch9

Mujtaba, M., Fernandes Fraceto, L., Fazeli, M., Mukherjee, S., Savassa, S. M., Araujo De Medeiros, G., Do Espírito Santo Pereira, A., Mancini, S. D., Lipponen, J., & Vilaplana, F. (2023). Lignocellulosic biomass from agricultural waste to the circular economy: A review with focus on biofuels, biocomposites and bioplastics. *Journal of Cleaner Production*, 402, 136815. https://doi.org/10.1016/j.jclepro.2023.136815

Nagarajan, D., Senthilkumar, G., Chen, C., Karmegam, N., Praburaman, L., Kim, W., & Dong, C. (2024). Sustainable bioplastics from seaweed polysaccharides: A comprehensive review. *Polymers for Advanced Technologies*, 35 (8), e6536. https://doi.org/10.1002/pat.6536

Nalini, S., & Sathiyamurthi, S. (2023). Bioplastic from microbial and agricultural biomass. In Inamuddin, Altalhi, T. & Alrooqi, A. (Eds.), *Green Sustainable Process for Chemical and Environmental Engineering and Science* (pp. 455–475). Elsevier. https://doi.org/10.1016/B978-0-323-95183-8.00010-X

Neeraj Siddiqui, S., Dalal, N., Srivastva, A., & Pathera, A. K. (2021). Physicochemical, morphological, functional, and pasting properties of potato starch as a function of extraction methods. *Journal of Food Measurement and Characterization*, 15 (3), 2805–2820. https://doi.org/10.1007/s11694-021-00862-5

Nguyen, T. K., That, N. T. T., Nguyen, N. T., & Nguyen, H. T. (2022). Development of starch-based bioplastic from jackfruit seed. *Advances in Polymer Technology*, 2022, 1–9. https://doi.org/10.1155/2022/6547461

Obadi, M., & Xu, B. (2021). Review on the physicochemical properties, modifications, and applications of starches and its common modified forms used in noodle products. *Food Hydrocolloids*, 112, 106286. https://doi.org/10.1016/j.foodhyd.2020.106286

Ochoa, S., & Osorio-Tobón, J. F. (2024). Isolation and characterization of starch from the purple yam (Dioscorea alata) anthocyanin extraction residue obtained by ultrasound-assisted extraction. *Waste and Biomass Valorization*, 15 (1), 379–389. https://doi.org/10.1007/s12649-023-02155-y

Oyeyinka, S. A., Adeloye, A. A., Olaomo, O. O., & Kayitesi, E. (2020). Effect of fermentation time on physicochemical properties of starch extracted from cassava root. *Food Bioscience*, 33, 100485. https://doi.org/10.1016/j.fbio.2019.100485

Padi, R. K., & Chimphango, A. (2020). Commercial viability of integrated waste treatment in cassava starch industries for targeted resource recoveries. *Journal of Cleaner Production*, 265, 121619. https://doi.org/10.1016/j.jclepro.2020.121619

Peixoto, A. M., Petronilho, S., Domingues, M. R., Nunes, F. M., Lopes, J., Pettersen, M. K., Grøvlen, M. S., Wetterhus, E. M., Gonçalves, I., & Coimbra, M. A. (2023). Potato chips byproducts as feedstocks for developing active starch-based films with potential for cheese packaging. *Foods*, 12 (6), 1167. https://doi.org/10.3390/foods12061167

Pham, D. T., Hoang, H. G., & Thuy, B. T. P. (2024). Avocado seed, coconut fiber, and vinegar residue blended composite biodegradable bioplastic film development: Agricultural waste management approach. *Biomass Conversion and Biorefinery*, 15 (10), 14779–14790. https://doi.org/10.1007/s13399-024-06195-9

Pires, J. B., Santos, F. N. D., Cruz, E. P. D., Fonseca, L. M., Siebeneichler, T. J., Lemos, G. S., Gandra, E. A., Zavareze, E. D. R., & Dias, A. R. G. (2024). Starch extraction from avocado by-product and its use for encapsulation of ginger essential oil by electrospinning. *International Journal of Biological Macromolecules*, 254, 127617. https://doi.org/10.1016/j.ijbiomac.2023.127617

Polman, E. M. N., Gruter, G.-J. M., Parsons, J. R., & Tietema, A. (2021). Comparison of the aerobic biodegradation of biopolymers and the corresponding bioplastics: A review. *Science of the Total Environment*, 753, 141953. https://doi.org/10.1016/j.scitotenv.2020.141953

Puri, A., Syukri, D. M., Silvia, E., Ladyani, F., Mohite, P., Ade, N., Munde, S., Chidrawar, V. R., Singh, S., & Shafi, S. (2024). Waste-to-value-added customized cationic banana starch for potential flocculant application. *Journal of Polymers and the Environment*. https://doi.org/10.1007/s10924-024-03349-8

Rahardiyan, D., Moko, E. M., Tan, J. S., & Lee, C. K. (2023). Thermoplastic starch (TPS) bioplastic, the green solution for single-use petroleum plastic food packaging – A review. *Enzyme and Microbial Technology*, 168, 110260. https://doi.org/10.1016/j.enzmictec.2023.110260

Rashwan, A. K., Younis, H. A., Abdelshafy, A. M., Osman, A. I., Eletmany, M. R., Hafouda, M. A., & Chen, W. (2024). Plant starch extraction, modification, and green applications: A review. *Environmental Chemistry Letters*, 22 (5), 2483–2530. https://doi.org/10.1007/s10311-024-01753-z

Raza, H., Xu, H., Zhou, Q., He, J., Zhu, B., Li, S., & Wang, M. (2023). A review of green methods used in starch–polyphenol interactions: Physicochemical and digestion aspects. *Food & Function*, 14 (18), 8071–8100. https://doi.org/10.1039/D3FO01729J

Rosenboom, J.-G., Langer, R., & Traverso, G. (2022). Bioplastics for a circular economy. *Nature Reviews Materials*, 7 (2), 117–137. https://doi.org/10.1038/s41578-021-00407-8

Roy, P., Mohanty, A. K., Dick, P., & Misra, M. (2023). A review on the challenges and choices for food waste valorization: Environmental and economic impacts. *ACS Environmental Au*, 3 (2), 58–75. https://doi.org/10.1021/acsenvironau.2c00050

Saba, B., Bharathidasan, A. K., Ezeji, T. C., & Cornish, K. (2023). Characterization and potential valorization of industrial food processing wastes. *Science of the Total Environment*, 868, 161550. https://doi.org/10.1016/j.scitotenv.2023.161550

Sadh, P. K., Chawla, P., Kumar, S., Das, A., Kumar, R., Bains, A., Sridhar, K., Duhan, J. S., & Sharma, M. (2023). Recovery of agricultural waste biomass: A path for circular bioeconomy. *Science of the Total Environment*, 870, 161904. https://doi.org/10.1016/j.scitotenv.2023.161904

Samer, M. (2019). Bioplastics production from agricultural crop residues. *Agricultural Engineering International: CIGR Journal*, 21 (3), 190–194.

Saputri, C. A., Julyatmojo, F. A., Harmiansyah, Febrina, M., Mahardika, M., & Maulana, S. (2024). Characteristics of bioplastics prepared from cassava starch reinforced with banana bunch cellulose at various concentrations. *IOP Conference Series: Earth and Environmental Science*, 1309 (1), 012006. https://doi.org/10.1088/1755-1315/1309/1/012006

Sasikala, K. L. N. S., Kondarage, Y. G., & Diyabalanage, S. (2022). Development of sweet potato starch-based thin films to produce biodegradable packaging materials. http://repository.kln.ac.lk/handle/123456789/25974

Shamim, M. Z., Mishra, A. K., Kausar, T., Mahanta, S., Sarma, B., Kumar, V., Mishra, P. K., Panda, J., Baek, K.-H., & Mohanta, Y. K. (2023). Exploring edible mushrooms for diabetes: Unveiling their role in prevention and treatment. *Molecules*, 28 (6), 2837. https://doi.org/10.3390/molecules28062837

Show, P., Tang, M., Nagarajan, D., Ling, T., Ooi, C.-W., & Chang, J.-S. (2017). A holistic approach to managing microalgae for biofuel applications. *International Journal of Molecular Sciences*, 18 (1), 215. https://doi.org/10.3390/ijms18010215

Singh, P., Pandey, V. K., Singh, R., Singh, K., Dash, K. K., & Malik, S. (2024). Unveiling the potential of starch-blended biodegradable polymers for substantializing the eco-friendly innovations. *Journal of Agriculture and Food Research,* 15, 101065. https://doi.org/10.1016/j.jafr.2024.101065

Souza, J. C. A. D., Macena, J. F. F., Andrade, I. H. P., Camilloto, G. P., & Cruz, R. S. (2021). Functional characterization of mango seed starch (Mangifera indica L.). *Research, Society and Development*, 10 (3), e30310310118. https://doi.org/10.33448/rsd-v10i3.10118

Taha, T. H., Abu-Saied, M. A., Elnouby, M., Hashem, M., Alamri, S., Desouky, E. A. E., & Morsy, K. (2024). Profitable exploitation of biodegradable polymer including chitosan blended potato peels' starch waste as an alternative source of petroleum plastics. *Biomass Conversion and Biorefinery*, 14 (1), 207–215. https://doi.org/10.1007/s13399-021-02244-9

Tecorralco-Bobadilla, A. L., Vázquez-Morillas, A., González-Cardoso, G., & Sotelo-Navarro, P. X. (2024). Life cycle assessment (LCA) of bioplastics. In Inamuddin, T. Altalhi, & V. C. Fernandes (Eds.), *Toxic Effects of Micro- and Nanoplastics* (1st ed., pp. 21–59). Wiley. https://doi.org/10.1002/9781394238163.ch2

Vannini, M., Marchese, P., Sisti, L., Saccani, A., Mu, T., Sun, H., & Celli, A. (2021). Integrated efforts for the valorization of sweet potato by-products within a circular economy concept: Biocomposites for packaging applications close the loop. *Polymers*, 13 (7), 1048. https://doi.org/10.3390/polym13071048

Venkatachalam, H., & Palaniswamy, R. (2020). Bioplastic world: A review. *Journal of Advanced Scientific Research*, 11 (03), 43–53.

Vigneswari, S., Kee, S. H., Hazwan, M. H., Ganeson, K., Tamilselvan, K., Bhubalan, K., Amirul, A.-A., & Ramakrishna, S. (2024). Turning agricultural waste streams into biodegradable plastic: A step forward into adopting sustainable carbon neutrality. *Journal of Environmental Chemical Engineering*, 12 (2), 112135. https://doi.org/10.1016/j.jece.2024.112135

Vilpoux, O. F., & Santos Silveira Junior, J. F. (2023). Global production and use of starch. In *Starchy Crops Morphology, Extraction, Properties and Applications* (pp. 43–66). Elsevier. https://doi.org/10.1016/B978-0-323-90058-4.00014-1

Weligama Thuppahige, V. T., Moghaddam, L., Welsh, Z. G., Wang, T., Xiao, H.-W., & Karim, A. (2023). Extraction and characterisation of starch from cassava (Manihot esculenta) agro-industrial wastes. *LWT—Food Science and Technology*, 182, 114787. https://doi.org/10.1016/j.lwt.2023.114787

Xiang, H., Li, Z., Liu, H., Chen, T., Zhou, H., & Huang, W. (2022). Green flexible electronics based on starch. *Npj Flexible Electronics*, 6 (1), 1–16. https://doi.org/10.1038/s41528-022-00147-x

Yaashikaa, P. R., Senthil Kumar, P., & Varjani, S. (2022). Valorization of agro-industrial wastes for biorefinery process and circular bioeconomy: A critical review. *Bioresource Technology*, 343, 126126. https://doi.org/10.1016/j.biortech.2021.126126

Zhang, R., Ma, S., Li, L., Zhang, M., Tian, S., Wang, D., Liu, K., Liu, H., Zhu, W., & Wang, X. (2021). Comprehensive utilization of corn starch processing by-products: A review. *Grain & Oil Science and Technology*, 4 (3), 89–107. https://doi.org/10.1016/j.gaost.2021.08.003

Zhang, X., Cheng, Y., Jia, X., Geng, D., Bian, X., & Tang, N. (2022). Effects of extraction methods on physicochemical and structural properties of common vetch starch. *Foods*, 11 (18), 2920. https://doi.org/10.3390/foods11182920

4 Roots, Tubers, and Bulbs
Beyond the Usual Suspects

Mine Koçyiğit

4.1 INTRODUCTION

In recent years, interest in the development of biodegradable polymers from renewable resources has increased, especially in the packaging industry. These studies are against the danger of microplastics, which has begun to reach dangerous levels due to environmental pollution. It is done with the aim of moving toward a greener environment (Schlemmer et al., 2010). The risks caused by plastics on living resources are causing concern. The packaging industry accounts for a significant portion of the plastic waste produced all over the world. Therefore, reducing the accumulation of plastics in the environment and its associated consequences has become an important problem for every country. Accordingly, edible plastics have been proposed as an alternative to traditional plastics. Edible plastics are plastics that can be consumed by humans, animals, and microorganisms without any harmful effects on health (Ghorpade et al., 2022). Natural fibers are used in structural applications due to their properties such as being biodegradable, light weight, renewable, and low cost (Awais et al., 2021; Ovalı & Sancak, 2022; Parbin et al., 2019; Dixit et al., 2017). Plant-based natural fibers can be made from different parts of plants, such as cotton for seed-derived fiber, agave for leaf-derived fiber, hemp for stem-derived fiber, and coconut for fruit-derived fiber (Li et al., 2020; Fortea-Verdejo et al., 2017; Gurunathan et al., 2015; Sanjay et al., 2015; Balaji et al., 2016). Due to their flexibility, petrochemical-based plastics can be easily formed into sheets and various shapes. In addition, the mechanical durability, oxygen and carbon dioxide (CO_2) barriers, and heat resistance of these petrochemical-based plastics make them a priority for use as packaging materials. However, their use has now decreased because they are very difficult to degrade in nature (Marsh & Bugusu, 2007). Additionally, foods are also contaminated due to the chemicals in these packages (Pocas & Hogg, 2007). To overcome these negativities, the use of biodegradable and nontoxic plastics produced from renewable resources should be encouraged instead of synthetic plastics. Biodegradable plastics are plastics designed to break down in natural aerobic (composting) and anaerobic (landfill) environments. Biodegradation of plastics occurs through the metabolization of microorganisms (Patel et al., 2011). Among biopolymers, starch is an inexpensive biopolymer that is completely biodegradable and can ultimately break down to CO_2 and water. Tuberous plants are also the most important sources of starch.

4.2 ROOTS AS POLYMER SOURCES

Interest in fibers obtained from plant roots is increasing day by day as natural plant fibers are accepted worldwide. Determining suitable plant fibers with certain properties is an important research area in the development of environmentally friendly, renewable, low-cost, and fiber-based materials (Moshi

DOI: 10.1201/9781003534914-4

et al., 2020). Natural plant fibers have been processed for different purposes due to their properties such as high cellulose content, low lignin content, high temperature stability, and high crystallinity (Indran & Raj, 2015). The quality of plant fibers is often related to certain properties such as mechanical, thermal, and surface properties. However, these characteristics may vary depending on the plant species and plant parts (Indran et al., 2014). Regarding plant fibers obtained from roots, the first plants that were the subject of research were *Glycyrrhiza glabra* L., *Ficus benjamina* L., and *Cissus quadrangularis* L. (Ganapathy et al., 2019; Prabhu et al., 2014).

4.2.1 *Glycyrrhiza glabra* L.

The fiber obtained from the roots of *G. glabra* has recently been recognized as a material that can be used in this field. The physiological and chemical properties of natural fibers vary depending on various factors, such as their composition/structure, and unpredictable weather conditions that occur during plant growth (Dittenber & GangaRao, 2012). Commonly used plant-based fibers must meet the required qualities. *G. glabra* is also known as a medicinal plant in different parts of the world, especially in Eastern Europe and Russia (Wang et al., 2020). The genus *Glycyrrhiza* belongs to the Fabaceae family. It is represented by 16 taxa in the world and 8 taxa in Türkiye. It is called "licorice" by local people in Türkiye.

Licorice is a perennial shrub that can grow up to approximately 120–150 cm. It is native to Russia and China. Its small, bluish, purple flowers are 0.8–1.2 cm tall. Its leaves are 7–15 cm long and root structure reaches 1 m in length underground (Figure 4.1). It grows in Mediterranean countries, southeastern Europe, and parts of Asia (Chamberlain, 1970). In a study conducted on *G. glabra* roots obtained from Türkiye, the elemental composition, surface appearance, molecular interactions, crystallization behavior, thermal degradation and mechanical behavior of these roots were examined. The roots of *G. glabra,* in comparison with the roots of other cellulosic reinforcement fibers, showed remarkable properties. The cellulose-based structure of *G. glabra* fibers gives them exceptional elongation and thermal stability properties. Its fibers were found to consist of cellulose (40.46%), hemicellulose (15.94%), lignin (12%), wax (1.3%), moisture (9.93%) and other components (Ovalı, 2023).

Cellulose molecules contain hydroxyl groups that form hydrogen bonds, resulting in highly crystalline structures and providing them with strong mechanical properties (Indran & Raj, 2015).

FIGURE 4.1 *Glycyrrhiza glabra* flowers, leaves, and roots.

Hemicelluloses are polysaccharides with short branched chains. This feature makes them hydrophilic. They are also more susceptible to biodegradation. Additionally, they have low thermal resistance. Due to their amorphous structure, the strength of hemicelluloses decreases and they are less resistant to chemical effects (Kalia et al., 2011). Lignin has the potential to be combined with biopolymers such as starch, cellulose, and polyhydroxybutyrate to create biomaterials with improved mechanical properties (Bourmaud et al., 2020). Increasing the wax content in cellulose-based materials reduces the compatibility between the material surface and the matrix in the composite (Raju et al., 2022). The density of *G. glabra* fibers was measured as 1.43 g/cm^3. Thermogravimetric analysis results showed that *G. glabra* fibers were thermally stable at 354.09 °C, which is the polymerization process temperature. Cellulose density and tensile strength have been found to be lower than other fibers, but can improve its use with pre-treatments. Thermal stability of the fibers was determined by thermogravimetric analysis up to 354.09 °C. This value is comparable to other lignocellulosic fibers. Due to these properties, *G. glabra* roots have great potential as a source of cellulosic fiber for various industries (Ovalı, 2023).

Nanofibers are widely accepted in many fields such as membranes, tissue engineering, layered fabrics, wound healing reinforcement, etc. Nanofibers are produced primarily by launching electrically charged jets at high voltage relative to the surface tension of the electrochemical solvent (Haider et al., 2014; Kwak et al., 2016; Gupta et al., 2014). The use of medicinal plants in the treatment of human skin against skin infections, such as wound and burn healing, and antiviral, antibacterial, and antifungal applications, has a deep history. Nontoxic and environmentally friendly natural antimicrobial compounds may be a better alternative to synthetic nanoparticles such as silver and zinc currently used in wound treatment. Natural compounds such as gelatin and chitosan are now being studied with suitable carrier polymers to synthesize nanomats for biomedical applications (Min et al., 2004; Abid et al., 2019; Sadeghianmaryan et al., 2020; Reneker et al., 2008). In this context, the nanofiber membrane obtained from *G. glabra* can be considered as a new milestone in the field of medicine (Topuz et al., 2019; Elsabee et al., 2012; De Freitas Figueiredo et al., 2020; Mazzei et al., 2020).

4.2.2 *Ficus benjamina* L.

F. benjamina, known as weeping fig, Benjamin fig, or simply ficus tree, belongs to the Moraceae family. It is native to Asia and Australia. It is generally grown as an ornamental plant in homes and gardens. *F. benjamina*, in particular, is the official tree of Bangkok, Thailand (Maji et al., 2015). *F. benjamina* grows up to 30 m tall in natural conditions, with drooping branches and 6–13 cm long, oval, shiny leaves with pointed tips (Pitchayya et al., 2021). It is represented by 880 species worldwide (POWO, 2024).

A study has revealed that aerial root fibers of *F. benjamina* L. species can be used as a potential reinforcement material for composite structures (Moshi et al., 2020; Sumesh et al., 2022). In the density analysis, it was found that the fiber had a low density (1,175 kg/m^3), and according to this feature, it is suggested that it can be used in areas that require lightness. The food packaging industry may be one of these areas of use. The high cellulose content (64.48 %) in the tested fibers is indicative of mechanical strength. It has also been determined that the fiber obtained from the roots of this plant can be used at high temperatures. It was revealed that the fiber provides good fiber-matrix interface bonding due to its surface roughness (Rathinavelu & Paramathma, 2022).

In this period when we need to use biological resources sustainably, finding new cellulosic fibers is of great importance for current environmental awareness, as they have the potential to develop composite materials. Cellulosic fibers obtained from plant roots contribute to reducing environmental impact as they are an environmentally friendly alternative to synthetic fibers. These fibers can be used to create strong, lightweight materials for a wide variety of applications, from construction to automotive and packaging industries.

4.2.2.1 *Cissus quadrangularis* L.

The genus *Cissus*, which belongs to the Vitaceae family, is represented by 280 species worldwide (POWO, 2024). *C. quadrangularis* is known for both its medicinal value and the rich fiber content in its roots and trunk. It is native to India and Sri Lanka, but is also found in Africa, the Arabian Peninsula, and Southern Asia. To date, little research has been conducted on fibers derived from this plant. It is a common, perennial ivy plant that can grow up to approximately 2.5 m and has quadrangular branches with internodes that are 1.5–1.8 cm wide and 10–15 cm long (Figure 4.2) (Anitha & Suji, 2012).

Findings from a study of the roots of this plant are promising for its potential use in fiber industries, composite production, and other applications. Its lightweight structure, together with its high cellulose content (77.17%) and minimum wax content (0.14%), gives the fiber excellent specific strength and strong bonding properties. According to electron microscope observations, it has a scaly honeycomb-like outer surface. A low microfibril angle was also revealed, contributing to its high modulus. In addition, thermogravimetric analysis shows that the fiber has good thermal stability, maintaining integrity up to 230 °C and within the temperature range required for polymerization processes. The characterization results clearly showed that the roots of this plant are a good alternative to conventional man-made hazardous fibers due to their outstanding physical, mechanical, and thermal degradation properties (Indrana et al., 2014).

In another study, *C. quadrangularis* fiber mixed with polylactic acid was used. In this study, composites were produced by adding three different fibers and polylactic acid, 20%, 30%, and 40%, respectively. In contrast, another composite with the same volume percentage was created by coating the fiber with calcium hydroxide to change its microcrystalline structure and improve its mechanical properties. Tests have been conducted to evaluate the improved properties. According to the preliminary study, the mechanical properties of the combination of *C. quadrangularis* fiber treated with polylactic acid were increased compared to untreated *C. quadrangularis* fiber and pure polylactic acid. Calcium hydroxide served as a fiber binding agent, increasing stress transmission in the matrix and increasing tensile strength and durability (Gummadavalli et al., 2023). This may be

FIGURE 4.2 *Cissus quadrangularis*: (a) aerial parts and (b) roots. (Designed from Indran & Raj, 2015 and Kumar et al., 2019.)

due to changes in the fiber surface. Mechanical contact with the matrix increases the roughness of the surfaces of the fibers and strengthens them (Hossain et al., 2013).

4.3 TUBERS AS POLYMER SOURCES

Starch can be obtained from grains such as rice, potatoes, and corn, as well as from various sources, including tubers (Mali et al., 2002). Among various biopolymers, starch is an inexpensive and fully biodegradable option that breaks down into CO_2 and water. Starch is an attractive choice for biodegradable materials due to its low cost, low density, and ability to be blended with other polymers to create composites with customized properties (Teseme, 2020). Biodegradable plastics are typically divided into photodegradable, oxidatively degradable, hydrolytically degradable, and biodegradable types (Averous et al., 2000). The biodegradability of plastics is affected by their specific properties. Factors such as surface area, hydrophilic or hydrophobic nature, chemical structure, molecular weight, glass transition temperature, melting point, elasticity, and crystallinity play key roles in the biodegradation process. Below are some starch-producing tubers.

4.3.1 *Ipomoea batatas* (L.) Lam.

The genus *Ipomea*, commonly known as sweet potato, belongs to the Convolvulaceae family and is represented by 635 species in the world. It spreads naturally in the Southern Hemisphere. It is grown in other countries for consumption as food (POWO, 2024). Sweet potato is an herbaceous perennial plant bearing triangular or palmately lobed leaves and medium-sized sympetal flowers. The stems often creep along the ground, forming roots at nodes. The petiole is 13–51 cm long. Leaf blades are highly variable, 5–13 cm long, heart-shaped, kidney-shaped, egg-shaped, and round or triangular in shape (Alam, 2021). The flowers open before sunrise and remain open for a few hours, closing later in the morning as they begin to fade. The edible tuberous root is smooth and can range in color from yellow and orange to red, brown, purple, and beige. Sweet potato varieties with white or pale-yellow flesh tend to be less sweet and moist than those with red, pink, or orange flesh (Figure 4.3) (Loebenstein & Thottappilly, 2009).

FIGURE 4.3 *Ipomoea batatas*: (a) aerial parts and (b) tubers. (Designed from Alam 2021.)

Sweet potatoes contain 50–80% starch by dry weight; starch consists of 70–80% highly branched amylopectin and 20–30% linear and slightly branched amylose (Zhou et al., 2015). Additionally, sweet potatoes are rich in starch, dietary fiber, minerals, vitamins, and antioxidants such as phenolic acids, anthocyanins, tocopherol, β-carotene, and ascorbic acid. If sweet potato starch-based plastics are used in food packaging, these compounds can migrate into the food and potentially increase the nutritional value of the food packaging (Issa et al., 2018). Biodegradable films made from sweet potato starch can be produced by mixing starch with glycerol in varying starch-to-glycerol weight ratios. Changing the starch–glycerol ratio can affect the properties of bioplastics, making sweet potato starch-based bioplastics suitable for food packaging applications. Bioplastics with a higher starch/glycerol ratio degrade faster. Sweet potato starch-based plastic packaging films show signs of deterioration after 7 days (Teseme, 2020).

4.3.2 *Solanum tuberosum* L.

Potato, one of the most grown tuberous plants, belongs to the Solanaceae family. In 2017, worldwide potato production exceeded 359 million tons (FAOSTAT, 2022). The genus *Solanum*, represented by 1,232 species in the world, carries tropane alkaloids in its pink or white flowers and leaves. Alkaloids are present only in its tubers (Figure 4.4). Potato tubers are an important source of staple food (Alvani et al., 2011), and also used for starch isolation (Waterschoot et al., 2015) and as raw material for processed foods (Reyniers et al., 2019). Starch constitutes approximately 70–85% of the dry matter of potato (Singh & Kaur, 2016). In the food industry, starch is often used due to its ability to bind large amounts of water and its excellent thickening and gel-forming capacity. Among natural polymers, starch is one of the most promising materials for biopolymer development (Ghanbari et al., 2018). Starch is a widely available and easily biodegradable natural resource. High levels of water or plasticizers are required to produce a starch-based film. These plasticized materials (by applying thermal and mechanical energy) are called thermoplastic starch and offer an alternative to synthetic polymers (Peelman et al., 2013).

The use of starch in biodegradable packaging is both environmentally and economically beneficial, as it is often wasted during production, despite its richness in fiber and bioactive compounds (Luchese et al., 2019). Research indicated that blending starch with natural fibers can result in

FIGURE 4.4 *Solanum tuberosum*: (a) aerial parts and (b) tubers.

polymer materials with enhanced properties. For instance, a hybrid composite was created by combining manioc pulp (a starch-rich material) and sugar palm fiber, utilizing fructose in the process (Edhirej et al., 2017; Machado et al., 2018). The incorporation of fibers notably influenced various physical properties, including an increase in thickness and density, as well as changes in water content, solubility, and water absorption (Kaisangsri et al., 2014; Stoffel et al., 2018).

In a study, by-products of the potato industry were used as raw materials, xanthan gum was added as a plasticizer, and natural oat fiber was added as reinforcement. A biodegradable foam was developed from this mixture by Thermo-pressing. Morphology, mechanical properties, and the biodegradability of the final product were investigated. Water absorption and solubility indices were found to be at highest level in the sample containing 20% plasticizer and 20% fiber. Hydrolysis was regarded as the basic mechanism of degradation which took place in 14–20 days. The combination of a by-product of potatoes with xanthan gum and natural oat fiber made it possible to produce a promising product (Rodrigues et al., 2020). In another study, waste potato peels were washed, dried, and then ground to isolate starch for the production of bioplastic. This starch was used with the addition of hydrochloric (HCl) and acetic acid. It has been observed that bioplastic products produced using HCl are darker in color than those produced using acetic acid. Bioplastics recorded densities between 1.2 and 1.7 g/cm^3; HCl-based bioplastics showed more dense properties than acetic acid-based ones. Bioplastic containing HCl has a higher elasticity than that containing acetic acid (Bello et al., 2024). According to the results of these studies, bioplastics obtained from the peels of waste tubers are cheap, they are renewable and do not threaten food safety.

4.3.3 *Dioscorea alata* L.

Dioscorea is represented by 634 species worldwide and is the genus that gives its name to the family – Dioscoreaceae family (POWO, 2024). *D. alata*, also known as purple yam (Figure 4.5), is one of the important food crops in Southeast Asia and the Philippines. Domesticated species are used as food in many countries. These include species such as *D. bulbifera* L., *D. hispida* Dennst., *D. esculenta* (Lour.) Burkill, *D. nummularia* Lam., *D. pentaphylla* L., and *D. transversa* R.Br. Among these, only *D. alata* and *D. esculenta* are regularly grown and consumed. Others contain

FIGURE 4.5 *Dioscorea alata*: (a) aerial parts and (b) tubers.

high levels of toxins called dioscorins and are used as food only in times of hunger, but it is also very important to prepare them correctly (Bevacqua, 1994).

Yam starch is a polymeric carbohydrate modification; it consists of anhydroglucose units. This is not a single unit; most starches contain two types of glucose polymers. One of these is a linear chain molecule called amylose. The other branched glucose polymer is called amylopectin (Falguera et al., 2011). Yam starch (*D. alata*) contains approximately 24–26% amylose; the amylose–amylopectin content is 0.32, which is significant, because the film-forming capacity of starches depends on their amylose content (Alves et al., 2007). In research conducted considering the sectors in which purple yam starch can be used, it was observed that it has a significant potential as a functional ingredient in various industries. After extraction, physical, chemical, and enzymatic methods were used to increase starch's functionality, and after these modifications, properties such as solubility, viscosity, gelatinization, and resistance to retrogradation were tested. This shows that starch may find application in bioplastics beyond food. This will allow processes to be scaled up for industrial use, ensuring sustainability and discovering new applications (Shrivastava et al., 2024).

4.4 BULBS AS POLYMER SOURCES

4.4.1 *Allium cepa* L.

The genus *Allium* is a large genus of monocotyledonous flowering plants, consisting of approximately 1,000 recognized species. This makes it the largest genus of the Amaryllidaceae family and one of the largest plant genera in the world (Rabinowitch & Thomas, 2022). Many species in this genus are edible; many of them, such as onions, garlic, green onions, shallots, leeks, and chives, have a long history of growing and using them as vegetables. As of 2023, onion is the second most grown vegetable in the world after tomatoes (Ray, 2024). Onion (*A. cepa* L.) has emerged as a promising source for producing pureed, edible films suitable for biodegradable packaging applications. From a physiological point of view, onion is a bulbous plant with modified underground roots, characterized by adventitious roots emerging from its lower surface. It also has fleshy, edible scaly leaves arranged concentrically on the upper part of the shortened stem. The bulb is covered with a layer of dry, membranous scales known as the tunic, which acts as a protective outer skin (Figure 4.6). The cell

FIGURE 4.6 *Allium cepa*: (a) inflorescence and (b) bulbs.

wall of the onion consists of cellulose, which provides structural support and protection to the cells (Koyuncu et al., 2023).

As interest in renewable bioplastics increases, efforts to find different methods to isolate them from nature also increased. In a study on the genus *Allium*, the production of flexible and optically translucent bioplastics containing only onion was reported. Onion bioplastics can be comprehensively characterized by targeting correlations between chemical structure, film formation protocol, and film properties. While the washing process removed all soluble carbohydrates, the approximately 14% soluble carbohydrates present in the unwashed samples caused the films to have a lower glass transition temperature. Hydrophobic/hydrophilic properties and water barrier and moisture absorption properties of bioplastics were also determined (Santos Dias et al., 2020). Optically translucent onion-based films made from raw or hydrothermally treated *A. cepa* puree were fabricated using a solution extension-evaporation method (solvent casting). These bioplastics are created through a fully aqueous processing technique that does not require the addition of fillers or functional additives for mechanical reinforcement, representing an environmentally friendly approach. This is because substances with excellent film-forming properties occur naturally in onions or are produced during the hydrothermal process. Depending on the specific film-forming process used, bioplastics have exhibited varying properties in terms of thermal stability, glass transition temperature, hydrogen bonding energy, hydrophilic/hydrophobic properties, crystallinity index, moisture absorption, water vapor permeability, and mechanical strength. The results highlight the potential to produce environmentally friendly bioplastics with diverse properties suitable for a variety of high-value applications. These properties can also be adjusted by changing the parameters used during film formation in accordance with the structure–processing–property relationships.

In another study, composite films were developed by combining waste onion peels and stems of *Allium* with gluten. The applicability of this film was evaluated based on the quality of the peeled shallot. The conditions of the onions during storage for 21 days were also examined (Figure 4.7). The research showed that the developed composite film is suitable for storing shallots at 4 °C for up to 21 days, offering a suitable solution for the onion processing industry to meet consumer demand. Additionally, this active composite film can also be applied to preserve other minimally processed fruits and vegetables. The technology also contributes to reducing the environmental impact of onion solid waste (Thivya et al., 2022).

4.5 CONCLUSION

Bioplastics play a crucial role in decreasing reliance on petroleum-based polymers, mitigating the ongoing accumulation of plastic waste, and offering better control over CO_2 emissions. Sustainable plastic products are those made from bio-based plastics with minimal environmental and social impacts. These products are characterized by a smaller carbon footprint, reduced waste generation, and lower pollution compared to conventional plastics derived from crude oil. Bio-based polymers not only lower the carbon footprint of plastics but also integrate well into existing recycling systems. Life cycle assessments indicate that bio-based plastics outperform petroleum-based plastics, especially in areas like greenhouse gas emissions, fossil fuel consumption, and eutrophication. The increasing emphasis on sustainability in materials science requires innovative approaches to fiber extraction, processing and application that ensure minimal waste and environmental damage. By exploring and using natural, renewable resources such as plant fibers, we can create more sustainable, biodegradable, and cost-effective solutions that align with the global need for environmentally responsible innovation. The use of starch in biodegradable packaging is both environmentally and economically beneficial, as it is often wasted during production, despite its richness in fiber and bioactive compounds. Research indicated that blending starch with natural fibers can result in polymer materials with enhanced properties. Recently, roots, tubers, and bulbs from plants such as *G. glabra*,

FIGURE 4.7 Application of developed composite film on quality of fresh-peeled onion. (Designed from Thivya et al., 2022.)

F. benjamina, *C. quadrangularis*, *I. batatas*, *S. tuberosum*, *D. alata*, and *A. cepa* have become the subject of increasing research in the field of bioplastics. Promising studies should continue to grow. We must take preventive measures against the consequences of global climate change before they lead to wars over hunger in the future.

REFERENCES

Abid, S., Hussain, T., Raza, Z. A., & Nazir, A. (2019). Current applications of electrospun polymeric nanofibers in cancer therapy. *Materials Science and Engineering: C*, *97*, 966–977. https://doi.org/10.1016/j.msec.2018.12.105

Alam, M. K. (2021). A comprehensive review of sweet potato (*Ipomoea batatas* [L.] Lam): Revisiting the associated health benefits. *Trends in Food Science & Technology*, *115*, 512–529. https://doi.org/10.1016/j.tifs.2021.07.001

Alvani, K., Qi, X., Tester, R. F., & Snape, C. E. (2011). Physico-chemical properties of potato starches. *Food Chemistry*, *125* (3), 958–965. https://doi.org/10.1016/j.foodchem.2010.09.088

Alves, V. D., Mali, S., Beléia, A., & Grossmann, M. V. E. (2007). Effect of glycerol and amylose enrichment on cassava starch film properties. *Journal of Food Engineering*, *78* (3), 941–946. https://doi.org/10.1016/j.jfoodeng.2005.12.007

Anitha, R., & Suji, P. (2012). Pharmacognostic evaluation of *Cissus quadrangularis* L. stem. *International Journal of Pharmaceutical Sciences and Research*, *3* (7), 2296.

Averous, L., Moro, L., Dole, P., & Fringant, C. (2000). Properties of thermoplastic blends: starch–polycaprolactone. *Polymer*, *41* (11), 4157–4167. https://doi.org/10.1016/S0032-3861(99)00636-9

Awais, H., Nawab, Y., Amjad, A., Anjang, A., Akil, H. M., & Abidin, M. S. Z. (2021). Environmental benign natural fibre reinforced thermoplastic composites: A review. *Composites Part C: Open Access*, *4*, 100082. https://doi.org/10.1016/j.jcomc.2020.100082

Balaji, A. N., Karthikeyan, M. K. V., & Vignesh, V. (2016). Characterization of new natural cellulosic fiber from kusha grass. *International Journal of Polymer Analysis and Characterization*, *21* (7), 599–605. https://doi.org/10.1080/1023666X.2016.1192324

Bello, T. K., Eze, E. C., Usman, M. S., & Isa, M. T. (2024). Characterization of bioplastics produced from yam and potato peels using hydrochloric and acetic acids. *Biomass Conversion and Biorefinery*, *14* (15), 18019–18030. https://doi.org/10.1007/s13399-023-04021-2

Bevacqua, R. F. (1994). Origin of horticulture in Southeast Asia and the dispersal of domesticated plants to the Pacific Islands by Polynesian voyagers: The Hawaiian Islands case study. *HortScience*, *29* (11), 1226.

Bourmaud, A., Shah, D. U., Beaugrand, J., & Dhakal, H. N. (2020). Property changes in plant fibres during the processing of bio-based composites. *Industrial Crops and Products*, *154*, 112705. https://doi.org/10.1016/j.indcrop.2020.112705

Chamberlain, D.F. (1970). *Glycyrrhiza*. In Davis (ed.), Flora of Turkey and the East Aegean Islands, vol. 3, p. 260, Edinburgh University Press, Edinburgh, UK.

de Fátima Pocas, M., & Hogg, T. (2007). Exposure assessment of chemicals from packaging materials in foods: A review. *Trends in Food Science & Technology*, *18* (4), 219–230. https://doi.org/10.1016/j.tifs.2006.12.008

de Freitas Figueiredo, F., Cechinel Filho, V., Damazo, A. S., Arunachalam, K., Colodel, E. M., Ribeiro, M., ... & de Oliveira Martins, D. T. (2020). Sorocea guilleminiana Gaudich.: Wound healing activity, action mechanisms, and chemical characterization of the leaf infusion. *Journal of Ethnopharmacology*, *248*, 112307. https://doi.org/10.1016/j.jep.2019.112307

Dittenber, D. B., & GangaRao, H. V. (2012). Critical review of recent publications on use of natural composites in infrastructure. *Composites Part A: Applied Science and Manufacturing*, *43* (8), 1419–1429. https://doi.org/10.1016/j.compositesa.2011.11.019

Dixit, S., Goel, R., Dubey, A., Shivhare, P. R., & Bhalavi, T. (2017). Natural fibre reinforced polymer composite materials-a review. *Polymers from Renewable Resources*, *8* (2), 71–78. https://doi.org/10.1177/204124791700800203

Edhirej, A., Sapuan, S. M., Jawaid, M., & Zahari, N. I. (2017). Cassava/sugar palm fiber reinforced cassava starch hybrid composites: Physical, thermal and structural properties. *International Journal of Biological Macromolecules*, *101*, 75–83. https://doi.org/10.1016/j.ijbiomac.2017.03.045

Elsabee, M. Z., Naguib, H. F., & Morsi, R. E. (2012). Chitosan based nanofibers, review. *Materials Science and Engineering: C*, *32* (7), 1711–1726. https://doi.org/10.1016/j.msec.2012.05.009

Falguera, V., Quintero, J. P., Jiménez, A., Muñoz, J. A., & Ibarz, A. (2011). Edible films and coatings: Structures, active functions and trends in their use. *Trends in Food Science & Technology*, *22* (6), 292–303. https://doi.org/10.1016/j.tifs.2011.02.004

FAOSTAT. (2022). The statistic division of Food and Agriculture Organization. Retrieved from http://fao.org.

Fortea-Verdejo, M., Bumbaris, E., Burgstaller, C., Bismarck, A., & Lee, K. Y. (2017). Plant fibre-reinforced polymers: Where do we stand in terms of tensile properties?. *International Materials Reviews*, *62* (8), 441–464. https://doi.org/10.1080/09506608.2016.1271089

Ganapathy, T., Sathiskumar, R., Senthamaraikannan, P., Saravanakumar, S. S., & Khan, A. (2019). Characterization of raw and alkali treated new natural cellulosic fibres extracted from the aerial roots of banyan tree. *International Journal of Biological Macromolecules*, *138*, 573–581. https://doi.org/10.1016/j.ijbiomac.2019.07.136

Ghanbari, A., Tabarsa, T., Ashori, A., Shakeri, A., & Mashkour, M. (2018). Thermoplastic starch foamed composites reinforced with cellulose nanofibers: Thermal and mechanical properties. *Carbohydrate Polymers*, *197*, 305–311. https://doi.org/10.1016/j.carbpol.2018.06.017

Ghorpade, C. V., Shetty, S. C., & Murthy, K. (2022). Bioplastics production and applications: A mini review. *Internaional Journal of Scientific Research and Development*, *10*, 229–238.

Gummadavalli, G., Schuster, J., & Shaik, Y. P. (2023). Development of "Cissus quadrangularis" composite material with elevated mechanical properties and thermal stability. *Journal of Composites Science*, *7* (11), 445. https://doi.org/10.3390/jcs7110445

Gupta, K. C., Haider, A., Choi, Y. R., & Kang, I. K. (2014). Nanofibrous scaffolds in biomedical applications. *Biomaterials Research*, *18* (1), 5. https://doi.org/10.1186/2055-7124-18-5

Gurunathan, T., Mohanty, S., & Nayak, S. K. (2015). A review of the recent developments in biocomposites based on natural fibres and their application perspectives. *Composites Part A: Applied Science and Manufacturing*, *77*, 1–25. https://doi.org/10.1016/j.compositesa.2015.06.007

Haider, A., Gupta, K. C., & Kang, I. K. (2014). Morphological effects of HA on the cell compatibility of electrospun HA/PLGA composite nanofiber scaffolds. *BioMed Research International*, *2014* (1), 308306. https://doi.org/10.1155/2014/308306

Hossain, S. I., Hasan, M., Hasan, M. N., & Hassan, A. (2013). Effect of chemical treatment on physical, mechanical and thermal properties of ladies finger natural fiber. *Advances in Materials Science and Engineering*, *2013* (1), 824274. https://doi.org/10.1155/2013/824274

Indran, S., & Raj, R. E. (2015). Characterization of new natural cellulosic fiber from *Cissus quadrangularis* stem. *Carbohydrate Polymers*, *117*, 392–399. https://doi.org/10.1016/j.carbpol.2014.09.072

Indran, S., Raj, R. E., & Sreenivasan, V. S. (2014). Characterization of new natural cellulosic fiber from Cissus quadrangularis root. *Carbohydrate Polymers*, *110*, 423–429. https://doi.org/10.1016/j.carbpol.2014.04.051

Issa, A. T., Schimmel, K. A., Worku, M., Shahbazi, A., Ibrahim, S. A., & Tahergorabi, R. (2018). Sweet potato starch-based nanocomposites: Development, characterization, and biodegradability. *Starch-Stärke*, *70* (7–8), 1700273.https://doi.org/10.1002/star.201700273

Kaisangsri, N., Kerdchoechuen, O., & Laohakunjit, N. (2014). Characterization of cassava starch based foam blended with plant proteins, kraft fiber, and palm oil. *Carbohydrate Polymers*, *110*, 70–77. https://doi.org/10.1016/j.carbpol.2014.03.067

Kalia, S., Dufresne, A., Cherian, B. M., Kaith, B. S., Avérous, L., Njuguna, J., & Nassiopoulos, E. (2011). Cellulose-based bio-and nanocomposites: A review. *International Journal of Polymer Science*, *2011* (1), 837875. https://doi.org/10.1155/2011/837875

Koyuncu M., Özhatay N., Koçyiğit Avcı M. & Ekşi Bona G. (2023): *Allium*. — In: Güner A., Kandemir A., Menemen Y., Yıldırım H., Aslan S., Çimen A.Ö., Güner I., Bona G.E. & Gökmen F.Ş. (eds.), *Resimli Türkiye Florası*, *3b* (66–530). ANG Vakfı Nezahat Gökyiğit Botanik Bahçesi Yayınları, İstanbul.

Kumar, D., Guna, R., Abbirami, E., Gayathri, N., Karthik, R., & Sivasudha, T. (2019). Anti-oxidant and anti-inflammatory properties of stem and root extracts of Cissus quadrangularis (Veldt Grape). *Asian Journal of Engineering and Applied Technology*, *8* (1), 15–19. https://doi.org/10.51983/ajeat-2019.8.1.1066

Kwak, S., Haider, A., Gupta, K. C., Kim, S., & Kang, I. K. (2016). Micro/nano multilayered scaffolds of PLGA and collagen by alternately electrospinning for bone tissue engineering. *Nanoscale Research Letters*, *11*, 1–16. https://doi.org/10.1186/s11671-016-1532-4

Li, M., Pu, Y., Thomas, V. M., Yoo, C. G., Ozcan, S., Deng, Y., ... & Ragauskas, A. J. (2020). Recent advancements of plant-based natural fiber–reinforced composites and their applications. *Composites Part B: Engineering*, *200*, 108254. https://doi.org/10.1016/j.compositesb.2020.108254

Loebenstein, G., & Thottappilly, G. (Eds.). (2009). *The Sweetpotato*. Springer Science & Business Media.

Luchese, C. L., Pavoni, J. M. F., Spada, J. C., & Tessaro, I. C. (2019). Avaliação da influência da incorporação de diferentes resíduos agro industriais em filmes à base de amido de mandioca. *Brazilian Journal of Animal and Environmental Research*, *2* (2), 802–812.

Machado, C. M., Longhi, E. M., Spada, J. C., & Tessaro, I. C. (2018). Effect of broken rice flour addition on cassava starch-based foams. *Starch-Stärke*, *70* (11–12), 1700191.

Maji, A. K., Mahapatra, S., Banerji, P., & Banerjee, D. (2015). Effects of standardized Wrightia tinctoria and Stereospermum suaveolens extracts on various immuno-inflammatory parameters. *Journal of Herbs, Spices & Medicinal Plants*, *21* (1), 75–85. https://doi.org/10.1080/10496475.2014.910583

Mali, S., Grossmann, M. V. E., Garcia, M. A., Martino, M. N., & Zaritzky, N. E. (2002). Microstructural characterization of yam starch films. *Carbohydrate Polymers*, *50* (4), 379–386. https://doi.org/10.1016/S0144-8617(02)00058-9

Marsh, K., & Bugusu, B. (2007). Food packaging—roles, materials, and environmental issues. *Journal of Food Science*, *72* (3), R39–R55. https://doi:10.1111/j.1750-3841.2007.00301.x

Mazzei, R., Leonti, M., Spadafora, S., Patitucci, A., & Tagarelli, G. (2020). A review of the antimicrobial potential of herbal drugs used in popular Italian medicine (1850s–1950s) to treat bacterial skin diseases. *Journal of Ethnopharmacology*, *250*, 112443. https://doi.org/10.1016/j.jep.2019.112443

Min, B. M., Lee, G., Kim, S. H., Nam, Y. S., Lee, T. S., & Park, W. H. (2004). Electrospinning of silk fibroin nanofibers and its effect on the adhesion and spreading of normal human keratinocytes and fibroblasts in vitro. *Biomaterials*, *25* (7–8), 1289–1297. https://doi.org/10.1016/j.biomaterials.2003.08.045

Moshi, A. A. M., Ravindran, D., Bharathi, S. S., Indran, S., & Priyadharshini, G. S. (2020). Characterization of surface-modified natural cellulosic fiber extracted from the root of Ficus religiosa tree. *International Journal of Biological Macromolecules, 156*, 997–1006. https://doi.org/10.1016/j.ijbiomac.2020.04.117

Ovalı, S. (2023). Characterization of lignocellulosic glycyrrhiza glabra fibers as a potential reinforcement for polymer composites. *Journal of Thermoplastic Composite Materials, 36* (11), 4241–4256. https://doi.org/10.1177/08927057231151928

Ovalı, S., & Sancak, E. (2022). Investigation of mechanical properties of jute fiber reinforced low density polyethylene composites. *Journal of Natural Fibers, 19* (8), 3109–3126. https://doi.org/10.1080/15440478.2020.1838999

Parbin, S., Waghmare, N. K., Singh, S. K., & Khan, S. (2019). Mechanical properties of natural fiber reinforced epoxy composites: A review. *Procedia Computer Science, 152*, 375–379. https://doi.org/10.1016/j.procs.2019.05.003

Patel, P. N., Parmar, K. G., Nakum, A. N., Patel, M. N., Patel, P. R., Patel, V. R., & Sen, D. J. (2011). Biodegradable polymers: An ecofriendly approach in newer millenium. *Asian Journal of Biomedical and Pharmaceutical Sciences, 1* (3), 23–39.

Peelman, N., Ragaert, P., De Meulenaer, B., Adons, D., Peeters, R., Cardon, L., ... & Devlieghere, F. (2013). Application of bioplastics for food packaging. *Trends in Food Science & Technology, 32* (2), 128–141. https://doi.org/10.1016/j.tifs.2013.06.003

Pitchayya Pillai, G., Manimaran, P., & Vignesh, V. (2021). Physico-chemical and mechanical properties of alkali-treated red banana peduncle fiber. *Journal of Natural Fibers, 18* (12), 2102–2111. https://doi.org/10.1080/15440478.2020.1723777

POWO. 2024. Plants of the World Online. Royal Botanic Gardens, Kew; https://powo.science.kew.org/ [accessed 05.11.2024].

Prabhu Stalin, J. R., Jenish, I., & Indran, S. (2014). Tribological charecterization of carbon epoxy composite materials with particulate silane treated SiC fillers. *Advanced Materials Research, 984*, 331–335. https://doi.org/10.4028/www.scientific.net/AMR.984-985.331

Rabinowitch, H., & Thomas, B. (Eds.). (2022). *Edible Alliums: Botany, Production and Uses*. CABI. Boston. ISBN 978-1-78924-997-2. https://doi.org/10.1079/9781789249996.0000

Raju, J. S. N., Depoures, M. V., Shariff, J., & Chakravarthy, S. (2022). Characterization of natural cellulosic fibers from stem of symphirema involucratum plant. *Journal of Natural Fibers, 19* (13), 5355–5370. https://doi.org/10.1080/15440478.2021.1875376

Rathinavelu, R., Arumugam, E. & Paramathma, B.S. (2022). Suitability examination of a new cellulosic fiber extracted from the stem of Ventilago maderaspatana plant as polymer composite reinforcement. *Polymer Composites, 43* (5), 3015–3028. https://doi.org/10.1002/pc.26596

Ray, R. C. (Ed.). (2024). *Roots, Tubers, and Bulb Crop Wastes: Management by Biorefinery Approaches*. Springer Nature Singapore, Imprint: Springer. ISBN 978-981-99-8266-0.

Reneker, D. H., & Yarin, A. L. (2008). Electrospinning jets and polymer nanofibers. *Polymer, 49* (10), 2387–2425. https://doi.org/10.1016/j.polymer.2008.02.002

Reyniers, S., De Brier, N., Matthijs, S., Brijs, K., & Delcour, J. A. (2019). Impact of mineral ions on the release of starch and gel forming capacity of potato flakes in relation to water dynamics and oil uptake during the production of snacks made thereof. *Food Research International, 122*, 419–431. https://doi.org/10.1016/j.foodres.2019.03.065

Rodrigues, N. H. P., de Souza, J. T., Rodrigues, R. L., Canteri, M. H. G., Tramontin, S. M. K., & de Francisco, A. C. (2020). Starch-based foam packaging developed from a by-product of potato industrialization (Solanum tuberosum L.). *Applied Sciences, 10* (7), 2235. https://doi.org/10.3390/app10072235

Sadeghianmaryan, A., Sardroud, H. A., Allafasghari, S., Yazdanpanah, Z., Naghieh, S., Gorji, M., & Chen, X. (2020). Electrospinning of polyurethane/graphene oxide for skin wound dressing and its in vitro characterization. *Journal of Biomaterials Applications, 35* (1), 135–145. https://doi.org/10.1177/088532822091686

Sanjay, M. R., Arpitha, G. R., & Yogesha, B. (2015). Study on mechanical properties of natural-glass fibre reinforced polymer hybrid composites: A review. *Materials Today: Proceedings, 2* (4–5), 2959–2967. https://doi.org/10.1016/j.matpr.2015.07.264

Santos Dias, D., Otoni, C. G., da Silva, R. R., Meneguin, A. B., Mattoso, L. H. C., da Silva Barud, H., & Ribeiro, C. A. (2020). Large scale manufacturing of puree-only edible films from onion bulb (Allium cepa L.): Probing production and structure–processing–property correlations. *Industrial Crops and Products*, *145*, 111847. https://doi.org/10.1016/j.indcrop.2019.111847

Schlemmer, D., Angélica, R.S., & Sales, M.J.A. (2010). Morphological and thermomechanical characterization of thermoplastic starch/montmorillonite nanocomposites. *Composite Structures*, *92* (9), 2066–2070. https://doi.org/10.1016/j.compstruct.2009.10.034

Shrivastava, A., Gupta, R. K., & Srivastav, P. P. (2024). Exploring novel frontiers of advancements in purple yam (*Dioscorea alata* L.) starch extraction, modification, characterization, applications in food and other industries. *Measurement: Food*, *15*, 100196. https://doi.org/10.1016/j.meafoo.2024.100196

Singh, J., & Kaur, L. (2016). *Advances in Potato Chemistry and Technology* (2nd ed.). Oxford: Academic Press.

Stoffel, F., Barreto, L. P., & Zeni, M. (2018). Preparation of cassava starch-based trays with glycerol, sorbitol and poly (vinyl alcohol): Properties and influence for use as food packaging. *Research and Reviews: Jornal of Food Science and Technology*, *4*, 45–52.

Sumesh, K. R., Kavimani, V., Rajeshkumar, G., Ravikumar, P., & Indran, S. (2022). An investigation into the mechanical and wear characteristics of hybrid composites: Influence of different types and content of biodegradable reinforcements. *Journal of Natural Fibers*, *19* (8), 2823–2835. https://doi.org/10.1080/15440478.2020.1821297

Teseme, W. B. (2020). Review on the manufacturing of biodegradable plastic packaging film from root and tuber starches. *American Journal of Nano Research and Applications*, *8* (1), 1–8. https://doi:10.11648/j.nano.20200801.11

Thivya, P., Bhosale, Y. K., Anandakumar, S., Hema, V., & Sinija, V. R. (2022). Study on the characteristics of gluten/alginate-cellulose/onion waste extracts composite film and its food packaging application. *Food Chemistry*, *390*, 133221. https://doi.org/10.1016/j.foodchem.2022.133221

Topuz, F., Satilmis, B., & Uyar, T. (2019). Electrospinning of uniform nanofibers of polymers of intrinsic microporosity (PIM-1): The influence of solution conductivity and relative humidity. *Polymer*, *178*, 121610. https://doi.org/10.1016/j.polymer.2019.121610

Wang, S., Wang, X., Liu, W., Zhang, L., Ouyang, H., Hou, Q., ... & Liu, X. (2020). Fabricating cellulose nanofibril from licorice residues and its cellulose composite incorporated with natural nanoparticles. *Carbohydrate Polymers*, *229*, 115464. https://doi.org/10.1016/j.carbpol.2019.115464

Waterschoot, J., Gomand, S. V., Fierens, E., & Delcour, J. A. (2015). Production, structure, physicochemical and functional properties of maize, cassava, wheat, potato and rice starches. *Starch-Stärke*, *67* (1–2), 14–29. https://doi.org/10.1002/star.201300238

Zhou, W., Yang, J., Hong, Y., Liu, G., Zheng, J., Gu, Z., & Zhang, P. (2015). Impact of amylose content on starch physicochemical properties in transgenic sweet potato. *Carbohydrate Polymers*, *122*, 417–427. https://doi.org/10.1016/j.carbpol.2014.11.003

5 Preparation, Structural Evaluation, and Potential Applications of Cornstarch Biodegradable Plastics

Linganaboina Srinivasa Rao, Jyotsna Cherukuri, C. D. Naidu, K. Haripriya, P. Srujana, and Sahith Battula

5.1 INTRODUCTION

The use of plastics has surged dramatically over the past 70 years, rising from approximately 0.5 million tons in 1950 to over 460 million tons worldwide in 2019. Plastics have become an important part of present-day life. As they offer significant benefits due to their various properties such as flexible size, inertness, transparency, lightweight, and low cost, these features make them the preferred materials for a wide range of applications in both households and businesses. Excessive utilization of plastics results in environmental pollution. Plastics cannot be degraded quickly in normal forms; they are left out as plastic waste, causing environmental issues (Global Plastics Outlook, 2022). Some plastics cannot be recycled, and reusing them poses challenges such as high energy consumption and issues with contaminants and fiber reinforcements. Moreover, the option of reusing is becoming increasingly unfeasible because of the production of complex, multilayered products. Hence, to address these challenges, it is essential to develop alternatives, such as biodegradable plastics. Typically, biodegradable plastics made from materials like starch, cellulose, poly-3-hydroxybutyrate (PHB), and polylactic acid (PLA) are recognized for their polymerization potential and moisture absorption capability (Ali et al., 2024).

In general, biodegradable plastics are not completely biodegradable and do not decompose more quickly than those made from fossil fuels. However, most starch-based bioplastics break down significantly faster than other types. Starch is preferred for plastic production because it is renewable, abundant, low-cost, and biodegradable, and also offers a high oxygen barrier when dry. Starch is a common biopolymer made up of two glucose polymer types: amylose and amylopectin that are linked together. Furthermore, because it includes roughly 30% amylose, a main linear polymer having alpha-1, 4 glycosidic chains, starch is an excellent raw material for quickly producing biodegradable plastic films. Such starch-based polymers have recently been created for commercial films, injection molding, and other applications (Moshood et al., 2021).

The journey of cornstarch-based bioplastics started in the early 1990s when the environmental impact of traditional plastics started to become increasingly apparent. The production of bioplastics from cornstarch has become significantly more efficient over time. Today, the process primarily involves gelatinization, which is the swelling and breaking of starch granules that release amylose and amylopectin molecules which then combine to form the bioplastic. At the molecular level, cornstarch bioplastics have some unique properties that set them apart from traditional plastics. The structure of cornstarch composed of amylose (a linear polymer of glucose units) and amylopectin

DOI: 10.1201/9781003534914-5

(a branched polymer) plays a big role in determining the properties of the resulting bioplastic (Marichelvam et al., 2019).

Cornstarch is widely used in the preparation of bioplastics due to its biodegradable nature and abundant availability. Cornstarch bioplastics offer an added advantage over the traditional plastics with significant environmental and economic benefits. The natural breakdown of cornstarch bioplastics into carbon dioxide, water, and biomass when exposed to environmental conditions helps to reduce plastic pollution in landfills and oceans. Unlike fossil fuel-based plastics, corn is a widely grown crop which can be harvested annually and contributes to a more sustainable supply chain in view of the dependency on petroleum products. Production of starch-based bioplastics lowers the greenhouse gas emissions and contributes toward mitigating climate change by reducing the carbon footprint. The production of cornstarch-based plastics requires less energy and hence contributes to energy conservation. They can be tailored to meet different mechanical and functional properties. Furthermore, their easy compost ability helps in returning back the nutrients to the soil. Hence, corn-based plastics are an ideal solution for tackling the global plastic waste problem while promoting sustainability in various industries. Studies have shown that cornstarch bioplastic bags can lose up to 43% of their weight after 90 days of incubation in compost soil at 25 °C. This rapid degradation rate is in line with global sustainability goals and offers a solution to the plastic pollution problem (Singh et al., 2024). Thus, we have chosen cornstarch ($(C_6H_{10}O_5)_n$) for the present study and synthesized three samples of the cornstarch degradable plastics. Attempts have been made to portray them with the aid of the X-ray diffraction (XRD), scanning electron microscope (SEM), energy dispersive X-ray analysis (EDAX), Fourier transform infrared (FTIR), and decomposition studies.

5.2 METHODOLOGY

5.2.1 Selection of Materials

The materials used for the production of biodegradable plastics are corn starch, water, glycerin, and vinegar.

Cornstarch: Cornstarch is vital in the synthesis of biodegradable plastics due to its renewable nature and properties. As a natural polymer, it serves as a sustainable alternative to petroleum-based plastics, forming a matrix that enhances structure and functionality. Its biodegradability allows for easier breakdown in the environment, while its moisture absorption and high oxygen barrier make it suitable for various applications, such as packaging and agricultural films. Overall, cornstarch contributes to the creation of environmentally friendly plastics with improved performance (Marichelvam et al., 2019).

Water: Water plays a significant role in the synthesis of cornstarch bioplastics by facilitating the gelatinization process. When cornstarch is heated in the presence of water, the starch granules absorb moisture and swell, leading to the disruption of their crystalline structure. This process creates a viscous gel that can be molded into various shapes. Additionally, water helps in dissolving plasticizers, such as glycerin, which enhance the flexibility and workability of the final product. By controlling the amount of water used, manufacturers can influence the texture and properties of the bioplastic, ultimately optimizing its performance for specific applications (Singh et al., 2024).

Vinegar: Vinegar plays a key role in the synthesis of cornstarch bioplastics by acting as an acid that helps break down the starch molecules. Cornstarch consisting of large cellulose molecules with long polymer chains can become less brittle by adding a small amount of vinegar. This addition helps to break some of the polymer chains, resulting in a more flexible plastic. The addition of vinegar into the composition enhances the flexibility and ductility of the cornstarch, making it easier to mold and shape. As a result, the modified cornstarch becomes more versatile

TABLE 5.1
Composition of prepared cornstarch biodegradable plastic (CSBP) samples

Sample	Corn Flour (g)	Water (ml)	Vinegar (ml)	Glycerin (ml)
CSBP-1	15	25	5	15
CSBP-2	15	25	10	10
CSBP-3	15	25	15	5

for various applications, particularly in biodegradable plastics, where improved elasticity is essential for performance (Enwere et al., 2024).

Glycerin: Glycerin is a plasticizer used in film production as it softens the material. Glycerin is commonly used as a plasticizer in bioplastics production due to its ability to soften cornstarch materials. By incorporating glycerin into the formulation, the rigidity of the plastic is reduced, enhancing its flexibility and workability. This makes it easier to process and shape the film, resulting in a more pliable final product. Additionally, glycerin helps improve the moisture retention of the film, which can further enhance its durability and performance in various applications. Overall, glycerin plays a vital role in achieving the desired properties of flexible films (Singh et al., 2024; Enwere et al., 2024).

To fabricate the cornstarch biodegradable plastics (CSBP) samples, the appropriate amounts of cornstarch, water, glycerin, and vinegar were used. Three samples (CSBP-1, CSBP-2, and CSBP-3) were prepared as per the chemical composition shown in Table 5.1.

5.2.2 Samples Preparation

Figure 5.1 shows various stages of solution casting method for preparing cornstarch bioplastics (Singh et al., 2024; Dai et al., 2019). The process involves the following steps:

- The initial materials in specific composition were taken into a beaker and stirred continuously for around 1 hour to form a homogeneous solution. This prolonged mixing allows the ingredients (i.e., cornstarch, water, glycerin, and vinegar) to blend thoroughly, breaking down any clumps and ensuring even distribution of each component. The continuous stirring promotes the gelatinization of the cornstarch, helping to dissolve it and create a uniform mixture. Achieving a homogeneous solution is crucial for the consistency and performance of the final bioplastic, as it affects the material's properties and functionality during subsequent processing and molding.
- The solutions of all samples were simultaneously heated at a constant temperature of 100 °C for 1 minute until the jelly-like mixture was obtained. This heating process is essential for gelatinization, where the heat causes the starch granules to swell and absorb water, leading to the thickening of the mixture. At this temperature, the starch molecules undergo structural changes, allowing them to disperse more uniformly throughout the solution. The result is a cohesive jelly-like mixture that serves as the foundation for the biodegradable plastics, enhancing their texture and preparing them for the next steps in the fabrication process.
- The jelly mixture was spread on a flat aluminum foil sheet, obtaining a small thickness (≈0.5 mm). This step is crucial for the subsequent drying and curing processes, as a consistent thickness ensures even moisture removal and helps in achieving uniform properties throughout the bioplastic. The aluminum foil provides a smooth, non-stick surface that facilitates easy

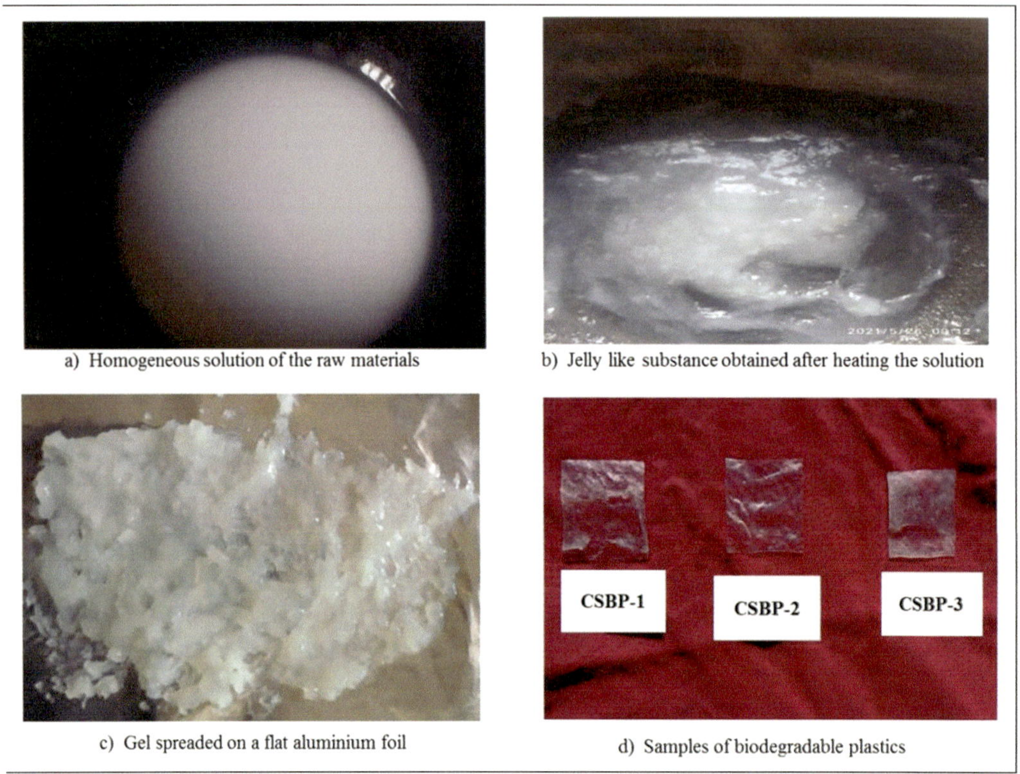

FIGURE 5.1 Various stages of solution casting method for preparation of cornstarch bioplastics.

handling and prevents the mixture from adhering too strongly, allowing for better removal once set. Spreading the mixture to this specific thickness also optimizes the mechanical properties of the final product, ensuring that it is neither too thick nor too thin for its intended applications.

- The products were dried for 24 hours in a hot air oven at 30 °C to convert the jelly mixture into solid plastic sheets. This drying process is essential for removing excess moisture, which allows the bioplastic to achieve the desired consistency and structural integrity. The controlled temperature of 30 °C ensures that the drying occurs gently, preventing any thermal degradation of the starch and maintaining the properties of the bioplastic. By drying for a full day, the moisture content is sufficiently reduced, resulting in flexible, durable plastic sheets that can be easily cut, shaped, and used in various applications. This step is vital for ensuring the functionality and longevity of the final product.

5.2.3 Characterization Techniques

5.2.3.1 XRD

The plastic nature of the samples was examined by XRD spectra (in the range of 10°–70°) recorded using X-pert[3]. XRD is a powerful analytical technique used to study the non-crystalline (amorphous), semicrystalline, and crystalline behavior of materials, including bioplastics. Bioplastics are not completely crystalline materials; instead, they often exhibit a combination of crystalline and amorphous regions. This partial crystallinity influences their physical properties, such as mechanical strength, flexibility, and degradation behavior. Thus, XRD technique allows for the detection of different phases present in the material, helping to understand how additives like plasticizers or fillers influence the overall structure (Rao et al., 2021).

XRD can determine the degree of crystallinity (X_c) in cornstarch biodegradable plastics. The degree of crystallinity can be determined using a two-phase model, where the sample consists of both crystalline and amorphous regions. The degree of crystallinity (X_c) is usually expressed as a percentage, indicating how much of the polymer is organized in a regular, repeating structure (crystalline) versus disordered regions (amorphous). Thus, the degree of crystallinity (X_c) can be calculated from XRD spectra using the following equation (Rao et al., 2021):

$$X_c = \frac{I_{crystalline}}{(I_{crystalline} + I_{amorphous})} \tag{5.1}$$

where $I_{crystalline}$ and $I_{amorphous}$ are intensities of crystalline and amorphous phases, respectively.

5.2.3.2 SEM and EDAX

The TESCAN VEGA SEM operating at an acceleration potential of 15 kV is used to capture high-resolution morphographs of the CSBP samples for carrying out comprehensive surface analysis (Marichelvam et al., 2019; Enwere et al., 2024). The elemental composition of CSBP biodegradable plastics was analyzed using EDAX (Oxford instruments). EDAX can provide a detailed elemental map and quantification of the constituent elements of biodegradable plastics for specific applications (Selvaraj et al., 2021).

5.2.3.3 FTIR

FTIR spectroscopy provides insights into the molecular structure of bioplastics by identifying functional groups and chemical bonds. The functional groups of the CSBP samples were monitored using FTIR spectrophotometer (Bruker-IFS 66 V) in the range 400–4000 cm^{-1}. Different bands can be identified in the FTIR spectra corresponding to specific functional groups (Rao et al., 2021; Enwere et al., 2024).

5.2.3.4 Biodegradability

A biodegradability test measures how quickly and completely a material breaks down in a natural environment through the action of microorganisms. The test typically involves exposing the material to specific conditions, such as soil, compost, or aquatic environments, where microbial activity can occur. The methodology for testing the volume degradability of the samples involves cutting a small part from each of the plastic film samples. The thickness and area of the test samples are measured before and after the test using a screw gauge. The samples were buried in the soil for a week and 10 ml of water was poured on the soil every day. After 7 days, the area and thickness of the samples were found to be significantly reduced. The volume degradability was measured using the following formula (Rao et al., 2021):

$$\text{Volume degradability} = \left| \frac{V_f - V_i}{V_i} \right| \times 100 \tag{5.2}$$

where V_i and V_f are initial and final volumes of the sample and $(V_f - V_i)$ is change in the volume.

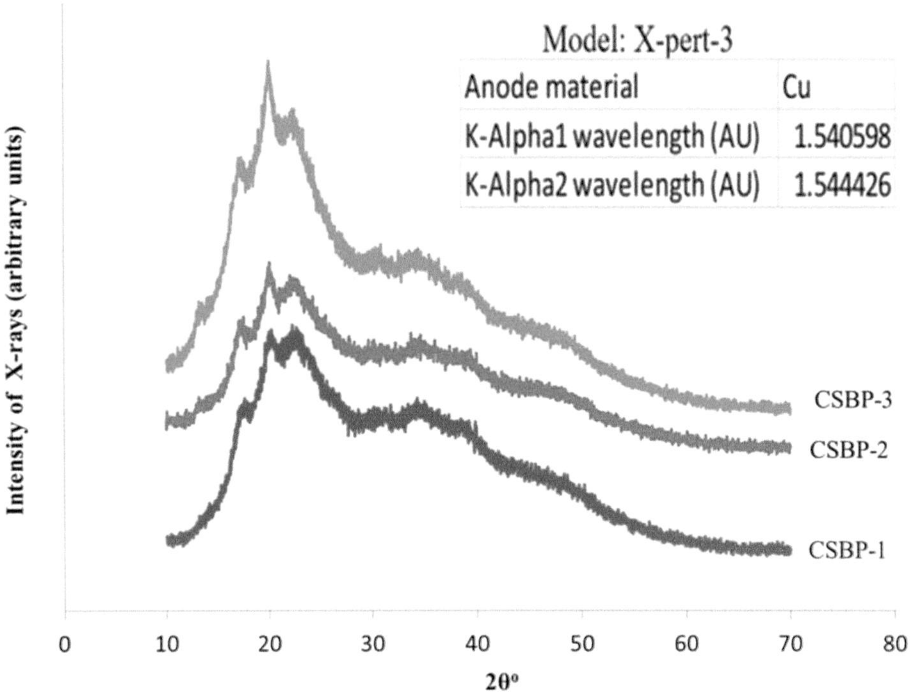

FIGURE 5.2 XRD spectra of the prepared cornstarch bioplastics.

5.3 RESULTS AND DISCUSSION

5.3.1 Degree of Crystallinity (X_c)

Figure 5.2 shows the XRD spectra of the CSBP samples. The XRD spectra of the samples have not displayed any prominent sharp peaks, which confirm their existence in a predominantly amorphous state. However, these samples have exhibited noticeable bumps and feeble peaks at around 13°, 17°, 21°, 22°, 30°, 33°, 39°, and 49° due to the presence of minute semicrystalline and crystalline phases associated in the amorphous phase (Rao et al., 2021; Mendes et al., 2015; Takeno et al., 2022). Such a response is accredited to the presence of granules of corn starches in terms of A-type crystals. These cereal granules are complex carbohydrates derived from cornstarches. They are primarily composed of two polysaccharides: amylose and amylopectin. These starches play a crucial role in various food and industrial applications because of their functional properties, particularly their high viscosity.

Type-A crystals are known for their relatively low density and are associated with enhanced mechanical properties, such as tensile strength and impact resistance. The presence of these crystals affects the thermal behavior of bioplastics, including their melting temperature and crystallization kinetics. Understanding the formation of Type-A crystals is crucial for optimizing the processing conditions as they play a key role in determining the overall performance and biodegradability of the plastic (Rao et al., 2021; Mendes et al., 2015; Takeno et al., 2022).

The degree of crystallinity (X_c) of the prepared cornstarch samples is presented in Table 5.2. The value of X_c of the samples is found to be in the order CSBP-2 < CSBP-1 < CSBP-3. The XRD spectra have exposed the feeble peaks and bumps, which confirm the modest crystalline nature of the present bioplastic samples by means of polymerization of the various constituents (especially

TABLE 5.2
Pertinent data on XRD spectra of the CSBP samples

	Degree of Crystallinity (X_c)		
2θ (degree)	**CSBP-1**	**CSBP-2**	**CSBP-3**
13	0.521	0.515	0.524
17	0.507	0.502	0.513
20	0.508	0.503	0.515
22	0.493	0.494	0.505
30	0.502	0.495	0.509
33	0.509	0.496	0.511
39	0.515	0.486	0.498
49	0.506	0.500	0.510

vinegar and glycerin) in the composition. Nevertheless, these samples have shown low crystalline nature due to formation of esters at the cost of the hydrogen bonds O–H and C–H bonds during the process of their synthesis at same quantities of vinegar and glycerin in composition of the sample CSBP-2. Thus, the established polymer chains are very short and therefore this sample (CSBP-2) has unveiled the dominant amorphous state instead of whole crystallinity (Rao et al., 2021; Ibrahim et al., 2019).

5.3.2 Morphographs and Elemental Analysis

The SEM morphographs of three CSBP samples and EDAX spectrum of one of the samples (CSBP-2) are presented in Figure 5.3. The SEM analysis reveals a distinct morphology of the samples, characterized by a combination of homogeneous amorphous regions and small, heterogeneous crystalline domains. This structural arrangement suggests that the bioplastics exhibit a favorable balance between flexibility and strength. The presence of amorphous regions likely contributes to the material's ductility, while the crystalline areas enhance its mechanical stability (Marichelvam et al., 2019; Enwere et al., 2024).

The EDAX analysis confirms the presence of key elemental constituents in the cornstarch-based biodegradable plastics, including carbon (C), oxygen (O), and calcium (Ca), along with other minerals as shown in Table 5.3. These elements are the basic building blocks of the polysaccharides amylose and amylopectin that make up cornstarch. The presence of these elements can vary based on the source of the corn and its processing methods. These elements reflect the organic nature of the bioplastics and their derived components, supporting the findings from the SEM and XRD analyses regarding the material's structural characteristics. The elemental composition indicates a significant presence of the fundamental constituents of cornstarch, suggesting a well-integrated matrix that contributes to both the mechanical properties and the biodegradability of the samples (Selvaraj et al., 2021).

Cornstarch consists mainly of two polysaccharides: amylose and amylopectin. Its elemental composition mirrors that of carbohydrates, making it an effective energy source. Cornstarch breaks down into glucose, providing quick energy, which is especially important for those with high energy needs, like athletes and laborers. It also aids in controlling blood sugar levels, lowering cholesterol, and improving digestion. By serving as a slow-release energy source and promoting digestive health through its resistant starch content, cornstarch supports overall health and wellness in a variety of dietary applications (Birt et al., 2013).

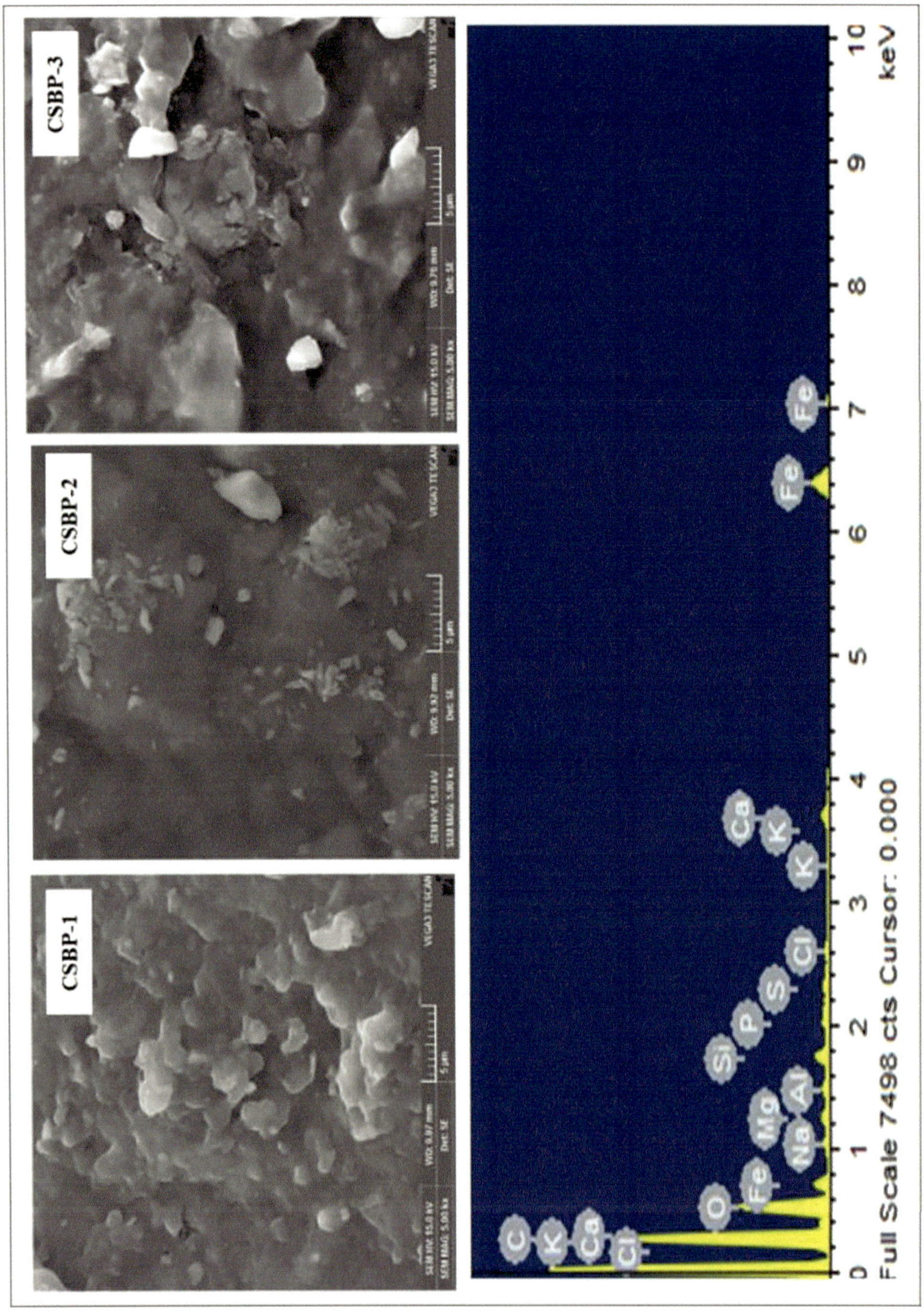

FIGURE 5.3 SEM morphographs of CSBP samples and EDAX spectrum of sample CSBP-2.

TABLE 5.3
Elemental analysis of the sample CSBP-2

Element	Weight %	Atomic %
C	50.44	63.28
O	33.32	31.38
Na	0.34	0.22
Mg	0.47	0.29
Al	0.26	0.15
Si	0.88	0.47
P	0.34	0.17
S	0.31	0.14
Cl	0.38	0.16
K	0.2	0.08
Ca	1.32	0.5
Fe	11.74	3.16
Total	100	100

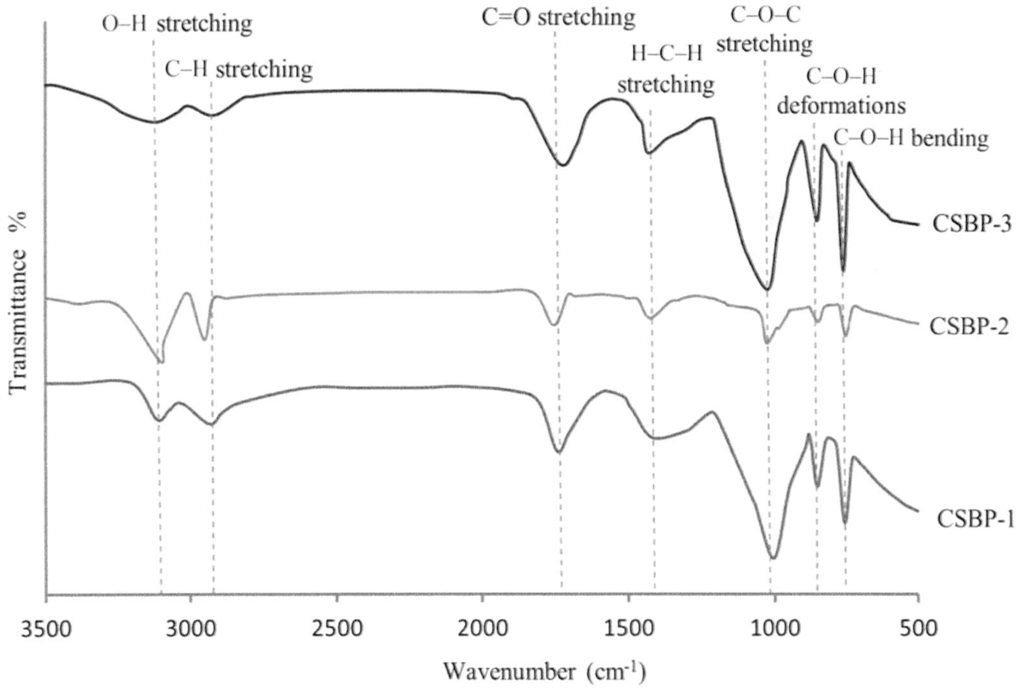

FIGURE 5.4 FTIR spectra of the prepared cornstarch biodegradable plastics.

5.3.3 Functional Groups

Figure 5.4 displays the FTIR spectra of the cornstarch bioplastics. The spectra show distinct bands at wave numbers corresponding to different functional groups within the polymers. These bands are aligning with those reported for starch-based degradable plastics (Singh et al., 2024, Enwere et al., 2024). A detailed attribution of various bands for the CSBP samples is provided in Table 5.4.

TABLE 5.4
Assignment of bands (cm^{-1}) in FTIR spectra of the cornstarch bioplastic samples

Sample-1 (CSBP-1)	Sample-2 (CSBP-2)	Sample-3 (CSBP-3)	Assignment of the Bands
758	755	762	C–O–H bending vibrations
854	851	856	Deformation of C–O–H rings
1005	1032	1024	C–O–C stretching vibrations
1416	1427	1434	$-CH_2$ stretching vibrations
1744	1755	1736	Stretching vibrations of C=O bonds
2934	2950	2918	Stretching vibrations of C–H bonds
3116	3102	3222	O–H stretching vibrations

TABLE 5.5
Volume degradability of the prepared cornstarch biodegradable plastic samples

Sample	Volume (V_i) (cm^3)	Volume (V_f) (cm^3)	Volume Degradability (%)
CSBP-1	0.297	0.1815	38.88
CSBP-2	0.342	0.122	64.36
CSBP-3	0.306	0.1884	38.43

The spectra revealed several characteristic vibrations, including the hydroxyl (O–H) stretching bonds (≈ 3100–3200 cm^{-1}), carbonyl C–H stretching bonds (≈2900 cm^{-1}), C=O stretching bonds (≈1750 cm^{-1}), $-CH_2$ stretching bonds (≈1420 cm^{-1}), C–O–C stretching bonds (≈1020 cm^{-1}), vibrations due to deformation of C–O–H rings (≈850 cm^{-1}), and C–O–H bending vibrations (≈750 cm^{-1}). These detected bands authenticate the plasticity of the prepared samples (Enwere et al., 2024; Rao et al., 2021). Notably, there is a noteworthy deviation in band positions and intensities for the second sample (CSBP-2) composed of equal quantity of both vinegar and glycerin compared to the other two samples, made from unequal quantity of them in the composition. Thus, this CSBP-2 sample would be more degradable than other two samples (Singh et al., 2024; Enwere et al., 2024).

5.3.4 Decomposition

The decomposition of the prepared cornstarch biodegradable plastics in the soil has been determined. Volume degradability of the CSBP samples is showcased in Table 5.5. Volume degradability of the samples is observed in the order: CSBP-2 > CSBP-1 > CSBP-3. Rate of degradation of sample-2 (CSBP-2) is identified to be the maximum as shown Figure 5.5. The molecular structure of the sample CSBP-2 may be more vulnerable to microbial attack. Its starch chains could be shorter or less crystalline than those in other samples, allowing easier access for enzymes produced by microorganisms, which accelerates the breakdown process (Omura et al., 2024). Moreover, the presence of vinegar and glycerin in equal quantities may create a synergistic effect, where the acidic conditions from vinegar and the moisture retention from glycerin work together to enhance degradation. Acidic conditions can hydrolyze ester bonds, leading to faster degradation of polymers that might be present in the sample CSBP-2. On the other hand, glycerin is a hygroscopic substance that

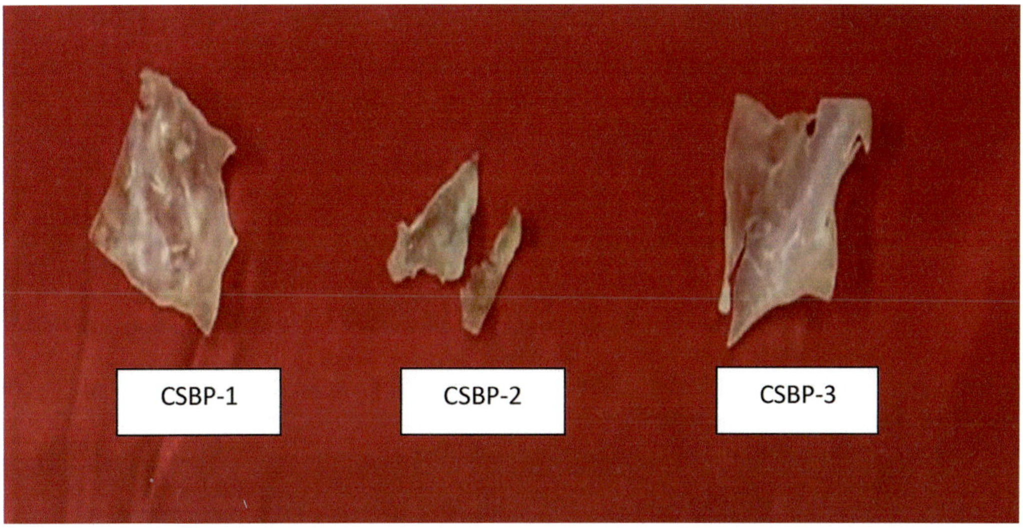

FIGURE 5.5 Decomposed cornstarch bioplastics in soil for seven days.

attracts water. Its presence can increase the moisture content around CSBP-2, facilitating hydrolytic degradation (Singh et al., 2024).

Biodegradability refers to the ability of cornstarch bioplastics to decompose naturally, contrasting with conventional plastics, which can take a long time to break down in the environment. Consequently, investigations are going on developing new degradable plastic products that can be decomposed quickly. Some of these products are labeled as "compostable" or "biodegradable." Typically, compostable materials are made from sources like rice starch, rice straw, cornstarch, potato starch, fibrous residue from sugarcanes, leaves of banana plant, and various agricultural and vegetable wastes. The use of these compostable materials not only helps to mitigate environmental issues related to plastic pollution but also supports agricultural waste management and resource efficiency. As demand for sustainable products grows, these materials are becoming increasingly important in the shift toward greener alternatives (Dejene & Geletaw, 2024; Routray et al., 2022).

5.4 POTENTIAL APPLICATIONS AND FUTURE SCOPE OF CORNSTARCH BIODEGRADABLE PLASTICS

5.4.1 Food Packing

Cornstarch-based bioplastics are nontoxic and safe for human contact, making them ideal for use in eco-friendly food packaging, particularly for single use items like plates and cutlery and bags. Improving their strength, flexibility, and barrier properties would result in competing with the conventional plastic food packaging. This reduces concerns about harmful chemicals leaching into food or the environment, as can happen with conventional plastics (Akshaykranth, Jayarambabu, Rao, et al., 2022; Anubha et al., 2022).

5.4.2 Compostable Trash Bags

Further starch-based bioplastics are used to manufacture compostable trash bags which decompose along with organic waste, making them ideal for home and industrial composting programs. This reduces reliance on conventional plastic garbage bags, which contribute to long-term environmental waste (Singh et al., 2024; López-Ibáñez et al., 2024).

5.4.3 Mulching in Agriculture

Corn starch-based films are used as mulch in agriculture as they help in suppressing the weed growth, maintaining soil moisture, regulating soil temperature. This would address the growing concerns about the intrusion of microplastics pollution from conventional mulch films which are not easily biodegradable and contribute to soil contamination (N. Yang et al., 2014).

5.4.4 Biomedical and Tissue Engineering

The biocompatibility and degradability of corn starch make it ideal and open for biomedical applications, including drug delivery systems, surgical implants, and tissue engineering scaffolds. These bioplastics degrade in the body without causing harm, thereby eliminating the need for removal surgery. The natural dissolvability of these materials within the body paves the way for innovative applications in medicine, particularly in the fields of implants and tissue engineering (Eslahi et al., 2019; Reshmy et al., 2021).

5.4.5 3D Printing

As three-dimensional (3D) printing becomes more widespread, especially in custom manufacturing and prototyping, cornstarch bioplastics could offer an eco-friendly alternative to petroleum-based filaments. Cornstarch-based bioplastics, especially polylactic acid derived from starch, are commonly used in 3D printing filaments for creating prototypes, models, and functional components (Chen et al., 2024).

5.4.6 Nonstructural Automotive Components

Starch-based bioplastics are being explored for use in nonstructural automotive components like interior panels and trims. These bioplastics offer a lightweight, biodegradable alternative to conventional plastic parts. As car manufacturers are seeking to reduce their carbon footprint, corn starch-based bioplastics could become a standard material for sustainable car interiors (Riedel & Nickel, 2002; Kouchakinejad et al., 2024).

5.4.7 Sustainable Electronics

With an increasing emphasis on sustainable electronics, starch-based materials could also help in reducing electronic waste by offering biodegradable alternatives for product packaging and casing. There is growing interest in biodegradable materials in the field of fashion, and starch-based bioplastics could contribute to reducing the environmental impact of textile waste (Kouchakinejad et al., 2024; Kuczynski & Boday, 2012).

5.4.8 Building Materials

Cornstarch bioplastics are emerging as innovative materials in the construction industry, offering sustainable alternatives to traditional building materials. Their lightweight nature, combined with good thermal insulation and moisture resistance, enhances their appeal for use in insulation panels, decorative elements, and even structural components. By incorporating cornstarch bioplastics into building practices, we can reduce our environmental footprint while promoting a circular economy in construction (Phadke & Rawtani, 2023).

In this way, the present work has portrayed cornstarch bioplastics as a promising alternative to traditional plastics, fostering the growth of global bioplastic industries in future (Chauhan et al., 2024).

5.5 CONCLUSION

Following successful synthesis and characterization with several analytical techniques, considerable information about the structure and characteristics of cornstarch degradable bioplastics was discovered. SEM morphographs indicated a mostly homogeneous amorphous nature interspersed with crystalline patches, whereas EDAX analysis proved the existence of key elements like calcium, carbon, and oxygen. The XRD analysis revealed semicrystalline and amorphous structures, indicating an optimal balance of flexibility and strength. Soil decomposition tests demonstrated the materials' notable degradation capacity, while FTIR spectra revealed a variety of functional groups that confirmed the synthetic materials' plastic nature. These findings show that prepared cornstarch-based bioplastics, created by combining water, vinegar, and glycerin, are a feasible, low-cost, and ecologically friendly alternative to standard plastics. They also provide a long-term answer to environmental challenges, including the growing worry over plastic waste pollution.

ACKNOWLEDGMENT

The authors would like to acknowledge the financial support and instrumentation facilities of VNRVJIET for carrying out this study under Seed Grant No.VNRVJIET/2018/01.

REFERENCES

Akshaykranth, A., Jayarambabu, N., Rao, T. V., Kumar, R. R., & Rao, L. S. (2022). Antibacterial activity study of ZnO incorporated biodegradable poly (lactic acid) films for food packaging applications. *Polymer Bulletin, 80* (2), 1369–1384. https://doi.org/10.1007/s00289-022-04126-0

Ali, S. A. S., Ilankoon, I. M. S. K., Zhang, L., & Tan, J. (2024). Understanding de-inking in packaging plastic recycling: Bridging the gap in resource conservation and establishing average hazard quotient. *Journal of Hazardous Materials, 479,* 135554. https://doi.org/10.1016/j.jhazmat.2024.135554

Anubha, M., Saranya, R., Chandrasatheesh, C., & Jayapriya, J. (2022). Effect of neem gum on water sorption, biodegradability and mechanical properties of thermoplastic corn starch-based packaging films. *Indian Chemical Engineer, 64* (5), 594–606. https://doi.org/10.1080/00194506.2022.2065368

Birt, D. F., Boylston, T., Hendrich, S., Jane, J., Hollis, J., Li, L., McClelland, J., Moore, S., Phillips, G. J., Rowling, M., Schalinske, K., Scott, M. P., & Whitley, E. M. (2013). *Resistant starch: Promise for improving human health. Advances in Nutrition, 4* (6), 587–601. https://doi.org/10.3945/an.113.004325

Chauhan, K., Kaur, R., & Chauhan, I. (2024). Sustainable bioplastic: A comprehensive review on sources, methods, advantages, and applications of bioplastics. *Polymer-Plastics Technology and Materials, 63* (8), 913–938. https://doi.org/10.1080/25740881.2024.2307369

Chen, K., Zhang, M., Bhandari, B., & Deng, D. (2024). 3D printed cinnamon essential oil/banana peel carbon dots loaded corn starch/gelatin bilayer film with enhanced functionality for food packaging application. *Food Chemistry, 448*, 139176. https://doi.org/10.1016/j.foodchem.2024.139176

Dai, L., Zhang, J., & Cheng, F. (2019). Effects of starches from different botanical sources and modification methods on physicochemical properties of starch-based edible films. *International Journal of Biological Macromolecules, 132*, 897–905. https://doi.org/10.1016/j.ijbiomac.2019.03.197

Dejene, B. K., & Geletaw, T. M. (2024). Development of fully green composites utilizing thermoplastic starch and cellulosic fibers from agro-waste: A critical review. *Polymer-Plastics Technology and Materials, 63* (5), 540–569. https://doi.org/10.1080/25740881.2023.2296649

Enwere, C. F., Okafor, I. S., Adeleke, A. A., Petrus, N., Jakada, K., Olosho, A. I., Ikubanni, P. P., Paramasivam, P., & Ayuba, S. (2024). Production of bioplastic films from wild cocoyam (Caladium bicolor) starch. *Results in Engineering*, 103132. https://doi.org/10.1016/j.rineng.2024.103132

Eslahi, N., Mahmoodi, A., Mahmoudi, N., Zandi, N., & Simchi, A. (2019). Processing and properties of nanofibrous bacterial cellulose-containing polymer composites: A review of recent advances for biomedical applications. *Polymer Reviews, 60* (1), 144–170. https://doi.org/10.1080/15583724.2019.1663210

Global Plastics Outlook. (2022). Policy scenarios to 2060. https://doi.org/10.1787/aa1edf33-en

Ibrahim, M., Sapuan, S., Zainudin, E., & Zuhri, M. (2019). Physical, thermal, morphological, and tensile properties of cornstarch-based films as affected by different plasticizers. *International Journal of Food Properties, 22* (1), 925–941. https://doi.org/10.1080/10942912.2019.1618324

Kouchakinejad, R., Lotfi, Z., & Golzary, A. (2024). Exploring azolla as a sustainable feedstock for eco-friendly bioplastics: A review. *Heliyon, 10* (20), e39252. https://doi.org/10.1016/j.heliyon.2024.e39252

Kuczynski, J., & Boday, D. (2012). Bio-based materials for high-end electronics applications. *International Journal of Sustainable Development & World Ecology, 19* (6), 557–563. https://doi.org/10.1080/13504509.2012.721404

López-Ibáñez, S., Quade, J., Wlodarczyk, A., Abad, M., & Beiras, R. (2024). Marine degradation and ecotoxicity of conventional, recycled and compostable plastic bags. *Environmental Pollution, 351*, 124096. https://doi.org/10.1016/j.envpol.2024.124096

Marichelvam, M. K., Jawaid, M., & Asim, M. (2019). Corn and rice starch-based bio-plastics as alternative packaging materials. *Fibers, 7* (4), 32. https://doi.org/10.3390/fib7040032

Mendes, J., Paschoalin, R., Carmona, V., Neto, A. R. S., Marques, A., Marconcini, J., Mattoso, L., Medeiros, E., & Oliveira, J. (2015). Biodegradable polymer blends based on corn starch and thermoplastic chitosan processed by extrusion. *Carbohydrate Polymers, 137,* 452–458. https://doi.org/10.1016/j.carbpol.2015.10.093

Moshood, T. D., Nawanir, G., & Mahmud, F. (2021). Sustainability of biodegradable plastics: A review on social, economic, and environmental factors. *Critical Reviews in Biotechnology, 42* (6), 892–912. https://doi.org/10.1080/07388551.2021.1973954

Omura, T., Isobe, N., Miura, T., Ishii, S., Mori, M., Ishitani, Y., Kimura, S., Hidaka, K., Komiyama, K., Suzuki, M., Kasuya, K., Nomaki, H., Nakajima, R., Tsuchiya, M., Kawagucci, S., Mori, H., Nakayama, A., Kunioka, M., Kamino, K., & Iwata, T. (2024). Microbial decomposition of biodegradable plastics on the deep-sea floor. *Nature Communications, 15* (1). https://doi.org/10.1038/s41467-023-44368-8

Phadke, G., & Rawtani, D. (2023). Bioplastics as polymeric building blocks: Paving the way for greener and cleaner environment. *European Polymer Journal, 199,* 112453. https://doi.org/10.1016/j.eurpolymj.2023.112453

Reshmy, R., Philip, E., Thomas, D., Madhavan, A., Sindhu, R., Binod, P., Varjani, S., Awasthi, M. K., & Pandey, A. (2021). Bacterial nanocellulose: engineering, production, and applications. *Bioengineered, 12* (2), 11463–11483. https://doi.org/10.1080/21655979.2021.2009753

Rao, L. S., Naidu, C., & Tiwari, S. (2021). Investigation on synthesis, structure and degradability of starch based bioplastics. *Materials Today Proceedings, 49*, 257–261. https://doi.org/10.1016/j.matpr.2021.01.917

Riedel, U., & Nickel, J. (2002). Applications of natural fiber composites for constructive parts in aerospace, automobiles, and other areas. *Biopolymers Online*. https://doi.org/10.1002/3527600035.bpola001

Routray, W., Chetry, R., & Jena, B. S. (2022). Drying of food industry and agricultural waste: Current scenario and future perspectives. *Drying Technology, 41* (5), 628–654. https://doi.org/10.1080/07373937.2022.2118767

Selvaraj, K., Vishvanathan, N., & Dhandapani, R. (2021). Screening, optimization and characterization of poly hydroxy butyrate from fresh water microalgal isolates. *International Journal of Biobased Plastics, 3* (1), 139–162. https://doi.org/10.1080/24759651.2021.1926621

Singh, P., Pandey, V. K., Singh, R., Singh, K., Dash, K. K., & Malik, S. (2024). Unveiling the potential of starch-blended biodegradable polymers for substantializing the eco-friendly innovations. *Journal of Agriculture and Food Research, 15,* 101065. https://doi.org/10.1016/j.jafr.2024.101065

Takeno, H., Shikano, R., & Kikuchi, R. (2022). Mechanical performance of corn starch/poly (vinyl alcohol) composite hydrogels reinforced by inorganic nanoparticles and cellulose nanofibers. *Gels, 8* (8), 514. https://doi.org/10.3390/gels8080514

Yang, N., Sun, Z., Feng, L., Zheng, M., Chi, D., Meng, W., Hou, Z., Bai, W., & Li, K. (2014). Plastic film mulching for water-efficient agricultural applications and degradable films materials development research. *Materials and Manufacturing Processes, 30* (2), 143–154. https://doi.org/10.1080/10426914.2014.930958

6 Yam-Starch-Based Biodegradable Bioplastic Film

Mohit Nagar, Yograj Bist, Yogesh Kumar, and D. C. Saxena

6.1 INTRODUCTION

Yams (*Dioscorea* spp.) are tuberous food crops mainly cultivated and consumed in West Africa. They rank as the fourth most produced tuber globally, following potato, cassava, and sweet potato (Zang et al., 2024). Therefore, they are a very important source of food, energy, and income in countries like Nigeria, Ghana, Côte d'Ivoire, and Benin. They can be consumed whole or processed into various forms, such as flour, powder, or dietary supplements. However, freshly harvested yams have very high moisture content (50–80% on wet basis) which reduce their shelf life (Ojediran et al., 2020). Consequently, a significant portion of yam is lost due to inadequate and inefficient storage facilities after harvesting. Drying methods like ovens, solar dryers, and hot air cabinets are often employed to reduce the moisture and convert the raw yam to more stable dried shape/powder. An alternative approach to reduce losses from freshly harvested tubers could involve extracting starch and other valuable components. This approach could also contribute to ease the burden on conventional starch sources in meeting the rising demand. Despite this potential, yams remain an underutilized starch source (Otegbayo et al., 2014). As shown in Figure 6.1, research interest in yam starches has only gained traction in the last decade.

The dry matter content in fresh yam varies between 20 and 50 g/100 g (wet basis), depending on cultivars, climate, and temperature (Akissoe et al., 2011). The dried yam powder contains approximately 60–80% starch, which can be extracted and used in various applications (Guan et al., 2019). The amylose content varies significantly between 4% and 29%, depending on the variety (Zou et al., 2022). High-amylose yam varieties could serve as a promising source for the development of biofilms. However, similar to other native starches, there are challenges in using yam starch for bioplastic film production, especially concerning its mechanical properties and water sensitivity. Native starch-based films are often brittle and have low moisture resistance, which limits their potential applications (Deep et al., 2024). To address these limitations, various modification techniques can be applied, such as blending yam starch with other biopolymers, incorporating plasticizers, or crosslinking starch molecules to enhance the film's properties (Thakur et al., 2023). Furthermore, the development of composite films by adding natural fibers or nanomaterials has also been investigated to enhance the mechanical strength, thermal stability, and water resistance of yam-starch-based bioplastic films (Muñoz-Gimena et al., 2023). Therefore, this chapter provides a comprehensive overview of yam-starch-based biodegradable bioplastic films, including preparation, properties, and applications. Several strategies like starch modification and reinforcement using additives to improve the properties of yam's starch-based films are also discussed.

DOI: 10.1201/9781003534914-6

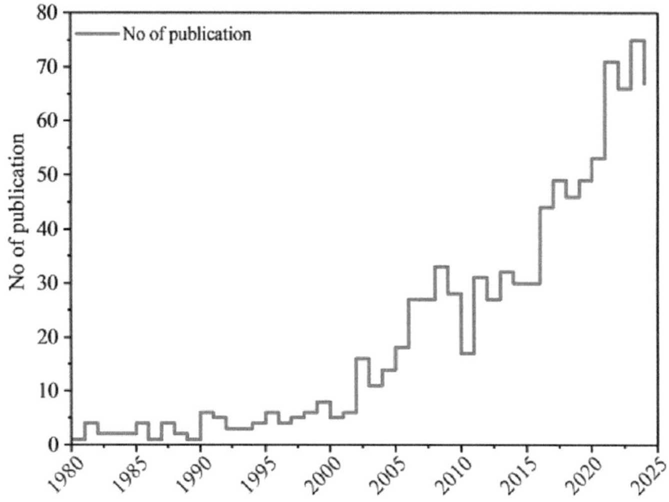

FIGURE 6.1 Number of publications on yam starch. (Based on a SCOPUS search conducted on 23 October 2024 using the keywords "yam" AND "Starch.")

6.2 GLOBAL PRODUCTION AND AVAILABILITY OF YAMS

Yams are a staple crop with a highly concentrated production in Africa, particularly in West Africa. According to FAOSTAT (2024), 97% of global yam production comes from Africa, and Nigeria alone contributes 67% of the world's total yam production (Figure 6.2). This heavy reliance on African countries for yam production highlights the importance of yams in the region, but it also indicates potential vulnerabilities in global supply. For instance, the volume of yams imported into Europe increased by 28% from 2018, reaching over 34,000 tons in 2022 (CBI, 2023). In the European market, 99% of the yam export is supplied by Ghana. Therefore, any disruption among the dominant yam producers can significantly affect the overall supply of yam. Despite this, the price of fresh yams remains relatively stable due to their niche consumption within ethnic communities. Additionally, the lack of widespread industrial applications and commercial utilization helps maintain a balance between supply and demand. However, as research and development on yams advance, commercialization on an industrial scale may become more dominant, requiring careful management of supply and price. These factors represent both strengths and risks associated with the fresh yam market. However, for starch production, the impact of global supply fluctuations can be mitigated due to the possibility of producing starch locally and exporting it. Compared to fresh yams, starch has a lower volume and moisture content, allowing for longer storage and easier transportation. Additionally, starch production can be increased during peak harvest seasons when prices of fresh yams are low and the risk of spoilage is high.

6.3 STARCH EXTRACTION AND PROPERTIES

Raw yams are rich in starch content which accounts for 15–40 g/100 g *(wet basis)* (Brunnschweiler et al., 2006). The extraction of starch from fresh yam involves a series of systematic steps to isolate the starch efficiently while minimizing impurities. Initially, the tubers are thoroughly washed and peeled to remove surface dirt and contaminants. They are then diced into smaller pieces and ground with distilled water in a blender. Sometimes, a sodium bisulfite solution (around 0.075%) or an alkaline solution is added before grinding to facilitate the separation of proteins and other impurities (Alves et al., 1999; Otegbayo et al., 2014). After grinding, the slurry is subsequently

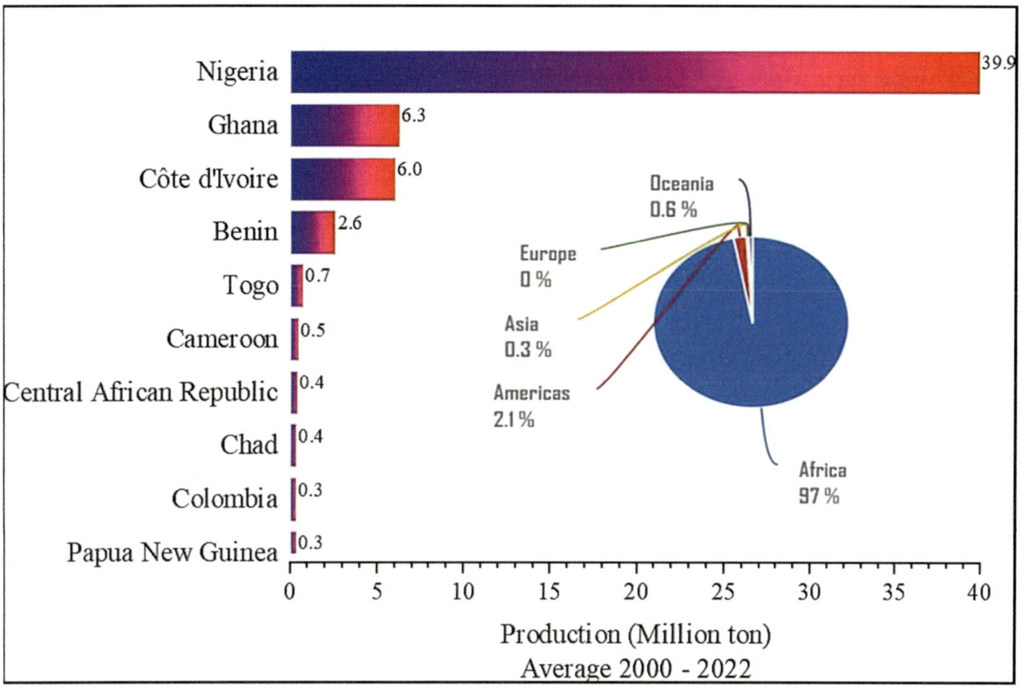

FIGURE 6.2 List of the top ten yam-producing countries. (Average between 2000 and 2022, FAOSTAT, 2024.)

filtered and sieved to eliminate larger impurities. Following filtration, sedimentation allows the starch granules to settle; the supernatant is discarded, and the sediment is re-suspended in water. This washing process is repeated multiple times to enhance the purity of the starch. Once the washing process is completed, the settled starch is hot air-dried. Some studies have also reported bubble method of starch extraction from yams (Lozano et al., 2018; Pérez et al., 2021). In this method, yams are converted into slurry and air is injected into the slurry, creating bubbles. This bubble formation enables phase separation, as mucilage-laden bubbles carry starch granules to the surface, while heavier impurities, such as fibers and proteins, remain at the bottom (Pérez et al., 2021). Once froth forms on the surface, it can be skimmed off, containing the concentrated starch. The collected starch is then washed to remove impurities and dried for storage. As a continuous process, this method not only enhances starch purity but also shows promise for large-scale production, making it a viable option for both laboratory and pilot-scale operations. The extracted yam starch has a high starch content and low impurities, including protein (<0.4%), ash (<0.4%), and lipids (<0.3%) (Zhu, 2015).

Yam starch can also be extracted from dried yam powder (Jiang et al., 2012). In this method, the dried powder is immersed in 95% ethanol for 24 h, allowing the extraction of proteins, pigments, lipids, and other low-molecular-weight components. The supernatant containing these impurities is then decanted, and the remaining sediment is resuspended in 95% ethanol. This process of settling and resuspension in ethanol is repeated 7–8 times to ensure thorough removal of non-starch components. After the ethanol washes, the remaining solid layer is suspended in distilled water. The mixture is allowed to settle, and the supernatant is decanted to eliminate any water-soluble impurities. The starch-containing sediment is repeatedly resuspended and rinsed with distilled water until the supernatant becomes transparent. The starch slurry is then centrifuged, with each cycle removing nonwhite layers and any remaining impurities in the sediment. Finally, the purified starch fraction is dried to obtain high-purity starch.

The native starches can be classified into four categories based on amylose content (Bhattacharya et al., 1982). Waxy starch contains 0–2% amylose and is mainly composed of amylopectin, making it ideal for products requiring a smooth texture, such as sauces and puddings. Low amylose starch has 10–20% amylose, which offers moderate gel strength, and is suitable for food applications that require a firmer texture. Intermediate amylose starch contains 20–25% amylose, which offers a balance between gel formation and viscosity, making it suitable for various culinary and industrial uses. Lastly, high amylose starch has more than 25% amylose and offers strong gelling properties, commonly used in biodegradable films and resistant starch products. The average amylose content of yam starch is around 23%, indicating that it is classified as intermediate amylose starch. However, many varieties have over 25% amylose, and some can reach up to 50% amylose (Rolland-Sabaté et al., 2003). The average size of native yam starch granules is about 25 μm, with a range of 1 to 92.22 μm (Daiuto et al., 2005; Riley et al., 2004; Otegbayo et al., 2014). Depending on the varieties of the yam, starch granules have various shapes like oval or elongated oval, round, angular ellipsoid, truncated sphere/oval ellipsoid, irregular triangular rod-like with a rough surface, and polygonal triangular ellipsoidal oval-shaped or roundel-shaped with an irregular surface (Figure 6.3; Zhu, 2015; Zou et al., 2022).

In addition to amylose content and morphology of starch granules, starch crystallinity also significantly influences the formation of filmogenic solutions. The degree of crystallinity affects the swelling behavior and solubility of granules during heating (Dome et al., 2020). Starches with high solubility tend to leach more amylose when heated in sufficient water, which contributes to the formation of a strong gel during retrogradation. The retrogradation process enhances the structural integrity of film, leading to improved mechanical properties and barrier functions. While most yam starches show B or C polymorphs, A-type starch is also present in some species like *Dioscorea dumetorum*, *Dioscorea zingiberensis,* and *Dioscorea nipponica* Makino (Figure 6.4; Li et al., 2011; Shujun et al., 2007; Amani et al., 2004; Zhou et al., 2013; Jiang et al., 2012). The crystallinity of yam starches is generally greater than 25%, although it has been reported to reach up to 50% in some species (Jiang et al., 2012). Higher crystallinity typically results in films with increased rigidity and better barrier properties, whereas lower crystallinity contributes to more flexible films. Since most yam starches exhibit B- or C-type crystallinity, their films are expected to maintain a balance between flexibility and mechanical strength, with the specific polymorph type and crystallinity level guiding the final properties of the starch-based films.

Other properties like gelatinization, retrogradation, and gelling are also important in the development of biodegradable films. Gelatinization of yam starches typically occurs at temperatures between 65 °C and 80 °C, where starch granules absorb water, swell, and lose their crystalline structure (Moorthy, 2002; Riley et al., 2006). It is influenced by the molecular organization of amylose and amylopectin in the starch, resulting in a relatively high gelatinization temperature compared to other starches like potato and cassava (Farhat et al., 1999; Wang et al., 2006). The gelatinized starch reassociates and forms a more ordered structure upon cooling by the retrogradation process. Yam starches show a tendency toward moderate to high retrogradation, where the amylose chains realign over time, resulting in firmer gels (Freitas et al., 2004; Otegbayo et al., 2014). The gelling properties of yam starches are marked by their ability to form strong, heat-resistant gels. The formation of firm gels upon cooling after gelatinization is a reflection of the strong network formed by amylose and amylopectin interactions. Overall, high gelatinization temperature, controlled retrogradation, and better gelling capacity of yam starches make them suitable for biodegradable film formation. Therefore, yam starch presents a promising alternative to conventional starch sources such as corn, wheat, and potatoes by offering comparable functional and gelling properties. Its better thickening and gelling characteristics make it suitable in enhancing food texture and stability. Yam cultivation is generally less resource-intensive than conventional starch sources which offer yam starch extraction and utilization, particularly in tropical and subtropical regions, a more sustainable option. Therefore,

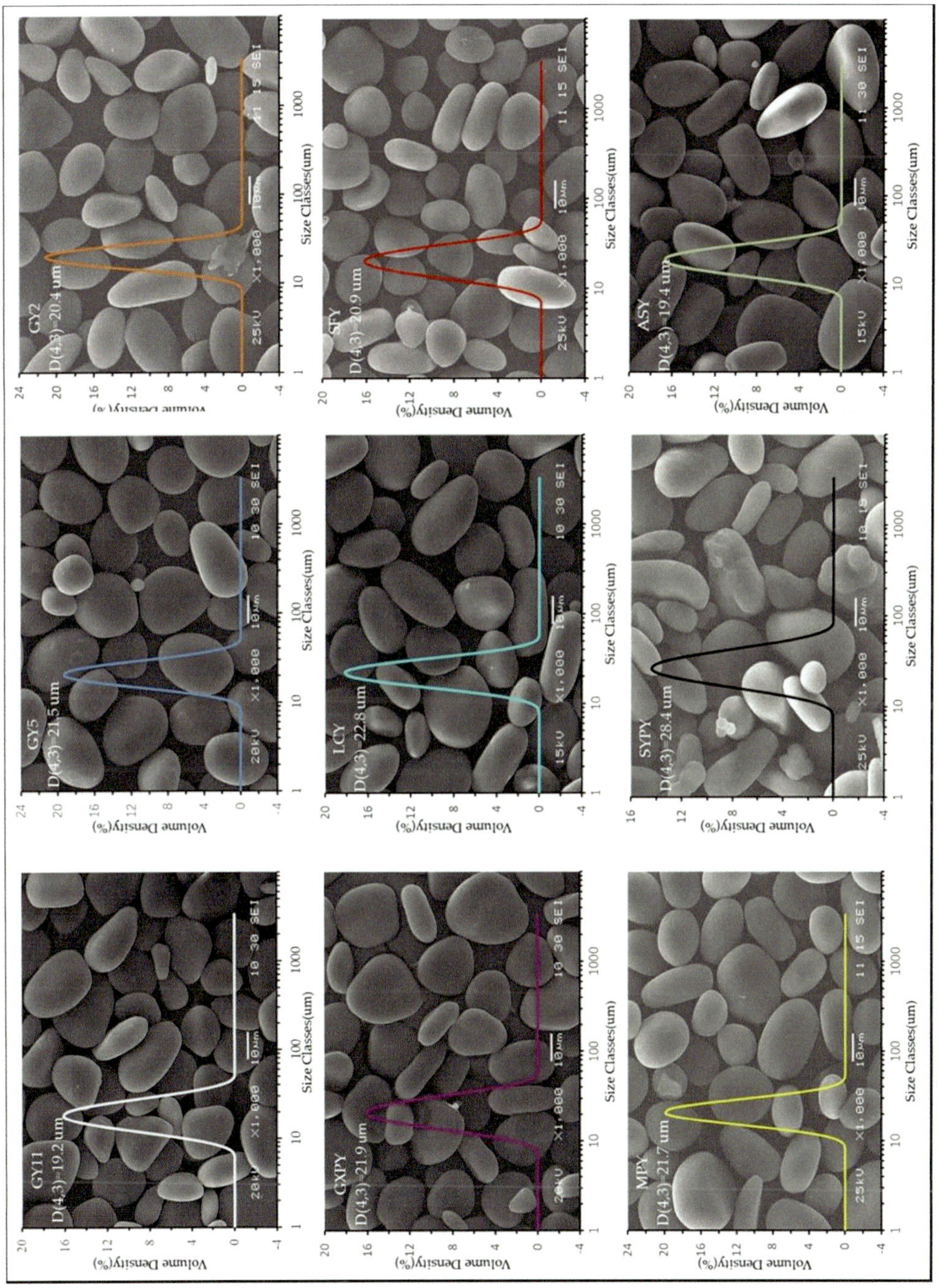

FIGURE 6.3 SEM micrographs and particle size distribution of Chinese yam starch. (Reprinted from Zou et al., 2022.)

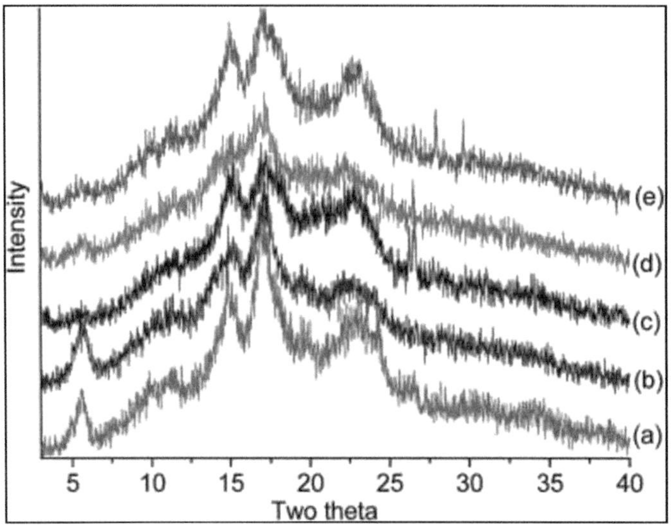

FIGURE 6.4 X-ray diffraction spectra of five *Dioscorea* L. species starches: (a) *Dioscorea opposita* Thunb., (b) *Dioscorea alata* Linn., (c) *Dioscorea nipponica* Makino, (d) *Dioscorea bulbifera* Linn., and (e) *Dioscorea septemloba* Thunb. (Reprinted from Jiang et al. (2012) with permission from Elsevier.)

exploring new applications by improving agricultural practices, advancing processing technologies, and exploring for yam starch, we can enhance its value and sustainability.

6.4 BIODEGRADABLE BIOPLASTIC FILMS FROM YAM STARCH

6.4.1 Film Formation Process Using Yam Starch

In general, two main techniques are used for the development of starch-based biodegradable edible films: wet methods (such as solution casting and drying) and dry methods (thermoplastic processing), with amylose and amylopectin acting as functional components (Paes et al., 2008). Most redundant studies on yam starch-based films have adopted the solution-casting method due to the poor thermoplastic properties of native starch compared to those used in dry methods. The wet process of yam-based starch film formation is divided into several key steps: solvent diffusion, starch gelatinization upon heating, uniform dispersion and homogenization (in the case of emulsions), casting onto Petri plates, drying, and peeling off the film. These steps are followed regardless of whether native, pregelatinized, or modified starch matrices are used, until the film formation process is completed. In the first stage, a yam starch-based filmogenic solution is prepared by dispersing yam starch 1–4% (w/v) (Mali et al., 2002; Narvaez et al., 2021) in 100 mL of distilled water with constant magnetic stirring and gentle heating. Additionally, common additives such as plasticizers are introduced to impart a plasticizing effect, while other additives, including triglycerides, fatty acids, and clays, are also incorporated into the film suspension before the gelatinization process begins.

In the second step, yam starch gelatinization occurs in the presence of water (>90% w/w). The film suspension is continuously heated at 95 °C for 30 minutes to achieve complete starch gelatinization, with intermittent stirring. The gelatinization temperature and time may vary depending on the source and type of starch, amylose content, interactions with water, and the type and content of plasticizer used. After gelatinization, the film suspensions are cooled to room temperature with gentle magnetic stirring to minimize air bubbles. During the casting or pouring process, an appropriate amount of the film suspension is transferred onto leveled Petri plates (made of polystyrene, polyethylene, or Teflon) after homogenization for cooling and drying. The drying of the yam starch film takes place under optimal temperature and relative humidity conditions to ensure the removal of excess water, making it easier to handle during the peeling stage. The films are carefully peeled off from the Petri plates and placed in a desiccator at optimal relative humidity for conditioning for 24 hours before further testing. The film formation process using native starch depends on several factors, including the type of native starch, the amylose and amylopectin content, the type of plasticizers used, any modifications made to the starch, and the desired film properties, which vary based on the intended application (Table 6.1).

6.4.2 Properties of Yam-Starch-Based Bioplastic Films

Several studies have investigated the physicochemical properties, gas and water vapor permeability (WVP), transparency, and mechanical properties of yam-starch-based bioplastic films. Mali et al. (2004) reported that yam-starch-based films exhibited homogeneous, colorless, odorless, and transparent characteristics (with an absence of insoluble particles) when prepared at different concentrations of starch (3.30–4%, w/w) and glycerol (1.30%–2%, w/w) in the filmogenic solution. Variations in film properties were observed as moisture content ranging from 23.63 to 27.72 (g water/100 g dried film), increased WVP (0.98–1.37 $\times 10^{-10}$ g m^{-1} s^{-1} Pa^{-1}) as starch concentration increased surpassing that of low-density polyethylene. Additionally, gas permeability increased (2.17–3.90 $\times 10^{-10}$ cm^3 m^{-1} s^{-1} Pa^{-1}) with rising glycerol concentration, improved puncture strength (6.03–15.96N), and increased deformation (3.44–4.78 mm) as film thickness increased. It was suggested that yam starch-based films with good mechanical and barrier properties for food products can be obtained at optimized parameters; a starch concentration (4%), glycerol concentration (1.3%) with thickness of 0.11 mm.

In another study, Nagar et al. (2020) developed elephant foot yam-starch-based film and revealed the effect of agar–agar and xanthan gum hydrocolloids on the film properties. The addition of hydrocolloids led to improved physiochemical properties compared to native elephant foot yam starch, including moisture content (23.66–26.42%, w/w), thickness (0.163–0.199 mm), density (1.73–2.02 g/cm^3) and transparency (20.27–33.57%). The addition of hydrocolloids also resulted in significant increase in other properties, including thermal properties (onset temperature 115.2–143.5 °C, melting temperature 121.8–150.6 °C, and end temperature 128.9–159.2°C), improved crystallinity (36.02%– 42.79%), and mechanical properties (Tensile strength 15.81–20.14 Mpa; Young's modulus 54.08–69.77 Mpa). Additionally, reductions were observed in the water vapor transmission rate (2386.67–1505.46 g/m^2.24 h) and oxygen transmission rate (0.053–0.017 cm^3/m^2.24 h). From the above studies it is clear that yam starch films alone are insufficient to impart desirable film characteristics needed for packaging of various food products. This highlights the need for modifications to yam starch, treatments, and the incorporation of additives to enhance its properties and achieve characteristics comparable to those of plastic materials.

Ulyarti et al. (2019) studied the properties of yam starch-based films made of two different yam cultivars: white and purple water yam using randomized block design at four levels of starch concentration: 1.33%, 2%, 2.67%, and 3.33%. The most desirable properties of edible yam starch film were observed at a starch concentration of 2.67% that exhibited a maximum thickness of 0.12 to 0.13 mm, comparatively lower transparency (12.42% to 13.24%), high tensile strength (48.10 to 49.35 N/m) and low water vapor transmission rate (1.16 – 1.08) x10^{-2} g/s/m^2.

6.4.3 Advantages of Yam Starch for Bioplastics Compared to Traditional Sources

The low cost, abundance, renewability, and favorable properties of yam starch make it a strong candidate to replace traditional starch sources in the development of biodegradable bioplastic films and coatings. Mali et al. (2006) developed biodegradable plasticized films from yam starch as well as traditional sources such as corn and cassava, examining their barrier, thermal, and mechanical

TABLE 6.1
Different preparation methods and properties of yam-starch-based biodegradable films

Starch Source	Plasticizer Used	Film Composition (Additives Added & Modification)	Gelatinization Condition, Drying, and Film Conditioning	WVTR (g/m^2. day) & WVP ($\times 10^{-10}$ g m^{-1} s^{-1} Pa^{-1})	Solubility (SOL; %) and Biodegradability (BD; %)	Elastic Modulus (EM; MPa)	Tensile Strength (TS; MPa)	Elongation (%E)	Reference
Water Yam starch	Glycerol	Edible film prepared from Water yam-cassava starch composite blended at different proportions (100:0, 90:10,80:20,70:30, 60:40,50:50 0:100).	80 °C for 10 min; dried at 35 °C & Conditioned at 25 °C at 52 RH for 72 h			EM ranged between (0.47 to 1.06) MPa, decreased as yam starch concentration increases	TS ranged between (0.28 to 0.60) MPa, Maximum for (50% yam:50%cassava starch) proportion film	%E ranged (11.34–32.20), Maximum for 100% yam proportion film	Adebowale et al. (2014)
Yam starch	Glycerol	Influence of glycerol concentration (1.65% & 2%) w/v and two surfactants span 80 and Tween 20 on yam starch film properties were investigated	80 °C for 10min; dried at 45 °C & Conditioned at 25 °C at 70% RH for 48 h	WVTR decreased (2105-2079) g/ m^2.day as glycerol concentration increased, showing strong plasticizing effect		Found highest for 2% glycerol concentration irrespective of surfactants	No significant change was observed	Maximum %E found for film contain 2% glycerol and 1.5 ml Tween 20	Rugchati et al. (2015)
Yam starch	Glycerol	Varying mass concentrations of both glycerol and sucrose (0%, 22%, 36%, 46%, 53% and 59%) are added to yam starch film solutions	Boiling with continuous stirring for 10 min			EM achieved maximum 49.3 MPa for both glycerol and sucrose at 0% concentration, then decreased	TS of yam starch film decreased (5.10–0.05) MPa with increasing sucrose and glycerol concentration	%E increased (11–42) as plasticizer concentration increases shown increase in ductility of the films	Omotoso et al. (2015)

EFY starch	Glycerol	Films were prepared using EFY starch alone (3%,4% and 5%) along with composite film was also prepared at optimized whey protein concentrate and Psyllium husk Composite (YS:3/ WPC:1.87/ PH:0.04/GLY:1.75)	90 °C for 30 min to ensure complete gelatinization of starch	Composite film had shown lowest WVP (0.32×10^{-10} g m^{-1} s^{-1} Pa^{-1}) than alone EFY starch-based film (1.73 – 2.09) $\times 10^{-10}$ g m^{-1} s^{-1} Pa^{-1}	Increased starch concentration reduced SOL (43.23–21.49) %, found lowest for composite film		TS of EFY starch-based film varied (0.8–1.83) MPa and found maximum 5.04MPa for composite film	Increased starch concentration (3–5%) reduced %E (36.56–30.84)	Sukhija et al. (2018)
Yam starch	Glycerol	Films were prepared by mixing the yam starch of three different species of *Dioscorea* yam tuber at different proportions.	At 90 °C till complete gelatinization occurs. Dried at 35 °C & placed in desiccator	WVP of films developed with native starch were higher than mixed starch film and ranged between (3.07–3.37) $\times 10^{-10}$ g m^{-1} s^{-1} Pa^{-1}	SOL ranged between (40.89–49.73) % and found higher for native starch species		TS of mixed starch film found to be higher (15MPa-22.8) MPa than native starch film (10.3–18) MPa	Whereas %E was on the higher side for native starch (17–43) than mixed starch films.	Hornung et al. (2018)
Iron yam starch	Lemon essential oil (LO)	Iron yam /maize starch-based biodegradable film were developed by incorporating lemon essential oil as plasticizer at (0.5%, 1%, 1.5% and 2%, v/v) concentration level in film forming solutions.	100 °C for 15 min. Stored at 25 °C & 53% RH for 7days in desiccator containing saturated magnesium chloride	Compared to control, addition of LO significantly) reduced WVP by 14% and 21% for film composition IYM/1.5%LO and IYM/ 2%LO respectively.	Addition of surfactant span 80 significantly enhanced SOL 55.23% (IYM/ 1LO/0.1S) than LO-based film		TS of the edible films decreased with increasing LO content as 11.8 MPa (control) to 7.60 MPa (IYM/1LO/0.1S) film.	%E was increased (31.86–38.25) with increasing LO content	Song et al. (2018)

(continued)

TABLE 6.1 (Continued)
Different preparation methods and properties of yam-starch-based biodegradable films

Starch Source	Plasticizer Used	Film Composition (Additives Added & Modification)	Gelatinization Condition, Drying, and Film Conditioning	WVTR (g/m^2. day) & WVP (×10^{-10} g m^{-1} s^{-1} Pa^{-1})	Solubility (SOL; %) and Biodegradability (BD; %)	Elastic Modulus (EM; MPa)	Tensile Strength (TS; MPa)	Elongation (%E)	Reference
Chinese yam starch	Glycerol and sorbitol	Active antimicrobial Yam starch film was developed using eugenol (1%, 3% and 5%, w/w) with two plasticizers (glycerol and sorbitol at (0.8%, 1.6%, 2.4%, 3.2%, and 4.0%, m/v) concentration level for Pork preservation.	At 90 °C till complete gelatinization occurs Dried at 25 °C for 48 h	Enhanced plasticizer concentration reduced WVP and found higher for glycerol (10.33– 5.87) × 10^{-10} g m^{-1} s^{-1} Pa^{-1} than sorbitol (6.83–2.51) × 10^{-10} g m^{-1} s^{-1} Pa^{-1}		Maximum EM was attained at 2.4% concentration level for both plasticizers; glycerol (4.50MPa) and sorbitol (7.53 MPa) respectively.	Similarly, TS was also found maximum at 2.4% plasticizers concentration.	Increased plasticizer concentration enhanced %E (15.02 → 60.61) for glycerol and (14.48 → 58.21) sorbitol respectively	Cheng et al. (2019)
White and purple water yam starch	Glycerol	Yam-starch-based edible films were developed using two yam cultivars (white and purple water yam) at different starch concentration (1.33%, 2%, 2,67%, and 3,33%) levels.	86 °C for 10 min. Conditioned at RT & 52 %RH for 48 h.	Increased starch concentration reduced WVTR for both varieties of starch	SOL of yam starch edible films are over 80% showed that the films are highly soluble		Starch at 2.67% concentration level reported having most appropriate properties, compressive strength (48.10–49.35) N/m^2 for both yam cultivars-based film		Ulyarti et al. (2019)

Purple yam starch	Glycerol	Purple yam starch based biodegradable film was developed by incorporating chitosan at (0%, 1% and 1.5%) concentration level in film forming solutions.	70 °C for 20–30 min. Conditioned in desiccator at RT & 75 %RH for 7Day.	Addition of chitosan significantly enhanced WVP, ranged between (1.31- 1.94) $\times 10^{-10}$ g m^{-1} s^{-1} Pa^{-1}					da Costa et al. (2020)
Iron Yam starch	Glycerol	Bilayer films were prepared by adding different concentrations of corn zein layer solution (0ml, 1ml, 2ml, and 3ml) in 10 ml Iron yam-pea starch forming solution.	80 °C for 30 min, dried at RT Stored at 25 °C & 54 %RH for 48 h.	As the proportion of corn zein increased, WVP of bilayer film decreased (0.88–0.55) $\times 10^{-10}$ g m^{-1} s^{-1} Pa^{-1}	SOL also decreased with an increase in zein content and reached minimum 34.99% in the IYP8/CZ3 film		TS was decreased (15.20–9.16) MPa significantly as the corn zein proportion was increased	% E at break was significantly increased (41.70–79.11) as the corn zein proportion increased	Song et al. (2020)
EFY starch	Glycerol	EFY starch-based biodegradable film developed using hydrocolloids (xanthan gum & agar-agar) at 0.5%, 1%, 1.5%, and 2% concentration level	95 ◦C for 30 min. Conditioned at 25 °C & 60 %RH for 24 h.	lowest WVTR (1494.54 g/ m^2.day) was observed for the film with 1% xanthan gum concentration level		YM of control EFY film was 54.08 MPa and increased (60.81–65.08) MPa and (56.15–69.77) MPa for agar-agar and xanthan gum respectively	TS increased significantly Incorporation of xanthan gum (19.27–19.48) MPa and agar-agar (17.30–20.14) MPa respectively	%E found maximum (23.96) for control EFY starch and reduced with addition of hydrocolloids	Nagar et al. (2020)

(continued)

TABLE 6.1 (Continued)
Different preparation methods and properties of yam-starch-based biodegradable films

Starch Source	Plasticizer Used	Film Composition (Additives Added & Modification)	Gelatinization Condition, Drying, and Film Conditioning	WVTR (g/m^2. day) & WVP ($\times 10^{-10}$ g m^{-1} s^{-1} Pa^{-1})	Solubility (SOL; %) and Biodegradability (BD; %)	Elastic Modulus (EM; MPa)	Tensile Strength (TS; MPa)	Elongation (%E)	Reference
Yam starch	Glycerol	Effect of chitosan (1%, 1.2%, 1.4%, 1.6%, and 1.8%) and clove oil (0.3%, 0.6%, 0.9%, 1.2%, and 1.5%) at five different concentration level on yam starch-based bioplastic properties evaluated	80 ◦C for 30 min. Stored at 25 °C & 52%RH for 48 h.	Lowest WVTR of film found for chitosan at 1% (45 g/ m^2.day) and for clove oil at 1.5% (21.32 g/ m^2. day) among the 5 different levels	Yam-starch–clove-oil bioplastics have a lower SOL (16.05–21.48) % than yam-starch–chitosan (69.34–77.48) %		Lower TS was observed for yam-starch-clove oil film than yam -starch-chitosan film		Ulyarti et al. (2021)
Yam starch	Glycerol	Film properties of dual modified (oxidized-NaOCl and crosslinked-STMP) yam starch at different concentration (5% and 15%) of glycerol & starch (1% and 2%) are studied	90 ◦C for 5 min. Stored at 25 °C & 52%RH for 48 h.	WVP of yam starch film found between (4.5–15.09) × 10^{-10} g m^{-1} s^{-1} Pa^{-1} and follow order of WVP as (dual modified= crosslink = native > oxidized starch)			TS ranged between (2.21–6.09) MPa and found to be decreased as glycerol concentration increased irrespective of starch.	%E found to be maximum (13.02–16.62) observed and minimum (9.10–12.66) for native starch and dual modified starch-based film respectively at higher glycerol concentration	Narváez-Gómez et al. (2021)

Lesser yam starch	Glycerol	Effect of glycerol concentration (17.5%, 22.5% and 27.5 % v/w) and crosslinking agent STPP (0.1%, 0.2% and 0.3 % w/v) were investigated on properties of yam starch based biodegradable film	Casting method	Enhancement in transmission of water vapor occurred when both STPP and glycerol applied at greater concentration	Film SOL found to be decreased and ranged (64.44–36.31) % as glycerol concentration increased irrespective of crosslinking with STPP.		Addition of STPP (0.1–0.3%) crosslinking agent significantly increased TS (1.23–1.90) MPa	Increased glycerol concentration enhanced %E and found maximum (21.18) at 27.5% glycerol concentration.	Warkoyo et al. (2021)
White yam starch	Glycerol	Epoxidized sesame oil (ESO:1.5% & 3%) incorporated in thermoplastic yam starch (TPS) and PLA polymer blends and film properties were evaluated	TPS obtained by extrusion process at 130 °C for 10 min. Conditioned at 53 % RH and 25 °C	PLA film shown lowest WVP, and also, incorporation of ESO in PLA/TPS blends reduced WVTR by 37%.	Increase in SOL (6.3–6.7) % of the composite films in comparison with the PLA (0.30%) ones was attributed to the presence of starch.	Addition of ESO decreased YM from (1123–1094) MPa for film based on PLA/TPS/ESO blends	TS found to be minimum 8.1MPa for TPS and maximum 48 Mpa for PLA films.	Addition of ESO increased %E (3.8–5.3) shown its coupling effect in blends.	Ortega-Toro et al. (2021)
Yam starch	Glycerol	Yam, cassava, jackfruit, and mango-kernel starch-based films were prepared at three starch concentrations (2%, 3%, and 3.5%) for each source.	90 °C for 15 min. Conditioned at 24 °C RT & 50 %RH for 48 h	Higher WVP was found for yam starch-based films ranged than other starches and enhanced with increasing starch concentration.	Lower SOL (11.56–18.61) % found for yam starch-based films, increased with reducing starch concentration.		The highest tensile strength (mean value of 36.63 MPa) was found for the 3.5% yam starch.	%E of yam starch film ranged (6.3 -10.39) comparatively higher than other starch films	Rodrigues et al. (2021)

(continued)

TABLE 6.1 (Continued)
Different preparation methods and properties of yam-starch-based biodegradable films

Starch Source	Plasticizer Used	Film Composition (Additives Added & Modification)	Gelatinization Condition, Drying, and Film Conditioning	WVTR (g/m^2. day) & WVP ($\times 10^{-10}$ $g\ m^{-1}\ s^{-1}\ Pa^{-1}$)	Solubility (SOL; %) and Biodegradability (BD; %)	Elastic Modulus (EM; MPa)	Tensile Strength (TS; MPa)	Elongation (%E)	Reference
Yam starch	Glycerol	Yam starch bioplastic films developed by incorporating Bentonite powder as filler material at (0.5%, 1% and 1.5% w/w) concentration level in film solution.	85 °C until gelation occurred. Dried at 40 °C for 24 h.	Increased bentonite concentration decreased WVTR for without bentonite (0.036 - 0.031 g/m^2. day) at 1.5% bentonite concentration level.	SOL% increased with increasing bentonite concentration. Bentonite enhanced biodegradability with maximum degradation of film at 1.5%.	Addition of bentonite improved EM of films ranged between (40.90–72.65) MPa	TS increased with bentonite concentration, found maximum 4.063 MPa at 1.5% bentonite.		Behera et al. (2022)
Water yam starch	Glycerol	Starch modified using ethanol precipitation at (1:5, 1:7.5, 1:10, 1:12.5, and 1:15) ratio. The ethanol vol. ratio (1:5) modified starch (0%, 5%, 10%, 15% and 20%) used for film development	Heated at 80 °C for 15 min. Dried at 50 °C & 52%RH for 24 h	WVTR of ethanol modified starch film decreased (871.68–258.96) g/ m^2. day as the concentration of modified starch increased	SOL of ethanol modified starch film decreased (54.29–29.86) % as the starch concentration increased		TS found to be increased with increasing ethanol modified starch concentration for 1:5 starch pastes to ethanol ratio.		Suseno et al. (2022)

EFY starch	Sorbitol	Nano crystalline cellulose (NCC) at different concentrations (3%, 5% and 7%) added to form EFY starch-based bio - nanocomposite film	Heated at 79.5 °C for 15 min: Dried at 50 °C for 20 h	WVTR decreased with increasing NCC concentration, found maximum 206.4 g/ m^2. day for control	Both SOL (47 – 33) % & BD (39.03–22.57) % decreased with increasing NCC concentration	Addition of NCC (3–7%) concentration enhanced EM	Similarly, NCC reinforcement also increased TS, found max. at 7%	%E of bio-nanocomposite film decreased with increased NCC up to 5%, then increased at 7%	Radiani Arifin et al. (2023)
Yam starch	Glycerol	Yam starch (64.22–71.43 wt.%) reinforced with TiO_2 (0.0–8.1 wt.%) and different glycerol concentration (26.17–32.00 wt.%) to develop film	Heated at 83.5 °C for 17–20 min	WVTR for optimized film (64.22 g of starch, 28.917 of glycerol, and 6.366 g TiO2) was observed 68.83×10^{-6} g/ m^2. day at 60 °C and 90%RH	The presence of TiO_2 slowed the biodegradation rate of bioplastic material		TS increased significantly with increasing TiO_2 concentration, found maximum 5.46 MPa at 6.8% TiO_2.	In opposite to TS, %E was found maximum 42.80 at 0% TiO_2	Chemiru et al. (2023)
Yam starch	Glycerol	Eggshells filler at (0%, 2%, 2.5%, 3%) concentration level added to yam starch bioplastic films	Heated at 90 °C and Dried at 40 °C	Egg shell reinforcement decreased WVTR, found lowest 96×10^{-6} g/ m^2.day at 3% reinforcement	Reduced film SOL and delay degradation rate reduced as a result of eggshell incorporation.	Increasing eggshell filler concentration enhanced EM (16.36–27.45) MPa.	Similarly, TS also improved, found highest 0.97Mpa at 3% eggshell reinforcement.		Sharif et al. (2023)

(continued)

TABLE 6.1 (Continued)
Different preparation methods and properties of yam-starch-based biodegradable films

Starch Source	Plasticizer Used	Film Composition (Additives Added & Modification)	Gelatinization Condition, Drying, and Film Conditioning	WVTR (g/m^2. day) & WVP ($\times 10^{-10}$ $g\ m^{-1}\ s^{-1}\ Pa^{-1}$)	Solubility (SOL; %) and Biodegradability (BD; %)	Elastic Modulus (EM; MPa)	Tensile Strength (TS; MPa)	Elongation (%E)	Reference
Yam starch	Glycerol	Starch modified using ethanol precipitation (starch paste to ethanol volume 1:5, 1:7.5, 1:10, 1:12.5, 1:15) and composite were made with smallest starch particle size		Composite film showed lower WVTR 442.08 g/ m^2. day than natural starch film 1010.64 g/ m^2. day.	Similarly, SOL also found to lower 27.38% for composite film		Higher TS observed for composite starch film than natural starch film.		Gunawan et al. (2023)
Chinese yam starch	Glycerol	Chitosan (CS) at (0.1 %, 0.2 %, 0.6 %, and 1 %, w/v) level dispersed in Gallic acid (GA) induced yam starch (YS) film for pork preservation	CS dissolved in 0.1 % acetum was mixed with GA/YS solution. Dried for 3 h at 55 ◦C				Lower TS 1.49MPa obtained for YS/CS1 Increased significantly to 14.74 MPa for GA/YS/CS3 film		Rong et al. (2023)
Mountain's yam starch		PLA/starch-based bio composites were developed at varying starch (7.5%, 15%, 22.5% & 30% db.) concentration level	Annealed at (100 bar pressure & 160 ◦C for 3 min) Stored at 25◦C for 3 days			EM declined as the starch content increases from 968.35MPa (pure polymer) to 849.45MPa for 30% starch bio composite.	TS decreased as starch level increases and found maximum for pure polymer bio composite (without starch) 79.11MPa		Estrada-Giron et al. (2024)

EFY starch	Glycerol	EFY Starch based (SB) bioplastic film amended with nano starch (NSB) and tannin coated nano starch (TNS) at 25% and 50% level	Heated at 100 °C till gel like consistency appeared Dried at 40 °C	WVTR significantly decreased (832–454 g/m^2.day) have minimum for TNSB 50 film	T-NSB film showed higher SOL compared to NS. SB film showed faster degradation % than NSB and T-NSB bio-film samples	Get higher EM for NSB compared to SB and TNSB film	T-NSB25 film showed the highest TS (0.62) MPa which was significantly higher compared to the other samples		Karmakar et al. (2024)
EFY starch	Glycerol	Effect of $CaCO_3$ as crosslinking agent (0.2 %, 0.4 %, 0.6 %, w/v) and EFY starch (3 %, 4 %, 5 %, w/v) on edible film characteristics are studied.	Heated at 85 °C for 5 min. Dried at 55 °C for 22 h	WVTR of EFY starch film were found to be lowest 23.8 g/m^2. day for 3% starch at 0.2% $CaCO_3$	SOL of EFY starch edible film ranged between (46.51–49.34) %, enhanced with starch concentration		Get best TS (0.064) MPa at 3% starch with 0.2% $CaCO_3$	%E for EFY starch film found maximum 26.73 % for 3% starch at 0.2% $CaCO_3$	Warkoyo et al. (2024)

Water vapor transmission rate; WVTR (24×g/m^2.h= g/m^2.day) & Water vapor permeability; WVP Conversion (2. 77×g.mm/kPa·h·m^2 = 10^{-10} g m^{-1} s^{-1} Pa^{-1}).

properties during storage. Compared to traditional sources like corn and cassava starch, yam starch has a higher amylose content (29%), which contributes to the development of yam starch-based bioplastic films with superior tensile strength (49 MPa) and higher gelatinization temperature range (72–83 °C). The water vapor permeability of yam starch bioplastic films, which ranges from 4.46 to 7.59×10^{-10} $g \cdot m^{-1} \cdot s^{-1} \cdot Pa^{-1}$ as glycerol concentration increases, is also lower than that of corn starch films.

In another study, Martelo-Guzman et al. (2020) developed both active edible coating material and film from yam and cassava starch, incorporating oregano essential oil for grape coating applications. The properties of these films and coatings showed notable differences depending on the starch source. Yam starch demonstrated higher film-forming dispersion stability, resulting in films with lower water absorption capacity, higher moisture content, enhanced antioxidant capacity, greater antimicrobial activity, and improved internal transmittance. The edible coating developed from yam starch showed great potential by maintaining better conditions for a longer duration and a slower rate of weight loss. Similarly, several other research studies have highlighted the immense potential of yam starch over traditional starch sources for developing bioplastic films, owing to its excellent biodegradability, superior mechanical and barrier properties, and its effectiveness as an active edible coating material. In the food industry, yam starch is also utilized as a thickening agent in soups, sauces, and syrups, and for microencapsulation of essential oils, which opens up applications in the cosmetics and pharmaceutical industries.

6.4.4 Need to Improve Properties of Native Yam Starch Bioplastic Film

In the initial phase of yam-starch-based bioplastic film development, researchers primarily focused on studying the application and properties of native yam-starch-based bioplastic films to investigate the potential of yam starch for forming biodegradable films from different sources, variety, and suitable availability of yam. Based on studies, starch concentration was found to be a primary factor influencing the properties of the film in the film-forming solution. Yam starch at 2.67% concentration level exhibited the most appropriate properties, like higher compressive strength, better transparency, good barrier properties, and color appearance for both yam cultivar-based films (Ulyarti et al., 2019). Similarly, properties of films developed from different nontraditional starch sources, including yam starch, were significantly affected by starch concentration and showed the greatest potential at optimized levels of 3% yam starch, 2% mango starch, 3% cassava starch, and 3.5% jack fruit seed starch (Rodrigues et al., 2021). Overall, native yam-starch-based bioplastic films exhibit low mechanical properties and a hydrophilic nature, which restricts their use compared to films made from nonstarch sources and synthetic polymers. This opens up opportunities for modifying starch and blending it with other starch and nonstarch polymers to enhance their suitability for food applications.

6.5 ENHANCING YAM STARCH BIOPLASTICS

Native yam starch films exhibit desirable properties such as transparency and being colorless, tasteless, and odorless. However, their hydrophilic nature leads to limitations in physicochemical, mechanical, and barrier properties, restricting their application in food compared to plastic films. To overcome these challenges, various strategies have been employed, including the modification of native starch, variations in film-forming conditions, and the incorporation of additives like plasticizers and polysaccharides. In addition, incorporation of functional additives, antioxidants, antimicrobial compounds, bioactive compounds, hydrocolloids, hydrophobic compounds, natural fillers, and compatible polymer combinations has been explored to enhance film characteristics (Majeed et al., 2023). Studies indicate significant improvements in the properties of yam starch films through these approaches.

6.5.1 The Role of Plasticizers and Other Additives

The functional additives utilized in the development of starch biocomposite films are majorly categorized as plasticizers (glycerol, sorbitol, mannitol, glucose, and sucrose), emulsifiers (polysorbate, lecithin, sodium lauryl sulfate, etc.), antioxidants, and antimicrobial compounds (Jost et al., 2016). Among these, plasticizers (nonvolatile, low molecular weight) are most frequently used due to their plasticizing effect. They play an important role in enhancing workability, flexibility, mechanical, barrier, and chemical properties while reducing brittleness of film due to their effective water binding capacity when added within prescribed limit, with no harmful effects. The incorporation of plasticizers in starch film forming solution reduces intermolecular hydrogen bonding, leading to increased intermolecular space between the starch layers. This led to improvement in chain mobility, extensibility, resistance to fracture, and enhanced film flexibility (Gracia et al., 2020).

Most researchers have widely used glycerol, sorbitol, and sugar compounds in the development of yam-starch-based film composites as a potential plasticizer. The ability of glycerol, due to the presence of (-OH) groups, to interfere between amylose molecules packed in starch layer makes it uniquely suited for working with starch polymers. In contrast, sorbitol has shown more intra-molecular hydrogen bonding with plant proteins, chitosan (containing-NH group) than starch polymer. Glycerol and sorbitol are the main primary plasticizers, widely used in the development of yam-starch-based bioplastic films. Their impact on mechanical, barrier, and optical properties of film is significantly observed at their different concentration levels. Chinese yam-starch-based antimicrobial film developed by Cheng et al. (2019) investigated the effect of two plasticizers, glycerol and sorbitol, at 0.8%, 1.6%, 2.4%, 3.2%, and 4.0% (m/v) concentration levels for pork preservation. Increased plasticizer concentrations reduced the WVP, with more pronounced effect shown for glycerol than sorbitol and also restricted oxygen permeability. Meanwhile, the mechanical properties including tensile strength and young modulus increased up to 2.4% concentration level then decreased. Similarly, interaction of STMP (crosslinking agent) and glycerol at varying concentration level has shown synergistic effect on properties of lesser yam starch-based films, such as improved water vapor transmission, enhanced transparency and reduced solubility, as the concentration of both STMP and glycerol increased (Warkoyo et al., 2021). In short, various other research studies have also proven the potential of both glycerol and sorbitol as plasticizing agents, by retaining moisture strongly when mixed with water.

Apart from plasticizers, various other additives including emulsifiers, high molecular weight polysaccharides with proteins (e.g., chitosan, pectin, carrageenan, soya, and corn zein protein), as well as hydrocolloids have been used to improve the film forming properties of yam starch bioplastic films. The lipophilic nature of commercial emulsifiers, including polysorbates (Span 20, 40, 60, and 80 and Tween 20, 40, 60, and 80), lecithin, monoglycerides, and glycerol monostearate improved water barrier and mechanical properties of films by creating emulsions as similar to their application in ice cream, mayonnaise, salad dressing, and bakery goods. The emulsifiers increase adhesiveness, decrease surface tension, and improve film forming capacity of starch-film forming solutions. Rugchati et al. (2015) reported that glycerol content significantly improved the flexibility and puncture deformation, while surfactants (Span 80 and Tween 80) enhanced film's surface properties by reducing stickiness and increasing smoothness. Additionally, antioxidants retard film oxidation and enhance stability during storage, while the inclusion of antimicrobial components prevents microbial degradation, resulting in improved shelf life of film and also the food product packaged by it. Nowadays, reinforcement of nano-filler materials in yam starch bioplastic films has increased significantly, as they are more effective in improving the specific film properties based on desired characteristics. The morphological and mechanical properties of yam starch film nanocomposites are significantly affected by incorporation of TiO_2 and nanocrystal cellulose as nanofiller materials, respectively, at varying glycerol concentration levels (Chemiru et al. 2023; Arifin et al., 2023).

6.5.2 Effect of Blending of Alternative Starches

Blending different starches has shown promising results in enhancing the barrier, mechanical, and sensory film properties and also improving their end use characteristics for various food applications. Adebowale et al. (2014) developed an edible film from a water yam-cassava starch composite blended at different proportions, where yam starch was successfully replaced by 60% cassava starch. This blend demonstrated better sensory and mechanical properties than 100% yam starch films. The addition of cassava starch affected the film structure and drying kinetics, resulting in greater thickness and improved mechanical properties. Hornung et al. (2018) studied the properties of bioplastic films prepared by mixing the yam starch from three different species of *Dioscorea* yam tuber in different proportions. Mixing yam starches improved crystallinity with no pinholes and cracks on surface, improved tensile strength by 200%, which supported biodegradability, forming flexible and strong films suitable for food applications. This innovation is not limited to starch alone; recent developments, such as synthesis and amendment of nano-starch material into yam starch films, have shown strong antimicrobial activity, enhanced mechanical strength with significant biodegradability (Karmakar et al., 2024).

6.5.3 Effect of Incorporation of Non-Starch Polysaccharides

In addition to starches, several non-starch based natural polymers like chitosan, chitin, dietary fiber, hydrocolloids, and gums have been studied to enhance biodegradability and antimicrobial properties as well as to improve the barrier and microstructure properties of yam-starch-based films. Rong et al. (2023) successfully developed an antioxidative and antimicrobial chitosan-yam-starch-based film. Chitosan, derived from deacetylation of chitin, served as a natural antimicrobial agent, preventing protein degradation and significantly increasing the shelf life of pork and other meat products. In this similar manner, Ulyarti et al. (2021) revealed that yam starch-chitosan-based plastic decreased the growth of *Staphylococcus aureus* in pempek (a fish product) while maintaining sensory and antimicrobial properties compared to plastic. Nagar et al. (2020) studied the interaction between elephant foot yam starch and hydrocolloids, including xanthan gum and agar-agar, which resulted in decreasing oxygen and water vapor permeabilities, as well as improved mechanical and thermal properties with the addition of hydrocolloids.

6.5.4 Effect of Starch Modification

Primarily chemical modification, including crosslinking, oxidation, hydroxypropylation and their combinations with suitable plasticizers (mainly glycerol and sorbitol) significantly enhanced the properties of native yam-starch-based edible films. These modifications improved water vapor permeability, wettability, color, appearance, and flexibility and addressed other deficiencies that limit the application of native starch in film formation. Warkoyo et al. (2021) investigated the effect of $CaCO_3$ as crosslinking agent and revealed the best film properties at 0.2% w/v $CaCO_3$ and 3% yam starch concentration. Addition of $CaCO_3$ strengthened the film matrix by introducing Ca^{2+} ions, which formed strong hydrogen bonds with starch. Continuing their work, Warkoyo et al. (2024) also used STPP as crosslinking agent, where phosphorylation of yam starch increased the attachment of phosphorus groups, enhancing hydrophilicity. This resulted in increasing thickness, film solubility, rate of water vapor transmission, and tensile strength of the yam starch film. Apart from single modifications, Narvaez-Gomez et al. (2021) studied the effect of single modifications (crosslinking and oxidation) as well as dual modifications, which include starch treatment with both kinds of chemical agents. The type of modifying agent, along with starch and glycerol concentration level used, showed a significant effect on the film properties of native and dual modified yam-starch-based films. Ethanol precipitation, as a novel chemical modification method, showed potential to improve yam starch film properties, including increased tensile strength, reduced rate of water vapor transmission, lower solubility, and enhanced transparency at an optimal starch paste to ethanol ratio (Gunawan et al., 2023; Suseno et al., 2022).

6.5.5 Effect of Reinforcement of Non-Starch Additives as Filler Materials

In addition to starch polysaccharides, the compatibility of non-starch additives including proteins (whey, corn zein, and pea protein), fiber, plant extracts, and essential oils has been studied in combination with synthetic biopolymers like polylactic acid (PLA) and polyhydroxyalkanoate (PHA). For instance, yam starch edible bioplastic films have been produced from pea starch blend with corn zein (Song et al., 2020), whey protein concentrate and psyllium husk (Sukhija et al., 2018), PLA bio-composite (Estrada-Giron et al., 2024), and epoxidized sesame oil as coupling agent in PLA-yam starch thermoplastics (Ortega-Toro et al., 2021). Incorporation of filler materials is introduced with plasticizers to fill the space within the starch matrix, improving the mechanical and barrier properties and enhancing the functionality of the shrunken, fragile native yam starch film. The successful reinforcement of bentonite as filler material in yam starch film, confirmed by SEM images, eliminated void spaces and smoothened the film surface. The uniform distribution of bentonite in starch matrix increased tensile strength, reduced water vapor transmission, and enhanced the soil degradation rate due to its biodegradable nature (Behera et al., 2022). Similarly, yam-starch-based nanocomposite films with uniformly distributed nanoparticle-based filler material such as TiO_2 (Chemiru et al., 2023), nanocrystalline cellulose (Radiani Arifin et al., 2023), and eggshell filler (Sharif et al., 2023) have improved surface smoothness and mechanical properties, leading to faster and consistent soil degradation.

6.6 APPLICATIONS OF YAM STARCH BIOPLASTIC FILMS

Yam starch bioplastic films offer a promising alternative to conventional synthetic packaging materials due to their natural origin, biodegradability, and excellent functional properties. In the food packaging industry, these films present a sustainable solution that addresses both environmental concerns and consumer demand for safer, eco-friendly packaging options (da Costa et al., 2020). With effective barrier properties against moisture and gases, yam starch films help to extend the shelf life of food products while maintaining quality (Sharif et al., 2023; Nagar et al., 2020). Additionally, their potential use in active and edible packaging systems places yam starch films at the forefront of innovation in sustainable food packaging. Some applications of yam-starch-based films are given in Table 6.2.

da Costa et al. (2020) developed biodegradable films using purple yam starch and chitosan for the coating of apples. The apples were dipped in the film solution for 2 seconds and dried at a controlled temperature (25 ± 2 °C), with weight measurements taken every five days over 30 days. Apples coated with yam starch and 0.5% chitosan exhibited the least weight loss, followed by those coated with yam starch and 1% chitosan, yam starch films without chitosan, and finally the uncoated apples, which had the highest weight loss. The internal appearance of the apples, assessed by slicing them into half, revealed that those coated with yam starch and 0.5% chitosan had a superior visual appearance, correlating with the minor weight losses (Figure 6.5).

Similarly, Rong et al. (2023) investigated the use of Chinese yam starch biodegradable films, incorporated with chitosan, for pork preservation. Fresh pork meat was cut into 2 cm portions, packaged in yam starch films, and stored at room temperature (25 ± 1 °C) for 24 hours. A control sample was packaged with polyethylene (PE). The color values (L*, a*, and b*) indicated that pork wrapped in yam starch films, particularly those with gallic acid and 0.6% chitosan, had better preservation outcomes compared to the PE-packaged control. The increased L* values and decreased a* and b* values suggested better antioxidant properties and lower oxygen transmissivity in the yam starch films. The pH evaluation revealed that the pork packaged with PE films had a pH above

TABLE 6.2
Application of yam starch based biodegradable films

Source	Preparation Method	Findings	Application	References
Chinese yam	Solvent casting yam starch suspension 4%, plasticizers-glycerol and sorbitol 0.8–4%, eugenol 1–5%	The film with 3% eugenol can extend the shelf life of pork beyond 50%, suggesting that the film has great potential in active packaging for extending the shelf life of foodstuffs.	Active packaging (Pork)	Chen et al. (2019)
Purple yam	solvent casting yam starch 2%, chitosan 0.5 and 1%, glycerol 2%	Films are coated on apples for 4 weeks and able to preserve the fruit quality and weight loss.	Biodegradable coating (apples)	da Costa et al. (2020)
Gallic acid-induced Chinese yam	Solvent casting gallic acid-induced Chinese yam starch 6%, chitosan 0.1–1%, glycerol 1.5%	The pork meat packaged with yam-starch-based film has an elastic surface without the condition of sticky and water penetration than control polyethylene films	Biodegradable films (Pork)	Rong et al. (2023)
Yam starch	Solvent casting yam starch 3%, glycerol 1.5%, bentonite 0.5–1.5%	SEM micrographs show that bentonite particles were homogeneously distributed in the starch matrix and hydrophilicity of the bioplastic was increased with the addition of bentonite	Biodegradable bioplastic film	Behera et al. (2022)
Yam starch	Solvent casting starch 0.88 g, glycerol 0.44 g	Mixtures of starch comprising starches from different botanical origins and species can improve the mechanical properties of films and coatings	Biodegradable films	Hornung et al. (2018)
Water yam (mun lued)	Solvent casting glycerol 1.65–2%, span 80 and tween 20 1–1.5 ml	Glycerol 1.65 and span 80 (1 ml/100 g) as desired filmogenic formulation. Film sample was molding into plastic bag with sealing machine and applied as agriculture bag for agriculture activity	Biodegradable Film	Rugchati (2015)
Yam (Dorceasea)	solvent casting starch 3%, glycerol 1.5% eggshells 0–3%	The hydrophilicity of films slightly decreased with an increase in the eggshell concentrations due to the covalent bonding of carbonate with the glycoside group and the tensile strength of films was improved with the increase in filler (eggshell) concentration	Biodegradable Bioplastics	Sharif et al. (2023)
Fresh yam (*Diospyros abyssinica*)	solvent casting starch 64.22–71.43 wt.%, glycerol 26.17–32 wt.% and titanium dioxide 0–8.1%	Optimal tensile strength (4.68 MPa) and elongation percentage (32.96%) were observed with the RSM model at starch (3.60 g), glycerol (1.40 ml), and TiO_2 nanofiller concentration (0.35 g). The biodegradability was 29% for first 15 min and antimicrobial *Escherichia coli* inhibition zone of 3.25 mm	Biofilms	Chemiru and Gonfa (2023)

TABLE 6.2 (Continued)
Application of yam starch based biodegradable films

Source	Preparation Method	Findings	Application	References
Elephant foot-yam	Solvent casting starch 3.5%, carboxymethyl cellulose 5%, nanocrystal cellulose (3,5,7%) Sorbitol 2%	7 wt.% NCC, showing tensile strength, elongation, WVTR, and thickness consistent with the Japanese Industrial Standard	Biodegradable nanocomposite film	Arifin et al. (2023)
Yam	Extrusion starch: glycerol proportion of 4:1, extrusion temp (80–130 °C), thermoplastic starch (0, 75, 100) polylactic acid (0, 25, 100), epoxidized sesame oil (0,1.5, 3)	Addition of the epoxidized oil improve the blend interface by increasing of the PLA–TPS molecular interactions and could be used to package low moisture foods such as flours, dry cereal grains, or cookies.	Thermoplastic film	Gómez-Contreras et al. (2021)

FIGURE 6.5 Photographs of the internal appearance of the apples coated with yam starch without chitosan (a), with 0.5% chitosan (b), with 1% chitosan (c), and without coating (d). (Reprinted from da Costa et al., 2020.)

6.5, indicating sub-fresh or rotten meat. During sensory evaluation, the samples in yam starch films remained fresh without any unpleasant odor. Moreover, pork packaged with PE films developed a sticky surface and exhibited water seepage, while those wrapped in yam starch films retained a firm, elastic surface without water penetration.

Cheng et al. (2019) explored active packaging by incorporating eugenol into yam starch films for pork preservation. Fresh pork (7 g) was covered with yam starch films containing 1%, 3%, or

5% eugenol, and then stored in sterile petri dishes at 25 °C and 65% relative humidity for 7 days. By the fourth day, bacterial colonies had formed on the control samples due to nutrient degradation and bacterial proliferation. Films without eugenol saw higher spoilage, whereas the films with 3% eugenol showed only minimal bacterial growth by the seventh day (Figure 6.6). Eugenol demonstrated a strong antimicrobial activity by reacting with lipids and proteins, disrupting bacterial cell membranes, and effectively inhibiting bacterial growth.

6.6.1 Biodegradability and Environmental Benefits

Biodegradable films, developed from renewable resources, are highly compatible with the environment and have gained traction in the food packaging industry as a means of reducing plastic waste. Additionally, biodegradable films have a low carbon footprint, incur lower energy costs during manufacturing, and offer improved compost ability compared to petroleum-based plastics (Ismail et al., 2016). Behera et al. (2022) prepared yam starch biodegradable films reinforced with 0.5–1.5% bentonite as filler. The addition of bentonite strengthened the O–H and Si–O–Si bonds within the bioplastic matrix, enhancing its mechanical properties. Films with 1.5% bentonite showed the highest soil degradation rate and the lowest water absorption, likely due to the silicate layers in the bentonite, which contributed to the film's structural integrity. Additionally, these films demonstrated resistance to degradation in salt and acidic conditions, though they were less effective in basic environments. In another study, Ismail et al. (2016) incorporated propan-triol as a plasticizer in yam starch films and tested them through a 5-day soil burial test, resulting in a 26% biodegradation rate. The weight loss observed in these films during soil burial indicated that microorganisms in the soil consumed the starch content, breaking down the polymer chains and facilitating the biodegradation process. This highlights the potential of yam- starch-based films to break down in natural environments without leaving harmful residues.

Sharif et al. (2023) developed eggshell-reinforced yam-starch-based bioplastics for packaging applications. The incorporation of 3% eggshell in the bioplastic matrix significantly enhanced the biodegradability of the films, with 56% degradation observed after 21 days of soil burial. The presence of calcium carbonate in the eggshell likely contributed to the accelerated breakdown of the polymer, making these films highly effective in reducing environmental pollution. The environmental benefits of yam-starch-based biodegradable films extend beyond their degradation rates. These materials contribute to the reduction of plastic waste, decrease dependency on fossil fuels, and promote the use of renewable agricultural resources. Their compatibility with existing composting processes also makes them an ideal candidate for closed-loop waste management systems, helping to reduce the environmental burden of packaging waste in both consumer and industrial sectors.

6.7 FUTURE PROSPECTS AND CHALLENGES

The future of yam-starch-based biodegradable films is promising as the demand for sustainable alternatives to traditional plastics continues to grow. With their biodegradability and low environmental impact, these films could play an important role in reducing plastic pollution. However, the natural hydrophilicity of starch makes these films susceptible to moisture, limiting their use in humid environments. Addressing this will require further research into treatments and additives that can improve water resistance without compromising biodegradability. Strategies like blending yam starch with other biopolymers, modifying yam starch structure to enhance durability, and adding bioactive compounds to improve antimicrobial properties of films have shown promising results and need attention to optimize the film for particular applications. However, research focus on yam starches are still very limited compared to conventional starch sources. Most of the studies on yam-starch-based biodegradable films are limited to exploring the properties of films. Therefore, there is huge scope for application-based studies.

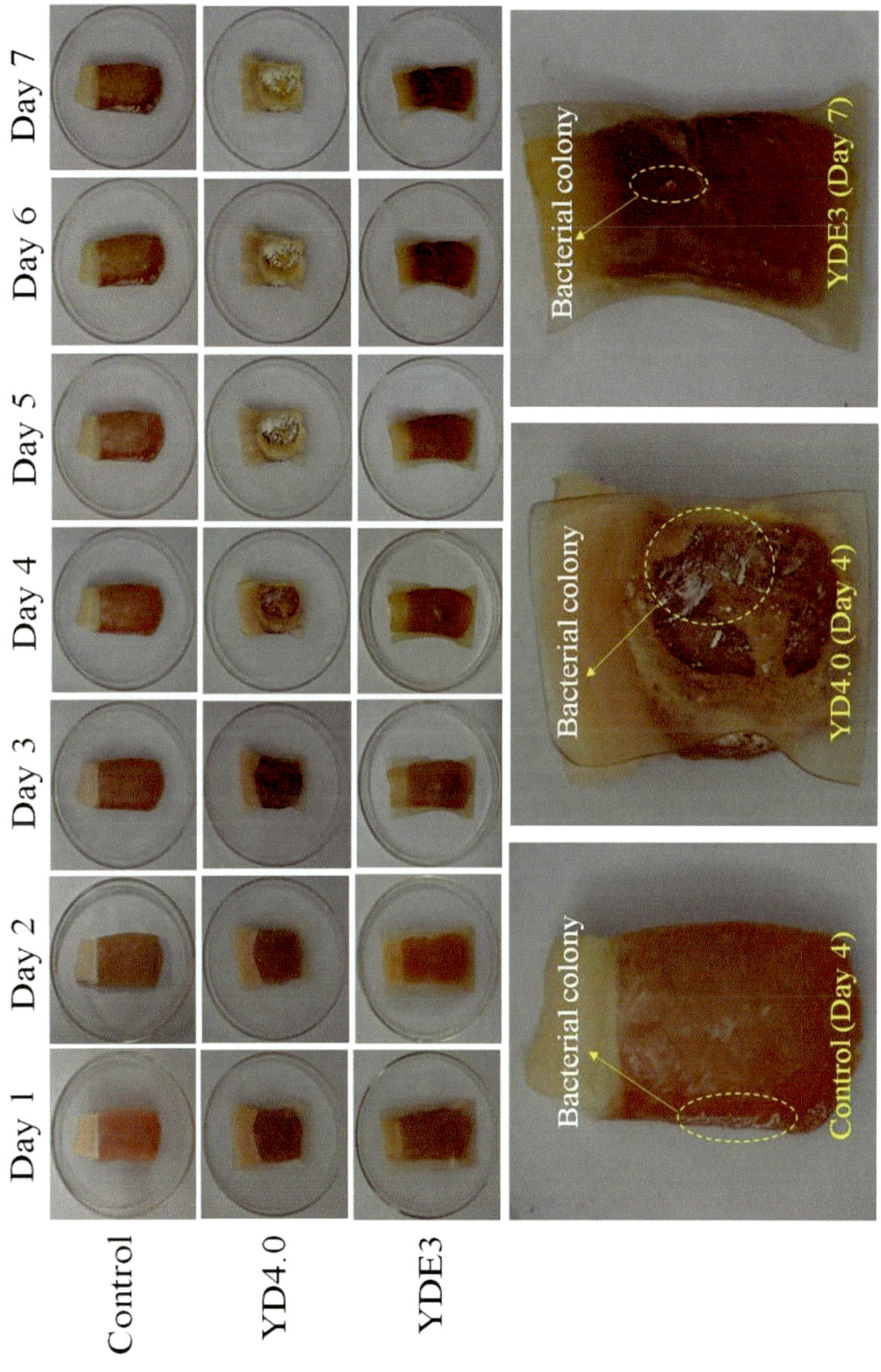

FIGURE 6.6 Application of antibacterial yam starch films on pork meat. (Reprinted from Cheng et al. (2019) with permission from Elsevier.)

6.8 CONCLUSION

Yam-starch-based bioplastic films represent an eco-friendly alternative to conventional plastics, harnessing the abundance and renewability of yam starch for sustainable material production. These films offer properties suitable for food packaging and other applications, especially with modifications to enhance mechanical, thermal, and barrier properties. However, challenges remain, particularly with moisture sensitivity and the need for cost-effective scaling beyond laboratory settings. Nevertheless, these limitations also open pathways for further research, aimed at optimizing these films' properties for broader commercial applications.

REFERENCES

Adebowale, A. A., Olatunde, O. O., Adegunwa, M. O., Asiru, W. B., & Sanni, L. O. (2014). Mechanical and sensorial characteristics of cassava and yam composite starch films. *Journal of Food Processing and Preservation*, *38* (4), 1994–1998. http://dx.doi.org/10.1111/jfpp.12175

Akissoe, N., Mestres, C., Handschin, S., Gibert, O., Hounhouigan, J., & Nago, M. (2011). Microstructure and physico-chemical bases of textural quality of yam products. *LWT-Food Science and Technology*, *44* (1), 321–329. https://doi.org/10.1016/j.lwt.2010.06.016

Alves, R. M. L., Grossmann, M. V. E., & Silva, R. S. S. F. (1999). Gelling properties of extruded yam (Dioscorea alata) starch. *Food Chemistry*, *67* (2), 123–127. https://doi.org/10.1016/S0308-8146(99)00064-3

Amani, N. G. G., Buléon, A., Kamenan, A., & Colonna, P. (2004). Variability in starch physicochemical and functional properties of yam (Dioscorea sp) cultivated in Ivory Coast. *Journal of the Science of Food and Agriculture*, *84* (15), 2085–2096. http://dx.doi.org/10.1002/jsfa.2160

Behera, L., Mohanta, M., & Thirugnanam, A. (2022). Intensification of yam-starch based biodegradable bioplastic film with bentonite for food packaging application. *Environmental Technology & Innovation*, *25*, 102180. https://doi.org/10.1016/j.eti.2021.102180

Bhattacharya, K. R., Sowbhagya, C. M., & Swamy, Y. I. (1982). Quality profiles of rice: A tentative scheme for classification. *Journal of Food Science*, *47* (2), 564–569. http://dx.doi.org/10.1111/j.1365-2621.1982.tb10124.x

Brunnschweiler, J., Mang, D., Farah, Z., Escher, F., & Conde-Petit, B. (2006). Structure–texture relationships of fresh pastes prepared from different yam (Dioscorea spp.) varieties. *LWT-Food Science and Technology*, *39* (7), 762–769. https://doi.org/10.1016/j.lwt.2005.05.011

CBI. (2023). The European market potential for yams. Centre for the Promotion of Imports from developing countries, European Unions. Retrieved September 24, 2024, from www.cbi.eu/market-information/fresh-fruit-vegetables/yams/market-potential

Chemiru, G., & Gonfa, G. (2023). Preparation and characterization of glycerol plasticized yam starch-based films reinforced with titanium dioxide nanofiller. *Carbohydrate Polymer Technologies and Applications*, *5*, 100300. https://doi.org/10.1016/j.carpta.2023.100300

Cheng, J., Wang, H., Kang, S., Xia, L., Jiang, S., Chen, M., & Jiang, S. (2019). An active packaging film based on yam starch with eugenol and its application for pork preservation. *Food Hydrocolloids*, *96*, 546–554. https://doi.org/10.1016/j.foodhyd.2019.06.007

da Costa, J. C. M., Miki, K. S. L., da Silva Ramos, A., & Teixeira-Costa, B. E. (2020). Development of biodegradable films based on purple yam starch/chitosan for food application. *Heliyon*, *6* (4). https://doi.org/10.1016/j.heliyon.2020.e03718

Daiuto, E., Cereda, M., Sarmento, S., & Vilpoux, O. (2005). Effects of extraction methods on yam (Dioscorea alata) starch characteristics. *Starch-Stärke*, *57* (3-4), 153–160. http://dx.doi.org/10.1002/star.200400324

Deep, D., Kumar, Y., Bist, Y., & Saxena, D. C. (2024). Valorization of guinea grass seed (Megathyrsus maximus): Synthesis and utilization of cellulose microfiber to reinforce esterified and cross-linked guinea starch films. *International Journal of Biological Macromolecules*, *263*, 130434. https://doi.org/10.1016/j.ijbiomac.2024.130434

Dome, K., Podgorbunskikh, E., Bychkov, A., & Lomovsky, O. (2020). Changes in the crystallinity degree of starch having different types of crystal structure after mechanical pretreatment. *Polymers*, *12* (3), 641. https://doi.org/10.3390/polym12030641

Estrada-Giron, Y., Fernandez-Escamilla, V. V. A., Martin-del-Campo, A., Gonzalez-Nunez, R., Canche-Escamilla, G., Uribe-Calderon, J., Tepale, N., Aguilar, J., & Moscoso-Sanchez, F. J. (2024). Characterization of polylactic acid biocomposites filled with native starch granules *from Dioscorea remotiflora* tubers. *Polymers*, *16* (7), 899. https://doi.org/10.3390/polym16070899

FAOSTAT. (2024). Crops and livestock products, Food and Agriculture Organization of the United Nations. Accessed on September 21, 2024 from www.fao.org/faostat/en/#data/QCL/visualize

Farhat, I. A., Oguntona, T., & Neale, R. J. (1999). Characterisation of starches from West African yams. *Journal of the Science of Food and Agriculture, 79* (15), 2105–2112. https://doi.org/10.1002/(SICI)1097-0010(199912)79:15%3C2105::AID-JSFA510%3E3.0.CO;2-R

Freitas, R. A., Paula, R. C., Feitosa, J. P. A., Rocha, S., & Sierakowski, M. R. (2004). Amylose contents, rheological properties and gelatinization kinetics of yam (Dioscorea alata) and cassava (Manihot utilissima) starches. *Carbohydrate Polymers*, *55* (1), 3–8. https://doi.org/10.1016/S0144-8617(03)00142-5

Garcia, A., Perez, L. M., Piccirilli, G. N., & Verdini, R. A. (2020). Evaluation of antioxidant, antibacterial and physicochemical properties of whey protein-based edible films incorporated with different soy sauces. *LWT*, *117*, 108587. https://doi.org/10.1016/j.lwt.2019.108587

Gómez-Contreras, P., Contreras-Camacho, M., Avalos-Belmontes, F., Collazo-Bigliardi, S., & Ortega-Toro, R. (2021). Physicochemical properties of composite materials based on thermoplastic yam starch and polylactic acid improved with the addition of epoxidized sesame oil. *Journal of Polymers and the Environment*, *29*, 3324–3334. https://doi.org/10.1007/s10924-021-02119-0

Guan, Q., Li, H., Qi, Y., Zhang, W., Liu, L., Qiu, B., ... & Zong, A. (2019). Changes in physicochemical properties and digestive characteristics of yam (Dioscorea opposita Thunb.) powder under different processing conditions using principal component analysis. *Drying Technology*, *37* (12), 1551–1562. https://doi.org/10.1080/07373937.2018.1518915

Gunawan, M. A., & Rahmi, S. L. (2023). Modification of yellow yam starch (Dioscorea alata) using various solvent ratio and its application for edible film: Starch modification; yellow yam starch; edible films; volume, starch composite. *Jurnal Bio-Geo Material Dan Energi*, *3* (1), 1–11. http://dx.doi.org/10.22437/bigme.v3i1.30914

Hornung, P. S., Avila, S., Masisi, K., Malunga, L. N., Lazzarotto, M., Schnitzler, E., ... & Beta, T. (2018). Green development of biodegradable films based on native yam (Dioscoreaceae) starch mixtures. *Starch-Stärke*, *70* (5-6), 1700234. https://doi.org/10.1002/star.201700234

Jiang, Q., Gao, W., Li, X., Xia, Y., Wang, H., Wu, S., ... & Xiao, P. (2012). Characterizations of starches isolated from five different Dioscorea L. species. *Food Hydrocolloids*, *29* (1), 35–41. https://doi.org/10.1016/j.foodhyd.2012.01.011

Jost, V., & Stramm, C. (2016). Influence of plasticizers on the mechanical and barrier properties of cast biopolymer films. *Journal of Applied Polymer Science*, *133* (2). https://doi.org/10.1002/app.42513

Karmakar, B., Sarkar, S., Chakraborty, R., Saha, S. P., Thirugnanam, A., Roy, P. K., & Roy, S. (2024). Starch-based biodegradable films amended with nano-starch and tannic acid-coated nano-starch exhibit enhanced mechanical and functional attributes with antimicrobial activity. *Carbohydrate Polymers*, 341, 122321. https://doi.org/10.1016/j.carbpol.2024.122321

Li, X., Gao, W., Wang, Y., Jiang, Q., & Huang, L. (2011). Granule structural, crystalline, and thermal changes in native Chinese yam starch after hydrolysis with two different enzymes–α-amylase and gluco-amylase. *Starch-Stärke*, *63* (2), 75–82. https://doi.org/10.1002/star.201000104

Lozano, E. J., Andrade, R. D., & Salcedo, J. (2018). Functional and rheological properties of yam (Dioscorea rotundata) mucilage. *Advance Journal of Food Science and Technology*, *15*, 134–142. http://dx.doi.org/10.19026/ajfst.14.5885

Majeed, T., Dar, A. H., Pandey, V. K., Dash, K. K., Srivastava, S., Shams, R., Jeevarathinam, G., Singh, P., Echegaray, N., & Pandiselvam, R. (2023). Role of additives in starch-based edible films and coating: A review with current knowledge. *Progress in Organic Coatings*, *181*, 107597. https://doi.org/10.1016/j.porgcoat.2023.107597

Mali, S., Grossmann, M. V. E., Garcia, M. A., Martino, M. M., & Zaritzky, N. E. (2002). Microstructural characterization of yam starch films. *Carbohydrate Polymers, 50*, 379–386. https://doi.org/10.1016/S0144-8617(02)00058-9

Mali, S., Grossmann, M. V. E., Garcia, M. A., Martino, M. N., & Zaritzky, N. E. (2004). Barrier, mechanical and optical properties of plasticized yam starch films. *Carbohydrate Polymers*, *56* (2), 129–135. http://dx.doi.org/10.1016/j.carbpol.2004.01.004

Mali, S., Grossmann, M. V. E., García, M. A., Martino, M. N., & Zaritzky, N. E. (2006). Effects of controlled storage on thermal, mechanical and barrier properties of plasticized films from different starch sources. *Journal of Food Engineering*, *75* (4), 453–460. https://doi.org/10.1016/j.jfoodeng.2005.04.031

Martelo-Guzman, J., Collazo-Bigliardi, S., & Ortega-Toro, R. (2020). Comparative study of active edible coatings based on yam starch and cassava starch. *Revista EIA*, *17* (34), 312–325.

Moorthy, S. N. (2002). Physicochemical and functional properties of tropical tuber starches: A review. *Starch-Stärke*, *54* (12), 559–592. https://doi.org/10.1002/1521-379X(200212)54:12%3C559::AID-STAR2222559%3E3.0.CO;2-F

Muñoz-Gimena, P. F., Oliver-Cuenca, V., Peponi, L., & López, D. (2023). A review on reinforcements and additives in starch-based composites for food packaging. *Polymers*, *15* (13), 2972. https://doi.org/10.3390/polym15132972

Nagar, M., Sharanagat, V. S., Kumar, Y., & Singh, L. (2020). Development and characterization of elephant foot yam starch–hydrocolloids based edible packaging film: physical, optical, thermal and barrier properties. *Journal of Food Science and Technology*, *57*, 1331–1341. https://doi.org/10.1007/s13197-019-04167-w

Narvaez-Gomez, G., Figueroa-Florez, J., Salcedo-Mendoza, J., Perez-Cervera, C., & Andrade-Pizarro, R. (2021). Development and characterization of dual-modified yam (Dioscorea rotundata) starch-based films. *Heliyon*, *7* (4). https://doi.org/10.1016/j.heliyon.2021.e06644

Ojediran, J. O., Okonkwo, C. E., Adeyi, A. J., Adeyi, O., Olaniran, A. F., George, N. E., & Olayanju, A. T. (2020). Drying characteristics of yam slices (Dioscorea rotundata) in a convective hot air dryer: Application of ANFIS in the prediction of drying kinetics. *Heliyon*, *6* (3). https://doi.org/10.1016/j.heliyon.2020.e03555

Omotoso, M. A., Adeyefa, O. S., Animashaun, E. A., & Osibanjo, O. O. (2015). Biogradable starch film from cassava, corn, potato and yam. *Chemistry and Materials Research*, *7* (12), 15–24.

Ortega-Toro, R., Lopez-Cordoba, A., & Avalos-Belmontes, F. (2021). Epoxidised sesame oil as a biobased coupling agent and plasticiser in polylactic acid/thermoplastic yam starch blends. *Heliyon*, *7* (2). https://doi.org/10.1016/j.heliyon.2021.e06176

Otegbayo, B., Oguniyan, D., & Akinwumi, O. (2014). Physicochemical and functional characterization of yam starch for potential industrial applications. *Starch-Stärke*, *66* (3–4), 235–250. https://doi.org/10.1002/star.201300056

Paes, S. S., Yakimets, I., & Mitchell, J. R. (2008). Influence of gelatinization process on functional properties of cassava starch films. *Food Hydrocolloids, 22*, 788–797. https://doi.org/10.1016/j.foodhyd.2007.03.008

Pérez, J., Arteaga, M., Andrade, R., Durango, A., & Salcedo, J. (2021). Effect of yam (Dioscorea spp.) starch on the physicochemical, rheological, and sensory properties of yogurt. *Heliyon*, *7* (1). https://doi.org/10.1016/j.heliyon.2021.e05987

Radiani Arifin, H., Adisti Mauliani, Z., Widyanti Harlina, P., Subroto, E., Nissa, R. C., & Nawaz, A. (2023). Characteristics of nanocomposite film based on elephant foot-yam starch (Amorphophallus paeoniifolius) with different nanocrystalline cellulose concentration. *International Journal of Food Properties*, *26* (2), 3512–3530. https://doi.org/10.1080/10942912.2023.2286897

Riley, C. K., Wheatley, A. O., & Asemota, H. N. (2006). Isolation and characterization of starches from eight Dioscorea alata cultivars grown in Jamaica. *African Journal of Biotechnology*, *5* (17).

Riley, C. K., Wheatley, A. O., Hassan, I., Ahmad, M. H., Morrison, E. Y. S. A., & Asemota, H. N. (2004). In vitro digestibility of raw starches extracted from five yam (Dioscorea spp.) species grown in Jamaica. *Starch-Stärke*, *56* (2), 69–73. https://doi.org/10.1002/star.200300195

Rodrigues, A. A. M., Costa, R. R. D., Santos, L. F. D., Silva, S. D. M., Britto, D. D., & Lima, M. A. C. D. (2021). Properties and characterization of biodegradable films obtained from different starch sources. *Food Science and Technology*, *41*, 476–482. https://doi.org/10.1590/fst.28520

Rolland-Sabaté, A., Georges Amani, N. G., Dufour, D., Guilois, S., & Colonna, P. (2003). Macromolecular characteristics of ten yam (Dioscorea spp) starches. *Journal of the Science of Food and Agriculture*, *83* (9), 927–936. https://doi.org/10.1002/jsfa.1410

Rong, L., Ji, X., Shen, M., Chen, X., Qi, X., Li, Y., & Xie, J. (2023). Characterization of gallic acid-Chinese yam starch biodegradable film incorporated with chitosan for potential use in pork preservation. *Food Research International*, *164*, 112331. https://doi.org/10.1016/j.foodres.2022.112331

Rugchati, O., & Thanacharoenchanaphas, K. (2015). Application of biodegradable film from yam (Dioscorea alata) starch in Thailand for agriculture activity. *International Journal of Environmental and Rural Development*, *6* (1), 28–33. https://doi.org/10.32115/ijerd.6.1_28

Sharif, N., Mohanta, M., & Thirugnanam, A. (2023). Eggshell reinforced yam starch-based bioplastic for packaging applications. *Journal of Packaging Technology and Research*, *7* (2), 75–86. https://doi.org/10.1007/s41783-023-00152-z

Shujun, W., Jinglin, Y., Wenyuan, G., Jiping, P., Hongyan, L., & Jiugao, Y. (2007). Granule structural changes in native Chinese Yam (Dioscorea opposita Thunb var. Anguo) starch during acid hydrolysis. *Carbohydrate Polymers*, *69* (2), 286–292. https://doi.org/10.1016/j.carbpol.2006.10.005

Song, X., Cheng, L., & Tan, L. (2018). Edible iron yam and maize starch convenient food flavoring packaging films with lemon essential oil as plasticization. *Food Science and Technology*, *39* (4), 971–979. https://doi.org/10.1590/fst.13118

Song, X., & Wang, Y. (2020). Development and characterization of edible bilayer films based on iron yam–pea starch blend and corn zein. *Food Science and Technology*, *41*, 684–694. https://doi.org/10.1590/fst.29820

Sukhija, S., Singh, S., & Riar, C. S. (2018). Physical, mechanical, morphological, and barrier properties of elephant foot yam starch, whey protein concentrate, and psyllium husk based composite biodegradable films. *Polymer Composites*, *39*, E407–E415. https://doi.org/10.1002/pc.24488

Suseno, R., Nursela, S., Megawati, H., & Rahmayani, I. (2022). Improving edible film quality using modified water yam (Dioscorea alata L) starch. *Jurnal Teknologi Pertanian*, *23* (2), 89–100. http://dx.doi.org/10.21776/ub.jtp.2022.023.02.1

Thakur, D., Kumar, Y., Sharanagat, V. S., Srivastava, T., & Saxena, D. C. (2023). Development of pH-sensitive films based on buckwheat starch, critic acid and rose petal extract for active food packaging. *Sustainable Chemistry and Pharmacy*, *36*, 101236. https://doi.org/10.1016/j.scp.2023.101236

Ulyarti, U., Lavlinesia, L., Surhaini, S., Lisani, L., & Nazarudin, N. (2021). Development of yam-starch-based bioplastics with the addition of chitosan and clove oil. *Makara Journal of Science*, *25* (5), 71–79. https://doi.org/10.7454/mss.v25i2.1155

Ulyarti, U., Maryana, E., Rahmayani, I., Nazarudin, N., Susilawati, S., & Doyan, A. (2019). The characteristics of yam (Dioscorea alata) starch edible films. *Jurnal Penelitian Pendidikan IPA*, *5* (1). http://dx.doi.org/10.29303/jppipa.v5i1.174

Wang, W. M., Lai, V. M. F., Chang, K. F., Lu, S., & Ho, H. H. (2006). Effect of amylopectin structure on the gelatinization and pasting properties of selected yam (Dioscorea spp.) starches. *Starch-Stärke*, *58* (11), 572–579. https://doi.org/10.1002/star.200600531

Warkoyo, W., Ruskamila, W., Wachid, M., Zekker, I., & Siskawardani, D. D. (2024). Characteristics of edible film elephant foot yam starch with calcium carbonate as cross–linking agent. In *BIO Web of Conferences* (Vol. 104, p. 00049). EDP Sciences. https://doi.org/10.1051/bioconf/202410400049

Warkoyo, W., Siskawardani, D. D., Siwi, A. A. P., Wachid, M., Zekker, I., & Onthong, J. (2021). Physical and mechanical characteristics of edible film based on lesser yam starch with addition of glycerol and sodium tripolyphosphate. *Sarhad Journal of Agriculture*, *37*, 144–152. http://dx.doi.org/10.17582/journal.sja/2021.37.s1.144.152

Zang, Z., Gong, X., Cao, L., Ni, H., & Chang, H. (2024). Resistant starch from yam: Preparation, nutrition, properties and applications in the food sector. *International Journal of Biological Macromolecules*, 133087. http://dx.doi.org/10.1016/j.ijbiomac.2024.133087

Zhou, Q., Shi, W., Meng, X., & Liu, Y. (2013). Studies on the morphological, crystalline, thermal properties of an under utilized starch from yam Dioscoreae zingiberensis CH Wright. *Starch-Stärke*, *65* (1–2), 123–133. https://doi.org/10.1002/star.201200058

Zhu, F. (2015). Isolation, composition, structure, properties, modifications, and uses of yam starch. *Comprehensive Reviews in Food Science and Food Safety*, *14* (4), 357–386. https://doi.org/10.1111/1541-4337.12134

Zou, J., Li, Y., Su, X., Wang, F., Li, Q., & Xia, H. (2022). Structure and processing properties of nine yam (Dioscorea opposita Thunb) starches from South China: A comparison study. *Molecules*, *27* (7), 2254. https://doi.org/10.3390/molecules27072254

7 Lotus Stem Starch
Composition, Properties, and Potential Applications

Chandrasekaran Sankaranarayanan

7.1 INTRODUCTION

Indian lotus (*Nelumbo nucifera*), popularly referred to as Sacred Lotus, belongs to Nelumbonaceae family that exists in two varieties, namely, *Nelumbo nucifera* var. *alba* (White type) and *Nelumbo nucifera* var. *rubra* (Red type). It is the national flower of India and is known by several names, which include Indian lotus, East Indian lotus, Oriental Lotus, Lily of Nile, Bean of India, and Sacred Water Lily. In India, varieties of lotus flowers with wide phenotypic diversity are found across the nation. The variations occur in their shapes, sizes, shade of flower, and in petal numbers. Of the three varieties reddish-white flowers are in demand than white ones. Lotus is mainly known for its ornamental, nutritional, and medicinal value. All parts of lotus that include rhizomes, leaves, seeds, and flowers are used for its ornamental and medicinal values (Zhang et al., 2015). Fresh lotus and its processed products are consumed all over the world as they are rich in carbohydrates, proteins, flavonoids, minerals, and vitamins. The therapeutic importance of lotus is attributed to the presence of phytoconsitutents which are documented in Indian and Chinese herbal medicine.

7.2 GEOGRAPHICAL DISTRIBUTION AND HABITAT

Lotus (*Nelumbo nucifera* Gaertn.) is an aquatic dicotyledonous flowering plant widely distributed in South-East Asia and is native to China, Japan, and India. Lotus is one of the most fascinating aquatic plants that regulate the ecosystem structure and functions of wetlands. It grows under tropical and subtropical climate and its cultivation requires heavy and muddy soil rich in organic matter. Based on the ecotypes in which they grow, lotus is classified into temperate and tropical, which differs in their geographic distribution, morphology, flowering time, genetics, gene expression, and the size of rhizome (Yang et al., 2015). Of the two species in the genus *Nelumbo*, *N. lucifera* Gaertn. is cultivated in many countries, such as China, India, Russia, and Australia, whereas the other species *N. lutea* Willd. is mostly found in Eastern and Southern parts of North America (Mukherjee et al., 2009). In India, China, Japan, and Australia lotus is cultivated for its economic value. It occupies an important position in the world due to its numerous purposes. Lotus cultivars are grouped into seed, rhizome, and flower lotus based on agricultural use and traits.

7.3 HISTORICAL AND RELIGIOUS BACKGROUND

N. nucifera has a rich historical background with cultural and spiritual values. It was first discovered in the ancient river beds of China and India, suggesting its origin in Asia. In Hinduism, lotus is considered as a sacred symbol representing purity, divinity, and enlightenment. Lord Vishnu, Brahma, and the goddesses Lakshmi and Saraswathi are depicted as seated or standing on a blooming

DOI: 10.1201/9781003534914-7

lotus, symbolizing spiritual transcendence and harmony. In Rigveda, lotus symbolizes the universe's unfolding from primordial waters. In Buddhism too, lotus occupies a central position, symbolizing the path to enlightenment. Even though lotus grows in muddy water it blooms clean and beautiful and thus represents spiritual growth and purity, unaffected by worldly desires. The color of lotus also carries specific meanings; white symbolizes spiritual perfection and blue for wisdom. Lotus signifies purity and virtue and is associated with Confucian ideals of integrity in Chinese culture. In modern spiritual movements and national symbols lotus continues to serve as a powerful symbol. It is the national flower of India and Vietnam. Through its cross-cultural presence, lotus has become one of the most enduring symbols of purity, renewal, and spiritual awakening in human history. Its historical significance reflects both natural beauty and philosophical depth, inspiring generations across the world.

7.4 LOTUS IN ALTERNATIVE SYSTEM OF MEDICINE

In the Indian system of medicine, every part of the lotus – roots, rhizomes, leaves, flowers, and seeds – has therapeutic properties and is used to improve physical, mental, and emotional health. In Ayurveda, lotus is considered as a "Tridosha-nashak" herb, as it helps to balance all three doshas: Vata, Pitta, and Kapha. As a coolant lotus flower relieves pitta and treats hyperacidity and polydipsia. Due to astringent properties it helps to manage bleeding disorders. It calms the mind and relieves stress and thus is used as an ingredient in formulations to treat anxiety and insomnia. In the Siddha system of medicine, lotus is considered as a rejuvenating herb that promotes overall well-being, enhances memory and cognition, and improves nervous system disorders. Similarly, in Unani medicine, lotus is beneficial for conditions related to heat and dryness as it has cold and moist properties. Lotus seeds improve cardiovascular function and are used to strengthen sexual health. It is an age-old practice in Asian traditional medicine to treat intestinal upsets, fever, and dizziness with a decoction of herbs containing leaves of lotus and rhizomes. According to *Compendium of Materia* written by Shizhen Li, "The lotus decocted with water could detoxify fungus poison and prevent blood collapse and blood drowning" (Zhao et al., 2022).

7.5 LOTUS PLANT – FAMILY AND MORPHOLOGY

Nelumbonaceae family, commonly known as the *Nelumbo* or lotus family, is a small family of flowering aquatic plants. The botanical features of this family include rounded leaves that either float or stand above water with long stalks. Further they have unique waxy surface that repels water, a phenomenon known as the "lotus effect." The flowers are large, showy with fragrance. Understanding the genetics of the Nelumbonaceae family highlights the plant's resilience and its potential applications in agriculture, medicine, and environmental studies. The family has only one genus, *Nelumbo*, which contains two species:

- *Nelumbo nucifera* (sacred or Indian lotus)
- *Nelumbo lutea* (American or yellow lotus)

The *Nelumbo* lineage diverged early from other eudicots, providing unique insights into plant evolution. This family's genetic makeup reflects a balance between evolutionary conservation and adaptation to aquatic environments. The taxonomic classification of lotus is given in Table 7.1.

7.5.1 Flower

The main feature of the plant is its huge, ostentatious flowers with 10–25 cm in diameter, which are primarily pink or white in hue. The lotus bud is shaped like a teardrop, with tightly packed petals

TABLE 7.1
Taxonomic classification of lotus

Phylum	Tracheophyta
Class	Magnoliopsida
Order	Proteales
Family	Nelumbonaceae
Genus	*Nelumbo*
Species	*Nelumbo nucifera; Nelumbo lutea*

and a pointy tip. The sepals, petals, and stamens are spirally arranged, passing gradually one into another. Each of the long stalks' single flowers has four sepals, as well as a large number of petals and stamens. Following them is a slightly cone-shaped torus that is 5–10 cm in diameter and has 10–30 carpels embedded into the upper surface. They are enormous, 15–25 cm across, quite showy, and exhibit a variety of pink colors. These carpels grow into tasty achenes that resemble ovoid nuts. The delicate petals frame the thalamus, a cone-shaped central female reproductive structure. The seeds are dark brown, oval, and hard.

7.5.2 Rhizomes and Roots

The rhizomes have a diameter of 0.5–2.5 cm and a length of 60–140 cm. The rhizome has nodes and internodes and is smooth longitudinally striated with brown patches. Lotus rhizomes and stems resemble wooden sausages and are often creamy-white in color. The lotus root system's terminal roots give rise to lotus rhizomes. The air tubes that run the length of the segment give the impression of a wagon wheel form in cross-section and are a unique characteristic of the Lotus rhizome. It has a tough, fibrous fracture and exudes mucilaginous juice. It also has a few large cavities encircled by several larger ones. Lotus rhizomes are the main commercial portion of the plant; they are edible and have a crisp, bland texture. On the basis of starch content in rhizome, there are two main categories of cultivars. The cultivar with high water, low starch, and crude fiber content has crispiness and considered as first category, while the cultivar with high water and low starch content is considered to be the second category as it is soft and ductile (Wattebled et al., 2002).

7.5.3 Fruits and Seeds

The fruit is a collection of indehiscent nutlets. The hard, smooth, brownish or grayish black pericarp of ripe nutlets is pedunculated, one-seeded, and ovoid, roundish, or oblongish, reaching 1.0 cm in length and 1.5 cm in width (Nagarajan et al., 1966). The ripe carpe is filled with seeds. Lotus seeds have been shown to last longer than those of any other blooming plant species, and their fruits have a remarkable ability to go dormant. The constituents and properties of lotus parts are presented in Table 7.2.

7.6 STARCH

Starch is a homopolysaccharide that consists of two different glucan chains connected by glycosidic bond: amylose and amylopectin. Amylose is composed of a straight chain of glucose molecules linked by an α-1,4 bond with slightly branched structure. In contrast, amylopectin is made up of glucose molecules forming highly branched chains connected by both α-1,4 and α-1,6 bonds. Starch granules range from 1 to 100 m in size and are stored as discrete semicrystalline granules in higher plants that vary in shape, size, amylose/amylopectin ratio, and functional properties. Differences

TABLE 7.2
Constituents and properties of lotus parts

Parts	Constituents	Properties	References
Flower	Quercetin, luteolin, luteolin glucoside, kaempferol, kaempferol-3-o-glucoside, isoquercitrin	Antipyretic, antimicrobial, vasodilative, antihypertensive and antiarrhythmic activities. Antioxidant, antibacterial, and free radical scavenging capacity	Ranchana et al. (2023)
Leaf	Flavonoids, alkaloids, phenolics, terpenoids, polysaccharides, steroids, tannins, and glycosides	Hematemesis, epistaxis, hemoptysis, and metrorrhagia	Wang et al. (2021)
Rhizome	Carbohydrates, fiber, vitamins, minerals, polyphenols, triterpenes, flavonoids, steroids	Diuretic, antidiabetic, anti-obesity, hypoglycemic, antipyretic, and antioxidant activities	Zhu et al. (2017)
Seed	Proteins, carbohydrates, minerals, vitamins, phospholipids, alkaloids, flavonoids, glycosides, polyphenols	Antidiabetic, anti-inflammatory, antipyretic, anti-arrhythmic, and cardioprotective activities	Bangar et al. (2022)

TABLE 7.3
Characteristics features and sources of allomorphic forms of starch

Starch Type	Characteristics	Sources
A	Tightly packed, dense, monoclinic crystalline structure	Cereals
B	Looser, more open structure compared to A-type hexagonal crystalline structures	Tubers, high amylase cereal seeds
C	Combination of A- and B-type characteristics, intermediate in packing density. Consist of both monoclinic and hexagonal crystalline structures	Lotus rhizomes and legumes

in starch composition and structure lead to different properties that often vary with region and variety and contribute to the various applications of starch. This semi-crystalline biopolymer serves as a storage carbohydrate for energy reserves in many plants, including cereals, roots, tubers, seeds, and fruits. Based on the peak distribution patterns in X-ray diffraction the polymorphic structure of starch can be classified into A, B, and C types, and this variation depends on starch composition, genetic diversity, and the molecular structure of amylose and amylopectin (Srichuwong & Jane, 2007). Many researchers have reported A- and B-type starches; however, there is little information on C-type starch which contains A- and B-type allomorphs within the same granule. Table 7.3 shows the characteristic features and sources of allomorphic forms of starch.

In addition, changes in structure and composition are generally related to the digestibility and glycemic index of starch. On the basis of glucose release and absorption across the gastrointestinal tract, starch is classified into rapidly digestible, slowly digestible, and resistant starch (RS) varieties. RS has gained attention in regulating metabolic homeostasis due to its unique characteristics that include poor digestibility in intestine with fermentation in large intestine. Experimental studies have documented that RS prevents colonic cancer, ameliorates fat metabolism, promotes growth of probiotics and exhibits hypoglycemic effects (Grubben et al., 2001). Due to high thermal stability and low water holding capacity RS are useful in food industries. The classification of RS is given in Table 7.4 with its characteristics and sources.

From the above it is clear that starch finds wide applications in culinary, pharmaceutical, textile, paper, cosmetic, and construction sectors and in food manufacturing industry as a thickener,

TABLE 7.4
Characteristic features and sources of resistant starches

Types of Resistant Starch	Characteristic Features	Sources
RS1	Physical encapsulation with plant cell wall or proteins	Grains and beans
RS2	Granules with B-type or C-type crystal nature	Bananas, raw rice, raw flour, and potatoes
RS3	Aged starch – retrograded starch	Under freezing conditions/food processing
RS4	Chemically modified starch – esterification, phosphorylation, ethyl esterification, or γ-ray irradiation	Frozen food/canned foods
RS5	Amylose-lipid complex, V-shaped crystal structure	Grains, soybeans, and corn

stabilizer, or gelling agent to provide consistency and generate attractive texture (Bemiller Whistler, 2009; Pérez & Bertoft, 2010).

7.7 NEED FOR NOVEL STARCH

As starch serves as a staple food for humans and also as raw material in food, pharmaceutical, textile, paper, and packaging industries there occurs a demand in the supply of starch-rich cereals and tubers. Thus, exploring alternate sources of starch that can supplement the conventional starch source becomes essential. Nowadays, researchers have paid attention to exploring new sources of starch, including underutilized plants, agricultural residues and wastes, and by-products of fruit and vegetables. Lotus starches are underutilized and represent novel starch resources. Lotus starch is an exotic source of starch as it grows in different environments compared to corn, wheat, potato, and tapioca (Chandak et al., 2022). The cultivation of aquatic plants for starch does not require land space and hence does not compete with humans and terrestrial animals. Further starch isolated from aquatic plants exhibit similarities with rice, potato, and maize starches, rendering their wide-scale industrial applications (Syed et al., 2012).

7.8 LOTUS STEM

Lotus rhizome is a type of modified stem and is the storage organ similar to those of other plant species, such as bamboo, onion, and potato. Lotus stem is a significant edible aquatic vegetable that is rich in nutrients such as carbohydrates, protein, dietary fiber, flavonoids, polyphenols, and minerals that offers a number of health benefits (Zaidi et al., 2021). Rhizomes measure 0.5–2.5 cm in diameter and are 60–140 cm long. They are creamish white, yellowish, or slightly yellowish brown and have a smooth surface with nodes and internodes, brown patches, and longitudinal striations. The growth characteristic is complex which follows an initial longitudinal growth followed by swelling to enlarge the girth (Zhang et al., 2022).

The development of rhizome falls into four stages:

- Initial stolon stage – elongation in single direction
- Initial swelling stage – cessation of longitudinal growth and increase in girth
- Middle swelling stage – gradual accumulation of starch and swelling of rhizome
- Later swelling stage – rapid accumulation of starch with cessation of rhizome enlargement

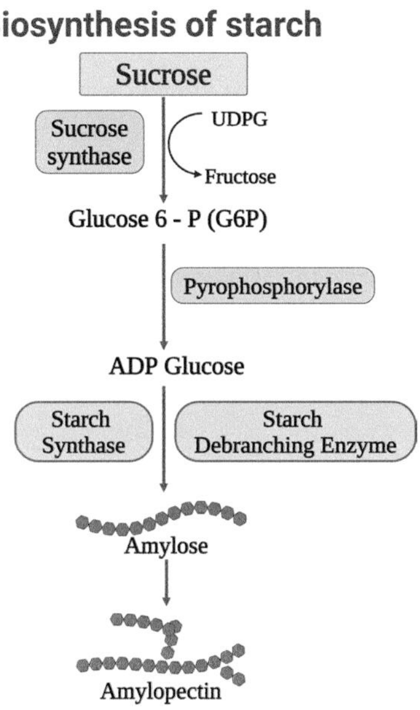

FIGURE 7.1 Biosynthesis of starch in lotus rhizome.

The development of rhizomes shows seasonal variations as it is evident that in early spring it elongates unidirectionally with the formation of few floating leaves. In late summer, rhizome stops longitudinal growth and starts swelling and subsequent rapid accumulation of starch to enable its survival during autumn (Yang et al., 2015).

7.9 LOTUS STEM STARCH

Starch is the main component of lotus rhizome and starch granules are oval and elongated with 10–50μm in long axis and 10–35μm in short axis. Molecular structures of amylose and amylopectin are of importance to the properties and functionalities of a starch. The A, B, and C types of polymorphism are produced based on the arrangement of granule crystals and it has been found that C-type starch which is a mixture of both A- and B-types is found in lotus rhizome (Yu et al., 2013). The center of starch granule contains a combination of both A- and B-types, the outer hilum contains A-type and the opposite part contains B-type of starch. The physicochemical characteristics of rhizome starch seem to be more similar to those of tuber and root starches rather than cereal starches. As amylose in starch resists solubility and swelling by forming lipid complexes, its concentration greatly affects the physicochemical and functional properties of starches. The fresh rhizome contains 10–20% starch and there is a significant disparity in the chemical makeup of lotus rhizome starches and their nutritional components which are attributed to plant genetics, geography, and climate. The biosynthesis of starch in lotus rhizome and its physicochemical properties are influenced by the regulation of enzymes involved in its synthesis and are illustrated in Figure 7.1. A comparison of the properties of lotus stem starch with other common starches is presented in Table 7.5.

7.10 MODIFICATION OF LOTUS STEM STARCH

Properties like low paste transparency, high paste viscosity and temperature, susceptibility to retrogradation and syneresis limits the broad applications of native starches. However, the physicochemical properties of lotus starch favor its applicability in food and non-food industries. To improve

TABLE 7.5
Comparison of lotus stem starch with other starches

Property	Corn Starch	Potato Starch	Cassava starch	Lotus Stem Starch
Amylose content	25–28%	20–22%	17–24%	20–30%
Granule size (µm)	5–20 µm	15–100 µm	5–35 µm	5–40 µm
Granule shape	Polygonal	Oval, spherical	Round shape	Oval, spherical
Gelatinization Temperature	62–72 °C	58–66 °C	62–74°C	60–80 °C
Retrogradation	Moderate	High	Low	Low
Water holding Capacity	Moderate	High	High	High
Viscosity	High	Very high	High	High
Pasting Stability	Moderate	Low	High	High
Freeze-Thaw Stability	Moderate	Low	Moderate freeze	High
Digestibility/glycemic index	Moderate, Medium glycemic index	High, higher glycemic index	High, higher glycemic index	Slow, moderate glycemic index
Applications	Soups, sauces, paper	Noodles, frozen foods	Biodegradable films, adhesives, textiles.	Food gels, noodles, bioplastics

the applicability and to minimize the undesirable properties modifications of starch are carried out (Sodhi & Singh, 2005). Lotus stem starches are modified by physical, chemical, and enzymatic means to improve their physicochemical/nutritional property that has further increased its applicability in various fields.

7.10.1 Chemical Modification

Chemical modifications disrupt starch granules and affect the water holding capacity and the re-association capacity during retrogradation. Some of the major chemical modification procedures are as follows:

Acetylation: This method involves the introduction of acetyl groups by converting hydroxyl groups into acetate ester groups. Such modifications enhance the properties of modified starch over the native form (El Halal et al., 2015). Lotus rhizome starch is acetylated using acetic anhydride. Acetylation endows lotus rhizome starch with improved physical and chemical properties rendering its potential applications in food industry as raw material and as additives. Investigations on the acetylated starch reveal an increase in paste transparency, swelling power, solubility, and the freeze-thaw stability and a decrease in pasting temperature, viscosity, and syneresis. Acetylated starch is used in the preparation of pudding and as additives in jellies, soups, and noodles.

Hydroxypropylation: Hydroxypropylation is a popular method of starch modification that introduces hydrophilic bulky groups to the starch backbone. Susceptibility to amylase with an increase in swelling factor and decreases in gelatinization temperatures, retrogradation, acid tolerance, pasting viscosity, and gel firmness were observed in hydroxypropylated starch. The introduction of hydroxypropyl groups destabilizes the granular structure of starch, increases water holding capacity, and hinders the reassociation of amylose chains during retrogradation. Due to the enhancement of these properties, hydroxypropylated starch has received great attention in food and pharmaceutical industries (Onofre et al., 2010).

Cross-linking: Cross linking is yet another modification of starch wherein the amylose chains are cross linked using phosphoryl chloride to reduce the mobility of chains and to stabilize the granules. Cross-linked starches are resistant to enzyme digestion and acid hydrolysis thereby exhibit decreased leaching, swelling and pasting viscosity (Gunaratne & Corke, 2007).

Dual modification: In order to mitigate the shortcomings of single modified starches, dual modifications are carried out to further optimize and increase their functional applicability (Bangar et al., 2021). In dual modifications, emphasis is given to the order, medium, reaction conditions and to the sources of starches. Such dual modifications of starches fall under two categories, viz.,

- Homo dual and
- Hetero dual modification.

Cross-linked hydroxypropylated rhizome starches possess the characteristics of both cross liking and hydroxylation. This dual modification has decreased gelatinization, leaching and pasting viscosity while increasing pasting stability (Sukhija et al., 2016).

Acid hydrolysis: Acid hydrolysis of rhizome starch changed the polymorphology from C to A type and reduced the swelling power, water binding capacity and solubility while increasing crystallinity and paste syneresis (Shujun et al., 2008).

7.10.2 Physical Modification

Physical methods of starch modifications are generally considered as safe in food applications. These modifications are carried out to produce pre-gelatinized starch by utilizing hydrothermal and nonthermal changes without incorporating any chemical compound into the starch molecules. These modifications include:

γ-Irradiation

Modification of starch by γ-irradiation is relatively a simple and faster technique that involves direct exposure to electromagnetic rays. γ-Irradiation in a dose-dependent manner induces granular damage and degrades starch chains via free radical mediated reactions. Further they increase amylose leaching and water absorption capacity while decreasing amylose content, syneresis, and the degree of crystallinity (Gani et al., 2013). However, the polymorph type of starch remains the same.

Ultrasonication

Modification of starch by ultrasonication is a green, nonthermal, effective, and innovative technology that changes the amorphous region of starch granules and enhances their physicochemical properties using physical forces. This method is advantageous over other methods due to less processing time and is environmentally safe.

High-Pressure Processing

High pressure changes the polymorph of lotus rhizome starch from C to B type (Chandak et al., 2022). As pressure forces water molecules into the granular structure, hydration and gelatinization increase with a decrease in gelatinization temperature and pasting profiles.

Heat-Moisture Treatment

Heat-moisture treatment is a simple hydrothermal and eco-friendly treatment that modifies the physicochemical properties of starch that makes starch granules to absorb more water causing swelling. It facilitates starch chain interactions and disrupts crystalline domains making it an alternative to other modification methods. Such modified starch increases the fiber content of processed food to improve their nutritional values.

Pre-gelatinization

Pre-gelatinized starch also termed as instant starch are produced by heating and by adopting mechanical shearing processes such as drying and grinding. This process increases the water solubility and viscosity of starch.

7.11 FOOD APPLICATIONS

In Asia, lotus rhizome is a popular dietary component that is consumed in various forms, such as fresh, salted, boiled and as decoctions (Hu & Skibsted, 2002). In Asian market various lotus-based food products such as herbal tea, juices, beverages, porridge, starch flours, pudding, baked and canned foods, cakes, biscuits, pickles, and chips are available (Wang & Zhang, 2010). The nutrients which are abundant in rhizome are amino acids, proteins, starch, fibers, minerals and vitamins, making it a functional food. In China, the products of rhizome are important dietary components and isolated starch is consumed as fast food, breakfast, and as food additives in traditional confectionary. In Japan, rhizome is served as pickles in vinegar, fried or sautéed. Lotus-based processed products such as salted and boiled rhizomes, drinks and teas are rich in starch, proteins, vitamins, and minerals; hence, they are popular and in great demand in many European countries (Liu et al., 2010). Modified starch finds extensive application as additives in the preparation of fruit jellies, noodle, soups, and other food products. The entire rhizome is consumed as a vegetable and supplementary food. The size of starch granules determines its applicability in food and non-food industries. Due to higher swelling and solubility properties than seed starch, lotus rhizome starch is used as thickeners, binders in gravies, soups, sauces, bread, pancakes, emulsifier, gelling agent, stabilizer, or binder bakery products, edible films, coatings, meat products, fat substitutes and in canned foods. Some of the value-added food products from different lotus starches are tabulated below (Table 7.6).

7.11.1 Pudding

Pudding was prepared by heating lotus rhizome starch, especially acetylated starch with milk, sugar, and vanilla followed by cooling (Figure 7.2). This results in the formation of pudding with a high sensory value and the degree of sensory scores increases with the extent of acetylation of rhizome starch (Sun et al., 2016). The appearance and quality of puddings prepared using modified starch enhanced attractiveness as evaluated by sensory analysis methods. Sensory analysis focusing on the taste, smell, consistency in appearance, adhesiveness, and acceptability revealed that pudding prepared using acetylated lotus rhizome starch exhibited better appearance and quality. They also exhibit good paste transparency, swelling power, freeze-thaw stability, and solubility.

TABLE 7.6
Value added food products from different lotus starches

Lotus Parts	Value Added Products	Functional Uses	References
Roots	Bread sticks	Nutraceutical bread	Thanushree et al. (2017)
	Unleavened flatbread	Nutraceutical flour	Saeed et al. (2021)
	Chinese steamed bread	Hypoglycemic, hypolipidemic and antioxidant activities	Li et al. (2021)
	Soup, sausage, mayonnaise, salad dressing, fat mimetic biscuits	Biscuits with antioxidants	Saeed et al. (2021)
	Tofu noodle	Diabetic supplements	Kim et al. (2021)
	Chicken nuggets	Provision of dietary fibers	Verma et al. (2024)
Seeds	Cookies	Improved biscuits	Dinkar & Mishra (2020)
	Noodles	Alternative product with health benefits for consumers	Jirukkakul & Sengkhampam (2018)
	Patties	Enhance fiber in meat products	Bharti et al. (2017)
Rhizome	Port patties	Higher antimicrobial and antioxidant activities	Qiu & Chin (2022)

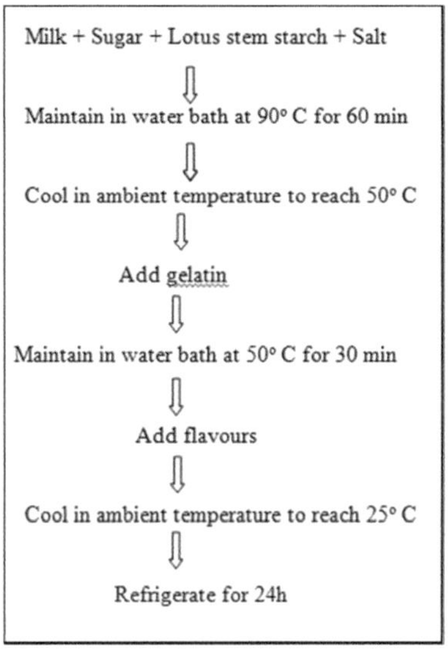

FIGURE 7.2 Preparation of lotus stem starch-based puddings.

7.11.2 Edible Films

Food preservation remains a challenging task in extending the shelf life of perishable food products. Preservation protocols deal with the use of edible films and coatings. Biodegradable edible films are thin layers of natural, consumable materials applied directly to food products or used as standalone wraps. These films improve food preservation, quality, and safety while minimizing environmental waste. Edible films are made from biopolymers such as starch, proteins, polysaccharides, and lipids, and may include plasticizers, antimicrobial agents, or antioxidants for added functionality. Lotus rhizome has a moderate amount of amylose ranging from 20% to 30% that exhibits strong film-forming property. Biodegradable films from lotus starch are cost-effective and possess improved physical, mechanical, and barrier properties. These films are substitutes for petroleum-based plastics. Lotus stem starch is increasingly being explored for edible film production due to its excellent film-forming properties (Liu et al., 2014). Currently, lotus rhizome starch is used for the preparation of edible films and coatings by mixing them with sodium alginate and sorbitol. These films have maximum tensile strength and it has been observed that such films increased the shelf life of pineapple by reducing microbial growth, rate of browning, and softening of fruit pulp. The physicochemical properties of lotus stem starch render its applicability in the preparation of edible films. The main features include:

- **High Amylose Content**: Improves film strength, flexibility, and transparency.
- **Good Gelatinization Properties**: Forms smooth, cohesive films when heated with water.
- **Antioxidant Activity**: Contains natural polyphenols, offering protection against oxidative damage.
- **Hydrophilic Nature**: Absorbs water easily, which may affect water resistance.

The key benefits of lotus stem starch-based films include:

- **Biocompatibility and Safety**: Safe for consumption and has minimal allergenic potential.
- **Good Mechanical Strength**: The starch forms smooth, transparent films with moderate flexibility.

- **Antioxidant and Antibacterial Potential**: Incorporating lotus starch with natural additives enhances film functionality.
- **Improved Shelf Life**: Reduces oxidative spoilage by limiting oxygen exposure.
- **Water Solubility**: Suitable for applications where rapid dissolution is needed, such as spice packaging or instant food systems.

7.12 NONFOOD APPLICATIONS

Lotus stem starch is widely used in the preparation of sponges, hydrogels, and in pharmaceutical and cosmetic industries.

7.12.1 Sponges

Starch-based sponges are eco-friendly alternatives to traditional synthetic sponges. These biodegradable sponges can be composted, unlike plastic sponges which contribute to microplastic pollution. Starch-based sponges are used for household cleaning, dishwashing, for wiping surfaces, in shock absorbing packaging, as wound dressings, or as bio-resorbable sponges in surgery. Lotus rhizome has high starch content, making it suitable for producing porous, absorbent materials. The fibrous structure of the lotus rhizome starch can enhance sponge texture because of its natural porosity. Thus, production of lotus rhizome starch-based sponges is an innovative approach to sustainable materials, leveraging its unique properties. Lotus rhizomes contain antioxidants, which may confer antimicrobial properties to the sponge, making it ideal for hygienic applications. TiO_2 sponges with mesopores and macropores are prepared by mixing titanium (IV) ammonium lactato dihydroxide with lotus rhizome starch which was freeze-dried and calcinated. These sponges have efficient photoconversion of carbon dioxide (Wang et al., 2014). The macroporous structure formed by the presence of starch promoted the diffusion of reagents and also improved the effect of multiple reflections within the sponge to capture more light.

7.12.2 Hydrogel

Hydrogels are three-dimensional networks of hydrophilic polymers that can absorb and retain large amounts of water or biological fluids. Due to their high water content and tunable properties, they are widely used in a range of applications, from biomedicine to agriculture. Hydrogels consist of cross-linked polymers, forming a mesh-like structure. The hydrophilic nature of the polymers allows them to absorb water without dissolution. Natural polymer hydrogels are made from materials like collagen, alginate, chitosan, or starch. Lotus rhizome starch-based hydrogels have reduced degree of crystallinity and are used in the field of drug delivery (Zhu, 2016).

Lotus rhizome starch-based hydrogels offer a biodegradable and sustainable alternative to synthetic polymers. Starch from lotus rhizomes can be cross-linked using nontoxic agents (like citric acid) to form hydrogel networks. These hydrogels have antimicrobial properties and could biodegrade without harming the environment. Lotus rhizome starch-based hydrogels are used as wound dressing agents, in agricultural practices as water retention agents or biodegradable seed coating and as moisturizing face masks and skincare products in cosmetics. Thus, hydrogels are versatile materials with a broad range of applications, and exploring lotus rhizome starch-based hydrogels can pave the way for eco-friendly innovations. With their natural origin, biodegradability, and potential health benefits, these hydrogels represent a promising development for sustainable technologies across healthcare, agriculture, and cosmetics.

7.12.2.1 In Pharmaceutical Industries

Starch is commonly used as a binder in drug formulations to provide structural integrity and ensure tablets maintain their shape. It also plays a vital role as a disintegrant, helping tablets dissolve efficiently once ingested. Starch-based binders are favored for being natural, biocompatible, nontoxic, and economical. The characteristics of lotus starch are investigated for possible application in the pharmaceutical industry as a binder. Large granular lotus rhizome starch is used as a binding, dissolution, disintegrating agent and as excipients in the preparation of tablets. Its antioxidant properties could help stabilize drugs prone to oxidation. Exploring lotus rhizome starch in pharmaceutical applications could introduce additional benefits, such as bioactivity and sustainability.

7.12.2.2 In Cosmetic Industry

The crystalline nature of starch has been attributed to the well-ordered structure of the amylopectin molecules inside the granules. Starch with small granular size isolated from lotus rhizome is ideal as aerosols and is used in talcum powder preparations (Moorthy, 2002).

7.13 CONCLUSION

The growing needs of starch in both food and nonfood industries emphasize the cultivation of starch-rich plant varieties. Urbanization, that is, decline in the cultivable area, presses the search for novel starch sources. Lotus stem starch is a valuable alternative with structural and physicochemical properties that render its wide applications in various industries. Lotus stem starch is modified further to enhance its applicability. Approaches to enhance yield and scientific validation in novel areas will open new avenues for lotus stem starch.

REFERENCES

Bangar, S. P., Ashogbon, A. O., Dhull, S. B., Thirumdas, R., Kumar, M., Hasan, M., & Pathem, S. (2021). Proso-millet starch: Properties, functionality, and applications. *International Journal of Biological Macromolecules, 190*, 960–968. https://doi.org/10.1016/j.ijbiomac.2021.09.064.

Bangar, S. P., Dunno, K., Kumar, M., Mostafa, H., & Maqsood, S. (2022). A comprehensive review on lotus seeds (*Nelumbo nucifera* Gaertn.): Nutritional composition, health-related bioactive properties, and industrial applications. *Journal of Functional Foods, 89*, 104937. https://doi.org/10.1016/j.jff.2022.104937.

BeMiller, J. N., & Whistler, R. L. (Eds.). (2009). *Starch: Chemistry and Technology*. Academic Press.

Bharti, S.K., Pathak, V., Goswami, M., Sharma, S., & Ojha, S. (2017). Quality assessment of Nelumbo nucifera supplemented functional muscle food. *Journal of Entomology and Zoology Study, 5* (4), 445–451.

Chandak, A., Dhull, S. B., Chawla, P., Fogarasi, M., & Fogarasi, S. (2022). Effect of single and dual modifications on properties of lotus rhizome starch modified by microwave and γ-irradiation: A comparative study. *Foods, 11* (19), 2969. https://doi.org/10.3390/foods11192969.

Dinkar, D. K., & Mishra, S. (2020). Preparation of biscuits by using lotus seed, pearl millets and multigrain wheat for elderly people. *International Journal of Research in Engineering, Science and Management, 3* (7), 413–415.

El Halal, S. L. M., Colussi, R., Deon, V. G., Pinto, V. Z., Villanova, F. A., Carreño, N. L. V., & da Rosa Zavareze, E. (2015). Films based on oxidized starch and cellulose from barley. *Carbohydrate Polymers, 133*, 644–653. https://doi.org/10.1016/j.carbpol.2015.07.024.

Gani, A., Gazanfar, T., Jan, R., Wani, S. M., & Masoodi, F. A. (2013). Effect of gamma irradiation on the physicochemical and morphological properties of starch extracted from lotus stem harvested from Dal lake of Jammu and Kashmir, India. *Journal of the Saudi Society of Agricultural Sciences, 12* (2), 109–115. https://doi.org/10.1016/j.jssas.2012.08.004.

Grubben, M. J. A. L., Braak, C. V. D., Essenberg, M. V., Olthof, M., Tangerman, A., Katan, M. B., & Nagengast, F. M. (2001). Effect of resistant starch on potential biomarkers for colonic cancer risk in patients with colonic adenomas. *Digestive Diseases and Sciences, 46*, 750–756.

Gunaratne, A., & Corke, H. (2007). Functional properties of hydroxypropylated, cross-linked, and hydroxypropylated cross-linked tuber and root starches. *Cereal Chemistry, 84*, 30e37. https://doi.org/10.1023/A:1010787931002.

Hu, M., & Skibsted, L. H. (2002). Antioxidative capacity of rhizome extract and rhizome knot extract of edible lotus (*Nelumbo nuficera*). *Food Chemistry, 76* (3), 327–333. https://doi.org/10.1016/S0308-8146(01)00280-1.

Jirukkakul, N., & Sengkhamparn, N. (2018). Physicochemical properties and potential of lotus seed flour as wheat flour substitute in noodles. *Songklanakarin Journal of Science & Technology, 40* (6), 1354–1360.

Kim, J. H., Kim, J. E., Kim, W. H., Tan, K. V., & Shim, S. M. (2021). Plant based protein products: Characterization and functionality of dried tofu noodles containing lotus root powder. *Food Bioscience, 43*, 101342. https://doi.org/10.1016/j.fbio.2021.101342.

Li, X., Lei, S., Liu, L., Zhang, Y., Zheng, B., & Zeng, H. (2021). Synergistic effect of lotus seed resistant starch and short-chain fatty acids on mice fecal microbiota in vitro. *International Journal of Biological Macromolecules, 183*, 2272–2281. https://doi.org/10.1016/j.ijbiomac.2021.06.016.

Liu, C. Q., Jin, J. C., Wang, N., Yao, Y. Q., & Huo, P. (2014). Optimization of lotus root starch and konjak composite edible films by response surface analysis. *Science and Technology of Food Industry, 35* (14), 333–338. DOI: 10.13386/j.issn1002-0306.2014.14.065

Liu, J., Zhang, M., & Wang, S. (2010). Processing characteristics and flavour of full lotus root powder beverage. *Journal of the Science of Food and Agriculture, 90* (14), 2482–2489.

Man, J., Cai, J., Cai, C., Xu, B., Huai, H., & Wei, C. (2012). Comparison of physicochemical properties of starches from seed and rhizome of lotus. *Carbohydrate Polymers, 88* (2), 676–683. https://doi.org/10.1016/j.carbpol.2012.01.016.

Moorthy, S. N. (2002). Physicochemical and functional properties of tropical tuber starches: A review. *Starch-Stärke, 54* (12), 559–592.

Mukherjee, P. K., Mukherjee, D., Maji, A. K., Rai, S., & Heinrich, M. (2009). The sacred lotus (*Nelumbo nucifera*) – phytochemical and therapeutic profile. *The Journal of Pharmacy and Pharmacology, 61* (4), 407–422. https://doi.org/10.1211/jpp.61.04.0001.

Nagarajan, S., Nair, A. G. R., Ramakrishnan, S., & Subramanian, S. S. (1966). Chemical examination of the flowers of *Nelumbium speciosum* wild. *Current Science, 35*(7), 176–-177.

Onofre, F. O., & Wang, Y. J. (2010). Hydroxypropylated starches of varying amylose contents as sustained release matrices in tablets. *International Journal of Pharmaceutics, 385* (1–2), 104–112. https://doi.org/10.1016/j.ijpharm.2009.10.038.

Pérez, S., & Bertoft, E. (2010). The molecular structures of starch components and their contribution to the architecture of starch granules: A comprehensive review. *Starch-Stärke, 62* (8), 389–420. https://doi.org/10.1002/star.201000013.

Qiu, Z. Z., & Chin, K. B. (2022). Evaluation of antioxidant activities of lotus rhizome root powder prepared by different drying conditions and its application to raw and cooked pork patties. *Food Packaging and Shelf Life, 33*, 100917.

Ranchana, P., Thamaraiselvi, S. P., Baranidharan, R., & Mangaiyarkarasi, R. (2023). Lotus (national flower of India): A review on floral biology, ornamental, nutritional and medicinal importance. *The Pharma Innovation Journal, 12* (5), 2810–2818.

Saeed, S. M. G., Ayesha, R., Ali, S. A., Ali, R., & Ahmed, R. (2021). Lotus root (Nelumbo nucifera Gaertn) flour a novel ingredient for the formulation of traditional unleavened flatbread: Rheological, physical and nutritional characteristics, and sensory attributes. *Journal of Food Processing and Preservation, 45* (12), e16078.

Shujun, W., Jinglin, Y., Jiugao, Y., Jiping, P., & Hongyan, L. (2008). Structure characterization of C-type starch granule by acid hydrolysis. *Food Hydrocolloids, 22* (7), 1283–1290. https://doi.org/10.1016/j.foodhyd.2007.06.012.

Sodhi, N. S., & Singh, N. (2005). Characteristics of acetylated starches prepared using starches from different rice cultivars. *Journal of Food Engineering, 70*, 117–127.

Srichuwong, S., & Jane, J. I. (2007). Physicochemical properties of starch affected by molecular composition and structures: A review. *Food Science and Biotechnology, 16* (5), 663–674.

Sukhija, S., Singh, S., & Riar, C. S. (2016). Physicochemical, crystalline, morphological, pasting and thermal properties of modified lotus rhizome (Nelumbo nucifera) starch. *Food Hydrocolloids, 60*, 50–58.

Sun, S., Zhang, G., & Ma, C. (2016). Preparation, physicochemical characterization and application of acetylated lotus rhizome starches. *Carbohydrate Polymers, 135*, 10–17. https://doi.org/10.1016/j.carbpol.2015.07.090.

Syed, N. H., Kalyna, M., Marquez, Y., Barta, A., & Brown, J. W. (2012). Alternative splicing in plants–coming of age. *Trends in Plant Science*, *17* (10), 616–623.

Thanushree, M. P., Sudha, M. L., & Crassina, K. (2017). Lotus (Nelumbo nucifera) rhizome powder as a novel ingredient in bread sticks: Rheological characteristics and nutrient composition. *Journal of Food Measurement and Characterization, 11*, 1795–1803. https://doi.org/10.1007/s11694-017-9561-y.

Verma, R., Yadav, A., Gupta, R. K., & Sanyal, I. (2024). Development of a high-frequency in vitro regeneration system in Indian lotus (Nelumbo nucifera Gaertn.). *Vitro Cellular & Developmental Biology-Plant, 60*(6), 1–12.

Wang, F., Zhou, Y., Li, P., Li, H., Tu, W., Yan, S., & Zou, Z. (2014). Formation of 3D interconnectively macro/mesoporous TiO2 sponges through gelation of lotus root starch toward CO2 photoreduction into hydrocarbon fuels. *RSC Advances*, *4* (81), 43172–43177. https://doi.org/10.1039/C4RA06565D.

Wang, J., & Zhang, G. (2010). The yield and chemical composition of lotus seed on different culture conditions. *Hubei Journal of Traditional Chinese Medicine, 32*, 75–76.

Wang, Z., Cheng, Y., Zeng, M., Wang, Z., Qin, F., Wang, Y., & He, Z. (2021). Lotus (*Nelumbo nucifera* Gaertn.) leaf: A narrative review of its Phytoconstituents, health benefits and food industry applications. *Trends in Food Science & Technology*, *112*, 631–650. https://doi.org/10.1016/j.tifs.2021.04.033.

Wattebled, F., Buléon, A., Bouchet, B., Ral, J. P., Liénard, L., Delvallé, D., Binderup, K., Dauvillée, D., Ball, S., & D'Hulst, C. (2002). Granule-bound starch synthase I. A major enzyme involved in the biogenesis of B-crystallites in starch granules. *European Journal of Biochemistry*, 269 (15), 3810–3820. https://doi.org/10.1046/j.1432-1033.2002.03072.x.

Yang, M., Zhu, L., Pan, C., Xu, L., Liu, Y., Ke, W., & Yang, P. (2015). Transcriptomic analysis of the regulation of rhizome formation in temperate and tropical lotus (Nelumbo nucifera). *Scientific Reports*, *5* (1), 13059. https://doi.org/10.1038/srep13059.

Yu, H., Cheng, L., Yin, J., Yan, S., Liu, K., Zhang, F., Xu, B., & Li, L. (2013). Structure and physicochemical properties of starches in lotus (Nelumbo nucifera Gaertn.) rhizome. *Food Science & Nutrition*, *1* (4), 273–283. https://doi.org/10.1002/fsn3.37.

Zaidi, A., Srivastava, A. K., & Parveen, K. (2021). Role of *Nelumbo nucifera* (Lotus) stem as immunity booster and anti-inflammatory food. *Era's Journal of Medical Research*, *8* (2). https://doi.org/10.24041/ejmr2021.34.

Zhang, Q., Xu, R., Wu, S., Wang, X., Cheng, D., Weng, B., & Guo, Y. (2022). Novel natural cellulosic fiber from lotus stem: Structure and performance characterization. *Journal of Natural Fibers*, *19* (17), 15704–15715. https://doi.org/10.1080/15440478.2022.2133051.

Zhang, Y., Lu, X., Zeng, S., Huang, X., Guo, Z., Zheng, Y., & Zheng, B. (2015). Nutritional composition, physiological functions and processing of lotus (*Nelumbo nucifera* Gaertn.) seeds: A review. *Phytochemistry Reviews*, 14, 321–334. https://doi.org/10.1007/s11101-015-9401-9.

Zhao, S., Ruan, F., Shen, W., Deng, K., Jiang, T., Wu, P., & Li, L. (2022). The effect of nitrogen fertilizer on rhizome quality and starch physicochemical properties in Nelumbo nucifera. *Agronomy*, *12* (4), 794. https://doi.org/10.3390/agronomy12040794.

Zhu, F. (2016). Impact of γ-irradiation on structure, physicochemical properties, and applications of starch. *Food Hydrocolloids*, *52*, 201–212. https://doi.org/10.1016/j.foodhyd.2015.05.035.

8 Fruit Wastes as Promising Sources of Starch

S. K. Ayiti, B. O. Olatunbosun, and E. I. O. Ajayi

8.1 INTRODUCTION

Could fruit waste starch, an alternative source of raw material for producing bioplastics, be the hoped-for solution to the problem and shortcomings posed by fossil-generated plastics and other chemically formed bioplastics? For almost a century, the waste generated from plastics and some chemically produced bioplastics has been quite alarming, and their unprecedented effects on the ecosystem are daunting. The advent of macro plastics has migrated to the application of both micro, nano-plastics and some bio-based but chemically formed bioplastics, and this has resulted in very high level of plastics and bioplastics waste that are non-compostable and non-biodegradable. Due to the nondegradability of these materials, research carried out on their presence in living organisms, especially human, showed that they can be ingested, inhaled and absorbed (Yin et al., 2022). Certain research to identify these materials in human body revealed the potential damaging effects they might cause to various organs in the body, especially to the mitochondria, the power house of the cell (Siyi et al., 2022). The task to explore alternatives to fossil-derived plastics and chemically formed bioplastics, therefore, becomes imperative.

Most bioplastics that are bio-based materials but chemically formed were found not to be degraded by microorganisms, thus making them non-biodegradable or non-compostable materials like their petroleum-derived counterparts, thereby, compounding the adverse effects of polluting the ecosystem (Carlota, 2019). However, less focus on biodegradable bioplastics is due to limitations in terms of their thermal stability and mechanical properties, making these bioplastics unsustainable; other properties, such as durability and high cost, are also major differences from their counterparts (Jaso, 2020). Research carried out on biodegradable polymers based on natural polysaccharides, such as starch and cellulose, as raw material to develop biodegradable films, revealed their capability of forming a continuous matrix that is renewable and abundant in nature (Jiang et al., 2020; Brandelero et al., 2011; Polnaya et al., 2021; Jiménez et al., 2012). Additionally, the renewability status of starch also confers other unique features on bioplastics that include low energy consumption involved in its production, its potential to add value to by-products from other industries, and its characteristic biodegradability, among others (Yu et al., 2006; Adamcová et al., 2019).

These properties can also be accounted for by the composition of starch, which entails a mixture of polysaccharides, such as amylose, amylopectin, and a small percentage of non-glycosidic conformation-D-glucose (Seung, 2020). Structurally, starches are built of crystalline branched polymer amylopectin that comprises poly-α-1,4-D-glucopyranoside and α-1,6-D-glucopyranoside and the amylose part, a linear polymer of poly-α-1,4-D-glucopyranoside that is present in the amorphous starch granule as shown in Figure 8.1 (Bertoft, 2017). These components are found in most native or normal starches, which have approximately 70–80% and 20–30% of amylopectin and amylose, respectively, by mass of starch granule (Seung, 2020; Bertoft, 2017).

DOI: 10.1201/9781003534914-8

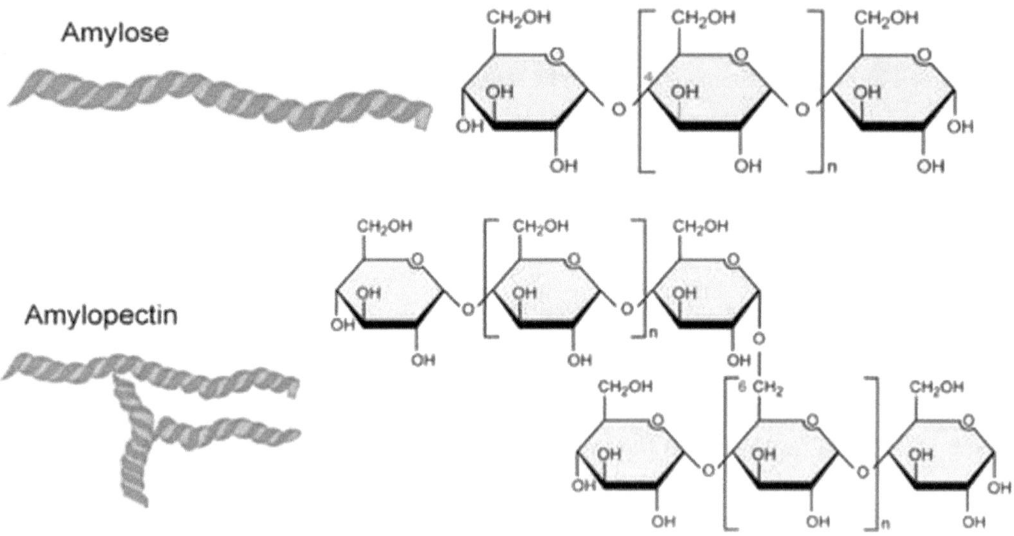

FIGURE 8.1 Structure of amylose and amylopectin in starch (Zia et al., 2015).

8.2 PROPERTIES OF FRUIT WASTE STARCH

The dynamic properties of fruit waste starch are generally determined by its two homopolysaccharides (amylose and amylopectin), their arrangements, molecular and structural composition, and percentage composition (Egharevba, 2019; Varghese et al., 2022).

These properties include thermal stability, gelatinization behavior, and rheological properties, as discussed next.

8.2.1 Thermal Stability

Thermal properties are crucial reference parameters for assessing the quality of starch during cooking and processing. The thermal stability of fruit waste starch refers to its capacity to retain structural and functional properties under high-temperature conditions in food processing. Research indicates that fruit waste starch with higher amylose content exhibits greater thermal stability than starches with lower amylose content (Lemos et al., 2018).

8.2.2 Gelatinization Behavior

Gelatinization involves breaking the intermolecular bonds of starch molecules using heat and water. During this process, amylose leaches into the water phase between granules while amylopectin becomes fully hydrated, causing the starch to swell and irreversibly burst in water. Heat disrupts the amylose chains, and water acts as a plasticizer (Chakraborty et al., 2022). Aside from water, other solvents can be used to promote gelatinization (Ulyarti et al., 2022). The key factor when using solvents is their ability to form hydrogen bonds with the molecules in the fruit waste starch granules (Chakraborty et al., 2022). The gelatinization process is influenced by the type of solvent, the ratio of fruit waste starch to solvent, as well as the time and temperature (Ulyarti et al., 2022).

8.2.3 Rheological Properties

Rheological properties describe the deformation and flow behaviors of starch in response to stress and other external factors and are often measured using a rheometer (Matignon et al., 2014). After heating above the gelatinization temperature, starch granules lose their crystalline structure, absorb water, swell, some disperse, and develop significant viscosity (Cornejo-Ramírez et al., 2018). This process of viscosity development is known as starch pasting. Although the rheological properties of starch depend on the level of modification (Bansal et al., 2016), differences in the rheological values of various fruit waste starch contents have been attributed to differences in their amylose contents and granular structures. Studies have shown that fruit starches with lower amylose content exhibit lower peaks compared with fruit waste starches with higher amylose content (Li et al., 2022; Chakraborty et al., 2022).

8.3 APPLICATION OF FRUIT WASTE STARCH

8.3.1 In The Food Industry

The application of fruit waste starch in the food industry encompasses a wide range of areas, from food processing to packaging. In food processing, using treatments such as drying and micronization, fruit waste starch has been transformed into various forms, such as powders and gels, which are used as functional ingredients like thickeners, binding agents, stabilizers, and sweeteners (Gurikar et al., 2021; Punia Bangar et al., 2021). Starches derived from fruit waste, such as pineapple stem starch with high amylose content, can also be utilized as resistant starch in low-calorie, reduced-fat food products, offering healthier and safer options (Nakthong et al., 2017). In food packaging, fruit waste starches are used to produce biodegradable packaging materials (Yaradoddi et al., 2021).

8.3.2 In Bioplastic Production

Plastics have become one of the biggest environmental factors affecting the globe. This is especially so because of its low degradability and pollution, which have unfavorably affected lands, oceans, and waterways. Studies have shown that plastic poses a significant health risk to humans, including the disruption of thyroid hormone levels and cancers (Andra & Makris, 2012; Govind & Nishitha, 2023). Due to its many negative effects, scientists have sought healthier and less harmful alternatives, including bioplastics that are biodegradable entirely or partially. Bioplastics are synthesized using fruit waste starch, such as banana peel and sugarcane. The high starch content in the peels makes them suitable for bioplastic production, with the pectin in them enhancing bioplastic strength (Mellinas et al., 2020). These bioplastics have shown similar characteristics and features to those of conventional plastics. When compared to conventional plastics, it is not only safer, but also more renewable, cheaper, and readily available (Punia Bangar et al., 2021).

8.3.3 In The Pharmaceutical Industry

Fruit wastes have been investigated for their potential in the production of various excipients and active pharmaceutical ingredients. They have been found to improve formulation properties and reduce manufacturing costs (Sarmah et al., 2023). It is a versatile and inexpensive polysaccharide that has received significant attention in drug delivery applications because it is hydrophilic, biodegradable, and biocompatible with tissues and cells. Polysaccharides, particularly cellulose and pectin extracted from fruit waste starch, have been widely utilized in the pharmaceutical industry as excipients (Choudhury et al., 2022). They are commonly used as binders, diluents, fillers, and disintegrants in the formulation of tablet dosage forms. For example, starch obtained from the seeds of *Mangifera indica* (mango) has been used as an excipient (binder and disintegrating agent) in the formulation of Paracetamol and Ibuprofen tablets (Sarmah et al., 2023). Given the advantageous nature of fruit waste starch, as extractable from their pulp, it is now also being used in the field of nanodrug delivery systems (Troncoso & Torres, 2020).

8.4 IMPORTANCE OF FRUIT WASTE STARCH SOURCES ON THE ENVIRONMENT

8.4.1 Bio Circularity

Bio circularity is essential for achieving environmental sustainability. It represents a regenerative cycle that enhances resource utilization and waste recovery while reducing emissions during manufacturing and extraction processes (Geissdoerfer et al., 2017). Studies on fruit waste starch have demonstrated the industrial possibility of converting this starch into a wide range of value-added chemicals and biofuels (Venkata Mohan et al., 2016; Gosavi et al., 2017). In converting renewable bioresources into high-value bio-based products, a sustainable bioeconomy requires more than just substituting renewable energy for fossil carbon; it also requires low-carbon energy inputs and sustainable supply chains, and studies have shown that fruit waste starch can be highly effective in achieving these objectives (Okuthe, 2024).

8.4.2 Reduction of Waste Disposal

Due to their high production, high biodegradability, and harmful effects on the ecosystem, fruit wastes pose a major threat to the environment, contributing to water and soil pollution, eutrophication, global warming, the greenhouse effect, and other health problems (Lee et al., 2024). Adequate use of fruit waste starch plays a significant role in waste reduction by providing an eco-friendly way of managing organic waste. By harnessing its renewability and the principles of "circular economy," these wastes have been valorized and repurposed into valuable products like biofuels, bioplastics, and other bio-based chemicals (Okuthe, 2024; Brunerová et al., 2017), thereby lessening the adverse effects of food waste on the environment and encouraging sustainable waste management practices.

8.4.3 Water Pollution Mitigation

Globally, an estimated 80% of industrial and municipal wastewater is discharged into the environment untreated, adversely affecting human health and the ecosystem (Lin et al., 2022), emphasizing the need for urgent intervention. Fruit waste peels are highly porous and abundant in cellulose, which is hydrophilic and therefore provides them with high chemical surface reform and high mechanical strength functionality, which makes them a promising source of absorbents (biosorbents) for water purification (Solangi et al., 2021). Fruit peel waste has shown remarkable adsorption potential for various water pollutants, including heavy metals, toxic dyes, and organic contaminants (Mallampati et al., 2015), serving as binders, chelators, and natural coagulants to remove turbidity (Haritha et al., 2024).

8.4.4 Decreasing Dependence on Conventional Starch Sources

With the growing demand for starch in different industries, the production of starch also increases globally, which is a major competition for human food sources. New sources of starch that do not compete as food sources for human populations are therefore exploited and focused on, including starch obtained from waste, including fruit waste (Tagliapietra et al., 2021). The structural and functional properties of starch isolated from some fruit wastes have been studied and found to exhibit very similar or higher functional characteristics in comparison with their conventional counterparts (Kringel et al., 2020; Abdelhai et al., 2023; Guo et al., 2018). These wastes, as unconventional starch sources, are a promising alternative to conventional starch. Because starch has a longer shelf life, post-harvest losses are avoided, and these starches can be used to generate a variety of innovative products (Phukan & Nongkhlaw, 2022), including bio-briquettes, an alternative source and clean cooking energy that can avert deforestation (Ferronato et al., 2022; United Nations Climate Change, 2023; Yirijor et al., 2024; Dugbazah et al., 2022; Jula et al., 2024).

Analysis of the starch content of different plants with scanning electron microscopy, Fourier-transform infrared spectroscopy, differential scanning, calorimetry, and thermogravimetric analysis showed that the starch with the highest crystallinity tends to have the highest thermal stability while the starch with the least crystalline content was said to have lower thermal stability (Pérez-Pacheco et al., 2024). Other studies of organic materials as sources of starch for the production of bioplastic, such as the work of Marichelvam et al. (2019) on corn and rice-based starch bioplastics, using food-based starch source indicated unsustainability, due to the fast-growing rate of the world population, and more food is required to feed the world's population.

Otoni et al. (2021) revealed that incorporating low-value or underutilized biomass, biocolloids, water-soluble biopolymers, polymerizable monomers, and nutrients into agri-food losses and waste to form functional bioplastics is a way of converting waste into useful products. The use of fruit and vegetable wastes as a component of biodegradable plastic was also explored with the addition of additives such as polyvinyl alcohol as binder, glycerin as plasticizer, soya oil, and glacial acetic acid as stabilizer. When subjected to different tests such as biodegradability test, chemical solubility test, air test, and tensile stress test, it was discovered that the products with the highest quantity of powdered waste (100 g) showed the largest tensile stress suggesting that the high starch content of the largest powdered waste may be due to the increased tensile strength with considerable biodegradation and chemical solubility (Orenia et al., 2018). Also, fruit wastes with no or low starch content were used for the production of bioplastic, and during this research, there was addition of other materials, such as paper and starch, to make the bioplastic more tensile and durable to handle, and glycerol was added to provide some level of water resistivity to the bioplastics. However, the bioplastics produced this way appeared to be firmer, more tensile and durable with high solubility in water and biodegradability (Delos et al., 2021).

Certain fruit waste that has proven to be effective in the production of bioplastic includes pomelo peels, watermelon and melon rinds, unripe and ripe mangos, lemon, tangerine, and orange peels, as well as recycled paper when combined in right proportions together with the proper amounts of additive components showed the possibility of giving rise to bioplastics that are naturally degradable (Delos et al., 2021). Rendón-Villalobos et al. (2022) studied the extraction of starch and micro-cellulose from mango waste used in the production of bioplastics. Results indicated that the amylose content of starch was found to be responsible for its film-forming ability and mango starch with its high amylose content (31.4%) can be processed to form biodegradable films or bioplastics. The study also revealed that banana inflorescence waste fiber contains insoluble fibers, such as lignocellulose, that serve as a starch matrix filler in the formation of bioplastic (Pongsuwan et al., 2022).

8.5 CONCLUSION

The sustainable use of industrial bioplastics for storing, packaging, and other purposes is a good move. Relying on their flexibility, tensile strength, biodegradability, recyclability, and compostability, the overwhelming effect of biomass generation and greenhouse gas emission can be mitigated to a great extent. The application of fruit waste starch in the production of these sustainable bioplastics and bioleathers takes advantage of the composition and properties of its starch. This not only helps to reduce the pressure of using the available food for the growing world population as a starch source, but also encourages the conversion of fruit waste to valuable products, thereby reducing the problem of waste disposal and also allowing the turning of waste to wealth, perhaps improving national development and encouraging bio-circular economy.

REFERENCES

Abdelhai, M., Boudries, N., Taibi, H., Blecker, C., Sinnaeve, G., & Sindic, M. (2023). Functional properties and potential utilization of starch isolated from chayote fruit. *Journal of Microbiology, Biotechnology and Food Sciences,* 13 (1), 1–8. https://hdl.handle.net/2268/303587

Adamcová, D., Zloch, J., Brtnický, M., & Vaverková, M. D. (2019). Biodegradation/desintegration of selected range of polymers: Impact on the compost quality. *Journal of Polymers and the Environment*, 27 (4), 892–899. http://dx.doi.org/10.1007/s10924-019-01393-3

Andra, S. S., & Makris, K. C. (2012). Thyroid disrupting chemicals in plastic additives and thyroid health. *Journal of Environmental Science and Health – Part C Environmental Carcinogenesis and Ecotoxicology Reviews*, 30 (2), 107–151. https://doi.org/10.1080/10590501.2012.681487

Bansal, P., Choudhary, S., Taneja, T., Sangwan, S., Gupta, B., Goyal, S., Kumar, R., & Sharma, P. (2016). We are IntechOpen, the world' s leading publisher of Open Access books built by scientists for scientists TOP 1 %. *Intech,* i (tourism), 15. www.intechopen.com/books/advanced-biometric-technologies/liveness-detection-in-biometrics

Bertoft, E. (2017). Understanding starch structure: recent progress. *Agronomy,* 7 (3), 56. http://dx.doi.org/10.3390/agronomy7030056.

Brandelero, R. P. H., Grossmann, M. V. E., & Yamashita, F. (2011). Effect of the method of production of the blends on mechanical and structural properties of biodegradable starch films produced by blown extrusion. *Carbohydrate Polymers*, 86 (3), 1344–1350. http://dx.doi.org/10.1016/j.carbpol.2011.06.045.

Brunerová, A., Roubík, H., Brožek, M., Herák, D., Šleger, V., & Mazancová, J. (2017). Potential of tropical fruit waste biomass for production of bio-briquette fuel: Using Indonesia as an example. *Energies*, 10 (12). https://doi.org/10.3390/en10122119

Carlota, V. (2019). Is PLA filament actually biodegradable. www.3dnatives.com/en/pla-filament- 230720194/

Chakraborty, I., Pooja, N., Mal, S. S., Paul, U. C., Rahman, M. H. & Mazumder, N. (2022). An insight into the gelatinization properties influencing the modified starches used in food industry: A review. *Food and Bioprocess Technology*, 15 (6), 1195–1223. https://doi.org/10.1007/s11947-022-02761-z

Choudhury, A., Sarma, S., Sarkar, S., Kumari, M. & Dey, B. K. (2022). Polysaccharides obtained from vegetables: An effective source of alternative excipient. *Journal of Pharmacopuncture*, 25 (4), 317–326. https://doi.org/10.3831/KPI.2022.25.4.317

Cornejo-Ramírez, Y. I., Martínez-Cruz, O., Del Toro-Sánchez, C. L., Wong-Corral, F. J., Borboa-Flores, J. & Cinco-Moroyoqui, F. J. (2018). The structural characteristics of starches and their functional properties. *CYTA – Journal of Food*, 16 (1), 1003–1017. https://doi.org/10.1080/19476337.2018.1518343

Delos Santos, K., Calitis, A. Y., Gonzales, M. L. S., Hussin, Y. A., Ilan, C. M. S. & Perin, C. F. J. C. (2021). Utilization of fruit waste and recycled paper for bioplastic synthesis. *SDCA Student Research Journal*, 10 (1), 37–45.

Dugbazah, J., Glover, B., Mbuli, B., Kungade, C. & Shikwambane, N. (2022). Preserving the lungs of Africa: Leveraging on briquettes from agricultural waste as an alternative fuel source. AU High-Level Panel on Emerging Technologies (APET) Secretariat and Calestous Juma Executive Dialogues (CJED). www.nepad.org/blog/preserving-lungs-of-africa-leveraging-briquettes-agricultural-waste-alternative-fuel-source

Egharevba, H. O. (2019). Chemical properties of starch and its application in the food industry. *Chemical Properties of Starch*, 9. doi:10.5772/intechopen.87777.

Ferronato, N., Mendoza, I. J. C., Portillo, M. A. G., Conti, F. & Torretta, V. (2022). Are waste-based briquettes alternative fuels in developing countries? A critical review. *Energy for Sustainable Development*, 68, 220–241.

Geissdoerfer, M., Savaget, P., Bocken, N. M. P. & Hultink, E. J. (2017). The circular economy – A new sustainability paradigm? *Journal of Cleaner Production,* 143, 757–768. https://doi.org/10.1016/j.jclepro.2016.12.048

Gosavi, P., Chaudhary, Y. & Durve-Gupta, A. (2017). Production of biofuel from fruits and vegetable wastes. *European Journal of Biotechnology and Bioscience European Journal of Biotechnology and Bioscience Online ISSN*, 5 (3), 2321–9122.

Govind, A. & Nishitha, K. (2023). Plastic and its side Effects on Humans–A review article. *Asian Pacific Journal of Environment and Cancer*, 6 (1), 81–85. https://doi.org/10.31557/APJEC.2023.6.1.81-85

Guo, K., Lin, L., Fan, X., Zhang, L. & Wei, C. (2018). Comparison of structural and functional properties of starches from five fruit kernels. *Food Chemistry*, 257, 75–82. https://doi.org/10.1016/j.foodchem.2018.03.004

Gurikar, C., Vandana, H. B., Netravati, B. P., Kumari, B. P. C., Gowda, N. A. N., Hanumantharaju, K. N. & Lokesh, A. C. (2021). Microbial fuel cells: An alternate approach for bioelectricity generation and waste management. *Journal of Pure and Applied Microbiology*, 15 (4), 1833–1845. https://doi.org/10.22207/JPAM.15.4.74

Haritha, M., Raju, G., Jacob, A. M. & Thampi, G. (2024). Remediation of waste water through natural coagulants such as lemon and banana peel. *E3S Web of Conferences*, 529, 1–8. https://doi.org/10.1051/e3sconf/202452903004

Jaso Sanchez, M. A. (2020). Bioplastics emergence: A technological niches study. *Actauniversitaria* 30, e2654. https://doi.org/10.15174/au.2020.2654

Jiang, T., Duan, Q., Zhu, J., Liu, H. & Yu, L. (2020). Starch-based biodegradable materials: Challenges and opportunities. *Advanced Industrial and Engineering Polymer Research*, 3 (1), 8–18. https://doi.org/10.1016/j.aiepr.2019.11.0033

Jiménez, A., Fabra, M. J., Talens, P. & Chiralt, A. (2012). Edible and biodegradable starch films: A review. *Food and Bioprocess Technology*, 5 (6), 2058–2076. https://doi.org/10.1007/s11947-012-0835-4

Jula, G., Kim, D.-G. & Nigatu, S. (2024). Potential of floriculture waste-derived charcoal briquettes as an alternative energy source and means of mitigating indoor air pollution in Ethiopia. *Energy for Sustainable Development* 79, 101390. https://doi.org/10.1016/j.esd.2024.101390

Kringel, D. H., Dias, A. R. G., Zavareze, E. da R. & Gandra, E. A. (2020). Fruit wastes as promising sources of starch: Extraction, properties, and applications. *Starch/Staerke*, 72 (3–4). https://doi.org/10.1002/star.201900200

Lee, A., Lan, J. C. W., Jambrak, A. R., Chang, J. S., Lim, J. W. & Khoo, K. S. (2024). Upcycling fruit waste into microalgae biotechnology: Perspective views and way forward. *Food Chemistry: Molecular Sciences*, 8 (February). https://doi.org/10.1016/j.fochms.2024.100203

Lemos, P. V. F., Barbosa, L. S., Ramos, I. G., Coelho, R. E. & Druzian, J. I. (2018). The important role of crystallinity and amylose ratio in thermal stability of starches. *Journal of Thermal Analysis and Calorimetry*, 131 (3), 2555–2567. https://doi.org/10.1007/s10973-017-6834-y

Li, B., Xie, B., Liu, J., Chen, X., Zhang, Y., Tan, L., Wang, Y., Zhu, L., Zhu, K. & Huang, C. (2022). A study of starch resources with high-amylose content from five Chinese mutant banana species. *Frontiers in Nutrition*, 9. https://doi.org/10.3389/fnut.2022.1073368

Lin, L., Yang, H. & Xu, X. (2022). Effects of water pollution on human health and disease heterogeneity: A review. *Frontiers in Environmental Science*, 10 (June). https://doi.org/10.3389/fenvs.2022.880246

Mallampati, R., Xuanjun, L., Adin, A. & Valiyaveettil, S. (2015). Fruit peels as efficient renewable adsorbents for removal of dissolved heavy metals and dyes from water. *ACS Sustainable Chemistry and Engineering*, 3 (6), 1117–1124. https://doi.org/10.1021/acssuschemeng.5b00207

Marichelvam, M. K., Jawaid M. & Asim M. (2019). Corn and rice starch-based bio-plastics as alternative packaging materials. *Fibers*. 7 (4), 32. https://doi.org/10.3390/fib7040032

Matignon, A., Ducept, F., Sieffermann, J. M., Barey, P., Desprairies, M., Mauduit, S. & Michon, C. (2014). Rheological properties of starch suspensions using a rotational rheometer fitted with a starch stirrer cell. *Rheologica Acta*, 53 (3), 255–267. https://doi.org/10.1007/s00397-013-0754-y

Mellinas, C., Ramos, M., Jiménez, A. & Garrigós, M. C. (2020). Recent trends in the use of pectin from agro-waste residues as a natural-based biopolymer for food packaging applications. *Materials*, 13 (3). https://doi.org/10.3390/ma13030673

Nakthong, N., Wongsagonsup, R. & Amornsakchai, T. (2017). Characteristics and potential utilizations of starch from pineapple stem waste. *Industrial Crops and Products*, 105 (March), 74–82. https://doi.org/10.1016/j.indcrop.2017.04.048

Okuthe, G. (2024). Valorizing fruit and vegetable waste: The untapped potential for entrepreneurship in sub-Saharan Africa—A systematic review. *Recycling*, 9 (3). https://doi.org/10.3390/recycling9030040

Orenia, M. R., Collado III, A., Magno, G. M. & Cancino, T. L. (2018). Fruit and vegetable wastes as potential component of biodegradable plastic. *Asian Journal of Multidisciplinary Studies*, 1 (1), 2651–6705.

Otoni, C. G., Azeredo, H. M., Mattos, B. D., Beaumont, M., Correa, D. S., & Rojas, O. J. (2021). The food–materials nexus: Next generation bioplastics and advanced materials from agri-food residues. *Advanced Materials*, 33 (43), 2102520. https://doi.org/10.1002/adma.202102520

Pérez-Pacheco, E., Ortiz-Fernández, A., Ríos-Soberanis, C. R. et al. (2024). Characterization of unconventional sources of starch: Physicochemical and thermal properties research square. PREPRINT (Version 1) *Polymer Bulletin* (In review). https://doi.org/10.21203/rs.3.rs-4745824/v1

Phukan, M. & Nongkhlaw, F. T. (2022). Starch: An overview of its conventional and unconventional sources, pharmaceutical and industrial applications. *Current Trends in Pharmaceutical Research*, 9 (1), 140–161. www.dibru.ac.in/ctprISSN:

Polnaya, F. J., Talahatu, J., Haryadi, H., & Marseno, D. W. (2021). Properties of biodegradable films from hydroxypropyl sago starches. *Asian Journal of Food and Agro-Industry*, 5 (3), 183–192.

Pongsuwan, C., Boonsuk, P., Sermwittayawong, D., Aiemcharoen, P., Mayakun, J. & Kaewtatip, K. (2022). Banana inflorescence waste fiber: An effective filler for starch-based bioplastics. *Industrial Crops and Products*, 180 (12), 1–8.

Punia Bangar, S., Kumar, M. & Whiteside, W. S. (2021). Mango seed starch: A sustainable and eco-friendly alternative to increasing industrial requirements. *International Journal of Biological Macromolecules*, 183, 1807–1817. https://doi.org/10.1016/j.ijbiomac.2021.05.157

Rendón-Villalobos, R., Lorenzo-Santiago, M. A., Olvera-Guerra, R., & Trujillo-Hernández, C. A. (2022). Bioplastic composed of starch and micro-cellulose from waste mango: Mechanical properties and biodegradation. *Polímeros: Ciência e Tecnologia,* 32 (3), e2022026. https://doi.org/10.1590/0104-1428.20210031

Sarmah, J., Choudhury, A., Deka, H., Ganguly, D., Baishya, D. & Jyrwa, R. (2023). A comprehensive overview of edible natural excipients and their potential use in pharmaceutical formulation development. *Journal of Young Pharmacists*, 15 (4), 589–594. https://doi.org/10.5530/jyp.2023.15.83

Seung, D. (2020). Amylose in starch: Towards an understanding of biosynthesis, structure and function. *New Phytologist*, 228 (5), 1490–1504. https://doi.org/10.1111/nph.16858

Siyi, L, Hongna, Z, Chen, W, Xiu-Li, S, Yuanyuan, S, Pengfei, W, et al. (2022). Metabolomics reveal nanoplastic-induced mitochondrial damage in human liver and lung cells. *Environmental Science & Technology,* 56 (17), 12483–12493. https://doi.org/10.1021/acs.est.2c03980.

Solangi, N. H., Kumar, J., Mazari, S. A., Ahmed, S., Fatima, N. & Mubarak, N. M. (2021). Development of fruit waste derived bio-adsorbents for wastewater treatment: A review. *Journal of Hazardous Materials*, 416 (April), 125848. https://doi.org/10.1016/j.jhazmat.2021.125848

Tagliapietra, B. L., Felisberto, M. H. F., Sanches, E. A., Campelo, P. H. & Clerici, M. T. P. S. (2021). Non-conventional starch sources. *Current Opinion in Food Science*, 39, 93–102. https://doi.org/10.1016/j.cofs.2020.11.011

Troncoso, O. P. & Torres, F. G. (2020). Non-conventional starch nanoparticles for drug delivery applications. *Medical Devices & Sensors*, 3 (6), 1–16. https://doi.org/10.1002/mds3.10111.

Ulyarti, U., Lisani, L., Surhaini, S., Lumbanraja, P., Satrio, B., Supriyadi, S. & Nazarudin, N. (2022). The application of gelatinisation techniques in modification of cassava and yam starches using precipitation method. *Journal of Food Science and Technology*, 59 (3), 1230–1238. https://doi.org/10.1007/s13197-021-05134-0

United Nations Climate Change (2023). Briquettes as an alternative to fuel wood and to prevent deforestation – Kigali, Rwanda. UN Global Climate Action Awards. https://unfccc.int/climate-action/momentum-for-change/activity-database/momentum-for-change-briquettes-as-an-alternative-to-fuel-wood-and-to-prevent-deforestation-%E2%80%93-kigali-rwanda

Varghese, S., Awana, M., Mondal, D., Rubiya, M. H., Melethil, K., Singh, A., Krishnan, V. & Thomas B. (2022). Amylose-amylopectin ratio: Comprehensive understanding of structure, physicochemical attributes, and applications of starch. In: Thomas S., AR A., Jose Chirayil C., Thomas B. (Eds.), *Handbook of Biopolymers*, Springer, Singapore. https://doi.org/10.1007/978-981-16-6603-2_48-1

Venkata Mohan, S., Nikhil, G. N., Chiranjeevi, P., Nagendranatha Reddy, C., Rohit, M. V., Kumar, A. N. & Sarkar, O. (2016). Waste biorefinery models towards sustainable circular bioeconomy: Critical review and future perspectives. *Bioresource Technology*, 215, 2–12. https://doi.org/10.1016/j.biortech.2016.03.130

Yaradoddi, J. S., Banapurmath, N. R., Ganachari, S. V., Soudagar, M. E. M., Sajjan, A. M., Kamat, S., Mujtaba, M. A., Shettar, A. S., Anqi, A. E., Safaei, M. R., Elfasakhany, A., Siddiqui, M. I. H. & Ali, M. A. (2021). Bio-based material from fruit waste of orange peel for industrial applications. *Journal of Materials Research and Technology*, *17*, 3186–3197. https://doi.org/10.1016/j.jmrt.2021.09.016

Yin, J, Ju, Y, Qian, H, Wang, J, Miao, X, Zhu, Y, et al. (2022). Nanoplastics and microplastics may be damaging our livers. *Toxics*. 10 (10), 586. https://doi.org/10.3390/toxics10100586.

Yirijor, J. & Bere A. A. T. (2024). Production and characterization of coconut shell charcoal-based bio-briquettes as an alternative energy source for rural communities. *Heliyon*m 10 (16), e35717.
Yu, L., Dean, K. & Li, L. (2006). Polymer blends and composites from renewable resources. *Progress in Polymer Science*, 31 (6), 576–602. https://doi.org/10.1016/j.progpolymsci.2006.03.002.
Zia, F., Zia, K. M., Zuber, M., Kamal, S. & Aslam, N. (2015). Starch based polyurethanes: A critical review updating recent literature. *Carbohydrate Polymers*, 134, 784–798. https://doi.org/10.1016/j.carbpol.2015.08.034

9 Starch-Based Bioplastic from Jackfruit Seed

Alice Singh, Aashita Bhardwaj, Sapna Kumari, Chetna Singh, and Tanya L. Swer

9.1 INTRODUCTION

A huge amount of food gets disposed of every day as waste due to poor handling and low shelf life. Improved preservation techniques would avoid this and lower the quantity of losses. Food product packaging is one of the many integral components of the food industry that helps improve food safety and stability while also making the handling and transportation of products much easier. In recent years, there have been growing social, political, and legislative pressures to develop new products based on renewable sources that can replace present petroleum plastics, whose recycling activity is not yet sufficiently efficient and cheap. Governments are trying to curb the landfill disposal of petroleum-based plastics by encouraging the development of new products as replacements. The growing concerns over synthetic plastic films regarding their negative impacts and pressure to curb their usage have stimulated several studies for the development of alternatives like biodegradable materials or bioplastics (Moreira et al., 2011). These environmentally friendly materials can decompose after fulfilling their intended purposes and can provide a potential solution to replace the traditional petroleum-based plastics (Zoungranan et al., 2020).

Among the biodegradable materials, edible films and coatings serve as the appropriate options for replacing plastics. These are defined as thin layers used as primary food packaging (Shahidi & Hossain, 2020). They do resemble plastic films where they are used as food coatings or barriers between the food and the environment, but they are also edible and are easily degradable (Indrianti et al., 2018). These bioplastics can regulate the biochemical processes within the food, like oxidation and transmission of moisture, aroma, and oil to or from the food, thus resulting in improved quality and enhanced shelf life (Bodana et al., 2024). Moreover, as they come in closest contact with the food, they act as primary packaging, therefore reducing the cost and complexity of packaging systems designed to protect foods (Aguirre-Joya et al., 2018).

The widely used and studied bioplastics are manufactured from biopolymers like cellulose, protein, and starch. Among these, starch, being the most abundant and inexpensive biopolymer in nature, has been gaining attention for fabricating biodegradable films (Singh et al., 2018). Starch-based films have been extensively studied over the decade due to their biodegradable nature as well as technological features suitable for applications in varied industries. The type of starch used, the plasticizers employed, the method for film production, and the incorporation of any additives and their concentrations can all have a big impact on the properties of the films (Janjarasskul et al., 2011). Starches from different sources show diverse characteristics that can significantly affect the film properties, such as barrier, mechanical, and thermal properties (Basiak et al., 2017). However,

DOI: 10.1201/9781003534914-9

a large portion of these starchy materials are often discarded and thrown away during production or consumption, leading to food waste that impacts the economy and environment. One of these by products is jackfruit seed.

The jackfruit seeds constitute about 8–15% of total jackfruit weight and are usually tossed away even when they contain a very high amount of starch. Several studies have shown that jackfruit seed starch can be used for the development of edible films or coatings that can have a wide range of uses in food and allied industries. Additionally, these films are potential carriers to hold antioxidants and antimicrobials, which can be used to improve food quality and safety (Singh et al., 2018), thus enhancing the functionality of the films. Thus, the development of bioplastics derived from the starch of jackfruit seed is a sustainable strategy to promote a circular economy through innovative value-added utilization of jackfruit waste that offers enhanced food safety and stability while simultaneously lowering food packaging costs.

This chapter focuses on jackfruit seed starch-based edible film and coating formulations and their applications. The starch extraction techniques, film development methods, properties and functionality determination, and the potential utilization in the food sector were described.

9.2 JACKFRUIT SEED STARCH

Jackfruit (*Artocarpus heterophyllus*) is one of the largest tropical fruits consumed for its edible flesh. However, its seeds also hold significant value as a promising source of starch, 70–85% db, with unique properties. The starch derived from jackfruit seeds comprise about 25–45% amylose and 45–80% amylopectin, contributing to the unique physicochemical and functional properties. The jackfruit seed starch exhibits desirable characteristics, viz., gelatinization, retrogradation, and water-holding capacity that can offer diverse applications as thickening, binding, and stabilizing agents in food and other allied industries. In addition, jackfruit starch contains resistant starch, a type of dietary fiber that acts as a prebiotic. The functional versatility and health benefits of jackfruit seed starch present it as a valuable resource that can be harnessed for the development of biodegradable plastics.

9.3 EXTRACTION OF STARCH FROM JACKFRUIT SEEDS

Numerous studies have been carried out in recent years to extract starch from jackfruit seeds. These methods adopted by several authors with differing complexity show varying efficiency and purity. The common methods for starch extraction are described below and a detailed look at the advantages and disadvantages of the potential methods for extracting starch from jackfruit seeds are given in Table 9.1.

9.3.1 Traditional Wet-Milling Method

The wet-milling method is one of the most frequently employed techniques used for extracting starch from various seeds (Singh & Singh, 2001). In this method, the jackfruit seeds are soaked in water to soften the seed coat, and thereafter, the outer layer of the seeds is removed by peeling. The seeds are then ground with water to form slurry. The slurry obtained is passed through a fine mesh or cloth for filtering. The starch granules are left to settle at the bottom by sedimentation. The liquid on top is decanted, leaving the starch, which is washed multiple times, followed by drying to a moisture level of 10% wb. This method is a simple and cost-effective process that does not require any complex instruments. However, the starch obtained may contain some residual protein and fiber impurities, thus making it less efficient than other modern techniques (Wong et al., 2021).

TABLE 9.1
Potential Methods for Extraction of Starch from Jackfruit Seeds

Method	Advantages	Disadvantages
Wet milling	Simple, low-cost, and widely used	Lower purity, requires multiple washings
Acidic extraction	High purity, fast process	Acidic waste is crucial to avoid environmental contamination
Alkaline extraction	High yield and purity	Possible chemical contamination, environmental concerns
Enzymatic extraction	High purity starch, efficient protein removal	Expensive enzymes, needs controlled conditions
Ultrasound-assisted extraction	Fast, less water use, increased efficiency	Requires specialized equipment, structural changes
Microwave-assisted extraction	Short extraction time, higher yield	Equipment cost, potential heat degradation

9.3.2 Alkaline Extraction Method

This method involves using an alkali, typically sodium hydroxide (NaOH), to facilitate the extraction of starch from jackfruit seeds (Bobbio et al., 1978). The seeds are peeled and ground into a slurry, which is further treated with dilute NaOH. The alkali causes the dissolution of the proteins and other nonstarch components. The mixture is then neutralized with a weak acid and subjected to filtration to obtain starch after sedimentation (Ma et al., 2005). The jackfruit seed starch is washed multiple times to remove any residual alkali, passed through a 150 µm mesh screen, and then dried to 11% wb moisture level. This process results in a high yield varying between 18.92 and 22.86%, with the purity above 99% (Zhang et al., 2020). However, El Halal et al. (2019) reported a higher yield of about 32% starch from jackfruit seed by the alkaline method.

9.3.3 Enzymatic Extraction Method

In this method, enzymes are employed to break down nonstarch components (such as proteins and lipids) in the jackfruit seed. Enzymatic hydrolysis of the nonstarch components results in enhanced recovery and purity of the starch. The seeds are cleaned, peeled, and ground into slurry. The slurry is treated with enzymes like a-amylase, proteases to break down proteins, or cellulases to degrade fibers for a predefined period. After enzymatic treatment, the slurry is filtered, and the starch is separated through sedimentation. The extracted starch undergoes further purification through washing before being dried. This is more costly due to the use of enzymes, and it requires carefully controlled conditions for effective enzyme activity. Protease was used to extract starch from jackfruit seeds resulting in better yield and purity than alkaline extraction (Noor et al., 2014).

9.3.4 Other Methods for Starch Extraction Methods

9.3.4.1 Acidic Extraction

The acid extraction method for isolating starch from any material uses hydrochloric acid (HCl) to break down nonstarch components, improving the separation and purity of starch granules (Wong et al., 2021). In this method, the material is treated with dilute HCl to hydrolyze proteins and fibers. After acid treatment, the solution is neutralized with sodium hydroxide (NaOH), and the starch is separated, washed, and dried (Singh & Singh, 2001). This method yields high-purity starch and accelerates extraction compared to traditional techniques. However, it requires careful control of acid concentration to avoid starch degradation and safe handling of corrosive chemicals to prevent safety risks and equipment corrosion (Palacios-Fonseca et al., 2013).

9.3.4.2 Ultrasound-Assisted Extraction

This method employs ultrasound waves to help break down cell walls and enhance starch release from the seed matrix. The slurry is subjected to ultrasonic treatment that generates cavitation bubbles, which cause disruption of the structure. Following this treatment, the slurry is filtered, and the starch is separated through sedimentation and washing. This technique improves starch extraction efficiency while reducing the extraction time. It provides a greener approach with minimal water and chemical requirements compared to traditional methods. However, it requires specialized equipment and may lead to structural changes in the starch due to the effects of cavitation (Resendiz-Vazquez et al., 2017).

9.3.4.3 Microwave-Assisted Extraction

This method employs microwave energy to speed up the extraction process by quickly heating the seed material, facilitating the breakdown of the seed matrix and the release of starch granules. Material will be suspended in water and followed by microwave treatment for a predesignated time. The starch is separated through filtration and sedimentation processes followed by repeated washing, and finally dried. This technique allows for a shorter extraction time compared to traditional methods and can yield higher starch amounts. However, it requires access to microwave extraction systems, and there is a risk of thermal degradation of starch if temperature control is not maintained (Araújo et al., 2020).

9.4 FORMULATION AND FORMATION OF BIOPLASTIC FILMS

The development of bioplastic films from jackfruit seed starch has gained attention in recent research due to the increasing demand for sustainable materials. Here are the formulation and formation processes, with examples of recent studies.

9.4.1 Formulation for Preparation of Starch-Based Bioplastic

The extraction process can impact the mechanical and barrier properties of the resulting bioplastic, making it essential to optimize the extraction process to obtain the highest starch quality from jackfruit seeds. Jackfruit seed starch is dispersed in distilled water with other additives like plasticizers for preparation of the formulation. Plasticizers such as glycerol, sorbitol, or citric acid are commonly added in bioplastic formulations using starch in order to improve flexibility and reduce brittleness. This is crucial, as pure starch films are often brittle without these additives. The concentration of plasticizer must again be optimized because it can influence the tensile strength, transparency, and water resistance of the bioplastic film formed. Studies have explored the influence of different plasticizers such as glycerol and sorbitol on jackfruit seed starch films and discusses optimal formulations for enhanced flexibility and water resistance of the film formed with glycerol (Lestari et al., 2020). The formulation is heated up to 95°C for approximately 5 min to gelatinize the starch and obtain a sol with a gel-like appearance (Santana et al., 2018).

9.4.2 Film Formation

Starch-based films are developed through several methods, each designed to optimize the film's properties for specific applications (Khan et al., 2016). Among these methods, solvent casting is the only technique employed so far by various authors for the development of jackfruit seed starch-based films at laboratory set-up.

9.4.2.1 Solvent Casting Technique

Solvent casting is one of the simplest and most common techniques for developing starch-based films. In this method, the sol obtained after heating the mixture of starch and plasticizer is poured onto a flat surface or mold and left to dry at a controlled temperature and humidity. The drying temperature and time influence the structural properties of the film, including thickness, density, and uniformity. A plasticizer (e.g., glycerol, sorbitol) is added to reduce brittleness (Rahmawati et al., 2023). This technique is easy and inexpensive, suitable for small-scale lab testing and film property evaluation. The problem faced is that the process is time-consuming with a prolonged drying process posing challenges to scale up for industrial production (Sanyang et al., 2015).

9.4.2.2 Extrusion

Extrusion is a commonly used method in the industrial production of starch-based films (Suhag et al., 2020). It involves heating and forcing the starch mixture through a die to form thin sheets or films. The starch is blended with plasticizers and water to create a dough-like mixture. The mixture is fed into an extruder where it undergoes high heat and pressure (Calderón-Castro et al., 2018). The melted starch is forced through a narrow die to form a thin film, which is then cooled and dried. It is suitable for large-scale production and is faster than casting method of extraction. It provides good mechanical properties. It requires specialized equipment as the high temperatures may degrade starch structure (Peressini et al., 2003)

9.4.2.3 Compression Molding

In this technique, starch mixture is placed in a mold and then compressed under high pressure and temperature to form films. The starch is mixed with water and plasticizers to form a dough. This mixture is placed in a mold and subjected to high temperature and pressure in a compression molding machine (Muller et al., 2017). After molding, the film is cooled to maintain its shape. It can produce thicker films with consistent properties and requires no solvent drying (Wu et al., 2023).

9.4.2.4 Thermoplastic Starch Processing

Thermoplastic starch (TPS) is developed by casting aqueous dispersion containing starch and water-soluble plasticizers under heat, converting it into a moldable material similar to conventional thermoplastics (Zhang, 2023). The TPS mixture can be processed into films through extrusion or injection molding (Roz et al., 2006). For this process parameters such as screw speed, moisture content of feed, barrel temperature etc. are important as they influence the TPS product. Starch is heated and mixed with plasticizers, such as glycerol, which lowers its glass transition temperature, allowing it to melt. The TPS extrudates are blown into films, sheets, or desired shapes (Thunwall et al., 2008). This process produces flexible, plastic-like films; compatible with industrial thermoplastic processing techniques. These are highly water sensitive and often needs to be blended with other water-soluble biopolymers for improved durability (Ma et al., 2005).

9.4.2.5 Electrospinning

Electrospinning is an emerging technique used for fabricating ultra-thin films and nanofibers from starch-based materials (Zhu et al., 2018). A starch solution is prepared and loaded into a syringe. A high-voltage electric field is applied to the solution, which is stretched to form nanofibers that are collected as a thin film (Thenmozhi et al., 2017). It produces very thin, uniform films with large surface area. It also has potential for various applications in food packaging. It requires precise control of conditions as the equipment and its setup is at costlier side (Kumar et al., 2019).

All these film formation methods have demonstrated potential in improving water resistance and mechanical strength. Among them, solvent casting and extrusion are the most widely used, as they offer comparatively better properties, making them more suitable for food packaging applications.

TABLE 9.2
Methods for Preparation of Starch-Based Bioplastics

Methods	Advantages	Disadvantages
Solvent casting	Simple, low-cost, suitable for lab testing	Slow drying; not ideal for large scale
Extrusion	High throughput; industrially scalable	Requires specialized equipment
Compression molding	Consistent thickness; no solvent needed	Requires high temperature and pressure
Thermoplastic starch (TPS)	Flexible, plastic-like films	High water sensitivity
Electrospinning	Produces ultra-thin films; high surface area	Expensive setup; requires precise conditions

However, other methods need further research to assess their practical effectiveness across different types of foods. The advantages and disadvantages of the common methods for preparing starch-based bioplastics are outlined in Table 9.2.

Each method provides unique advantages for creating starch-based films with specific characteristics, depending on the intended application, desired properties, and scale of production. Researchers continue to explore these techniques to optimize starch-based films for sustainable alternatives in packaging, agriculture, and biomedical fields.

9.5 PROPERTIES OF THE STARCH-BASED BIOPLASTICS FROM JACKFRUIT SEEDS

Jackfruit starch-based bioplastics have drawn enough attention because of their environmental impact and sustainability (Lestari et al., 2020). Understanding the properties of starch-based bioplastics from jackfruit seeds is important because of the growing need for sustainable alternatives to petroleum-based plastics. By understanding the properties of the bioplastics, it is possible to characterize their quality attributes, potential, and impact. This would help to improve the functional limitations of starch-based bioplastics, such as low water resistance and mechanical strength. Starch-based films lack these qualities when compared with other polymers; hence, several modifications have been done to improve the desired features of the film, such as starch modification (Bangar et al., 2022), plasticizer addition (Sharma et al., 2023), and the addition of essential oils (Singh et al., 2022), nanoparticles (Hashem et al., 2023), and solvents (Gupta et al., 2021). Better optimization of bioplastics based on jackfruit seed starch will be beneficial for several useful applications, such as food packaging, disposable cutlery, and even medical applications like biodegradable films or capsules. Some key properties of starch-based bioplastics made from jackfruit seeds include thermal, mechanical, and barrier properties.

9.5.1 Thermal Properties

The thermal properties of bioplastics, such as heat resistance, glass transition (T_g), melting temperature (T_m), thermal degradation, thermal conductivity (k), and heat capacity (C_p) are crucial in determining their performance, elasticity, and strength at various temperatures. Thermal properties are usually assessed by differential scanning calorimetry (DSC) and thermogravimetric analysis (TGA). TGA measures the thermal stability of films at higher temperatures, whereas DSC provides information about the melting temperature, glass transition, and crystallization temperature of starch-based films. These properties help assess the suitability of bioplastics for food packaging applications, especially for temperature-sensitive products requiring thermal stability (Santhosh et al., 2024).

Thermal degradation of starch-based films occurs in three distinct stages: the first stage is attributed to volatilization of water and glycerol, second stage to degradation of starch and other components and the third and final stage to complete decomposition of film matrix, usually above

400°C (Kahar et al., 2019; Wu et al., 2023). Chandran et al. (2023) observed that jackfruit starch blends had maximum degradation temperature around 295.2°C with about 50% weight loss. These results indicate that Jackfruit starch films show excellent thermal properties, suggesting their potential for commercial applications. Incorporation of pomegranate peel extract (Bodana et al., 2024), fillers-*Boswellia serrata* (Reshmy et al., 2021), and zinc nanoparticles (Santhosh & Sarkar, 2022) into jackfruit starch films significantly improved their thermal properties. Similar trends for thermal degradation was observed in bioplastics prepared with rice starch (Piyada et al., 2013) and corn starch (Ates & Kuz, 2020). Incorporation of nano-chitosan and plasticizer were found to improve the thermal stability of potato starch-guar gum nanocomposite films (Islam et al., 2020).

9.5.2 Mechanical Properties

The mechanical properties, such as tensile strength, elongation at break, young's modulus, and puncture/tear resistance are some of the important mechanical properties of bioplastics concerning their commercial applications. They determine the strength, durability and structure of the material (Santhosh et al., 2024). Tensile strength of starch-based bioplastics is generally lower as compared to conventional plastics, but their mechanical properties can be improved or modified by the use of different biopolymers, gums, plasticizers, and fillers (Santana et al., 2024; Arismendi et al., 2013).

Mechanical properties of starch-based films are influenced by starch content, plasticizers, gelatinization degree, and processing methods. Plasticizers like glycerol and sorbitol can affect the tensile strength and elongation at break of the films by decreasing intermolecular strength and increasing chain mobility (Santana et al., 2018; Lestari et al., 2020). Jackfruit-based bioplastics have shown remarkable mechanical properties. Santana et al. (2018) observed that the addition of 30% glycerol presented the highest tensile strength (3.12 MPa). The addition of polyvinyl alcohol (PVA) and sorbitol plasticizer to jackfruit seed starch films resulted in high tensile strength (Lestari et al., 2020). Adding more filler to the film was also found to increases tensile strength as starch molecules form compact bonds with fillers, making it less stretchable (Lubis et al., 2018). Similar results were observed by Lopez and Garcia (2012) for Ahipa and cassava starch films with similar glycerol concentrations. Incorporation of nanofillers and plasticizers enhanced the mechanical properties of bioplastics derived with rice starch (Piyada et al., 2013), potato starch (Islam et al., 2020), and corn starch (Ates & Kuz, 2020) making them suitable for food packaging applications.

9.5.3 Barrier Properties

Barrier properties of starch-based films, such as water vapor transmission rate (WVTR) and oxygen transmission rate (OTR), determine their suitability in various food packaging applications. In order to enhance food safety, these attributes provide protection against moisture, gases, light, and contaminants. Barrier properties of bioplastics are highly influenced by the use of starch concentrations, plasticizers, and additives (Santhosh et al., 2024). The water vapor permeability of starch-based films is relatively higher, which limits their use in applications requiring moisture resistance (Santana et al., 2018; Santana et al., 2024).

Studies on jackfruit starch bioplastics have shown that water vapor permeability is affected by starch and glycerol concentrations. Water molecules are transferred through the starch films primarily by adsorption, diffusion, and desorption. The microstructure formed by the chemical composition, crystallinity, and plasticizer influences these mechanisms (Santana et al., 2018; Lubis et al., 2017). Incorporation of lysozyme was also found to increase the permeability of the bioplastic (Bonomo et al., 2018). Few studies stated that water permeability of jackfruit starch bioplastic was affected by starch and glycerol concentrations (Santana et al., 2018; Lubis et al., 2017). Similar trends were observed by Piyada et al. (2013) and Islam et al. (2020) for rice starch and potato starch films, where the addition of starch nanoparticles and different types of plasticizers and nanoclays reduced the water permeability of films, enabling their use as food packaging materials.

9.6 IMPROVING THE FUNCTIONALITY OF STARCH-BASED BIOPLASTRICS FROM JACKFRUIT SEED

As a natural polysaccharide, jackfruit starch serves as a sustainable and biodegradable substitute for conventional plastics. However, considerable advancements in their functional characteristics are imperative to enable jackfruit starch bioplastics to compete successfully with synthetic counterparts. Enhancing the functionality of jackfruit starch bioplastic requires a multifaceted approach aimed at improving its characteristics, including tensile strength, elongation at break, durability, and many more. The following strategies outline potential avenues for improvement.

9.6.1 Types of Starch

Different types of starch are added in film formulations to improve mechanical, barrier, and transparency of the bioplastics. The types of starch that have been employed by different researchers for the development of films include cross-linked, oxidized and esterified starch. Cross-linked starch films were opaque and had better water vapor resistance, tensile strength, and young modulus among all the modifications due to high intermolecular interaction in the polymer matrix (Dai et al., 2019). In the study investigated by Kahar et al. (2019), different starch from jackfruit seeds such as native starch (NS), TPS, and cross-linked starch was used to formulate bioplastics. Cross-linking was done by citric acid, known as citric acid starch (CAS), and was combined with recycled high-density polyethylene (rHDPE) through a melt blending technique. The maximum tensile strength recorded was 23.3 MPa for CAS at a ratio of 90% rHDPE and 10% starch. This is attributed to the modification of TPS with 1% citric acid which facilitated cross-linking by forming ester bonds that led to excellent interfacial adhesion in the matrix. For the similar reason, elongation at break value also seemed to increase. On increasing the starch concentration, tensile strength decreased due to agglomeration in the matrix. The 60/40 ratio of rHDPE and starch was characterized by the highest Young's modulus and improved thermal stability.

9.6.2 Plasticizers

Plasticizers are added in film formulations to improve the flexibility and processability of the films (Vieira et al., 2010). The plasticizers that have been employed by different researchers for development of films from jackfruit seeds include glycerol, sorbitol, citric acid, and sodium bicarbonate polyvinyl acetate (Badona et al., 2024; Nguyen et al., 2022; Rahmawati et al., 2022; Lestari et al., 2020). In the work investigated by Nguyen et al. (2022), different plasticizers such as water, glycerol, citric acid, and sodium bicarbonate in varying concentrations were utilized to produce four distinct types of bioplastics, enabling an investigation into the influence of the plasticizers and the characterization of the resultant bioplastic properties. The findings demonstrated that as the glycerol concentration increased, the films became softer and more pliable while on decreasing the concentration of glycerol the films became firmer and less flexible. Augmenting the citric acid content enhanced the hardness but diminished the plasticity of the films. If the thickness of the film exceeded the optimal levels, it would likely experience cracking as a consequence of shrinkage. Lestari et al. (2020) on the other hand, utilized different plasticizers such as PVA, glycerol, and sorbitol for the formulation of starch films derived from jackfruit seed. Their findings indicated that increasing the glycerol and sorbitol concentrations significantly reduces the intermolecular strength of the films which further decreases their tensile strength. Increased PVA concentration, on the other hand, greatly improved the tensile strength of the films due to the intensive hydrogen bonding between molecules. On the contrary, glycerol and sorbitol inclusion improved the elongation value resulting in superior flexibility of films while PVA inclusion had a negative impact on elongation value reducing the flexibility. While increasing the sorbitol concentration, elongation increased due to reduced mobility between molecules attributable to low intermolecular forces. Overall, inclusion of plasticizers improved the hydrophilicity and biodegradability of the films.

9.6.3 Nanoparticles

Nanoparticles added to the edible films do not only improve the properties of the films but also improve the shelf life of the foods. Zinc oxide nanoparticles (ZNPs), FDA approved GRAS, have shown to improve the mechanical strength, barrier properties and thermal stability while also enhancing the antimicrobial and antioxidant activity of jackfruit seed starch-based composite films (Santhosh & Sarkar, 2022). The authors prepared a composite film from jackfruit seed starch by incorporating xyloglucan from tamarind kernels and ZNPs ranging from 1 to 5%. The incorporation of ZNPs led to an increase in thickness and roughness of the films with decreasing moisture content as it exhibited increased hydrophobic properties. This contributed to an enhancement in tensile strength. ZNPs also reduced the light transmittance of films, providing good barrier properties for food products. Varying concentrations of ZNPs had a significant effect on the interfacial adhesion, mobility, and rigidity of polymers, thus affecting the viscosity and rheology of the film's solution as well as the water solubility and glass transition temperature of the films. Furthermore, the incorporation of ZNPs exhibited a significant contribution to the antimicrobial properties against *Staphylococcus aureus* and *Escherichia coli* with the highest zone of inhibition obtained at 5% ZNPs.

9.6.4 Blending with Natural Polymers

Natural polymers are added in the film matrix to enhance mechanical and barrier properties, increase elastic modulus, and decrease elongation at break (Tabasum et al., 2018). The natural polymers that have been employed by different researchers for development of bioplastics from jackfruit seeds include microcrystalline cellulose (MCC) and chitosan (Kane et al., 2016; Lestari et al., 2020; Lubis et al., 2018). In the study conducted by Lubis et al. (2018) bioplastics were developed from jackfruit seed starch incorporating MCC sourced from cocoa pod husk using glycerol as plasticizer. A jackfruit seed starch to MCC ratio of 8:2, incorporating 20% glycerol, resulted in an improved tensile strength because of a high degree of polymerization between the MCC and starch molecules, which also contributed to the homogeneity of bioplastic. Kane et al. (2016) on the other hand utilized chitosan as filler for the formulation of starch bioplastic derived from jackfruit seeds. Chitosan and sorbitol as a plasticizing agent were combined in different concentrations to formulate bioplastics. As the concentration of chitosan increased, tensile strength increased due to the formation of hydrogen bonds, a more compact polymer structure and improved homogeneity. The optimal conditions were identified at a starch-to-chitosan ratio of 8:2 (w/w), with 25% concentration of sorbitol resulting in a tensile strength of 13.524 MPa.

9.6.5 Blending with Bioactive Compounds

9.6.5.1 Antimicrobials

The incorporation of antimicrobials into edible films enhances barrier, optical, and mechanical properties. It also plays a significant role in preserving fresh produce and safeguarding against spoilage (Bhatia & Bharti, 2015; Nizam et al., 2021). An innovative study by Bonomo et al. (2018) integrated lysozyme into the development of bioplastic derived from jackfruit seed starch, employing glycerol as a plasticizing agent. The biodegradable films exhibited an appealing appearance and demonstrated antimicrobial properties against *Micrococcus lysodeikticus*. The incorporation of lysozyme resulted in increased thickness, enhancement of the permeability with prominent effect on tensile strength and elongation at break of the biodegradable films. The findings indicated that jackfruit starch bioplastics supplemented with lysozyme exhibit significant promise for application in the packaging industry. Incorporating pomegranate peel in jackfruit seed bioplastic inhibits the growth of gram-positive and gram-negative bacteria due to the presence of bioactive compounds like polyphenols (Ali et al., 2019).

9.6.5.2 Antioxidants

Antioxidants added to the edible films and coatings enhance antioxidant activity, barrier properties, especially against UV radiation, surface hydrophobicity, and elastic modulus (Piñeros-Hernandez et al., 2017). Rahmawati et al. (2023) incorporated extracts from the rosella flower (1.5–3%) in jackfruit seed film utilizing sorbitol as a plasticizer. The rosella flower extracts imparted antioxidative activity of the bioplastic while augmenting its resistance to water. The study advocates for the incorporation of rosella flower extract into jackfruit starch bioplastic as a strategy to mitigate oxidation in high moisture food. Ali et al. (2019) demonstrated enhanced stiffness, young modulus, and tensile strength of the jackfruit seed films by incorporating pomegranate peel. In the work investigated by Bodana et al. (2024) varying concentration of extract from pomegranate peel (0.02–0.1 g/mL) was utilized in the formation of bioplastic from jackfruit starch 5% (w/w) with glycerol (3% w/w) as a plasticizer. Increasing concentration of the extract resulted in a corresponding increase in thickness, oxygen and water vapor permeability, and oxidative activity while reducing the solubility index. The film containing 0.04 g/mL extract demonstrated superior thermal stability and tensile strength, beyond which it declined due to insufficient interaction between the polymer matrix.

9.6.5.3 Anthocyanins

The addition of anthocyanins to edible films contributes to the improvement of the mechanical and barrier properties and provides a mean to monitor the freshness of food items. Apart from this, anthocyanins provide an array of potential health benefits, which include antioxidant, anticancer, antimicrobial, and anti-inflammatory activities (Qin et al., 2020). A study conducted by da Costa et al. (2020) incorporated starch from jackfruit seed and anthocyanins from black grape skin to gauge the freshness of fish and shrimp acting as a potential smart film to be used in the commercial market. The film changed its color from reddish pink to green during the storage of the sea foods due to the release of amines during storage which caused a color change due to pH variation as they react with anthocyanin in the film (Huang et al., 2019). However, increased concentration of anthocyanin enhanced the solubility owing to the presence of hydroxyl groups in anthocyanins, which facilitates interaction with water molecules through hydrogen bonding. This negatively effects the barrier properties and tensile strength due to the propensity for dissolution.

9.7 STARCH-BASED BIOPLASTICS FROM JACKFRUIT SEED FOR FOOD PACKAGING APPLICATION

A detailed analysis of recent scientific studies shows that majority applications of the jackfruit seed starch-based bioplastics have only been carried out in laboratory settings rather than at an industrial scale. Figure 9.1 summarizes the key functionalities of jackfruit seed starch-based bioplastics in food packaging whereas Table 9.3 describes the effects of jackfruit seed starch-based packaging for food applications at lab scale level.

The coating of white grapes with jackfruit seed starch containing 0.1 g/mL of pomegranate peel extract reduced the weight loss of about 26.05%. Further, the coating enhanced the shelf life of the grapes up to 8 days under 30°C and 70% humidity storage conditions while maintaining the firm texture, color profile, and sensory attributes of the grapes (Bodana et al., 2024). Further, guava fruits coated with jackfruit seed starch (2%) incorporated with chitosan (2%) and alginate (2%) showed

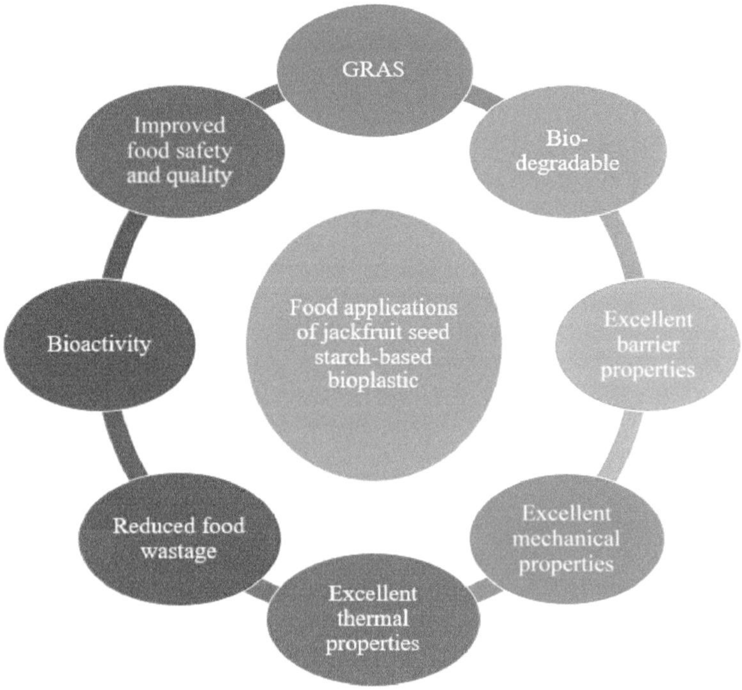

FIGURE 9.1 Functionalities of jackfruit seed starch-based bioplastic for food applications.

TABLE 9.3
Application of Jackfruit Seed Starch-Based Film in Several Food Products

Product	Other additives	Effects	References
Grapes	Pomegranate peel extract	Enhanced the shelf life of grapes at ambient conditions for up to 8 days of storage period	Bodana et al. (2024)
Guava	Chitosan and alginate	Reduced the weight loss of fruit by 2.5% after 20 days at 20°C	Rodrigues et al. (2018)
Mango	-	Reduced fruit weight loss by 6.11% on day 12 and also deferred yellow hue of fruit skin	Rodrigues et al. (2020)
Potato	Rosella flower extract	The coating prolonged the shelf life up to 8 days at 16°C and 6 days at 30°C	Ningsih et al. (2021)
Tomato	Tamarind kernel xyloglucan and zinc oxide nanoparticles	The combination of these two additives decreased the weight loss and augmented the firmness up to 15 days	Santhosh & Sarkar (2022)
Fish and shrimps	Anthocyanins	The film changed color after 2 days which acted as an indicator of freshness of the samples	da Costa et al. (2020)

reduced weight loss significantly by 2.5% than with jackfruit seed starch alone. This might be due to enhanced association between chitosan and alginate which made the coating less permeable to water loss, gases, and thereby reduced the metabolic rate of fruits. Fruits coated with starch or chitosan alone enhanced the firm texture and maintained the color attributes, hence increased consumer acceptability even after 16 days away from the refrigerated condition (Rodrigues et al., 2018). The jackfruit seed starch (3.5%) coating of mango stored at 24°C and 87% RH reduced the weight loss and ripening by reducing the respiration rate which resulted in lagged transition to yellow tint of the fruit (Rodrigues et al., 2020). Jackfruit seed starch incorporated with xyloglucans from tamarind kernel and ZNPs (3%) were coated on tomatoes which resulted in barrier property and increased shelf life of up to 15 days. The coated sample exhibited a lower weight loss, i.e. 40% lower than the uncoated tomato, with retention of firmness and lycopene throughout storage study (Santhosh & Sarkar, 2022).

The jackfruit seed film containing rosella flower extract increased the shelf life of potato slices for up to 8 days at 16°C and 6 days at 30°C. The findings suggested potato slices wrapped in film had a higher shelf life than unwrapped or packaged with parchment paper. The extract of rosella flower contains antioxidants that inhibit the process of oxidation occurring in slices. The slices appeared to be slightly wrinkled but there were no molds or any other discoloration (Ningsih et al., 2021). Jackfruit seed starch infused with anthocyanins from black grape was used as a covering material in the beaker containing fish and shrimps. The films acted as smart films as it changed its color indicating the freshness of the samples. The color change from reddish pink to green or blue due to the release of volatile compounds like trimethylamine, ammonia, and other amines which have a basic pH. The film containing fish turned green and the film with shrimps turned blue after 10 days of storage period as these basic compounds react with anthocyanins to form chalcone. These smart films have great potential in getting commercial because of their cost-effectiveness and convenience to check the freshness of food products (da Costa et al., 2020).

9.8 CONCLUSION AND FUTURE PROSPECTS

The jackfruit seed starch-based bioplastics have demonstrated promising alternative for packaging food applications. The several studies reviewed on chemical and physical modifications, the development of composites, the use of additives to create active and/or intelligent films, etc., have all claimed to significantly improve the quality of the jackfruit seed starch-based films. However, further research and development are essential to achieve the properties comparable to those of the conventional plastic materials. Such is the use of novel plasticizers, namely, deep eutectic solvents, which can improve the processibility and properties of jackfruit seed starch-based bioplastics. In contrast to the traditional plasticizers like glycerol or sorbitol, etc., these green plasticizers have been demonstrated to offer several advantages including better processability under extrusion conditions, increased starch dissolution and plasticization, resistance to retrogradation with improved film flexibility. Further, it is essential to introduce new cutting-edge technological methods to develop inventive jackfruit seed starch-based packaging. Emerging technologies such as cold plasma, pulsed electric field, ultrasound, and HPP can be used to modify starch-based raw materials, creating environmentally friendly packaging that maintains product quality and extends shelf life. Further, preparation techniques like 3D printing, electrospinning/electrospraying, reactive extrusion, and nanotechnology are also promising methods for development of starch-based active packaging using jackfruit seed.

REFERENCES

Aguirre-Joya, J. A., De Leon-Zapata, M. A., Alvarez-Perez, O. B., Torres-León, C., Nieto-Oropeza, D. E., Ventura-Sobrevilla, J. M., Aguilar, M. A., Ruelas-Chacón, X., Rojas, R., Ramos-Aguiñaga, M. E., & Aguilar, C. N. (2018). Basic and applied concepts of edible packaging for foods. In *Elsevier eBooks* (pp. 1–61). https://doi.org/10.1016/b978-0-12-811516-9.00001-4

Ali, A., Chen, Y., Liu, H., Yu, L., Baloch, Z., Khalid, S., Zhu, J., & Chen, L. (2019). Starch-based antimicrobial films functionalized by pomegranate peel. *International Journal of Biological Macromolecules*, *129*, 1120–1126. https://doi.org/10.1016/j.ijbiomac.2018.09.068

Araújo, R. G., Rodríguez-Jasso, R. M., Ruiz, H. A., Govea-Salas, M., Rosas-Flores, W., Aguilar-González, M. A., Pintado, M. E., Lopez-Badillo, C., Luevanos, C., & Aguilar, C. N. (2020). Hydrothermal–microwave processing for starch extraction from Mexican avocado seeds: Operational conditions and characterization. *Processes*, *8* (7), 759. https://doi.org/10.3390/pr8070759

Arismendi, C., Chillo, S., Conte, A., Del Nobile, M. A., Flores, S., & Gerschenson, L. N. (2013). Optimization of physical properties of xanthan gum/tapioca starch edible matrices containing potassium sorbate and evaluation of its antimicrobial effectiveness. *LWT – Food Science and Technology*, *53* (1), 290–296. https://doi.org/10.1016/j.lwt.2013.01.022

Ateş, M., & Kuz, P. (2020). Starch-based bioplastic materials for packaging industry. *Journal of Sustainable Construction Materials and Technologies*, *5* (1), 399–406. https://doi.org/10.29187/jscmt.2020.44

Bangar, S. P., Whiteside, W. S., Suri, S., Barua, S., & Phimolsiripol, Y. (2022). Native and modified biodegradable starch-based packaging for shelf-life extension and safety of fruits/vegetables. *International Journal of Food Science & Technology*, *58* (2), 862–870. https://doi.org/10.1111/ijfs.16219

Basiak, E., Lenart, A., & Debeaufort, F. (2017). Effect of starch type on the physico-chemical properties of edible films. *International Journal of Biological Macromolecules*, *98*, 348–356. https://doi.org/10.1016/j.ijbiomac.2017.01.122

Bhatia, S., & Bharti, A. (2015). Evaluating the antimicrobial activity of nisin, lysozyme and ethylenediaminetetraacetate incorporated in starch based active food packaging film. *Journal of Food Science and Technology*, *52* (6), 3504–3512. https://doi.org/10.1007/s13197-014-1414-7

Bobbio, F., El Dash, A., Bobbio, P., & Rodrigues, L. (1978). Isolation and characterization of the physicochemical properties of the starch of jackfruit seeds (*Artocarpus heterophyllus*). *Cereal Chemistry*, *55*, 505–511.

Bodana, V., Swer, T. L., Kumar, N., Singh, A., Samtiya, M., Sari, T. P., & Babar, O. A. (2024). Development and characterization of pomegranate peel extract-functionalized jackfruit seed starch-based edible films and coatings for prolonging the shelf life of white grapes. *International Journal of Biological Macromolecules*, *254* (P3), 127234. https://doi.org/10.1016/j.ijbiomac.2023.127234

Bonomo, R. C. F., Santos, T. A., Santos, L. S., da Costa, R. I. F., Rodrigues, L. B., dos Santos Pires, A. C., Veloso, C. M., Gandolfi, O. R. R., & Bonomo, P. (2018). Effect of the incorporation of lysozyme on the properties of jackfruit starch films. *Journal of Polymers and the Environment*, *26* (2), 508–517. https://doi.org/10.1007/s10924-016-0902-4

Calderón-Castro, A., Vega-García, M. O., De Jesús Zazueta-Morales, J., Fitch-Vargas, P. R., Carrillo-López, A., Gutiérrez-Dorado, R., Limón-Valenzuela, V., & Aguilar-Palazuelos, E. (2018). Effect of extrusion process on the functional properties of high amylose corn starch edible films and its application in mango (*Mangifera indica* L.) cv. Tommy Atkins. *Journal of Food Science and Technology*, *55* (3), 905–914. https://doi.org/10.1007/s13197-017-2997-6

Chandran, A. S., M., Ushamani, M., M. V., Krishnaraj (2023). On the mechanical, thermal and biodegradation of jackfruit seed starch bioplastic. *International Journal of Advanced Research in Science Communication and Technology*, *3* (1),15–22. https://doi.org/10.48175/ijarsct-11203

da Costa, L. A., Diógenes, I. C. N., Oliveira, M. D. A., Ribeiro, S. F., Furtado, R. F., Bastos, M. D. S. R., Silva, M. a. S., & Benevides, S. D. (2020). Smart film of jackfruit seed starch as a potential indicator of fish freshness. *Food Science and Technology*, *41* (2), 489–496. https://doi.org/10.1590/fst.06420

Dai, L., Zhang, J., & Cheng, F. (2019). Effects of starches from different botanical sources and modification methods on physicochemical properties of starch-based edible films. *International Journal of Biological Macromolecules*, *132*, 897–905. https://doi.org/10.1016/j.ijbiomac.2019.03.197

El Halal, S. L. M., Kringel, D. H., Zavareze, E. D. R., & Dias, A. R. G. (2019). Methods for extracting cereal starches from different sources: A review. *Starch – Stärke*, *71* (11–12), 1900128. https://doi.org/10.1002/star.201900128

Gupta, V., Thakur, R., & Das, A. B. (2021). Effect of natural deep eutectic solvents on thermal stability, syneresis, and viscoelastic properties of high amylose starch. *International Journal of Biological Macromolecules*, *187*, 575–583. https://doi.org/10.1016/j.ijbiomac.2021.07.099

Hashem, A. H., El-Naggar, M. E., Abdelaziz, A. M., Abdelbary, S., Hassan, Y. R., & Hasanin, M. S. (2023). Bio-based antimicrobial food packaging films based on hydroxypropyl starch/polyvinyl alcohol loaded with the biosynthesized zinc oxide nanoparticles. *International Journal of Biological Macromolecules*, *249*, 126011. https://doi.org/10.1016/j.ijbiomac.2023.126011

Huang, S., Xiong, Y., Zou, Y., Dong, Q., Ding, F., Liu, X., & Li, H. (2019). A novel colorimetric indicator based on agar incorporated with *Arnebia euchroma* root extracts for monitoring fish freshness. *Food Hydrocolloids*, 90, 198–205. http://dx.doi.org/10.1016/j.foodhyd.2018.12.009

Indrianti, N., Pranoto, Y., & Abbas, A. (2018). Preparation and characterization of edible films made from modified sweet potato starch through heat moisture treatment. *Indonesian Journal of Chemistry*, 18 (4), 679–687.

Islam, H. B. M. Z., Susan, Md. A. B. H., & Imran, A. B. (2020). Effects of plasticizers and clays on the physical, chemical, mechanical, thermal, and morphological properties of potato starch-based nanocomposite films. *ACS Omega*, *5* (28), 17543–17552. https://doi.org/10.1021/acsomega.0c02012

Janjarasskul, T., Min, S. C., & Krochta, J. M. (2011). Storage stability of ascorbic acid incorporated in edible whey protein films. *Journal of Agricultural and Food Chemistry*, *59* (23), 12428–12432. https://doi.org/10.1021/jf201629r

Kahar, A. W. M., Lingeswarran, M., Amirah Hulwani, M. Z., & Ismail, H. (2019). Plasticized jackfruit seed starch: A viable alternative for the partial replacement of petroleum-based polymer blends. *Polymer Bulletin*, *76* (2), 747–762. https://doi.org/10.1007/s00289-018-2402-2

Kane, S. N., Mishra, A., & Dutta, A. K. (2016). Preface: International Conference on Recent Trends in Physics (ICRTP 2016). *Journal of Physics: Conference Series*, *755* (1), 1–7. https://doi.org/10.1088/1742-6596/755/1/011001

Khan, B., Niazi, M. B. K., Samin, G., & Jahan, Z. (2016). Thermoplastic starch: A possible biodegradable food packaging material—a review. *Journal of Food Process Engineering*, *40* (3). https://doi.org/10.1111/jfpe.12447

Kumar, T. S. M., Kumar, K. S., Rajini, N., Siengchin, S., Ayrilmis, N., & Rajulu, A. V. (2019). A comprehensive review of electrospun nanofibers: Food and packaging perspective. *Composites Part B Engineering*, *175*, 107074. https://doi.org/10.1016/j.compositesb.2019.107074

Lestari, R. A., Kasmiyatun, M., Dermawan, K., Aini, A. N., Riyati, N., & Putri, F. R. (2020, April). Bioplastic from jackfruit seeds and rice. In *IOP Conference Series: Materials Science and Engineering* (Vol. 835, No. 1, p. 012035). IOP Publishing.

López, O. V., & García, M. A. (2012). Starch films from a novel (*Pachyrhizus ahipa*) and conventional sources: Development and characterization. *Materials Science and Engineering: C*, *32* (7), 1931–1940.

Lubis, M., Gana, A., Maysarah, S., Ginting, M. H. S., & Harahap, M. B. (2018). Production of bioplastic from jackfruit seed starch (*Artocarpus heterophyllus*) reinforced with microcrystalline cellulose from cocoa pod husk (*Theobroma cacao* L.) using glycerol as plasticizer. *IOP Conference Series: Materials Science and Engineering*, *309* (1), 1–10. https://doi.org/10.1088/1757-899X/309/1/012100

Lubis, M., Harahap, M. B., Manullang, A., Ginting, M. H. S., & Sartika, M. (2017). Utilization starch of jackfruit seed (*Artocarpus heterophyllus*) as raw material for bioplastics manufacturing using sorbitol as plasticizer and chitosan as filler. In Journal of Physics: Conference Series (Vol. 801, No. 1, p. 012014). IOP Publishing.

Ma, X., Yu, J., & Kennedy, J. F. (2005). Studies on the properties of natural fibers-reinforced thermoplastic starch composites. *Carbohydrate Polymers*, *62* (1), 19–24. https://doi.org/10.1016/j.carbpol.2005.07.015

Moreira, M. D. R., Pereda, M., Marcovich, N. E., & Roura, S. I. (2011). Antimicrobial effectiveness of bioactive packaging materials from edible chitosan and casein polymers: assessment on carrot, cheese, and salami. *Journal of Food Science*, 76 (1), M54–M63.

Muller, J., González-Martínez, C., & Chiralt, A. (2017). Poly (lactic) acid (PLA) and starch bilayer films, containing cinnamaldehyde, obtained by compression moulding. *European Polymer Journal*, *95*, 56–70. https://doi.org/10.1016/j.eurpolymj.2017.07.019

Nguyen, T. K., That, N. T. T., Nguyen, N. T., & Nguyen, H. T. (2022). Development of starch-based bioplastic from jackfruit seed. *Advances in Polymer Technology*. https://doi.org/10.1155/2022/6547461

Ningsih, P., Rahmawati, S., Santi, N. M. N., Suherman, & Diah, A. W. M. (2021). Making edible film from jackfruit seed starch (*Artocarpus heterophyllus*) with the addition of Rosella flower extract (*Hibiscus sabdariffa* L.) as antioxidant. *International Journal of Design and Nature and Ecodynamics*, *16* (6), 691–699. https://doi.org/10.18280/ijdne.160611

Nizam, N. H. M., Rawi, N. F. M., Ramle, S. F. M., Aziz, A. A., Abdullah, C., Rashedi, A., & Kassim, M. H. M. (2021). Physical, thermal, mechanical, antimicrobial and physicochemical properties of starch based film containing aloe vera: A review. *Journal of Materials Research and Technology*, *15*, 1572–1589. https://doi.org/10.1016/j.jmrt.2021.08.138

Noor, F. (2014). Physicochemical properties of flour and extraction of starch from jackfruit seed. *International Journal of Nutrition and Food Sciences*, *3* (4), 347. https://doi.org/10.11648/j.ijnfs.20140304.27

Palacios-Fonseca, A., Castro-Rosas, J., Gómez-Aldapa, C., Tovar-Benítez, T., Millán-Malo, B., Del Real, A., & Rodríguez-García, M. (2013). Effect of the alkaline and acid treatments on the physicochemical properties of corn starch. *CyTA – Journal of Food, 11* (sup1), 67–74. https://doi.org/10.1080/19476337.2012.761651

Peressini, D., Bravin, B., Lapasin, R., Rizzotti, C., & Sensidoni, A. (2003). Starch–methylcellulose based edible films: Rheological properties of film-forming dispersions. *Journal of Food Engineering, 59* (1), 25–32. https://doi.org/10.1016/S0260-8774(02)00426-0

Piñeros-Hernandez, D., Medina-Jaramillo, C., López-Córdoba, A., & Goyanes, S. (2017). Edible cassava starch films carrying rosemary antioxidant extracts for potential use as active food packaging. *Food Hydrocolloids, 63*, 488–495. https://doi.org/10.1016/j.foodhyd.2016.09.034

Piyada, K., Waranyou, S., & Thawien, W. (2013). Mechanical, thermal and structural properties of rice starch films reinforced with rice starch nanocrystals. *International Food Research Journal, 20* (1), 439.

Qin, Y., Xu, F., Yuan, L., Hu, H., Yao, X., & Liu, J. (2020). Comparison of the physical and functional properties of starch/polyvinyl alcohol films containing anthocyanins and/or betacyanins. *International Journal of Biological Macromolecules, 163*, 898–909. https://doi.org/10.1016/j.ijbiomac.2020.07.065

Rahmawati, S., Anggraeni, P., Nuryanti, S., Suherman, Sakung, J. M., Santoso, T., & Afadil. (2022). The making and characterization of edible film from jackfruit seeds (*Artocarpus heterophyllus* L.). *International Journal of Design and Nature and Ecodynamics, 17* (3), 407–415. https://doi.org/10.18280/ijdne.170311

Rahmawati, S., Mustapa, K., Suherman, N., Diah, A. W. M., Supriadi, N., Jura, M. R., & Wulan, M. R. (2023). Jackfruit (Artocarpus Heterophyllus) seed starch with sorbitol as a plasticizer and rosella flower antioxidant in the making edible film (*Hibiscus sabdariffa*). *International Journal of Design & Nature and Ecodynamics, 18* (2), 443–448. https://doi.org/10.18280/ijdne.180223

Resendiz-Vazquez, J. A., Ulloa, J. A., Urías-Silvas, J. E., Bautista-Rosales, P. U., Ramírez-Ramírez, J. C., Rosas-Ulloa, P., & González-Torres, L. (2017). Effect of high-intensity ultrasound on the technofunctional properties and structure of jackfruit (*Artocarpus heterophyllus*) seed protein isolate. *Ultrasonics Sonochemistry, 37*, 436–444. https://doi.org/10.1016/j.ultsonch.2017.01.042

Reshmy, R., Philip, E., Vaisakh, P. H., Raj, S., Paul, S. A., Madhavan, A., Sindhu, R., Binod, P., Sirohi, R., Pugazhendhi, A., & Pandey, A. (2021). Development of an eco-friendly biodegradable plastic from jack fruit peel cellulose with different plasticizers and *Boswellia serrata* as filler. *Science of the Total Environment, 767*, 144285. https://doi.org/10.1016/j.scitotenv.2020.144285

Rodrigues, A. a. M., De Melo Silva, S., Dantas, A. L., Da Silva, A. F., Da Silva Santos, L., & Moreira, D. D. N. (2018). Physiology and postharvest conservation of 'Paluma' guava under coatings using Jack fruit seed-based starch. *Revista Brasileira De Fruticultura, 40* (2). https://doi.org/10.1590/0100-29452018352

Rodrigues, A. A. M., Dos Santos, L. F., da Costa, R. R., Félix, D. T., Nascimento, J. H. B., & de Lima, M. A. C. (2020). Characterization of starch from different non-traditional sources and its application as coating in 'palmer' mango fruit. *Ciencia e Agrotecnologia, 44*, 1–12. https://doi.org/10.1590/1413-7054202044011220

Roz, A., Carvalho, A., Gandini, A., & Curvelo, A. (2006). The effect of plasticizers on thermoplastic starch compositions obtained by melt processing. *Carbohydrate Polymers, 63* (3), 417–424. https://doi.org/10.1016/j.carbpol.2005.09.017

Santana, R. F., Bonomo, R. C. F., Gandolfi, O. R. R., Rodrigues, L. B., Santos, L. S., dos Santos Pires, A. C., de Oliveira, C. P., da Costa Ilhéu Fontan, R., & Veloso, C. M. (2018). Characterization of starch-based bioplastics from jackfruit seed plasticized with glycerol. *Journal of Food Science and Technology, 55* (1), 278–286. https://doi.org/10.1007/s13197-017-2936-6

Santana, R. F., Mariana, C., Alves, A. N., Maniglia, B. C., Anjos, L. dos, Clarissa, A., Santos, L. S., Rodrigues, L. B., Rafael, Teixeira, J., Raquel, Melo, D., & Cristina, R. (2024). Sorbitol-plasticized jackfruit starch-based films: investigation of the effect of the starch and plasticizer concentration on the film properties. *Polymer Bulletin, 81* (14), 13089–13111. https://doi.org/10.1007/s00289-024-05316-8

Santhosh, R., & Sarkar, P. (2022). Jackfruit seed starch/tamarind kernel xyloglucan/zinc oxide nanoparticles-based composite films: Preparation, characterization, and application on tomato (*Solanum lycopersicum*) fruits. *Food Hydrocolloids, 133* (June), 107917. https://doi.org/10.1016/j.foodhyd.2022.107917

Santhosh, R., Ahmed, J., Thakur, R., & Sarkar, P. (2024). Starch-based edible packaging: rheological, thermal, mechanical, microstructural, and barrier properties – A review. *Sustainable Food Technology, 2* (2), 307–330. https://doi.org/10.1039/D3FB00211J

Sanyang, M., Sapuan, S., Jawaid, M., Ishak, M., & Sahari, J. (2015). Effect of glycerol and sorbitol plasticizers on physical and thermal properties of sugar palm starch based films. *Recent Advances in Environment, Ecosystems and Development*, 157–162.

Shahidi, F., & Hossain, A. (2020). Preservation of aquatic food using edible films and coatings containing essential oils: A review. *Critical Reviews in Food Science and Nutrition*, *62* (1), 66–105. https://doi.org/10.1080/10408398.2020.1812048

Sharma, R., Gautam, N., D. Giribabu, Kumar, R., & Yadav, S. (2023). Synergistic effect of plasticizer and cross-linking agent on the properties of finger millet starch-based film. *Starch Stärke*, *75* (9-10). https://doi.org/10.1002/star.202200251

Singh, G. P., Bangar, S. P., Yang, T., Trif, M., Kumar, V., & Kumar, D. (2022). Effect on the properties of edible starch-based films by the incorporation of additives: A review. *Polymers*, *14* (10), 1987. https://doi.org/10.3390/polym14101987

Singh, J., & Singh, N. (2001). Studies on the morphological, thermal and rheological properties of starch separated from some Indian potato cultivars. *Food Chemistry*, *75* (1), 67–77. https://doi.org/10.1016/S0308-8146 (01)00189-3

Singh, P., Chatli, A. S., & Mehndiratta, H. K. (2018). Development of starch based edible films. *International Journal of Development Research*, 8 (10), 23501–23506.

Suhag, R., Kumar, N., Petkoska, A. T., & Upadhyay, A. (2020). Film formation and deposition methods of edible coating on food products: A review. *Food Research International*, *136*, 109582. https://doi.org/10.1016/j.foodres.2020.109582

Tabasum, S., Younas, M., Zaeem, M. A., Majeed, I., Majeed, M., Noreen, A., Iqbal, M. N., & Zia, K. M. (2018). A review on blending of corn starch with natural and synthetic polymers, and inorganic nanoparticles with mathematical modeling. *International Journal of Biological Macromolecules*, *122*, 969–996. https://doi.org/10.1016/j.ijbiomac.2018.10.092

Thenmozhi, S., Dharmaraj, N., Kadirvelu, K., & Kim, H. Y. (2017). Electrospun nanofibers: New generation materials for advanced applications. *Materials Science and Engineering: B*, *217*, 36–48. https://doi.org/10.1016/j.mseb.2017.01.001

Thunwall, M., Kuthanová, V., Boldizar, A., & Rigdahl, M. (2008). Film blowing of thermoplastic starch. *Carbohydrate Polymers*, 71 (4), 583–590. https://doi.org/10.1016/j.carbpol.2007.07.001

Vieira, M. G. A., Da Silva, M. A., Santos, L. O. D., & Beppu, M. M. (2010). Natural-based plasticizers and biopolymer films: A review. *European Polymer Journal*, *47* (3), 254–263. https://doi.org/10.1016/j.eurpol ymj.2010.12.011

Wong, K. T., Poh, G. Y. Y., Goh, K. K. T., Wee, M. S. M., & Jeyakumar Henry, C. (2021). Comparison of physicochemical properties of jackfruit seed starch with potato and rice starches. *International Journal of Food Properties*, *24* (1), 364–379. https://doi.org/10.1080/10942912.2021.1885439

Wu, H., Li, T., Peng, L., Wang, J., Lei, Y., Li, S., Li, Q., Yuan, X., Zhou, M., & Zhang, Z. (2023). Development and characterization of antioxidant composite films based on starch and gelatin incorporating resveratrol fabricated by extrusion compression moulding. *Food Hydrocolloids*, *139*, 108509. https://doi.org/10.1016/j.foodhyd.2023.108509

Zhang, Y., Li, B., Xu, F., He, S., Zhang, Y., Sun, L., Zhu, K., Li, S., Wu, G., & Tan, L. (2020). Jackfruit starch: Composition, structure, functional properties, modifications and applications. *Trends in Food Science & Technology*, *107*, 268–283. https://doi.org/10.1016/j.tifs.2020.10.041

Zhu, H., Zhang, Y., Tian, J., & Chu, Z. (2018). Effect of a new shell material—Jackfruit seed starch on novel flavor microcapsules containing vanilla oil. *Industrial Crops and Products*, *112*, 47–52. https://doi.org/10.1016/j.indcrop.2017.10.060

Zoungranan, Y., Lynda, E., Dobi-Brice, K. K., Tchirioua, E., Bakary, C., & Yannick, D. D. (2020). Influence of natural factors on the biodegradation of simple and composite bioplastics based on cassava starch and corn starch. *Journal of Environmental Chemical Engineering*, 8 (5), 104396.

10 Unlocking the Potential of Plantain Starch

Extraction, Properties, and Applications

Poonam Choudhary, Ravi Prakash, Sumedha Arora Kapoor, and Sandeep Mann

10.1 INTRODUCTION

Starch is a complex carbohydrate that acts as a primary source of energy in plants and is an essential component of the human diet. It is made up of glucose units joined together, principally in two forms: amylose (linear) and amylopectin (branched). Plants, particularly cereals, tubers, and roots, are the primary sources of starch. Maize (corn), wheat, and rice are important cereal sources, as they are all staple meals around the world. Tuber crops such as potatoes and cassava are also prominent sources of starch, especially in Europe, Africa, and South America. Other sources include legumes such as peas and beans, which contain starch in lower amounts than grains and tubers. Starch is also commonly employed in the food sector as a thickening, stabilizing, texturizing, and gelling agent and in industrial applications such as food processing, pharmaceutical industry, packaging, film formation, papermaking, and textiles. The global demand for starch has been growing over the years, reaching USD 57.21 billion in 2024 and expected to reach USD 74.41 billion by 2029, with a compounded annual growth rate (CAGR) of 5.40% during 2024–2029 (Mordor Intelligence Research & Advisory, 2023). Therefore, in order to meet out the growing demand for starch in future, particularly in food and industrial applications, the researchers are looking for the unconventional sources of starch.

Plantain (*Musa paradisiaca*) belongs to the genus *Musa* of the family *Musaceae*. Plantains are closely related to bananas and originated in Southeast Asia, and grown extensively in tropical and subtropical regions. Many hybrids and polyploids of plantains are produced by crossing with two diploid wild species, *viz*., *Musa acuminate* COLLA and *Musa balbisiana* COLLA. However, all the crosses are scientifically known as *Musa paradisiaca*. The main genome groups are AA, AB, AAA, AAB, and ABB (Robinson and Sauco, 2010). Unlike dessert bananas, the plantains are typically larger, with thicker skin. The Food and Agriculture Organization (FAO) and International Institute of Tropical Agriculture mention the sweet *Musa* species, consumed uncooked as "banana", and *Musa* species with higher starch content, which require cooking before consumption termed as "plantain". Plantains are recognized by their higher starch content than bananas and cooking bananas. Like bananas, plantain are also seedless and consist of pulp and peel. Based on the color of pulp, plantain cultivars are classified into two groups: white and yellow cultivars.

The plantain fruit is an important food crop and is consumed widely in various cuisines across the tropical regions of Africa, Latin America, and Caribbean. It can be consumed in its ripe and unripe stages in soups, chips, cake, mash, etc. (Quintero-Castaño et al., 2020a). Plantains are an excellent source of nutrients and minerals. They are rich in complex carbohydrates mainly starch. The composition of starch and amount of resistant starch is dependent on the ripeness, cultivars, and

DOI: 10.1201/9781003534914-10

several other factors. Therefore, this chapter critically reviews and discusses the up-to-date studies on chemical composition, physico-chemical, and functional properties of plantain starch along with the different modification interventions of plantain starch and their utilization in food and industrial applications.

10.2 NUTRITIONAL PROFILE OF PLANTAINS

Plantains are a rich source of complex carbohydrates mainly starch, dietary fibre, vitamins A and C, and minerals such as potassium, calcium, and iron (Sojinu et al., 2021). The nutritional composition of plantains varies among cultivars, ripening stages, and cooking methods with 4.59–6.85% moisture content, 2.01–2.56% ash content, 2.54–3.47% protein, and 0.95–1.37% fiber content on dry weight basis (Honfo et al., 2022). Plantain contain total sugar in the range of 57.12% to 86.27%, depending on ripeness (Agbemafle et al., 2017). The ash, protein, fat, fiber, carbohydrate, and total sugar contents were ranged from 6.04% to 22.70%, 1.54% to 4.52%, 3.25% to 4.83%, 0.07% to 16.02%, 1.26% to 5.57%, 57.12% to 86.27%, and 4.06% to 14.14%, respectively. The mineral contents were 17.07 to 28.44 mg/100g, 4.92 to 9.36 mg/100g, 364.80 to 487.55 mg/100g, 0.10% to 0.39% and 0.16% to 0.42% for iron, zinc, potassium, magnesium, and calcium, respectively (Agbemafle et al., 2017).

10.3 CHEMICAL COMPOSITION OF PLANTAIN STARCH

The chemical composition of plantain starch extracted from various cultivars of plantain is presented in Table 10.1. Starch extracted from plantain cultivars shows significant variation in the moisture content, ranging from 7.53% to 13.01%. It also contains 0.03% to 1.09%, 0.325% to 2.53%, and 0.02% to 0.45% ash, protein, and lipid content, respectively. The lower amount of fat, protein, and ash contents suggests that the plantain starches are extremely pure. The crude fiber and amylose contents of plantain starch are ranged between 0.04% and 0.72% and 16.00 and 40.00%, respectively.

TABLE 10.1
Chemical Composition of Plantain Starches Extracted from Various Cultivars

Variety/Plant Part Used	Moisture (%)	Ash (%)	Lipid (%)	Protein (%)	Fiber (%)	Amylose (%)	References
Tanduk (Unripe fruit)	8.83	0.66	0.45	2.31	NA	26.08	Shittu et al. (2016)
Nangka (Unripe fruit)	9.84	0.53	0.382	0.84	NA	31.79	Shittu et al. (2016)
Kluai Namwa (Cull Peel)	9.12	0.18	0.10	1.28	0.20	32.58	Nasrin et al. (2015)
Kluai Namwa (Cull pulp)	9.76	0.12	0.02	0.90	0.10	39.80	Nasrin et al. (2015)
Agbagba (Unripe fruit)	11.20	0.62	0.44	2.53	0.72	38.79	Olatunde et al. (2017)
Dominico-Harton (unripe fruit)	7.53	0.08	0.015	1.30	0.04	16.00	Vega-Rojas et al. (2021)
Dominico-Harton (Fruit)	10.13	0.33	0.15	1.75	0.18	40.00	González et al. (2019)
Ogede fufun (White plantain fruit)	8.44	1.09	0.276	0.640	NA	26.01	Nwokocha & Williams (2009)
Ogede pupa (Yellow plantain fruit)	8.62	0.95	0.403	0.325	NA	26.13	Nwokocha & Williams (2009)
Agbagba (fruit pulp)	13.01	1.05	0.26	1.26	0.12	24.85	Otegbayo et al. (2010)
Terra (unripe fruit pulp)	9.27	0.03	0.02	0.97	0.28	35.00	Pelissari et al. (2012)

NA, not available.

González et al. (2019) reported 10.13%, 0.33%, 0.15%, 0.18%, and 1.75% moisture, ash, fat, fiber, and protein, respectively in the unripe *Dominico-Harton* plantain starch. The significant differences in chemical composition might be due to the genetic variation among cultivars, agronomic and climatic variations, and different extraction methods used for extraction of starch.

10.4 EXTRACTION OF STARCH FROM PLANTAIN PULP AND PEEL

Extraction of starch is a fundamental step that influences its utility in food and industrial applications. The extraction of starch involves its separation and purification from other plant components such as fiber, proteins, and lipids to produce the pure product. The method of starch extraction may vary depending upon the plant source; however, it usually involves steps like cleaning, grinding, slurry formation, separation, settling and decanting, washing, and drying to yield the pure starch powder. The ripe, unripe, rejected unripe fruits, and culled plantains are peeled and sliced into small pieces after removing the injured section of culled plantains. The peel of the plantain is then cut down into small pieces. The sliced pieces are then steeped into the water containing citric acid, ascorbic acid, or sodium meta-bisulfite solution to prevent the oxidation and discoloration. After steeping, the soaked pieces are blended into the pulp which is further mixed with water and strained through the fine mess sieve or muslin cloth to separate the starch-rich solution from the fibrous residues. The collected filtrate is allowed for settling the starch granules to the bottom. The top clear liquid is decanted, and the starch is collected. The obtained starch is washed several times with water to remove any residues and purify it further. The wet starch thus obtained is then allowed to dry using hot air oven and tray dryer to obtain dry, pure and powdered plantain starch. Nasrin et al. (2015) utilized 1% citric acid solution for extraction of starch from the culled "*Kluai Namwa*" plantain peel and pulp collected at 100–110 days after anthesis and to disintegrate the slices further, 0.05% sodium hydroxide (in ratio of 1:2, v/v) solution was used. Shittu et al. (2016) observed that *Nangka* cultivar contained higher amylose than the *Tanduk* cultivar. *Dominico-Harton* and FHIA 21 plantains starches contained 14% and 18% amylose content, respectively, and reported that esterification process improved the amylose content in modified starches (Quintero-Castaño et al., 2020a). Moreover, total starch content or purity in the isolated starch from plantain peel and pulp also varies according to the cultivar's genetic make-up and extraction method employed. The starch isolated from *Dominico-Harton* plantain showed 91% starch purity as compared to the other cultivar, FHIA 21, which has 96% purity (Quintero-Castaño et al., 2020a). However, the starch purity has been decreased during the starch modification process using esterification reaction. Vega-Rojas et al. (2021) also reported 16% amylose and 91% purity of starch in the isolated starch from unripe *Dominico-Harton* plantains. They also observed that the isolated starch had minerals such as potassium, phosphorus, calcium, magnesium, and silica. However, potassium is predominant mineral found in plantain starch which affects the pasting properties of starch. Similar results were reported by González et al. (2019) who reported that addition of potassium and magnesium ions to plantain starch improved the viscosity and gelatinization temperature.

10.5 PROPERTIES OF PLANTAIN STARCH

10.5.1 Morphological Properties

The morphological properties of starch are studied using scanning electron microscopy (SEM). SEM analysis provides information about the size, shape, and surface characteristics of starch granules. The differences in starch granule size are reflected in functional properties such as gelatinization temperature, viscosity, solubility, water-binding capacity, and swelling power (Otegbayo et al., 2010). The starch isolated from unripe fruits of *Dominico-Harton* plantain showed lenticular, elliptical, and semi-spherical morphologies with a diameter of 10–15 μm, length between 15 μm and 32 μm, and widths of 0–15 μm, and 8–12 μm, respectively (Vega-Rojas et al., 2021). The starch granules of *Nangka* and *Tanduk* cultivars were compact, irregular shapes varying from elliptical to round and oval (Shittu et al., 2016). Moreover, the *Tanduk* plantain starch showed bimodal size distribution with an average particle size of 23.91 μm, whereas *Nangka* cultivar showed unimodal distribution with an average particle size of 38.97 μm. Similarly, the average granule size of white variety of plantain ranged between 3.74 and 7.00 μm with smooth irregular shaped bimodal granules, whereas the average granule size of yellow variety of plantain ranged from 11.22 to 41.00 μm with elliptical smooth granules (Nwokocha & Williams, 2009). The starch granules were smooth surfaced of uneven size ranging between 9 and 25 μm with oval and elongated shapes in *Agbagba* variety of plantain (Otegbayo et al., 2010). The granule surface morphology is also affected by the ripening stages of the plantain fruits. Figure 10.1, demonstrates the SEM images of plantain starch granules of different types, *viz.*, "*Tanduk*" and "*Nangka*" starch, at 500× and 1000× magnifications.

10.5.2 Thermal Properties

The various unit operations and processing of starch in different food preparations and industrial applications depend on the thermal properties of starch. The gelatinization and retrogradation are most common thermal characteristics of starch. Starch gelatinization is a process in which starch granules are subjected to high temperature in the presence of water. As a result, the crystalline structure of starch is disrupted, leading to significant changes in its textural and functional properties, such as increased solubility and viscosity. Starch retrogradation is the process that occurs when gelatinized starch molecules begin to re-associate after cooling. The thermal and retrogradation transitions are represented as onset temperature (T_o), peak temperature (T_p), conclusion temperature (T_c), thermal transition temperature ($T_c - T_o$) and enthalpy of transition (ΔH), which can be measured using differential scanning calorimeter (DSC). The thermal properties of starch granules are affected by starch composition, shape or size of starch granule, crystallinity index, amylose content, and chain length and molecular structure of amylopectin (Wani et al., 2010; Nwokocha & Williams, 2009). DSC analysis also showed that the gelatinization temperatures (T_o, T_p, and T_e) of *Tanduk* starch were higher than the *Nangka* Starch (Shittu et al., 2016). The larger granule size of *Nangka* cultivar as compared to *Tanduk* cultivar showed the lower gelatinization due to increased swelling of *Nangka* starch granules. Similar results were also reported by other researchers, suggesting that the melting properties of starch are directly related to the granule size and granular structure (Steeneken & Woortman, 2008; Nwokocha & Williams, 2009).

10.5.3 Pasting and Rheological Properties

Each starch demonstrates the distinctive rheological and pasting properties including peak and break down viscosity during the heating process that depends on the starch source, starch structure, starch extraction and modification methods, amylose to amylopectin ratio, and amount of resistant starch which includes the changes in the viscosity of starch paste and gel formation ability and gel characteristics (Wang & Ren, 2020). Heating of starch in presence of water leads to swelling of starch granules by absorption of water which results in increased viscosity of starch/water mixture, reaching a peak viscosity before starch initiates to break down under continued heating and stirring. The pasting and rheological properties of starch are determined by Rapid Visco-Analyzer (RVA), Brabender Visco-Amylograph and Rheometer, respectively. The RVA provides information on pasting temperature, peak viscosity, breakdown, setback, and final viscosity of starch. The pasting profile of plantain starch is shown in Figure 10.2. The peak viscosity of plantain starch is also affected by the ratio of starch to water suspensions and the ratio of 3:18 showed the best rheological condition for analyzing the apparent viscosity of starch (Vega-Rojas et al., 2021). Modification of starch using pregelatinization, oxidation, and acetylation reduces all paste viscosities of plantain

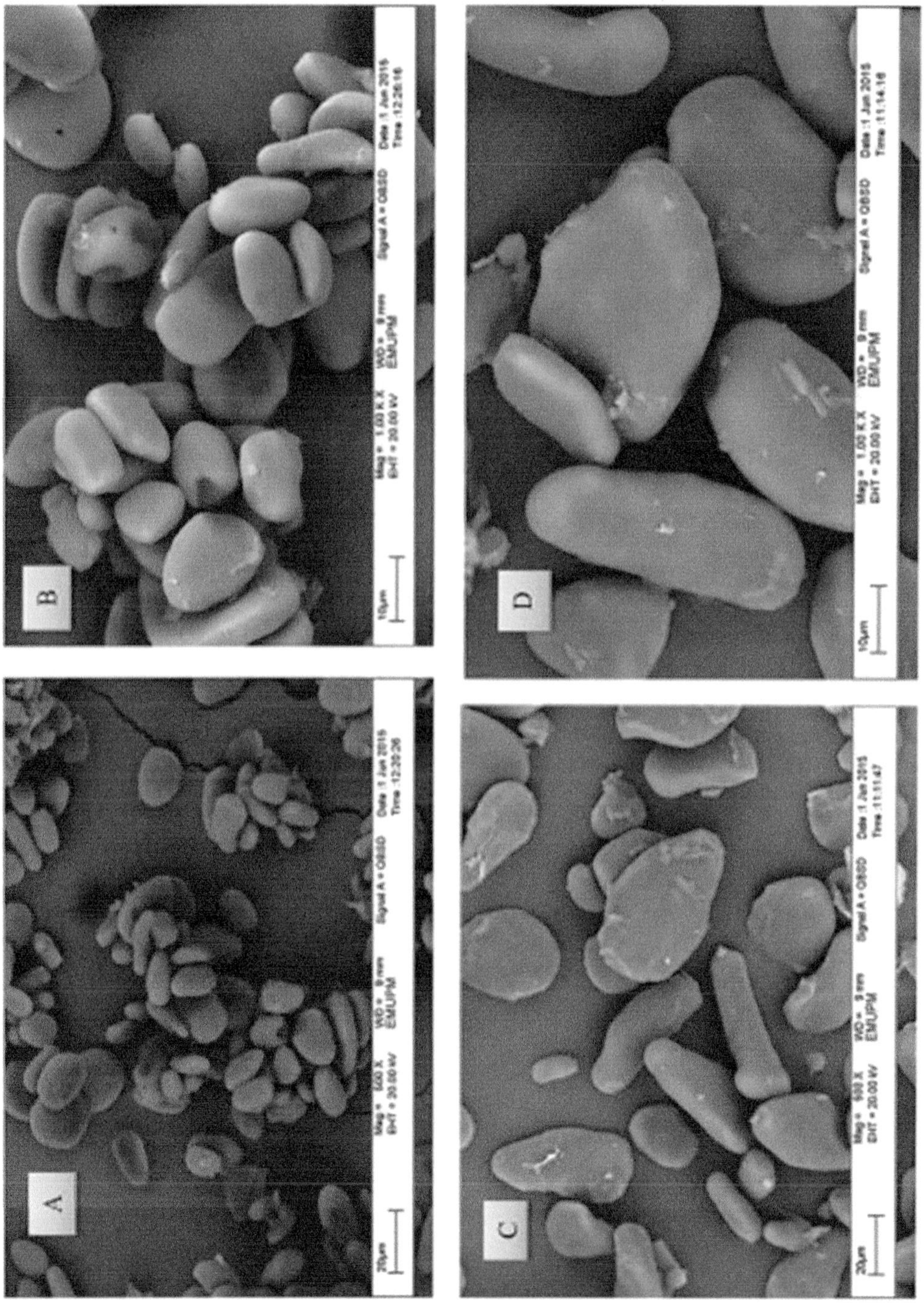

FIGURE 10.1 SEM images of plantain starch granules: (A, B) "*Tanduk*" starch at 500× and 1000×; (C, D) "*Nangka*" starch at 500× and 1000× (Shittu *et al.*, 2016).

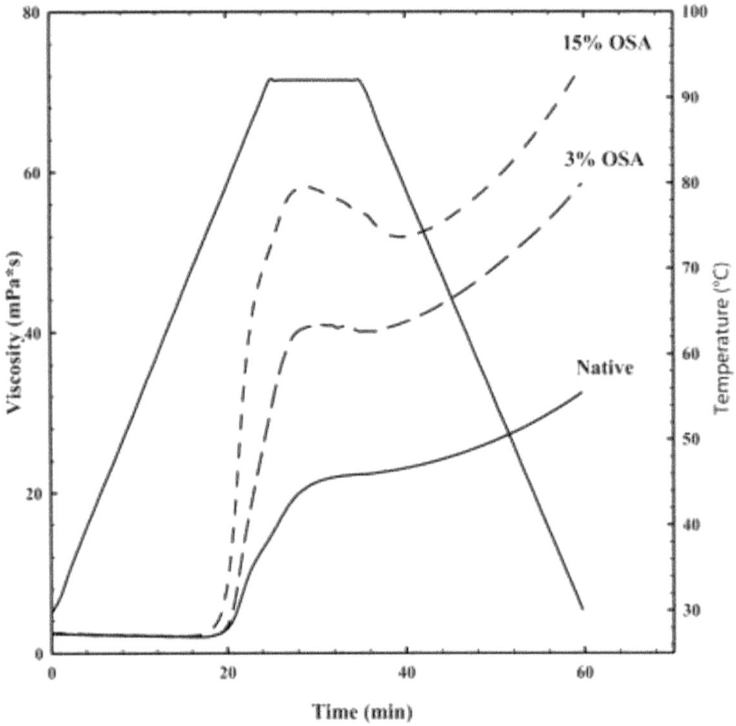

FIGURE 10.2 Pasting profile of native and OSA modified plantain starch (Bello-Pérez et al., 2015).

starch as compared to the native starch (Olatunde et al., 2017). The higher peak and final viscosities of plantain and other *Musa Spp.* starches at high temperatures make them suitable for application as thickening agent in sauces and baby food products (De la Torre-Gutiérrez et al., 2008).

10.5.4 X-Ray Crystallography

X-ray diffraction (XRD) technique is used to characterize the structure of starch, crystalline orientation and crystallinity. The plantain starch generally showed C-type of crystalline pattern which is a combination of A- and B-type, indicating a complex crystalline structure (Pelissari et al., 2012). *Dominico-Harton* unripe plantain starch showed an orthorhombic and hexagonal crystalline phase structure under X-ray analysis, indicating a C-type of pattern (Vega-Rojas et al., 2021). The starch isolated from unripe fruits of *Nangka* and *Tanduk* plantain cultivars showed three prominent peaks at diffraction angles of 15.0°, 17.5°, and 23.0° which specifies as an A-type diffraction pattern (Shittu et al., 2016). The relative crystallinity index *Nangka* starch (61.1%) was higher than *Tanduk* starch (52.1%) due to differences in the proportion of long and short chain amylopectin and amylose. Figure 10.3 shows a typical X-ray powder diffraction pattern of *Tanduk* and *Nangka* plantain starches with their relative crystallinity.

10.5.5 Digestibility

The digestibility of starch is generally known as ability of starch granules to be broken down into glucose by digestive enzymes and absorbed in the gastrointestinal tract. The starch digestibility is mainly affected by the molecular structure of starch, amylose to amylopectin ratio, cooking methods and interaction with other food components such as lipids, dietary fibers, and presence of

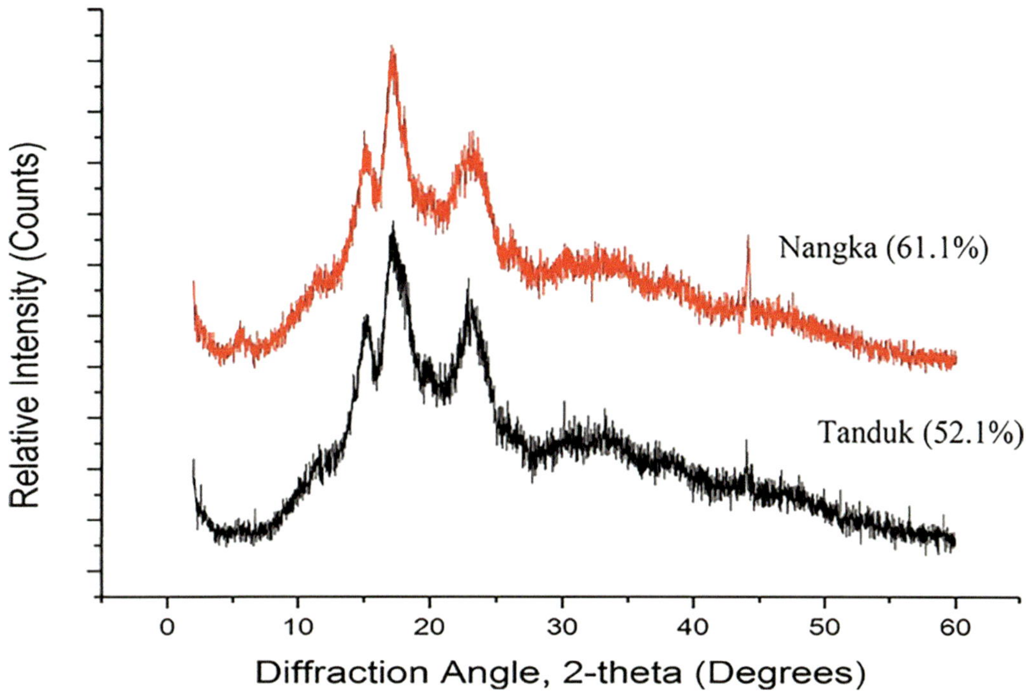

FIGURE 10.3 X-ray powder diffraction pattern of *Tanduk* and *Nangka* plantain starches with their relative crystallinity (Shittu et al., 2016).

hydrocolloids (Rong et al., 2024; Rostamabadi et al., 2024; Ma et al., 2024). The digestibility of plantain starch is also affected by the ripening stage of pulp as the unripe plantain starch contains higher resistant starch as compared to the ripe plantain. The structure of plantain starch changed with the ripening process and the amount of resistant starch also decreased gradually which resulted in improving the in vitro digestibility of plantain starch (Bi et al., 2019). The modified plantain starch with precooking and annealing exhibited changes in starch structure and showed a significant reduction in starch hydrolysis rates from 62.3% to 44.9% (De la Rosa-Millán et al., 2015).

10.6 MODIFICATION OF PLANTAIN STARCH

The commercial use of starch derived from both conventional and nonconventional sources is somewhat restricted due to several unfavorable characteristics inherent in its native form. These include its insolubility, low mechanical strength, and instability under conditions of high temperature, pH, and shear during processing. To broaden the range of applications and improve the versatility of native starch, these adverse properties must be addressed, which can be achieved through physical, enzymatic, or chemical modifications. The starch modification is an evolving sector which offers numerous opportunities to develop novel starches with new functional and value-added properties. The primary goal of starch modification is to alter its cooking characteristics, enhance its stability during freeze–thaw cycles and processing, reduce retrogradation, and gelling properties, and improve its film-forming capabilities (Kaur et al., 2012).

10.6.1 Physical Modification

Physical methods of starch include modification of starch through heat–moisture treatment (HMT), annealing, pregelatinization, ultra-sonication, and pressure treatment. These treatments are safe and less expensive methods of starch modification compared to the chemical and enzymatic methods. Physical methods of starch modification are more preferred because they can change functional properties of starch without using chemicals, making them a more natural and clean-label food processing choice (Fonseca et al., 2021).

10.6.1.1 Heat–Moisture Treatment

HMT is a physical modification method in which starch is subjected to high temperatures (between 80 and 120°C) at a low moisture content (10–30%) for a specific duration. This treatment alters the crystalline structure of the starch granules without distrusting the structure, leading to changes in its functional properties. HMT modification altered the crystalline type of banana starch from B-type to A-type. Due to higher ordered granule structure and tight packing in A type of starch, the modified starch showed lower affinity to hydrolytic enzyme, resulting in higher content of slowly digestible starch (SDS) (Cahyana et al., 2024). Similarly, the bioplastics obtained by HMT modified green plantain starch demonstrated improved mechanical properties such as elongation, tensile strength, force and puncture deformation and reduced water vapor permeability (Viana et al., 2022). Moreover, the treatment also decreases the pasting viscosities, which correlates with improved mechanical performance. HMT increases the gelatinization temperature, improves paste stability, reduces swelling power, and enhances resistance to enzymatic hydrolysis in high amylose corn and rice starch (Gu et al., 2024; Lee & Chung, 2022). HMT treatment increases the crystalline content in the native starch from 28.0 to 31.5% and enhances the content of slowly digestible and resistant starch (Hoyos-Leyva et al., 2015). The X-ray diffraction pattern of HMT modified starch also changed due to changes in the crystalline structure (Cordeiro et al., 2018). Starch modification through HMT method demonstrated significant reduction in the swelling power and an increase in the gelatinization peak temperature in *Prata* green banana starch. Moreover, the Rapid Visco Analyser showed reductions in viscosities such as peak, trough, breakdown, and final viscosity and increased the pasting temperature and time to peak viscosity, suggesting improvement in thermal stability in the modified starch (dos Santos Costa et al., 2019). These modifications make heat-moisture treated starches suitable for applications requiring improved thermal stability, controlled viscosity, and better textural properties in food products, as well as in nonfood industrial applications.

10.6.1.2 Annealing

Another method of starch modification is annealing that involves treating starch samples with excessive water (more than 40–70%) at the temperature above the glass transition temperature and below the gelatinization temperature of starch for a period ranging from minutes to days. Like HMT, this process also alters the crystalline structure of starch and improves functional properties without affecting the moisture content and gelatinization (Punia, 2020; BeMiller & Huber, 2015). This process improves the starch digestibility and reduces the glycemic responses. De la Rosa-Millán et al. (2015) reported that annealing decreased the susceptibility of modified starch to α-amylase hydrolysis which results in reduction in digestibility. The annealed starch is more suitable food ingredient for diet conscious and persons suffering from glycaemia. Moreover, annealing has been reported to improve the relative crystallinity index and gelatinization temperature in potato starch (Thomann et al., 2024).

10.6.1.3 Pregelatinization

The most common physical method of starch modification is pregelatinization. In this process, starch is subjected to heating and mechanical shearing before drying and grinding which make it more soluble, enhances water absorption and swelling power and easily digestible in cold water. Pregelatinized starch is also called as instant starch which can be easily dissolved in water below the gelatinization temperature (Bemiller, 2016). This process improves the viscosity of starch which makes it suitable as thickening agent (Zia-ud-Din et al., 2017; Hong & Liu, 2018). Spray drying, extrusion, and drum drying are most frequently used techniques to produce the pregelatinized starch. The modified starch produced by extrusion and drum drying showed greater destruction in granular shape as compared to the spray dried starch which still maintained granules in good shape at higher temperatures (Ma et al., 2022). Moreover, the water solubility and cold paste viscosity of extruded and drum-dried starch were higher than those of spray dried pregelatinized starch, due to the complete loss of crystallinity. Pregelatinized plantain starch exhibited increased water and oil absorption capacities and emulsion ability (Olatunde et al., 2017). Similarly, unripe banana starch modified using pregelatinization and annealing increases pasting properties (Akubor & Igba, 2019).

10.6.2 Chemical Modification

Chemical modification of starch comprises tweaking the molecular structure of starch through various chemical treatments to improve or add specific properties. These alterations can increase the functionality of starch by increasing its resilience to heat, acidity, and shear, increasing its solubility, decreasing retrogradation, or improving its binding and thickening properties. Esterification, etherification, crosslinking, and oxidation are common chemical modifications that modify starch to specific industrial applications.

10.6.2.1 Octenyl Succinic Anhydride

The octenyl succinic anhydride (OSA)-modified starches are produced by chemical esterification with OSA. This alteration imparts hydrophobic groups to the starch molecules. Regulatory agencies such as Food and Drug Administration (FDA) have considered OSA-modified starches to be generally recognized as safe (GRAS) and have approved their usage up to 3% in food products (Food and Drug Administration, 2001; Olawoye et al., 2023; Nurmilah & Subroto, 2021). Recently, Quintero-Castaño et al. (2020a) prepared OSA-modified starch in the starch extracted from the two unripe plantains of Colombia, viz., Dominico-Harton plantain (DH) and FHIA 21 (tetraploid plantain). The results showed that OSA modification altered the hexagonal and monoclinic structures with increased amylose content as compared to the native starch. Also, the peak viscosity of modified starches was decreased with higher degree of substitution. This modification enhances the capacity of starch to produce stable emulsions and serve as a thickening or gelling agent in food preparations (Agama-Acevedo & Bello-Perez, 2017; Bello-Pérez et al., 2015).

Drying of OSA-modified starch can affect the functionality and digestive properties of plantain starch (Chavez-Salazar et al., 2022). The modified starch dried with convection drying method showed an increase in slowly digestible starch (SDS) and resistant starch as compared to the spray dried starch. The spray dried OSA-modified starch exhibited slight loss in organization of the granular structure which resulted in higher water absorption index, solubility power, pasting peak viscosity, and lower gelatinization temperature. The OSA-modified plantain starch showed a slow enzymatic degradation of the starch chains on the granule surface thereby decreasing the rapidly digestible starch content and reducing the digestibility of starch (Quintero-Castaño et al., 2020b). Similar results were reported by Zhang et al. (2017) in various botanical starches that showed decreased content of rapidly digestible fractions after OSA modification. Moreover, the enzymatic breakdown of starch is positively correlated with the degree of substitution. The OSA-modified banana starches with higher degree of substitution were digested slowly (Bi et al., 2017). The OSA acts as a cross-linking agent and upon modification of starch, the apparent molecular weight of starch increased, and the modified starch showed greater resistance to the hydrolyzing enzymes (Lopez-Silva et al., 2020). The process of starch modification using OSA was optimized using response surface methodology and artificial neural network (ANN) for *Cardaba* banana starch to yield SDS, which ranged from 31.88 to 44.68% (Olawoye et al., 2020). The optimal conditions to obtain the highest yield of SDS were reported as: 3.69% OSA concentration, reaction time 43.75 min at pH of 9.8.

10.6.2.2 Oxidation

Another starch modification method is oxidation, in which starch molecules are treated with oxidizing agents, such as sodium hypochlorite, ozone, and hydrogen peroxide, to introduce carbonyl and carboxyl functional groups into the native starch structure. The addition of functional groups through an oxidation process is capable of modifying the solubility, mechanical, barrier, and thermal properties (Pham et al., 2024). The oxidized starches are more whitish in color and have limited retro-gradation. The oxidation improved the emulsifying ability and water absorption capacity of plantain starches as compared to the native starch. However, the oil absorption capacity was declined (Olatunde et al., 2017). The oxidized plantain starch exhibited significant changes in its physicochemical properties and increasing its utilization in various sectors. The banana starch, modified through an oxidation process, was blended with polyvinyl alcohol to develop composite film that showed improvements in mechanical and barrier properties (Palma-Rodríguez et al., 2012). The plantain starch modified by oxidation showed increased moisture content and decreased swelling power as compared to the native starch (Haruna et al., 2023). However, the solubility of oxidized starch was increased due to structural alteration in the native starch. The breakdown viscosity of oxidized banana was higher as compared to the native starch, whereas the acetylated starches showed reduced breakdown viscosity (Ogunbusola et al., 2023). Oxidation with ozonation significantly enhanced the freeze–thaw stability and pasting properties, suggesting a reduced syneresis extent (Cahyana et al., 2018). Likewise, oxidation with sodium hypochlorite increased its qualities by increasing moisture and starch content, as well as improving its thermal, mechanical, and solubility properties, making it ideal for possible usage in biodegradable sheet applications (Alanís-López et al., 2011). Oxidized starches can be used in food applications as thickening and binding agent.

10.6.2.3 Cross-linking

Cross-linking is a starch modification method in which starch molecules are chemically linked together using cross-linking chemicals like phosphorus oxychloride, sodium tri-metaphosphate, sodium tripolyphosphate, citric acid, boric acid, adipic acid, epichlorohydrin, etc. The cross-linking improves the internal structure of starch, making it more resistant to breakdown under high temperatures, acidic environments, and mechanical stress, due to development to intra- and inter-molecular bonds. Cross-linked starches have enhanced viscosity, stability during cooking, and resistance to retrogradation, making them ideal for use in processed foods and pharmaceutical sectors that require consistent texture and durability, as well as for industrial uses such as adhesives and coatings (Punia et al., 2024). The cross-linked starches showed improvement in thermal and mechanical properties.

Plantain starch cross-linked with urea has been used as a matrix for formulation of ambroxol hydrochloride tablets, which show improved mechanical properties and better drug release profile (Okunlola & Adedeji, 2021). Strach extracted from *Musa balbisiana Var. Abu* was modified using cross-linking agents such as sodium trimetaphosphate and sodium tripolyphosphate (99:1), and the process was optimized using response surface methodology. The maximum degree of cross-linking (81.73%) was obtained at 50.57°C at 9.52 pH. The degree of cross-linking affects the peak viscosity of the modified starch and shows a declining trend as the cross-linking exceeds 75%. The swelling power and solubility of cross-linked starches were decreased (Nor Sham et al., 2024a). Cross-linking with sodium tripolyphosphate and phosphorus oxychloride in plantain starches demonstrated increased moisture content, water absorption capacity, swelling capacity, solubility power, and peak viscosity. However, the pasting temperature and break down viscosity of modified starches were reduced (Ibikunle et al., 2022). The cross-linking modification process reduces the size of starch granules without affecting the crystallinity index of unripe banana starch (Nor Sham et al., 2024b). The plantain starch cross-linked with epichlorohydrin exhibited significant improvement in the resistant starch content as compared to the native starch. Moreover, the gelatinization temperature and enthalpy of modified starch were decreased, suggesting lower thermal stability during gelatinization (Aparicio-Saguilán et al., 2024).

10.6.3 Enzymatic Modification

The enzymatic starch modification method includes alteration in molecular mass, distribution of branch chain length, and ratio of amylose and amylopectin by starch-modifying enzymes such as α-amylase, branching enzyme (glucan transferase), cyclodextrin glycosyltransferase, β-amylase, amylo-sucrase, α-glucanotransferase, and de-branching enzymes (pullulanase and isoamylase) to improve the functional properties of modified starch (Bangar et al., 2022; Zhang & Bao, 2023; Park & Kim, 2021). Enzymatic and physical methods of starch modification are considered as environment-safe and clean-label methods because the final products are free of artificial and synthetic ingredients (Zhang & Bao, 2023; Park & Kim, 2021). Furthermore, the specificity and selectivity of enzymes offer the production of desired end products in the reaction and prevent the undesirable side reactions. Enzymatically modified plantain starch demonstrated an increase in the SDS content from 10.9 to 18.5% using β-amylase-transglucosidase enzyme, whereas the modified starch using β-amylase enzyme showed decline in SDS content from 10.9 to 7.1% (Casarrubias-Castillo et al., 2012). However, the resistant starch content increased in both modified starches, and a reduction in molar mass, and gyration radius was observed.

10.6.4 Dual Modification

Dual modification of starch includes combining two different modification techniques such as chemical, physical, or enzymatic methods to achieve a starch with enhanced or modified properties that cannot be obtained through a single modification process. Dual-modified starches can exhibit improved textural properties, greater resistance to retrogradation, enhanced film-forming properties, and other desirable traits, depending on the combination of applied modifications (Kumari & Sit, 2023). The cross-linking with phosphorus oxychloride and microwave irradiation decreases the solubility, swelling power, and water absorption capacity of modified banana starch. Moreover, the modified starch with higher degree of cross-linking exhibited reduced retrogradation properties and freeze–thaw stability (Surendra Babu et al., 2018). However, in another study, the degree of crystallinity has been reduced in the banana starch modified by cross-linking of sodium trimetaphosphate (16.32%) and dual modification (by cross-linking and dry heat treatment) (13.91%) as compared to the native starch which, showed 26.88% crystallinity index (Sharma et al., 2023). Moreover, the cross-linked starch had smoother surface as compared to the starch modified by dual modification process. However, the pasting temperature of both modified starches was reported to be higher than that of the native starch. The cross-linked starch demonstrated better stability in the emulsion as compared to dual-modified starch, which showed creaming and poor performance.

10.7 APPLICATION OF PLANTAIN STARCH

10.7.1 Food and Functional Food Sector

Plantain starch is a versatile ingredient with diverse applications in different industries. Plantain starch has a high level of resistant starch that is indigestible by digestive enzymes and is instead fermented by bacteria in the colon (Vaidya & Sheth, 2011), thereby does not spike blood glucose levels (Marta et al., 2022). Consequently, plantain starch is recommended as a functional food ingredient for diabetic patients due to its glycemic index management properties. Thus, flour made from plantain is a healthier alternative to wheat flour for bread and pasta (Burkhart et al., 2022; Munir et al., 2024). In addition, cake, bread, cookies, and biscuits can be prepared from plantain flour

(Marta et al., 2022). Also, this plantain flour would benefit patients with celiac disease, a food-sensitive enteropathy in humans. High glycemic indices in traditional noodles alongside strong carb content diets are the primary reasons for obesity and overweight problems in individuals. Plantain flour and pasta provide fewer calories without any gluten content (Yu et al., 2020).

Starch modified through HMT is used in noodle preparation. HMT starch has reduced swelling and minimal cooking loss with increased SDS, providing additional health benefits (Marta *et al.*, 2022). SDS helps to regulate satiety, support physical performance, and reduce the risks associated with metabolic disorders by controlling blood lipid levels and improving insulin sensitivity, which is crucial for glucose homeostasis (Björck et al., 2000). Furthermore, plantain starch is useful in various culinary applications, such as refrigerated meals because of its freeze–thaw stability, demonstrating its functional versatility in innovative food technology and product formulations (Marta et al., 2019; Fontes et al., 2017; Munir et al., 2024).

Plantain starch can act as a stabilizer and emulsifier in food formulations because of its remarkable ability to thicken and form a gel-like structure. Rheological properties such as high viscosity and a strong tendency to rearward account for its thickening and gel-like consistency (Munir et al., 2024). This stabilizing effect is essential for restraining ingredient separation, retaining consistency, and enhancing product quality. Plantain starch can withstand an array of processing conditions, such as heat, acidity, and shear forces stress, which makes it adaptable for preparing sauces, soups, puddings, flans, dairy products like yogurt, and baked food items (Bezerra et al., 2013a; Bezerra et al., 2013b; Munir et al., 2024; Rashwan et al., 2024). Also, plantain starch stabilizes emulsions; examples include mayonnaise and salad dressings, owing to its pseudo-plastic properties at both low and ambient temperatures (Izidoro et al., 2008; Mohapatra et al., 2009).

Green banana or plantain biomass is an effective fat replacer used as a fat addition in chicken mortadella (Auriema et al., 2021; Munir et al., 2024). The biomass provides a wholesome blend of nutrients, including 4.16% resistant starch, a good amount of dietary fiber, and antioxidants (Auriema et al., 2021). Thus, biomass can be blended with fat-based meat products to produce low-fat versions with good sensory acceptability. In addition, biomass can be employed as a partial fat replacer in *requeijao cremoso* processed cheese formulations to enhance the products' functional value (Pivetta et al., 2019; Munir et al., 2024).

10.7.2 Pharmacological Sector

Plantain is a good source of various bioactive compounds, making it ideal for the pharmaceutical industry. Plantain flour has excellent binding properties, making it an ideal alternative to corn starch and, hence can be employed in paracetamol tablet production. Faster disintegration and lamination are major concerns in paracetamol tablets, which are resolved by using plantain flour (Akin-Ajani et al., 2005). In the study by Shrestha et al. (2018) an autoclaved treated linearized banana starch and soy protein isolate conjugate emulsion system was utilized, which can bio-fortify food and medicinal formulations such as astaxanthin. Also, the resistant starch in plantain has weight management properties evidenced by a study by Fu et al. (2021) in which obese rats fed plantain starch showed improvements in metabolism and intestinal flora, resulting in weight reduction. Furthermore, the SDS in plantain is advantageous for maintaining satiety and supporting better physical performance. It also lowers blood lipid or fat levels and can improve insulin sensitivity, contributing to glucose homeostasis (Björck et al., 2000).

10.7.3 Nonfood Applications

Plantain starch has utility in nonfood items, including wood adhesives, bioplastics, textiles, and bio-based film, and coatings (Wang et al., 2022; Rashwan et al., 2024). For bio-based packaging films, plantain starch is preferred with some modifications (Pongsuwan et al., 2022). These modifications are necessary to ameliorate the poor barrier properties of the plantain starch. To improve barrier properties, a large amount of fiber from banana inflorescence debris is added as filler to starch-based bioplastics. Thus, bioplastics with better physical properties, water, and thermal resistance can be produced (Ponguswan et al., 2022). Studies supported the incorporation of fillers and modifications for the manufacturing of starch-based biofilms (Hernández et al., 2008). Also, plantain flour has the potential to produce filmogenic solutions and is hence, used in various industrial applications (Salazar et al., 2022). Plantain starch films are aesthetically pleasing because of their transparent appearance and are highly acceptable to consumers. In addition, because of their high glycerol content, starch films are preferred for packaging light-sensitive foods such as fruits and rolls (Alimi et al., 2022).

Moreover, starch is considered a promising polymer of natural origin because of its exceptional biodegradability, widespread availability, and ability to be modified or combined with other polymers (Jiang et al., 2020; Punia et al., 2020; Bangar et al., 2021). Starch-based edible films and coatings are gaining significant popularity for their potential to address environmental pollution (Bangar et al., 2021). Plantain starch stands out among cereal crops with approximately 40–60% amylose content, making it favorable for developing resistant starch-based biofilms (Hernández et al., 2008; Mohapatra et al., 2009). To further increase, tensile strength, Young's modulus, opacity, and crystallinity cellulose nanofibers can also be extracted from banana peel and incorporated into the starch film. This addition enhances the water resistance of the resulting film (Pelissari et al., 2012; Marta et al., 2022). Plantain starch can also be used in environmentally friendly capacitors as porous carbon material. Plantain starch can serve as the main component of electrode materials or gel electrolytes for supercapacitors, which are highly effective in conducting energy (Kasturi et al., 2019; Kumar et al., 2021; Marta et al., 2022).

10.8 CONCLUSION

Plantain starch has high potential as a food ingredient. Due to its high-resistant starch content, it is suitable for use in functional foods, including those for diabetics and babies. Furthermore, certain *Musa* cultivars possess distinct functional features, such as good stability at low and frozen temperatures (high freeze–thaw stability), making them suitable as ingredients in frozen dishes. However, natural plantain starch has several limitations, thus modifications are required to broaden its application. Studies have shown that modifying plantain starch, physically, chemically, or enzymatically improves its functional properties, including thermal stability and digestibility. Diabetic individuals can consume foods with a high SDS concentration.

Further investigations are needed to improve plantain starch properties using nonthermal approaches, such as ultrasonication, pulse-electric fields, and high-pressure treatments. Plantain starch (in flour form) can be used to replace wheat flour in gluten-free products for people with special diets, such as those with celiac disease and autism.

REFERENCES

Agama-Acevedo, E., & Bello-Perez, L. A. (2017). Starch as an emulsions stability: the case of octenyl succinic anhydride (OSA) starch. *Current Opinion in Food Science*, *13*, 78–83. https://doi.org/10.1016/j.cofs.2017.02.014

Agbemafle, R., Aggor-Woananu, S. E., & Hayford, D. (2017). Effect of cooking methods and ripening stages on the nutritional compositions of plantain (*Musa paradisiaca*). *International Journal of Food Science and Biotechnology*, 2 (4), 134–140. doi:10.11648/j.ijfsb.20170204.17

Akin-Ajani, O. D., Itiola, O. A. & Odeku, O. A. (2005). Effects of plantain and corn starches on the mechanical and disintegration properties of paracetamol tablets. *AAPS Pharmaceutical Science and Technology*, *6* (3), 57. https://doi.org/10.1208/pt060357

Akubor, P. I., & Igba, T. (2019). Effect of pre gelatinization and annealing on the chemical composition, functional and pasting properties of starch prepared from unripe banana fruits. *South Asian Journal of Food Technology and Environment*, *5*, 807–816. http://dx.doi.org/10.46370/sajfte.2019.v05i01.08

Alanís-López, P., Pérez-González, J., Rendón-Villalobos, R., Jiménez-Pérez, A., & Solorza-Feria, J. (2011). Extrusion and characterization of thermoplastic starch sheets from "Macho" banana. *Journal of Food Science*, *76* (6), E465–E471. https://doi.org/10.1111/j.1750-3841.2011.02254.x

Alimi, B. A., Workneh, T. S., & Zubair, B. A. (2022). Microstructural and physicochemical properties of biodegradable films developed from false banana (*Ensete ventricosum*) starch. *Heliyon*, *8* (3) e09148. https://doi.org/10.1016/j.heliyon.2022.e09148

Aparicio-Saguilán, A., Páramo-Calderón, D. E., Vázquez-León, L. A., Aguirre-Cruz, A., García-Muñoz, M. A., Valera-Zaragoza, M., ... & Ramírez-Hernández, A. (2024). Resistant starch from plantain (Macho *Musa paradisiaca* L.) and banana (Roatan *Musa sapientum* L.) varieties crosslinked with epichlorohydrin. *Food Production, Processing and Nutrition*, *6* (1), 66. https://doi.org/10.1186/s43014-024-00243-7

Auriema, B. E., Corrêa, F. J. B., de Toledo Guimarães, J., dos Santos Soares, P. T., Rosenthal, A., Zonta, E., Rosa, R. C., Luchese, R. H., Esmerino, E. A., & Mathias, S. P. (2021). Green banana biomass: physicochemical and functional properties and its potential as a fat replacer in a chicken mortadella. *LWT*, *140*, 110686. https://doi.org/10.1016/j.lwt.2020.110686

Bangar, S. P., Ashogbon, A. O., Singh, A., Chaudhary, V., & Whiteside, W. S. (2022). Enzymatic modification of starch: a green approach for starch applications. *Carbohydrate Polymers*, *287*, 119265. https://doi.org/10.1016/j.carbpol.2022.119265

Bangar, S. P., Purewal, S. S., Trif, M., Maqsood, S., Kumar, M., Manjunatha, V., & Rusu, A. V. (2021). Functionality and applicability of starch-based films: an eco-friendly approach. *Foods*, *10* (9), 2181. https://doi.org/10.3390/foods10092181

Bello-Pérez, L. A., Bello-Flores, C. A., del Carmen Nuñez-Santiago, M., Coronel-Aguilera, C. P., & Alvarez-Ramirez, J. (2015). Effect of the degree of substitution of octenyl succinic anhydride-banana starch on emulsion stability. *Carbohydrate Polymers*, *132*, 17–24. https://doi.org/10.1016/j.carbpol.2015.06.042

BeMiller, J. N. (2016). Reference module in food science. London: Elsevier.

BeMiller, J. N., & Huber, K. C. (2015). Physical modification of food starch functionalities. *Annual Review of Food Science and Technology*, *6* (1), 19–69. https://doi.org/10.1146/annurev-food-022814-015552

Bezerra, C. V., Amante, E. R., de Oliveira, D. C., Rodrigues, A. M. C., & da Silva, L. H. M. (2013a). Green banana (*Musa cavendishii*) flour obtained in spouted bed – effect of drying on physicochemical, functional and morphological characteristics of the starch. *Industrial Crops and Products, 41*, 241–249. https://doi.org/10.1016/j.indcrop.2012.04.035

Bezerra, C. V., Rodrigues, A. M. d. C., Amante, E. R., & Silva, L. H. M. d. (2013b). Nutritional potential of green banana flour obtained by drying in spouted bed. *Revista Brasileira de Fruticultura*, *35*, 1140–1146. https://doi.org/10.1590/S0100-29452013000400025

Bi, Y., Zhang, Y., Gu, Z., Cheng, L., Li, Z., Li, C., & Hong, Y. (2019). Effect of ripening on in vitro digestibility and structural characteristics of plantain (Musa ABB) starch. *Food Hydrocolloids*, *93*, 235–241. https://doi.org/10.1016/j.foodhyd.2019.02.043

Bi, Y., Zhang, Y., Jiang, H., Hong, Y., Gu, Z., Cheng, L., Li, Z., & Li, C. (2017). Molecular structure and digestibility of banana flour and starch. *Food Hydrocolloids*, *72*, 219–227. http://dx.doi.org/10.1016/j.food hyd.2017.06.003

Björck, I.; Liljeberg, H.; Ostman, E. (2000). Low glycemic index foods. *British Journal of Nutrition*, *83* (Suppl. S1), S149–S155. https://doi.org/10.1017/S0007114500001094

Burkhart, S., Underhill, S., & Raneri, J. (2022). Realizing the potential of neglected and underutilized bananas in improving diets for nutrition and health outcomes in the Pacific Islands. *Frontiers in Sustainable Food Systems*, *6,* 1–9. https://doi.org/10.3389/fsufs.2022.805776

Cahyana, Y., Annisa, N. D. N., Khoerunnisa, T. K., Sulastri, S., Marta, H., Rialita, T., ... & Şumnu, G. (2024). Banana starch modified by heat moisture treatment and annealing: study on digestion kinetics and enzyme affinity. *International Journal of Biological Macromolecules*, *258*, 128771. https://doi.org/10.1016/j.ijbiomac.2023.128771

Cahyana, Y., Titipanillah, R., Mardawati, E., Sukarminah, E., Rialita, T., Andoyo, R., ... & Handarini, K. (2018). Non-starch contents affect the susceptibility of banana starch and flour to ozonation. *Journal of Food Science and Technology*, *55*, 1726–1733. https://doi.org/10.1007/s13197-018-3085-2

Casarrubias-Castillo, M. G., Hamaker, B. R., Rodriguez-Ambriz, S. L., & Bello-Pérez, L. A. (2012). Physicochemical, structural, and digestibility properties of enzymatic modified plantain and mango starches. *Starch-Stärke*, *64* (4), 304–312. https://doi.org/10.1002/star.201100142

Chavez-Salazar, A., Alvarez-Barreto, C. I., Hoyos-Leyva, J. D., Bello-Pérez, L. A., & Castellanos-Galeano, F. J. (2022). Drying processes of OSA-modified plantain starch trigger changes in its functional properties and digestibility. *LWT*, *154*, 112846. https://doi.org/10.1016/j.lwt.2021.112846

Cordeiro, M. J. M., Veloso, C. M., Santos, L. S., Bonomo, R. C. F., Caliari, M., & Fontan, R. D. C. I. (2018). The impact of heat-moisture treatment on the properties of *Musa paradisiaca* L. starch and optimization of process variables. *Food Technology and Biotechnology*, *56* (4), 506. https://doi.org/10.17113/ftb.56.04.18.5490

De la Rosa-Millan, J., Lin, A. H. M., Osorio-Díaz, P., Agama-Acevedo, E., Hamaker, B. R., & Bello-Perez, L. A. (2015). Influence of annealing flours from raw and pre-cooked plantain fruit on cooked starch digestion rates. *Starch-Stärke*, *67* (1–2), 139–146. http://dx.doi.org/10.1002/star.201400136

De la Torre-Gutiérrez, L., Chel-Guerrero, L. A., & Betancur-Ancona, D. (2008). Functional properties of square banana (*Musa balbisiana*) starch. *Food Chemistry*, *106*, 1138–1144. https://doi.org/10.1016/j.foodchem.2007.07.044

dos Santos Costa, S., de Medeiros Almeida, M. C. B., Almeida, E. L., & Cavalcanti, M. T. (2019). Effects of low heat-moisture treatment in prata green banana starch (Musa AAB-Prata). *Food and Bioprocess Technology*, *12*, 1938–1944. https://link.springer.com/article/10.1007/s11947-019-02352-5

Fonseca, L. M., El Halal, S. L. M., Dias, A. R. G., & da Rosa Zavareze, E. (2021). Physical modification of starch by heat-moisture treatment and annealing and their applications: a review. *Carbohydrate Polymers*, *274*, 118665. https://doi.org/10.1016/j.carbpol.2021.118665

Fontes, S. D. M., Cavalcanti, M. T., Candeia, R. A., & Almeida, E. L. (2017). Characterization and study of functional properties of banana starch green variety of Mysore (Musa AAB-Mysore). *Food Science and Technology*, *37*, 224–231. http://dx.doi.org/10.1590/1678-457x.18916

Food and Drug Administration. (2001). Secondary direct food additives permitted in food for human consumption. *Federal Register*, *66* (123), 33829–33830.

Fu, J., Wang, Y., Tan, S., & Wang, J. (2021). Effects of banana resistant starch on the biochemical indexes and intestinal flora of obese rats induced by a high-fat diet and their correlation analysis. *Frontiers in Bioengineering and Biotechnology*, *9*, 575724. https://doi.org/10.3389/fbioe.2021.575724

González, N. M., Gómez, P. P., & García, M. E. R. (2019). Effect of the addition of potassium and magnesium ions on the thermal, pasting, and functional properties of plantain starch (*Musa paradisiaca*). *International Journal of Biological Macromolecules*, *124*, 41–49. https://doi.org/10.1016/j.ijbiomac.2018.11.051

Gu, Z., Cheng, G., Sha, X., Wu, H., Wang, X., Zhao, R., ... & Jiang, H. (2024). Heat-moisture treatment of freshly harvested high-amylose maize kernels improves its starch thermal stability and enzymatic resistance. *Carbohydrate Polymers*, *340*, 122303. https://doi.org/10.1016/j.carbpol.2024.122303

Haruna, M.S, Ezeanayanaso, C. S., Olagoke, M. F., Idris, R., & Lawal, O. S. (2023). Characterization of oxidized plantain (*Musa paradisiaca*) starch as influenced by reaction temperature. *IOSR Journal of Applied Chemistry (IOSR-JAC)*, 16 (3), 40–50. doi:10.9790/5736-1603014050

Hernández, O., Emaldi, U., & Tovar, J. (2008). In vitro digestibility of edible films from various starch sources. *Carbohydrate Polymers*, *71* (4), 648–655. http://dx.doi.org/10.1016/j.carbpol.2007.07.016

Honfo, F. G., Togbe, E. C., Dekker, M., Akissoe, N., & Ahohuendo, B. C. (2022). Nutritional and functional characterisation of flour from six plantain (Musa spp.) cultivars grown in Benin. *International Food Research Journal*, *29* (5), 1101–1109. https://doi.org/10.47836/ifrj.29.5.12

Hong, Y., Liu, X. (2018). Pre-gelatinized modification of starch. In: Sui, Z., Kong, X. (eds) Physical Modifications of Starch. Springer, Singapore. https://doi.org/10.1007/978-981-13-0725-6_4

Hoyos-Leyva, J. D., Bello-Pérez, L. A., Agama-Acevedo, E., & Alvarez-Ramirez, J. (2015). Optimising the heat moisture treatment of Morado banana starch by response surface analysis. *Starch-Stärke*, *67* (11-12), 1026–1034. https://doi.org/10.1002/star.201500149

Ibikunle, A. A., Yussuf, S. T., Sanyaolu, N. O., Ogundare, S. A., Oderinu, O. F., Atewolara-Odule, O. C., & Ogunmoye, A. O. (2022). Effects of phosporylation and cross-linking on starches obtained from banana (*Musa acuminata*) and plantains (*Musa paradisiaca* and *Musa balbisiana*). *FUW Trends in Science & Technology Journal*, *7* (3), 283–290.

Izidoro DR, Scheer AP, Sierakowski MR and Haminiuk CWI (2008). Influence of green banana pulp on the rheological behaviour and chemical characteristics of emulsions (mayonnaises). *LWT – Food Science and Technology*, *41* (6): 1018–1028. https://doi.org/10.1016/j.lwt.2007.07.009

Jiang, T., Duan, Q., Zhu, J., Liu, H., & Yu, L. (2020). Starch-based biodegradable materials: challenges and opportunities. *Advanced Industrial and Engineering Polymer Research*, *3* (1), 8–18. https://doi.org/10.1016/j.aiepr.2019.11.003

Kasturi, P. R., Ramasamy, H., Meyrick, D., Sung Lee, Y., & Kalai Selvan, R. (2019). Preparation of starch-based porous carbon electrode and biopolymer electrolyte for all solid-state electric double layer capacitor. *Journal of Colloid and Interface Science*, *554*, 142–156. https://doi.org/10.1016/j.jcis.2019.06.081

Kaur, B., Ariffin, F., Bhat, R., & Karim, A. A. (2012). Progress in starch modification in the last decade. *Food Hydrocolloids*, *26* (2), 398–404. https://doi.org/10.1016/j.foodhyd.2011.02.016

Kumar, S., Dhapola, P. S., Pandey, S. P., Singh, P. K., & Chauhan, M. (2021). Corn-starch based porous carbon and IL based electrolyte for high efficient supercapacitor. *Materials Today: Proceedings*, *34*, 842–845. https://doi.org/10.1016/j.matpr.2020.06.050

Kumari, B., & Sit, N. (2023). Comprehensive review on single and dual modification of starch: methods, properties and applications. *International Journal of Biological Macromolecules*, 253, 126952. https://doi.org/10.1016/j.ijbiomac.2023.126952

Lee, C. S., & Chung, H. J. (2022). Enhancing resistant starch content of high amylose rice starch through heat–moisture treatment for industrial application. *Molecules*, *27* (19), 6375. http://dx.doi.org/10.3390/molecules27196375

Lopez-Silva, M., Bello-Perez, L. A., Castillo-Rodriguez, V. M., Agama-Acevedo, E., & Alvarez-Ramirez, J. (2020). In vitro digestibility characteristics of octenyl succinic acid (OSA) modified starch with different amylose content. *Food Chemistry*, *304*, 125434. https://doi.org/10.1016/j.foodchem.2019.125434

Ma, H., Liu, M., Liang, Y., Zheng, X., Sun, L., Dang, W., ... & Liu, C. (2022). Research progress on properties of pre-gelatinized starch and its application in wheat flour products. *Grain & Oil Science and Technology*, *5* (2), 87–97. https://doi.org/10.1016/j.gaost.2022.01.001

Ma, M., Gu, Z., Cheng, L., Li, Z., Li, C., & Hong, Y. (2024). Effect of hydrocolloids on starch digestion: a review. *Food Chemistry*, 138636. https://doi.org/10.1016/j.foodchem.2024.138636

Marta, H., Cahyana, Y., Djali, M., & Pramafisi, G. (2022). The properties, modification, and application of banana starch. *Polymers*, *14* (15), 3092. https://doi.org/10.3390/polym14153092

Marta, H., Cahyana, Y., Djali, M., Arcot, J., & Tensiska, T. (2019). A comparative study on the physicochemical and pasting properties of starch and flour from different banana (Musa spp.) cultivars grown in Indonesia. *International Journal of Food Properties*, *22* (1), 1562–1575. http://dx.doi.org/10.1080/10942912.2019.1657447

Mohapatra, D., Mishra, S., & Meda, V. (2009). Plantains and their postharvest uses: an overview. *Stewart Postharvest Review*, *5* (5), 1–11. http://dx.doi.org/10.2212/spr.2009.5.4

Mordor Intelligence Research & Advisory. (2023, June). Industrial Starch Market Size & Share Analysis – Growth Trends & Forecasts (2024 – 2029). Mordor Intelligence. Retrieved August 22, 2024, from www.mordorintelligence.com/industry-reports/industrial-starches-market

Munir, H., Alam, H., Nadeem, M. T., Almalki, R. S., Arshad, M. S., & Suleria, H. A. R. (2024). Green banana resistant starch: a promising potential as functional ingredient against certain maladies. *Food Science & Nutrition* 12, 3787–3805. https://doi.org/10.1002/fsn3.4063

Nasrin, T. A. A., Noomhorm, A., & Anal, A. K. (2015). Physico-chemical characterization of culled plantain pulp starch, peel starch, and flour. *International Journal of Food Properties*, *18* (1), 165–177. https://doi.org/10.1080/10942912.2013.828747

Nor Sham, N. N. I., Yusoff, A., & Ab Mutalib, S. R. (2024b). Effect of cross-linking and pre-gelatinization on morphology and crystallinity characteristics of unripe banana starch (*Musa balbisiana*). *Science Letters (ScL)*, *18* (1), 88–96. doi:10.24191/sl.v18i1.2461

Nor Sham, N.N.I., Yusoff, A., Ahmad, N., & Mutalib, S. R. A. (2024a). Response surface methodology for optimising cross-linking conditions of unripe banana (*Musa balbisiana* VAR. Abu) starch. *Malaysian Journal of Analytical Sciences*, *28* (1), 127–140.

Nurmilah, S., & Subroto, E. (2021). Chemical modification of starch for the production of resistant starch type-4 (RS4): a review. *International Journal of Engineering Trends Technology*, *69*, 45–50. https://doi.org/10.14445/22315381/IJETT-V69I7P206

Nwokocha, L. M., & Williams, P. A. (2009). Some properties of white and yellow plantain (*Musa paradisiaca*, Normalis) starches. *Carbohydrate Polymers*, *76* (1), 133–138. https://doi.org/10.1016/j.carbpol.2008.10.012

Ogunbusola, E. M., Alabi, O. O., Sanni, T. A., Adubiaro, H. O., Araoye, K. T., Jaiyeoba, C. N., & Oni, K. (2023, April). Physicochemical, functional and viscosity profile of starches from three banana cultivars as affected by oxidation and acetylation. In *2023 International Conference on Science, Engineering and Business for Sustainable Development Goals (SEB-SDG)* (Vol. 1, pp. 1–8). https://doi.org/10.1109/SEB-SDG57117.2023.10124550

Okunlola, A., & Adedeji, A. T. (2021). Cross linked plantain starch-urea as a polymer in matrix tablets of ambroxol hydrochloride. *British Journal of Pharmacy*, *5* (1). https://doi.org/10.5920/bjpharm.774

Olatunde, G. O., Arogundade, L. K., & Orija, O. I. (2017). Chemical, functional and pasting properties of banana and plantain starches modified by pre-gelatinization, oxidation and acetylation. *Cogent Food & Agriculture*, *3* (1), 1283079. http://dx.doi.org/10.1080/23311932.2017.1283079

Olawoye, B., Fagbohun, O. F., Gbadamosi, S. O., & Akanbi, C. T. (2020). Succinylation improves the slowly digestible starch fraction of cardaba banana starch. A process parameter optimization study. *Artificial Intelligence in Agriculture*, *4*, 219–228. https://doi.org/10.1016/j.aiia.2020.09.004

Olawoye, B., Jolayemi, O. S., Akinyemi, T. Y., Nwaogu, M., Oluwajuyitan, T. D., Popoola-Akinola, O. O., ... & Akanbi, C. T. (2023). Modification of Starch. In *Starch: Advances in modifications, technologies and applications* (pp. 11–54). Cham: Springer International Publishing. https://doi.org/10.1007/978-3-031-35843-2_2

Otegbayo, B., Lana, O., & Ibitoye, W. (2010). Isolation and physicochemical characterization of starches isolated from plantain (*Musa paradisiaca*) and cooking banana (*Musa sapientum*). *Journal of Food Biochemistry*, *34* (6), 1303–1318. https://doi.org/10.1111/j.1745-4514.2010.00354.x

Palma-Rodríguez, H. M., Aguirre-Álvarez, G., Chavarría-Hernández, N., Rodríguez-Hernández, A. I., Bello-Pérez, L. A., & Vargas-Torres, A. (2012). Oxidized banana starch–polyvinyl alcohol film: partial characterization. *Starch-Stärke*, *64* (11), 882–889. http://dx.doi.org/10.1002/star.201200035

Park, S., & Kim, Y. R. (2021). Clean label starch: production, physicochemical characteristics, and industrial applications. *Food Science and Biotechnology*, *30*, 1–17. https://doi.org/10.1007/s10068-020-00834-3

Pelissari, F. M., Andrade-Mahecha, M. M., Sobral, P. J. D. A., & Menegalli, F. C. (2012). Isolation and characterization of the flour and starch of plantain bananas (*Musa paradisiaca*). *Starch-Stärke*, *64* (5), 382–391. https://doi.org/10.1002/star.201100133

Pham, T. T. H., Nguyen, T. T., Nguyen, V. K., Nguyen, T. D., Pham, T. T., Nguyen, T. H., & Pham, N. A. (2024). Effect of oxidation on physicochemical characteristics of thermoplastic canna starch. *Vietnam Journal of Chemistry*, *62* (S1), 15–22. http://dx.doi.org/10.1002/vjch.202300260

Pivetta, F. P., Silva, M. N. D., Tagliapietra, B. L., & Richards, N. S. D. S. (2019). Addition of green banana biomass as partial substitute for fat and encapsulated *Lactobacillus acidophilus* in requeijão cremoso processed cheese. *Food Science and Technology*, *40*, 451–457. https://doi.org/10.1590/fst.03919

Pongsuwan, C., Boonsuk, P., Sermwittayawong, D., Aiemcharoen, P., Mayakun, J., & Kaewtatip, K. (2022). Banana inflorescence waste fiber: an effective filler for starch-based bioplastics. *Industrial Crops and Products*, *180*, 114731. https://doi.org/10.1016/j.indcrop.2022.114731

Punia, B. S., Sunooj, K. V., Navaf, M., Phimolsiripol, Y., & Whiteside, W. S. (2024). Recent advancements in cross-linked starches for food applications-a review. *International Journal of Food Properties*, *27* (1), 411–430. https://doi.org/10.1080/10942912.2024.2318427

Punia, S. (2020). Barley starch modifications: physical, chemical and enzymatic-a review. *International Journal of Biological Macromolecules*, *144*, 578–585. https://doi.org/10.1016/j.ijbiomac.2019.12.088

Punia, S., Sandhu, K. S., Dhull, S. B., Siroha, A. K., Purewal, S. S., Kaur, M., & Kidwai, M. K. (2020). Oat starch: physico-chemical, morphological, rheological characteristics and its applications-a review. *International Journal of Biological Macromolecules*, *154*, 493–498. https://doi.org/10.1016/j.ijbiomac.2020.03.083

Quintero-Castaño, V. D., Castellanos-Galeano, F. J., Álvarez-Barreto, C. I., Lucas-Aguirre, J. C., Bello-Pérez, L. A., & Rodríguez-Garcia, M. E. (2020a). Starch from two unripe plantains and esterified with octenyl succinic anhydride (OSA): partial characterization. *Food Chemistry*, *315*, 126241. https://doi.org/10.1016/j.foodchem.2020.126241

Quintero-Castaño, V. D., Castellanos-Galeano, F. J., Álvarez-Barreto, C. I., Bello-Pérez, L. A., & Alvarez-Ramirez, J. (2020b). In vitro digestibility of octenyl succinic anhydride-starch from the fruit of three Colombian Musa. *Food Hydrocolloids*, *101*, 105566. https://doi.org/10.1016/j.foodhyd.2019.105566

Rashwan, A. K., Younis, H. A., Abdelshafy, A. M., Osman, A. I., Eletmany, M. R., Hafouda, M. A., & Chen, W. (2024). Plant starch extraction, modification, and green applications: a review. *Environmental Chemistry Letters*, *22*, 2483–2530. doi:10.1007/s10311-024-01753-z

Robinson, J. C., & Saúco, V. G. (2010). *Bananas and plantains* (Vol. 19). Cabi.

Rong, L., Fei, W., Wang, Z., Chen, X., Wen, H., Xie, J., & Shen, M. (2024). Modulation of starch digestibility using non-thermal processing techniques: a review. *Grain & Oil Science and Technology*, 7 (3), 209–218. https://doi.org/10.1016/j.gaost.2024.05.002

Rostamabadi, H., Demirkesen, I., Mert, B., Barua, S., Colussi, R., Frasson, S.F., Wang, Y., & Falsafi, S. R. (2024). Starch digestibility: how single, double, and multiple physicochemical modifications change nutritional attributes of starch?. *Food Frontiers*, 5, 1410–1444. https://doi.org/10.1002/fft2.396

Salazar, D., Arancibia, M., Lalaleo, D., Rodríguez-Maecker, R., López-Caballero, M. E., & Montero, M. P. (2022). Physico-chemical properties and filmogenic aptitude for edible packaging of Ecuadorian discard green banana flours (*Musa acuminanta* AAA). *Food Hydrocolloids*, *122*, 107048. http://dx.doi.org/10.1016/j.foodhyd.2021.107048

Sharma, A., Bist, Y., Chakkingal, F., Kumar, Y., & Saxena, D. C. (2023). Preparation, characterization, and utilization of crosslinked and dual-modified banana starch for stabilizing pickering emulsion. *Starch-Stärke*, *75* (11–12), 2300084. https://doi.org/10.1002/star.202300084

Shittu, R., Lasekan, O., Karim, R., & Sulaiman, R. (2016). Plantain-starch: microstructural, physicochemical, and morphological characteristics of two cultivars grown in Malaysia. *Starch – Stärke, 68* (11-12), 1187–1195. https://doi.org/10.1002/star.201500262

Shrestha, S., Sadiq, M. B., & Anal, A. K. (2018). Culled banana resistant starch- soy protein isolate conjugate based emulsion enriched with astaxanthin to enhance its stability. *International Journal of Biological Macromolecules*, *120*, 449–459. https://doi.org/10.1016/j.ijbiomac.2018.08.066

Sojinu, O. S., Biliaminu, N. T., Mosaku, A. M., Makinde, K. O., Adeniji, T. H., & Adeboye, B. M. (2021). The implications of ripening agents on chemical compositions of plantain (*Musa paradisiaca*). *Heliyon*, *7* (6). https://doi.org/10.1016/j.heliyon.2021.e07123

Steeneken, P. A., & Woortman, A. J. (2008). Surface effects in the acetylation of granular potato starch. *Carbohydrate Research*, *343* (13), 2278–2284. https://doi.org/10.1016/j.carres.2008.04.034

Surendra Babu, A., Mohan Naik, G. N., James, J., Aboobacker, A. B., Eldhose, A., & Jagan Mohan, R. (2018). A comparative study on dual modification of banana (*Musa paradisiaca*) starch by microwave irradiation and cross-linking. *Journal of Food Measurement and Characterization*, *12* (3), 2209–2217. https://doi.org/10.1007/s11694-018-9837-x

Thomann, J. A., Polhuis, M., Lasorsa, A., Heeres, H. J., & Heeres, A. (2024). Effects of mineral elements and annealing on the physicochemical properties of native potato starch. *ChemEngineering*, *8* (3), 60. https://doi.org/10.3390/chemengineering8030060

Vaidya, R. H., & Sheth, M. K. (2011). Processing and storage of Indian cereal and cereal products alters its resistant starch content. *Journal of Food Science and Technology*, *48* (5), 622–627. https://doi.org/10.1007/s13197-010-0151-9

Vega-Rojas, L. J., Londoño-Restrepo, S. M., & Rodriguez-García, M. E. (2021). Study of morphological, structural, thermal, and pasting properties of flour and isolated starch from unripe plantain (*Musa paradisiaca*). *International Journal of Biological Macromolecules*, *183*, 1723–1731. https://doi.org/10.1016/j.ijbiomac.2021.05.144

Viana, E. B. M., Oliveira, N. L., Ribeiro, J. S., Almeida, M. F., Souza, C. C. E., Resende, J. V., ... & Veloso, C. M. (2022). Development of starch-based bioplastics of green plantain banana (*Musa paradisiaca* L.) modified with heat-moisture treatment (HMT). *Food Packaging and Shelf Life*, *31*, 100776. https://doi.org/10.1016/j.fpsl.2021.100776

Wang, J., Euring, M., Ostendorf, K., & Zhang, K. (2022). Biobased materials for food packaging. *Journal of Bioresources and Bioproducts*, *7* (1), 1–13. https://doi.org/10.1016/j.jobab.2021.11.004

Wang, S., & Ren, F. (2020). Rheological, pasting, and textural properties of starch. In: Wang, S. (eds) Starch structure, functionality and application in foods. Springer, Singapore. 121–129. http://dx.doi.org/10.1007/978-981-15-0622-2_7

Wani, I. A., Sogi, D. S., Wani, A. A., Gill, B. S., & Shivhare, U. S. (2010). Physico-chemical properties of starches from Indian kidney bean (*Phaseolus vulgaris*) cultivars. *International Journal of Food Science & Technology*, *45* (10), 2176–2185. http://dx.doi.org/10.1111/j.1365-2621.2010.02379.x

Yu, A. H. M., Phoon, P. Y., Ng, G. C. F., & Henry, C. J. (2020). Physicochemical characteristics of green banana flour and its use in the development of konjac- green banana noodles. *Journal of Food Science*, *85* (10), 3026–3033. https://doi.org/10.1111/1750-3841.15458

Zhang, B., Mei, J. Q., Chen, B., & Chen, H. Q. (2017). Digestibility, physicochemical and structural properties of octenyl succinic anhydride-modified cassava starches with different degree of substitution. *Food Chemistry*, *229*, 136–141. https://doi.org/10.1016/j.foodchem.2017.02.061

Zhang, Z., & Bao, J. (2023). Recent advances in modification approaches, health benefits, and food applications of resistant starch. *Starch-Stärke*, *75* (9-10), 2100141. http://dx.doi.org/10.1002/star.202100141

Zia-ud-Din, Xiong, H., & Fei, P. (2017). Physical and chemical modification of starches: a review. *Critical Reviews in Food Science and Nutrition*, *57* (12), 2691–2705. https://doi.org/10.1080/10408398.2015.1087379

11 Utilization of Mango Seed Starch in the Manufacture of Polyester-Based Bioplastics

Edla Maria Bezerra Lima, Rossana Mara da Silva Moreira Thiré, Antonieta Middea, Renata Nunes Oliveira, Ana Carolina Corrêa, Reiner Neumann, and José Manoel Marconcini

11.1 INTRODUCTION

The circular economy has emerged as a dynamic approach in production and environmental protection, in which the reuse of residues from the processing of major commodities provides new raw materials for other industrial applications. In this context, the production, storage, safety, and environmentally responsible disposal of food waste are strategic objectives of sustainable economies. Mango is one of the world's major commodities, and its processing generates large amounts of agro-industrial waste. Native to South Asia, particularly Myanmar and eastern India, mango is widely distributed across the world's tropical regions. The fruit consists of three distinct parts: the pericarp (peel), the fleshy mesocarp (pulp), and the endocarp (a hard, woody stone containing the seed kernel), as shown in Figure 11.1.

Broadly speaking, mangoes can be classified based on the size and nature of their seeds, including seedless varieties. The "*Mahachanok*" variety is a hybrid cultivar of the pink mango and a wild mango known as "*Nang klanwan*", found in the Chiang Mai region of northern Thailand. Its seed is long and thin, and some fruits may even be seedless, making nearly 100% of the fruit edible. However, from an economic standpoint, its production is less competitive than seeded varieties such as *Mangifera indica L.*, *Tommy atkins*, *Van dyke*, *Manzano*, *Keitt*, *Azúcar*, *Criollo*, *Farchis*, *Yulina*, and others, which are among the most commercially popular in countries such as India (26 million tons), Indonesia (4.1 million tons), China (3.8 million tons), Pakistan (2.8 million tons), and Mexico (2.5 million tons). Brazil, in sixth place, produced 1.382 million tons in 2020 and is projected to produce 2.383 million tons by 2030 (Kist, 2023). Continuous production, driven by domestic and international demand, is maintained through irrigation (Figure 11.2).

Several technologies have been developed to enhance productivity. These include efficient water management, pruning techniques throughout different crop stages, floral induction to enable year-round production, and mineral nutrition to support plant and fruit development. Additionally, they focus on controlling the major diseases affecting each producing country and on production losses, which contribute to issues with residual biomass disposal. However, in countries with inadequate production techniques and controls, approximately 30% of mango production is lost before reaching consumers (Xavier et al., 2009). From the remaining 70%, the juice and fruit pulp manufacturers discard around 40% of the fruit's weight as organic waste, including peels, seeds, and pulp (Nascimento Filho & Franco, 2015; Lobo & Silva, 2003), with an annual production of 123,000 metric tons (Bangar et al., 2021). Recycling industries are now processing much of this waste into animal feed

 DOI: 10.1201/9781003534914-11

FIGURE 11.1 Structure of mango fruit (illustration by authors).

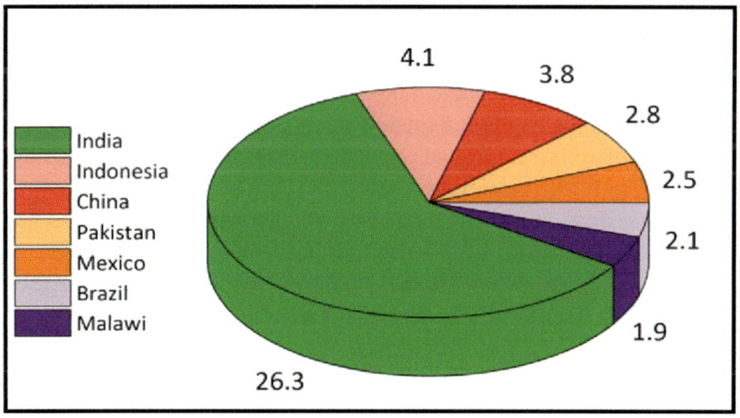

FIGURE 11.2 Mango's production in the world in 2022 (million tons) (based on Kist, 2023).

and raw materials for new products or bioproducts, aiming to create opportunities for production, profit, and reduced environmental impact.

11.2 MANGO SEED CHARACTERIZATION

The chemical characterization of mango has been extensively studied by numerous research teams, particularly in areas such as planting, harvest/postharvest management, agro-industrial processing, nutrition, pharmaceuticals, cosmetics, and the use of waste for developing new materials. These efforts aim for the total, rational use of the fruit, minimizing losses and environmental impact.

Mango varieties exhibit a wide range of colors, from deep green to pinkish yellow, and shapes, from elongated to spherical. Among the various mango varieties, it is important to note that mango's chemical composition varies due to climatic, geographic, and cultivation differences, including factors such as moisture content, topography, soil types, and planting and postharvest techniques. Comparative studies on different varieties (Maldonado et al., 2019) showed that the average nutritional composition includes fiber and carbohydrates as macronutrients (10–32% in ripe pulp), proteins (0–5%), and amino acids (alanine, arginine, glycine, serine, leucine, and isoleucine). It also contains lipids (0.75–1.7%) and organic acids, with citric acid being the major one (0.13%–0.71% FW). Mangoes are also a rich source of micronutrients, including vitamins such as vitamin C (9.79–186 mg/100 g of mango pulp), vitamin A (1,000–6,000 IU), and smaller amounts vitamins E and K. No vitamin D has been detected in any cultivars to date. Except for biotin, all other B vitamins have been found in mango fruit. Moreover, mangoes are an important source of polyphenols, including catechins, quercetin, kaempferol, and mangiferin. Carotenoids, organic acids, and volatile compounds are valuable for medicinal applications and serve as indicators of fruit quality. These concentrations vary depending on the ripeness of the mango pulp and peel.

Although biodegradable, mango biomass requires time to decompose, posing environmental pollution risks. Agro-industrial wastes, which are rich in polyphenolic compounds (González et al., 2007; Cataneo et al., 2008), also contain valuable nutrients. Mango peels are rich in calcium, sodium, potassium, iron, magnesium, manganese, fiber, vitamin C, proteins, carbohydrates, and pectin (França, 2014). Efficient utilization of these residues could create employment opportunities and add value to agro-industrial by-products while also preventing environmental pollution (Malacrida et al., 2007).

When producing mango juice, jams, preserves, and the processing of by-products, waste in the form of peels and seeds accounts for 40% of the fruit's total weight (Manhongo et al., 2021). Improper disposal of this leads to uncontrolled decomposition, promoting the growth of microorganisms, pathogens, and zoonoses that can contaminate the environment, particularly the water provision. This, in turn, contributes to the spread of diseases, contamination from leachate, and the proliferation of rodents. These issues are exacerbated during the rainy season, which increases the dispersion of contaminants. Alternatively, in rural areas, mango peels can be used as cattle feed, although whole seeds pose a choking hazard. Ground or whole seeds can also be used as fertilizer, adding value to the production chain with minimal waste.

The mango seed kernel makes up about 20% of the whole fruit, though depending on the variety, it can constitute 45–75% of the mango seed. Its composition, on a dry weight basis, is approximately 6–7.5% protein, 11–11.4% lipids, 1–2% ash, 67.8–77% carbohydrates, and 2–2.2% fiber (Yatnatti et al., 2014; El-Bagoury & Kassab, 2005; Choudhary et al., 2023), highlighting its potential for reuse. Based on world production, it is estimated that around 123,000 metric tons (Bangar et al., 2021) of waste are generated annually. In the industry, juice processing plants and other facilities are typically far from mango production areas, creating significant volumes of peels and seeds that cannot be stored due to rapid deterioration. This requires swift and efficient disposal to prevent environmental impact. The utilization of these residues can be optimized through a process resembling a biorefinery, as described by Silva et al. (2021), which builds on the work of Arora et al. (2018) and Ahlgren et al. (2013). In the framework of the circular economy, this approach focuses on extracting bioproducts of economic and biological interest (such as chemicals and pharmaceuticals) from mango waste. Thus, a bioproduct can be derived from a co-product using different processing methods, as demonstrated in Table 11.1, where the amount of raw material processed from the kernel by isolated extraction was 80 times higher than integrated extraction. The processing of the integument using alternative equipment processed 16% more raw material (Silva et al., 2021).

Specifically, the extraction and concentration of starch from the mango seed comprise the depulping process, with the following steps: cleaning, heat treatment, Figure 11.3, pulping using an extractor, drying, and storage (Lima et al., 2021). The chemical, thermal, and microstructural

TABLE 11.1
Quantity of Raw Material Processed and Bioproducts Obtained (adapted from Ahlgren et al., 2013)

Processed Co-Product	Bioproducts	Extraction Process	Quantity of Raw Material Processed per Month (kg/month)
Kernel	Starch	Isolated extraction	10.8
		Integrated extraction	0.13
Integument	Lignin and cellulose	Acetosolv extraction with conventional equipment	0.31
		Acetosolv extraction with alternative equipment	0.36
Peel	Pectin and phenolics	Soxhlet/ultrasound	1.8
		ESP Juice Extractor/ ESP Juice Extractor	0.28
		Soxhlet/ESP Juice Extractor	1.8
		ESP juice extractor/ultrasound	0.28

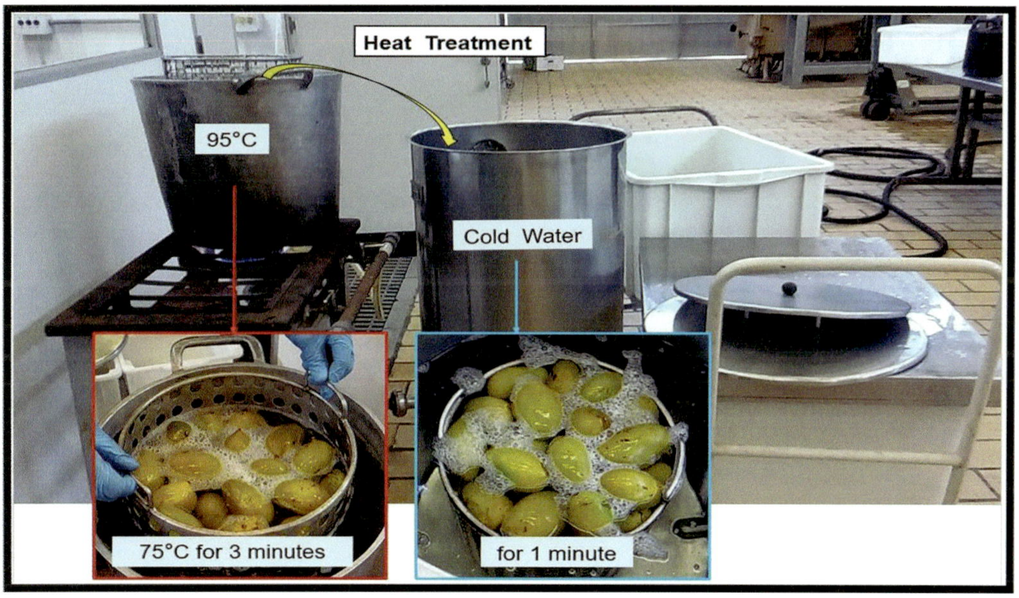

FIGURE 11.3 Pulping – heat treatment stage (photographs by the authors).

characterizations of *Mangifera indica L.*, which is the main type of mango produced in Brazil, were carried out by Lima et al. (2021) and are summarized in Table 11.2.

Starch and cellulose are present throughout the fruit, especially in the mango seed, which has two very distinct phases: the kernel, rich in starch, and the integument, rich in cellulose. Other constituents such as lipids, proteins, oils, and others are present in the amorphous phase of seed. In vegetables, as well as in mango, native starch has semicrystalline structure and is mainly composed by of two polysaccharides: amylose, mostly linear, and amylopectin, a branched molecule. Native starch granules can present three types of crystalline pattern (A, B, or C) according to the source. A-type is usually present in cereals such as wheat and corn. Tubers as potato present starch with B-type, while C-type, a superposition of the structures of A + B is found in some vegetables as

TABLE 11.2
Characterization of *Mangifera indica L.* Mango's Kernel and Integument (adapted from Lima et al., 2021a)

FIGURE 11.17 Seed.

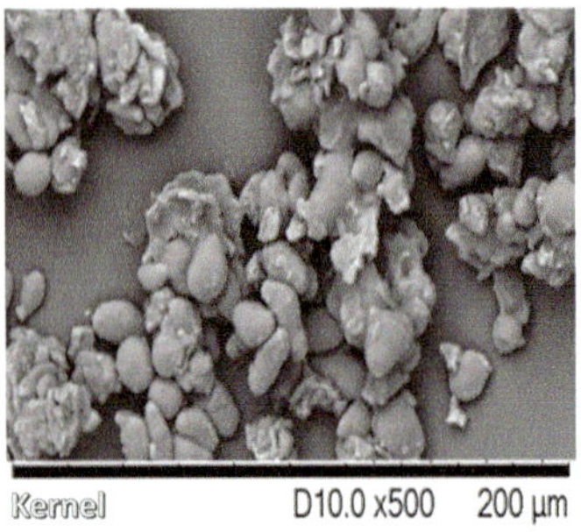

FIGURE 11.18 Kerel.

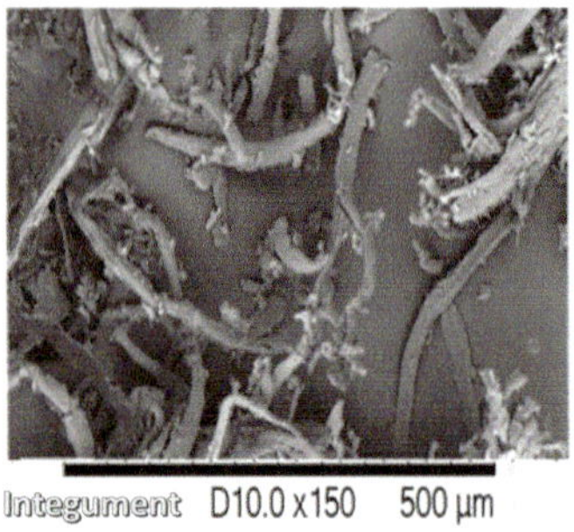

FIGURE 11.19 Integument.

Chemical properties (g/100g)	Moisture = 5.85; Ash = 1.03; Protein = 1.09; Ether extract = ND Carbohydrate = 92.03	Moisture = 7.72; Ash = 2.35; Protein = 4.43; Ether extract = 7.96 Carbohydrate = 77.44
Acid detergent fiber ADF (%)	5.76	65.98
Neutral detergent fiber (NDF) (%)	18.04	70.74
Density (g cm^{-3})	1.47	1.55
Average particle size	D_1 was around 44 µm	D_1 was around 207 µm
X-ray diffraction (XRD)	Presented semi-crystalline starch as a main component	Composed by semi-crystalline cellulose fibers
Fourier transform infrared spectroscopy (FTIR)	Starch (predominant), lignin and cellulose (less)	Cellulose (predominant), lignin, hemicelluloses and starch (less)
Scanning electron microscopy (SEM)	Exhibits granules due to starch	Presents a fibrous (cellulose) morphology
Thermal analysis	Cellulose thermal degradation was observed in the mango's integument Dynamic scanning calorimetry (DSC) analysis identified the probable presence of cellulose and starch both in mango's kernel and integument Higher amounts of starch in mango's kernel and higher amounts of cellulose in mango's integument	

pea (Zobel, 1988; Hizukuri, 1985). The X-ray diffraction of the mango kernel (Figure 11.4) shows that starch has a semi-crystalline structure, whose crystalline regions are packed according to an A-type pattern, meaning that the double helix of amylopectin chains have a regular and dense arrangement. This crystalline pattern could be identified since the diffractogram show strong peaks in the interplanar spaces (d) at 5.9; 5.2, and 3.8 Å due to angular distance observed in 2θ at 15°, 17°, 18°, and 23° 1988 (Zobel, 1988; Mestres et al., 1996; Lima et al., 2017; Santos Jr. et al., 2018).

Cellulose is concentrated in the mango seed, mainly in the integument, showing type I cellulose peaks located at 2θ = 14°, 16°, 23°, and 35° (Figure 11.4), which positions are the reflections of the crystallographic planes (101), (100), (002), and (040), unlike other polymorphs of types II, III, and IV. It is noteworthy that cellulose type II, obtained from man-made alkaline treatment, presents peaks at 2θ = 11°, 20°, 22°, and 37°, which correspond to the diffraction of the crystallographic planes (101), (100), (002), and (040).

The characterization of the functional groups of the mango seed integument and kernel was carried out by Fourier transform infrared spectroscopy (FTIR) analysis (Lima et al 2021a,

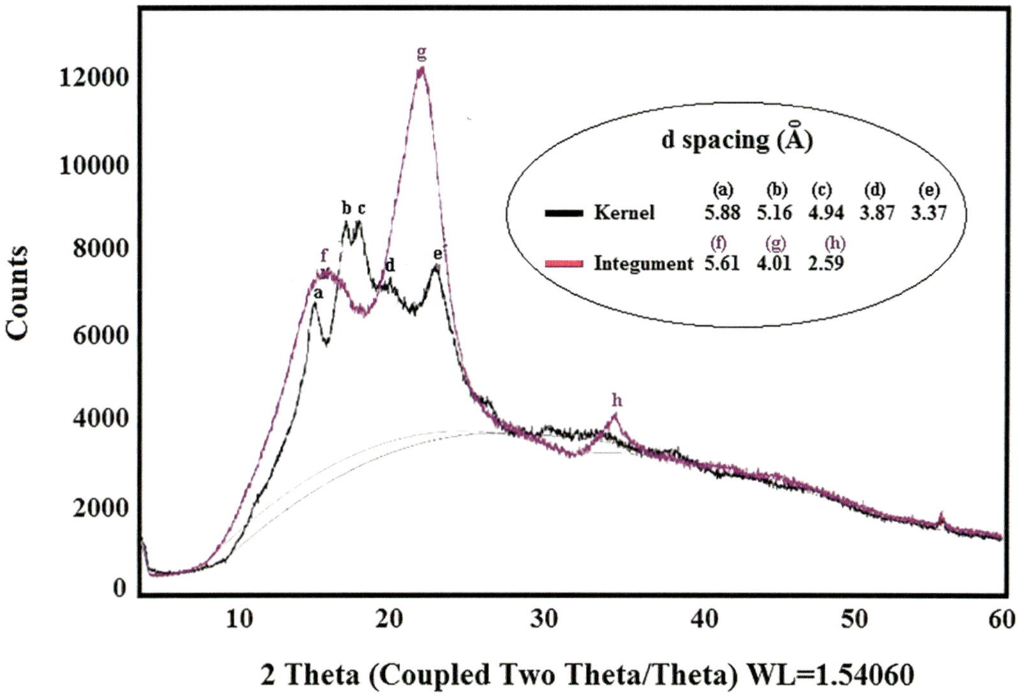

FIGURE 11.4 2θ XRD of Mango Kernel (Type A starch) and Mango Integument (Type I cellulose).

Figure 11.5). In both samples, the FTIR spectra show the cellulose bands. The OH stretching vibration, a broad band at 3342 cm^{-1} (Figure 11.5a) and 3297 cm^{-1} (Figure 11.5b), indicates the presence of water, as well as alcohols, phenols, and carboxylic acids (García et al., 2015; Ambrose et al., 2016), and was present in the integument sample, while the bands around 1370 cm^{-1}, 1155 cm^{-1} (Figure 11.5a) and 1336 cm^{-1}, 1151 cm^{-1} (Figure 11.5b) suggest the presence of carbonyl groups in both the integument and kernel. However, the bands at 1100–1150 cm^{-1} may be related to C–O, C–C, and C–O–H stretching of starch (Warren et al., 2016). The vibrations at 2916 cm^{-1} and 2850 cm^{-1} of the kernel sample are attributed to symmetric and asymmetric C–H stretching of aliphatic acids. According to Warren et al. (2016), the bands around 2900–3000 cm^{-1} may be attributed to C–H stretching of starch. The band at 1471 cm^{-1} can be attributed to C–H vibration (Carrot & Carrot, 2007; Lazzari et al., 2016; Warren et al., 2016). There is also a typical absorption band around 1730 cm^{-1} in both spectra, attributed to the stretching of the C=O bond in the aliphatic carboxylic, aryl ester, and acetyl groups in the xylan component of hemicelluloses and lignin (Iqbal et al., 2009; Meireles et al., 2020; Neto et al., 2013). The vibrations at 1638 cm^{-1} (Figure 11.5a) and at 645 cm^{-1} (Figure 11.5b) are mainly attributed to the angular deformation of the absorbed water (Cordeiro et al., 2014), while those at 1595 cm^{-1} of the shell and 1611 cm^{-1} of the kernel are probably due to the presence of carboxylic groups (Oun & Rhim, 2016). The bands at 1507 cm^{-1} and 1535 cm^{-1} are indicative of the presence of lignin and are associated with the C=C stretching vibration in the aromatic rings (Henrique et al., 2013), and those around 1453 cm^{-1} to 1417 cm^{-1} in both spectra correspond to the CH_2 vibration attributed to cellulose (Shankar & Rhim, 2016). The bands near 1239 cm^{-1} (Figure 11.5a) and 1249 cm^{-1} (Figure 11.5b) may be associated with C–O stretching in lignin, as well as with the asymmetric axial stretching of =C–O–C, which is observed in ether, ester, and phenol groups, while the vibrations around 1076 cm^{-1} and 862 cm^{-1} are associated with cellulose (Henrique et al., 2013; Souza Junior et al., 2015). Cellulose, lignin, and starch were found in both the integument and the kernel. More bands, probably related to starch, were found in the kernel

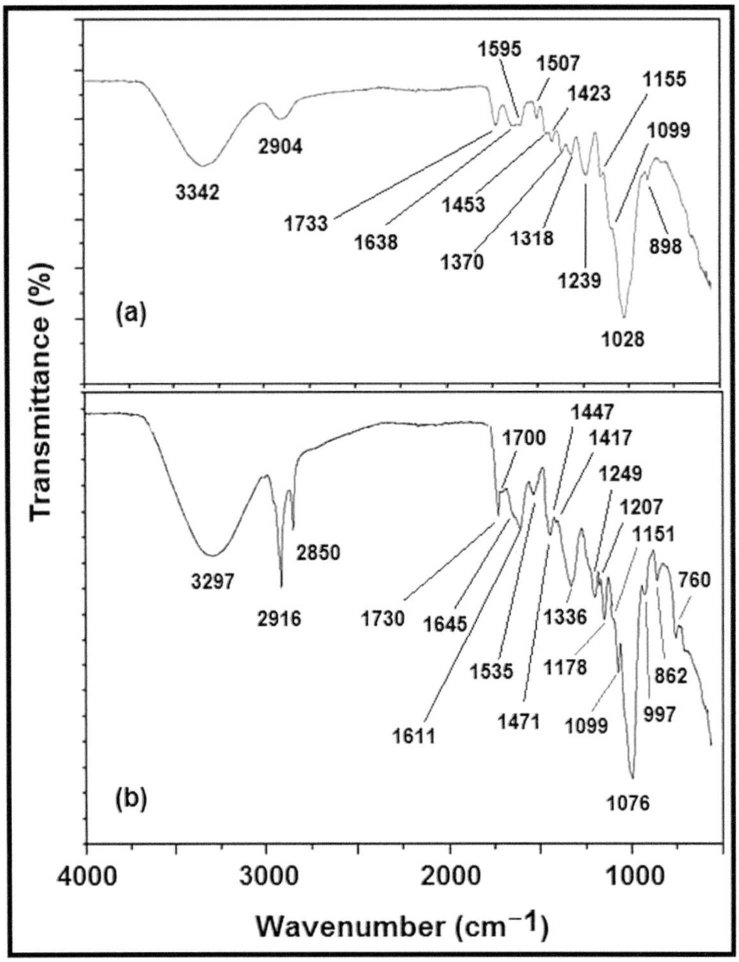

FIGURE 11.5 FTIR spectra of mango seed: (a) integument and (b) kernel (adapted from Lima et al., 2021a).

(1151 cm^{-1} and 2916 cm^{-1}) than in the integument (1155 cm^{-1}), while more cellulose bands were found in the integument (1423 cm^{-1} and 1318 cm^{-1}) than in the kernel (1417 cm^{-1}). Hemicellulose (1507 cm^{-1}) was found only in the integument.

The thermal stability of the mango seed integument and kernel, as analyzed by TGA/DTG, is demonstrated in Figure 11.6. The first event occurred in the temperature range of 50–100°C and is attributed to water evaporation due to surface tension (natural humidity), with low weight loss, around 7%. The second event, around 350°C, is attributed to the cellulose degradation process (depolymerization, dehydration, and decomposition of glycosyl units), where a large weight loss (73% by weight) is confirmed by the DTG curve (Gonzalez & Montoya, 2007; Henrique et al., 2013; Neto et al., 2013; Andrade et al., 2016). Figure 11.6b shows the TGA/DTG curves of the mango seed kernel. Thermal decomposition of kernels occurs in three main stages. The first stage refers to the loss of surface water at around 75°C, with a loss of ~8%. The second stage of degradation occurred at around 300°C, corresponding to a large weight loss (49%), and may be related to starch degradation, which occurs between 228°C and 393°C, where the scission of starch chains and oxidation of glucose rings occur due to thermal degradation (Solis-Fuentes et al., 2007; Ismail et al., 2017). The last event, at around 425°C, is related to the decomposition of cellulose polysiloxanes and the weight loss was ~19.0% (Cordeiro et al., 2014; Lazzari et al., 2016; Loof et al., 2016).

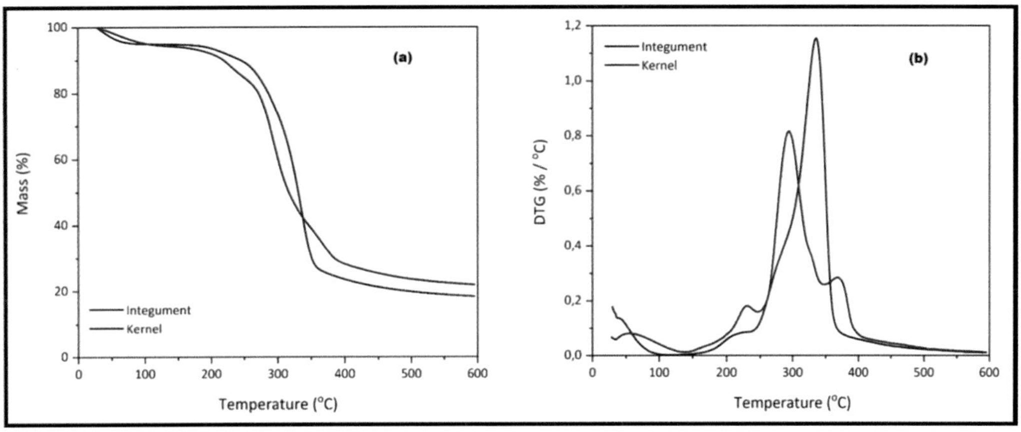

FIGURE 11.6 (a) TGA and (b) DTG analyses of the integument and kernel of the mango seed (analysis conducted by the authors).

The characterization of mango seed by DSC presents for the kernel (Figure 11.7) two T_g (Lima et al., 2021), one at ~50°C and the other at ~0°C. The T_g of starch varies according to the amount of water in the sample, since water acts as a plasticizer, and can be located between 30°C and 80°C (Zeleznak & Hoseney, 1987). In addition, three endothermic peaks were found around 15°C, 27°C, and at 128°C. In relation to the peak at 128°C, the T_m of starch, according to the water content in the sample, varies from 100°C to 140°C. Furthermore, there is cellulose in the kernel, which presents an endothermic peak at 50–150°C, where T_m varies with hydration and the degree of amorphization of the sample (Florêncio et al., 2012; Zeleznak & Hoseney, 1987). The peak at 128°C presents ΔH~51.71 J/g, which is greater than that of starch, ~10 J/g (Genkina et al., 2014), while the fusion enthalpy of cellulose is >25 J/g (Ciolacu et al., 2011). A similar profile, where more than one endothermic peak is found, is presented by starch, where the first peak (~70°C) would be related to the fusion of amylopectin crystals (or disorganization of the double helices of the chains). The second peak (~95°C) would be due to the fusion of the amylose–lipid complex. The T_m of starch varies according to the amount of amylose in the sample, number and size of crystals and granules, among other factors (Genkina et al., 2014; Huang et al., 2015). The T_g of the kernel at ~50°C and the T_m at ~128°C would be related to starch (Zeleznak & Hoseney, 1987). The peaks at ~15°C and ~27°C would be related to other low melting temperature components present in the kernel.

DSC analysis of the integument illustrated in Figure 11.7a revealed T_g ~58°C, T_m ~138°C and an exothermic peak at ~196°C. Both T_g and T_m are in the starch transition temperature range. Since FTIR demonstrates that the mango seed integument probably has less starch and more cellulose than the kernel, the endothermic peak may be related to cellulose, T_m range 50–150°C (Ciolacu et al., 2011). The exothermic peak may be related to oxidation, likely degradation of the integument (Neto et al., 2013; Pardauil et al., 2011; Rodrigues et al., 2012; Yamoneka et al., 2015).

It is important to highlight that depending on the method of preparation of the mango seed by grinding for application as a composite or biocomposite, the integument of the mango seed may present a two-dimensional morphology and the kernel a spheroidal morphology if a hammer-knife mill is used (Lima et al., 2021). However, if a high-impact disc mill is used and the grinding process is monitored to avoid high heating of the samples, the morphology produced will be uniform and one-dimensional for both the integument and the kernel, as shown in Figure 11.8 (Lima et al., 2024).

These behaviors are observed by optical and electron microscopy techniques. Microprobe analysis by energy dispersive X-ray spectrometry (EDS) coupled to the electron microscope do not present many answers, except in the case of contamination, since the composition of these samples is basically carbon (C) and hydrogen (H).

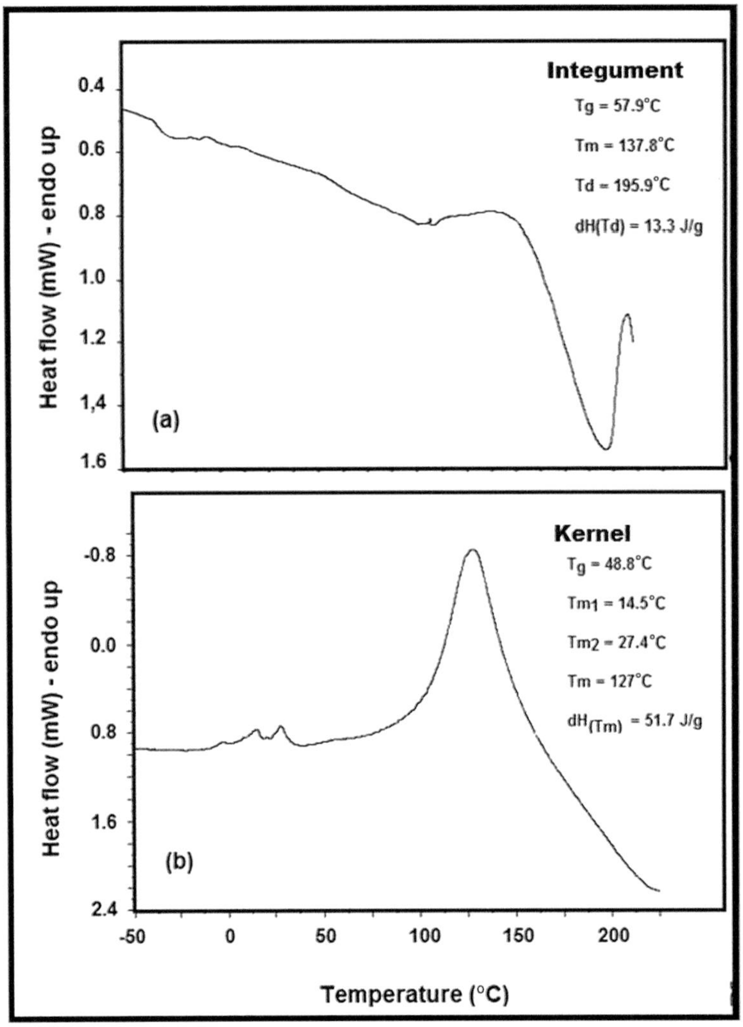

FIGURE 11.7 DSC analysis of mango seed (a) integument and (b) kernel (analysis conducted by the authors).

The reduction in the size of the particles present in the mango seed (integument and kernel) reduces the mechanical resistance of the biocomposites because they have the same charge as the biopolymer associated with the increase in the specific contact surface. The shape and size of the particles greatly influence biodegradability and mechanical resistance (Lima et al., 2021, 2024). On the other hand, a large size of the integument can favor anchoring in the polymer mass, increasing its resistance, in addition to being able to form percolation ducts for fluids (such as water, air etc.).

11.3 PLA/MANGO SEED RESIDUE BIOCOMPOSITES

Another possible application for these fractions would be as reinforcement "fillers", where the representative fraction of starch and cellulose are added to biopolymers, such as polylactic acid (PLA). Given that these residue fractions exhibit thermal stability exceeding typical processing temperatures such as those involved in compression molding, extrusion, or injection molding and considering that the polymer to be reinforced is water-soluble, the residue fractions must also possess sufficient chemical stability to endure processing methods like casting. These composites are environmentally

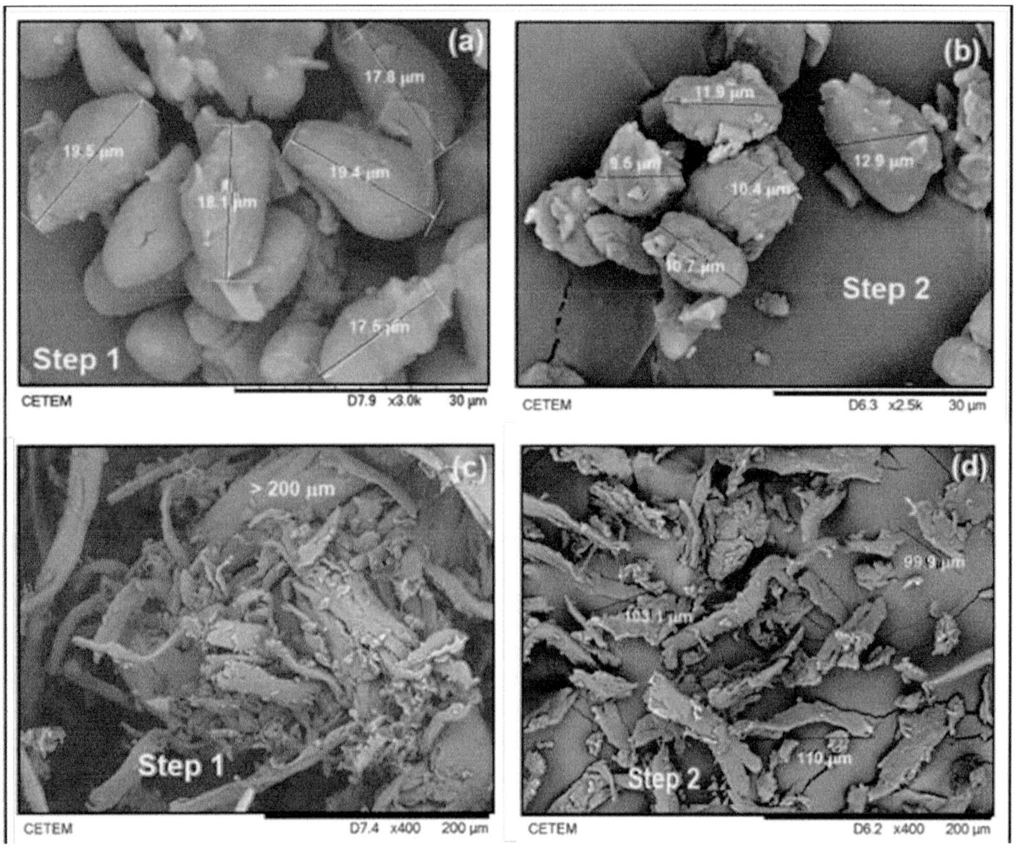

FIGURE 11.8 SEM backscattered electron images of Kernel: (a) Step 1, (b) Step 2 and Integument: (c) Step 1, (d) Step 2 (analysis conducted by the authors).

friendly solutions that can replace petroleum-derived polymers (Lima et al., 2019, 2021, 2022 and 2024). This challenge has motivated research into new materials that have the same mechanical resistance and cost–benefit ratio. In this way, natural polymers are studied in more detail in their nanometric size (Lima et al., 2021).

Recent studies on PLA-based biocomposites (Lima et al., 2019, 2020, 2021, 2021a, 2024) involved agricultural inputs based on tegument and kernel from the mango seed added with montmorillonite in the PLA matrix using casting/extrusion/injection techniques. From these studies, it is possible to observe some characteristics present in these biocomposites, such as the hydrophobic character measured by the contact angle of the PLA polymer of the matrix and of the integument and kernel of the mango seed. It is noted that the greater the volume of these materials, the greater the hydrophobic character presented (Figure 11.9).

The thermal analysis of PLA/integument/kernel biocomposites from mango seeds was presented by Lima et al. (2021). In Figure 11.10, it can be observed that all samples containing integument or kernel showed mass loss at 100°C, attributed to water evaporation, and the decomposition of the biocomposites is completed at 350°C. The highest degradation rate occurs for PLA (T_{onset}) at approximately 340°C, while for the biocomposites the T_{onset} is a little lower, that is, for PLA/integument, approximately 334°C, for PLA/kernel, at 328°C, and for PLA/integument/kernel, 324°C. This suggests that the addition of mango integument to PLA decreases the thermal degradation temperature of the biocomposites, likely due to the presence of cellulose and lignin in the residues (Lima

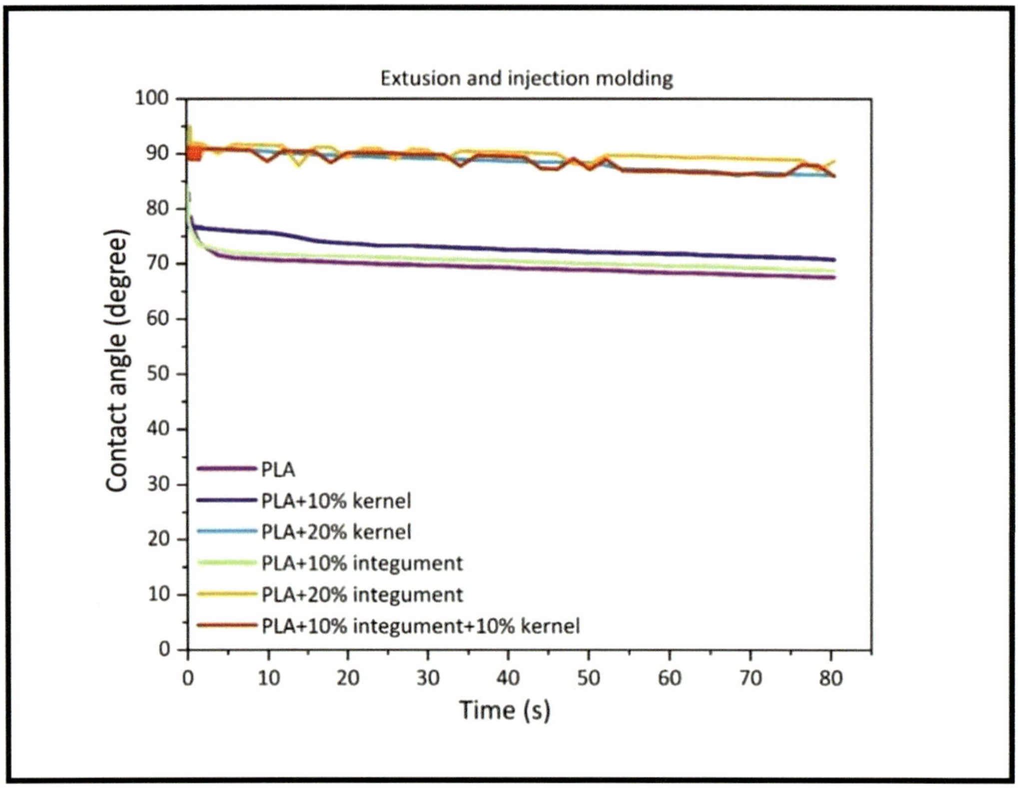

FIGURE 11.9 Hydrophobicity *versus* time showing greater hydrophobic character in biocomposites with 20% filler (analysis conducted by the authors).

et al., 2021), whose thermal stabilities are lower than that of the PLA matrix. Furthermore, with the increase in lignocellulosic components in the matrix, there was a more expressive decrease in the thermal stability of the biocomposites, presenting behaviors like those of the integument or kernel of the mango seed alone.

Regarding the production of biocomposites with mineral fillers, in the last 20 years, a special attention has been given to the generation of biocomposites through polymer nanocomposites with organoclay (PCN). In these materials, the synergistic interaction between organic and inorganic phases can give rise to materials with new properties that are not found in original materials. This interaction occurs due to its high specific surface area, which produces materials with better mechanical resistance, greater thermal stability, resistance to scratches and damage, gas barrier, and flame resistance.

Many studies use synthetic clays such as laponite, a hectorite with composition $Na_{0.7}$ $[(Mg_{5.5}Li_{0.4})$ Si_8O_{20} $(OH)_4]$, which according to some studies has some advantages over natural clays, such as the absence of impurities (silica, iron oxides, carbonates, etc.) and a well-defined chemical formula (Aouada et al., 2012). This synthetic clay has been employed in diverse areas such as surface coatings, paper and polymer films, agriculture, household, and personal care products. The work of Perotti et al. (2013) shows the development of transparent and homogeneous biomaterials from nanocomposites based on starch reinforced with hectorite synthetic clay (industrial waste) plasticized with invert sugar in an aqueous medium (casting), where it is observed that small amounts of clay (5–15 wt%) mainly produce exfoliated nanocomposites that initially present greater resistance, but if in large quantities (30 wt%) they promote the formation of intercalated nanocomposites.

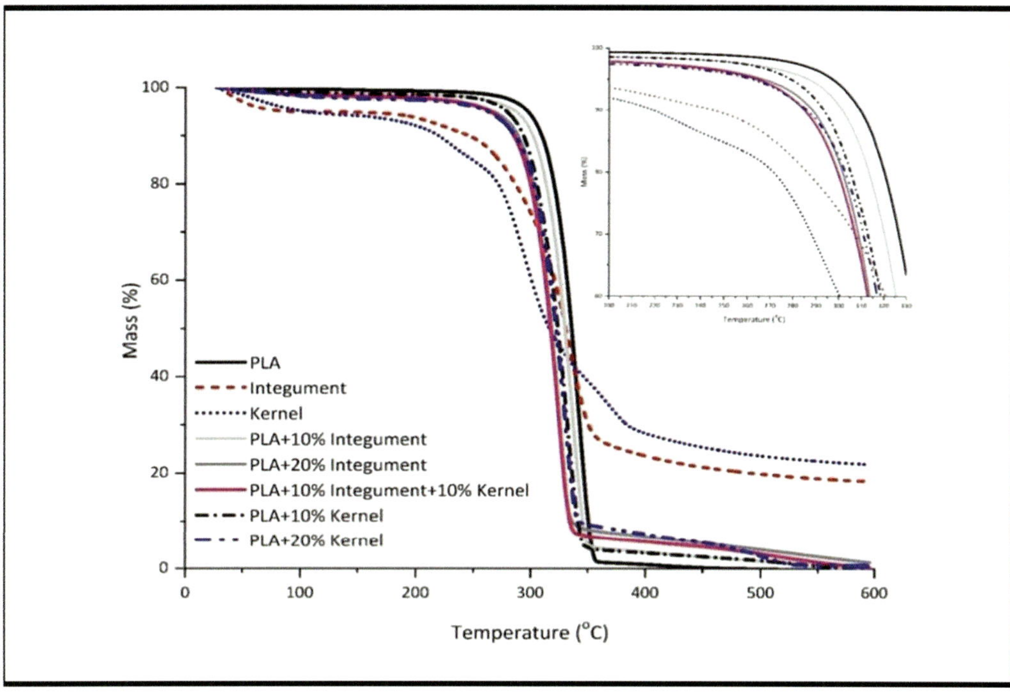

FIGURE 11.10 Thermogravimetric analysis (TGA) of PLA and mango's integument and/or kernel biocomposites (analysis conducted by the authors).

In this way, a higher content of fine material (greater than 10 wt%) does not present substantial improvements in the elongation properties or mechanical resistance of the developed biomaterials. Within the scope of the use of natural clays, the use of clay minerals from the 2:1 group stands out, where the use of bentonite can be highlighted. This is a generic technical term for clay composed predominantly of the clay mineral montmorillonite (55–70%), from the smectite group, regardless of its origin or occurrence. The main characteristics of smectites are high swelling power, up to 20 times their initial volume, reaching interplanar spaces of up to 100 Å, high surface area (up to 800 m^2/g), cation exchange capacity (CTC) in the range of 60 at 170 meq/100g and thixotropy. These characteristics give to bentonite (trade name of montmorillonite) very specific properties, which have justified a wide range of applications in the most diverse segments (Odom, 1984).

The development of biocomposites that increase shelf life of food by giving it a better appearance is of great importance. They can be through edible films as a coating, as they isolate the food from direct contact with oxygen in the air, delaying its deterioration. This edible polymeric coating is a film made up of different natural substances that, when polymerized, form barriers that can be consumed without risk to health and can be enriched with other elements, such as vitamin C (Amaral et al., 2010; Lorevice et al., 2009; Nascimento et al., 2010).

In addition to this previously mentioned work, we can list others that apply to the development of packaging obtained by casting and extrusion techniques, such as Cruz, Bassinello, and Ascheri (Cruz et al., 2010) and the work of Fakhouri et al. (2010). We can also mention the work developed to increase the post-harvest quality of "*Tommy atkins*" mangoes (Vieira et al., 2009; Santos et al., 2011). Technically, establishing a use for this volume of waste generated by mango farming has been a challenge that researchers have faced in Brazil and around the world. These PLA-mango residue and PLA-montmorillonite-based composites were characterized, and their PLA-based biocomposites were evaluated (Lima et al., 2021a, 2021b). It was observed that the addition of mango seed components and organo-nano-montmorillonite to the PLA matrix physically interfered with the freedom of movement of the PLA bonds when subjected to radiation in the mid-infrared wavelength range.

The presence of the mango seed peel altered the mobility of the PLA chains, as the samples exhibited higher glass transition and melting temperatures. Thus, the mango components could act (effect dependent on cellulose content) as nucleation agent of PLA crystallites, which contribute to greater mechanical resistance of the materials (Khoo et al., 2016). PLA films containing tegument and seed core (rich in starch) presented transition temperatures intermediate between those of PLA-tegument and PLA-seed core samples, i.e., there was a competitive effect between cellulose fibers (crystallite nucleators) and starch, resulting in lower mechanical resistance, as confirmed by texturometer analysis (Aragão et al., 2017).

The addition of organo-nano-montmorillonite to PLA increased the transition temperatures by an effect like that of cellulose fibers, nucleators of PLA crystallites, which has an affinity for silicates (Chang et al., 2016; Paul et al., 2003). It is worth mentioning that the PLA-mango samples, to which organo-nano-montmorillonite originating from the BOFE layer were added, did not allow the formation of perfect PLA crystals. While the incorporation of organo-nano-clays originating from the Chocolate appear layer to have physically bonded to the amorphous PLA chains and served as crystalline nucleation sites for PLA (Gorrasi et al., 2013), although it may have interfered with the physical knots in the amorphous PLA chains, which justifies the lower mechanical resistance of these samples. SEM analysis revealed that PLA has a greater affinity for the cellulose fibers in the mango seed coat (continuous interface) than for the starch in the nucleus, where there was interfacial discontinuity between these phases (Gil-Castell et al., 2016; Jaszkiewicz et al., 2016). The samples containing organo-nano-montmorillonite presented an internal polygonal morphology, apparently characteristic of the presence of the clay mineral (montmorillonite).

The samples showed relative degradation in soil and higher degradation in water over a 3-month experiment, though degradation by hydrolysis alone was not measurable due to microbial presence. Such degradation was observed through FTIR vibration band shifts in degraded soil samples. Other possibility of using these residues would be related to their texture due to their polymeric fillers need to be well dispersed and distributed in the matrix before and after processing, generating plastic materials with uniform color and texture and mechanical resistance, which would influence the biodegradability (Lima et al., 2024). The use of biopolymers for new materials is gaining increasing attention due to their major benefits over plastics, including polymer modification, nanoparticle reinforcement, blending of polymer matrices, multilayer strategies, biodegradability, environmentally friendly properties, and biocompatibility (Lima et al., 2019, 2020).

11.3.1 Comparison of Casting and Extrusion Processes

Casting and extrusion processes are well-known and used in the manufacture of plastics. However, they present several differences in their results. Most studies use the laboratory-scale casting process, as it is a well-known and simple method. However, there are challenges to be addressed, such as the long processing time and the need to evaporate large quantities of solvents (La Fuente et al., 2022). Extrusion, on the other hand, involves forcing specific materials through a die to create a specific shape or profile. In summary, casting involves pouring molten specific materials into a mold to create a shape, while extrusion involves forcing them through a die to create a specific profile. The slow drying in casting and the fast drying in extrusion process, promoted different molecular arrangements, which affected the material performance.

As a result, the current study aims to understand the impact of the casting and extrusion processes on the microstructure of the biocomposites studied by the team of Lima et al. (2019, 2020, 2021, 2021a, and 2024), focusing on comparing both technologies and correlating the materials

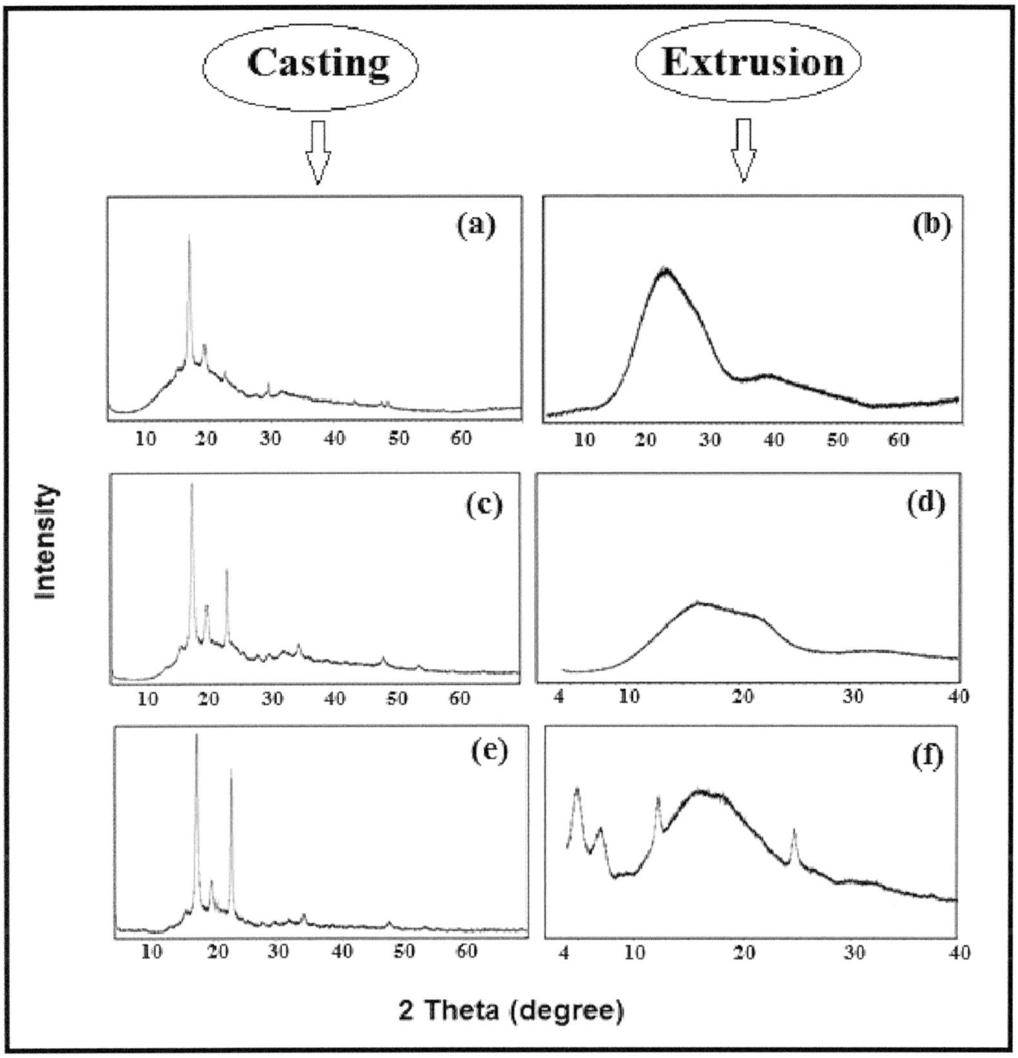

FIGURE 11.11 XRD patterns by Casting and Extrusion of (a) and (b) – neat PLA; (c) and (d) – PLA/Kernel/Integument; (e) and (f) PLA/BOFE organoclay showing the loss of crystallinity during the extrusion process (analysis conducted by the authors).

characteristics of the same parts of mango's seed. Figure 11.11 demonstrates the semicrystalline characteristics of neat PLA, PLA/mango seeds and PLA/organoclay by X-ray diffraction patterns of the casting films and extrusion sheets (Lima et al., 2019, 2020, 2021). The decrease in the number of peaks from casting to extrusion is probably caused by the destruction of the crystalline structure of starch granules during extrusion (Gonzales-Seligra et al., 2017).

Additionally, in Table 11.3, it could be seen the loss of crystallinity during the extrusion process. This is likely due to the different drying conditions. The sheets were quickly dried at high temperatures and high shear stress during extrusion, and the films were slowly dried at rest by casting at low temperatures. The casting process helps to promote more effective recrystallization in the matrix, while rapid drying at high temperatures and shear that occurs during the extrusion process causes the destruction of this crystalline structure. The properties of biocomposites may undergo changes due to the differences that occur between these two processes. Figure 11.12 shows

TABLE 11.3
Samples' Degree of Crystallinity and Amorphous for Casting and Extrusion Processes (Analysis Conducted by the Authors)

	Crystallinity (%)		Amorphous (%)	
Samples	**Casting**	**Extrusion**		**Casting**
Neat PLA	16.4	3.8	Neat PLA	16.4
PLA/Kernel/Integument	45.8	6.5	PLA/Kernel/Integument	45.8
PLA/Organoclay	42.1	18.8	PLA/Organoclay	42.1

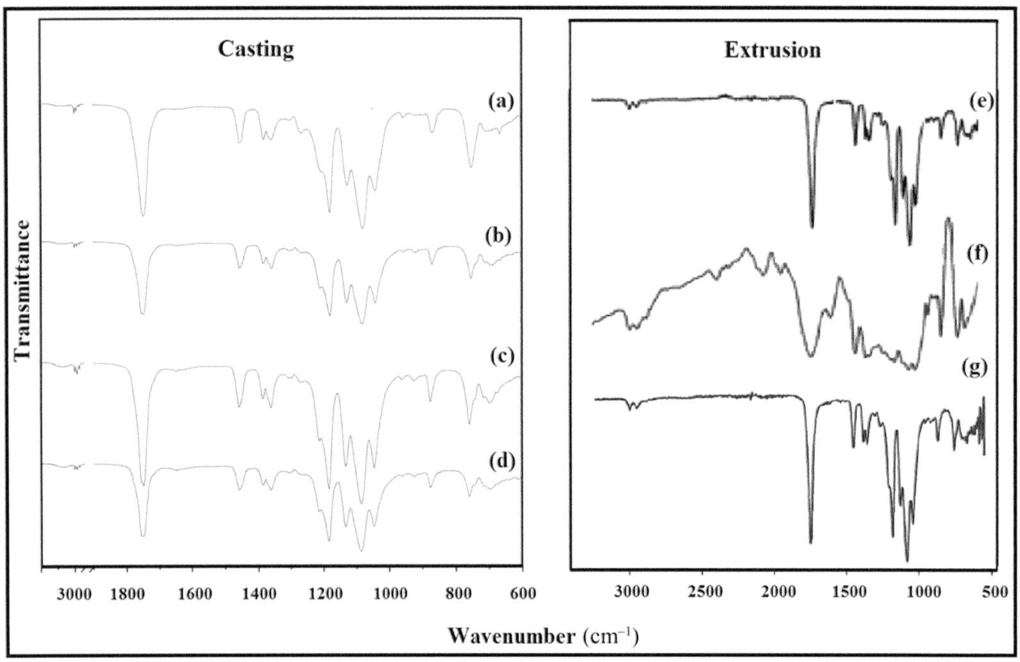

FIGURE 11.12 Biocomposites FTIR spectra of films (casting) and sheets (extrusion): (a), (e) neat PLA; (b), (f) PLA/Kernel/Integument; (c), (g) PLA/organoclay and (d) PLA/Kernel/Integument/organoclay (analysis conducted by the authors).

the FTIR results of casting films (Lima et al., 2019; Lima et al., 2020) and extrusion sheets (Lima et al., 2021) obtained from PLA biocomposites.

In general, both spectra showed a typical band behavior for PLA's biocomposites. The principal bands are shown in Table 11.4 (Lima et al., 2019; Lima et al., 2021a). These materials presented a region close to 1650 cm^{-1} that could be attributed to the bending vibration of H–O–H water molecules absorbed in the amorphous region, while the region between 1200 and 900 cm^{-1} bands could be associated with C–O, C–C, and C–O–H vibrations (La Fonte et al., 2022). Anyway, no differences were observed between the two processes (casting and extrusion), probably because the changes that occurred in the biocomposites were not sufficient to be identified by FTIR.

The thermogravimetric and differential analysis (TGA/DTA) for the two processes (casting and extrusion) described in previous work (Lima et al., 2019, 2021a) showed in general, no remarkable

TABLE 11.4
PLA Biocomposites' FTIR Principal Bands by Casting and Extrusion (Analysis Conducted by the Authors)

Wavenumber (cm^{-1})	Vibrational Mode
3342 and 3297	Bands of cellulose – stretching vibration of OH
2996 and 2945	Asymmetric stretching vibrations of CH_3
2921 and 2850	(stretching vibration of C—H bonds related to CH_2 groups) could be related to the effectiveness of the organophilization process
2921 and 2850	(stretching vibration of C—H bonds related to CH_2 groups) could be related to the effectiveness of the organophilization process
1748	Stretching and bending vibrations of C=O
1638 and 1645	Angular deformation of the absorbed water
1611 and 1595	Presence of carboxylic groups
1507 and 1525	Presence of lignin – stretching C=C vibration in the aromatic rings
1453-1417	CH_2 vibration attributed to cellulose
1382 and 1359	Symmetric deformation vibrations of CH_3
1303, 1266, and 1206	Deformation vibrations of tertiary carbon
1239 and 1249	C–O stretch in lignin and axial asymmetric strain of =C–O–C
1180	C–O stretching vibrations of lactide units or to C–O–C asymmetric stretching
1127	C-H (of CH_3 groups) rocking mode
1081 and 1043	Vibrations of CH_3
956	Stretching of C–C
919 and 690	Vibration of Si—O of clay
869 and 752	to–CH_2– rocking vibrations

changes in the thermal properties due to the kind of biocomposite preparation can be seen, and all materials obtained can be considered stable at normal temperatures of storage. Thermal decomposition occurred in three main steps. The first stage of degradation around 80°C corresponds to volatilization of free water, a second stage around 350°C is related to the cellulose and/or starch degradation process and the last transition near 500°C is attributed to dehydroxylation of structural OH units of clay minerals or cellulose's polysiloxanes. It can be also observed that the casting films and extrusion sheets degrade at the same rate showing that those processes do not interfere with the thermal degradation of the obtained biocomposites.

DSC data regarding the transition temperatures of the samples as well as their degree of crystallinity are displayed in Table 11.5. The addition of mangos starch and tegument separately to PLA increased the samples' T_m, T_c, and T_g. The addition of both mango components together lowered the samples' T_m and T_g toward those of PLA, while the T_c remained relatively high. The addition of clays increased the T_m, T_c and T_g of the samples, indicating that the clays act as a substrate for PLA crystallization.

Cellulose thermal degradation was observed in the mango's integument. However, the DSC analysis identified the probable presence of cellulose and starch both in mango's kernel and integument, although higher amounts of starch in mango's kernel and higher amounts of cellulose in mango's integument could be observed.

Morphology of cryo-fractured surfaces of materials obtained by SEM backscattered electrons by casting and extrusion are presented in Figure 11.13. In general, the biocomposites present compact and smoother structures on the surfaces and cross-sections, with typical composite characteristics showing low interfacial interaction. As can be seen in the two processes (casting and extrusion), there is no significant compatibility between the starch and PLA phases and limited adhesiveness between PLA and cellulose fibers (Lima et al., 2019, 2020). Additionally, PLA/Kernel/Integument/

TABLE 11.5
DSC Results for Some PLA Biocomposites by Casting (Cast) and Extrusion (Extr) (Analysis Conducted by the Authors)

	T_g (°C)		T_c (°C)		ΔH_c (J/g)		ΔH_m (J/g)	
Samples	**Cast**	**Extr**	**Cast**	**Extr**	**Cast**	**Extr**	**Cast**	**Extr**
PLA	54.4	64.3	-	93.4	23.1	7.1	34.0	48.2
PLA/Kernel	62.2	61.2	161.5	87.8	7.53	15.1	29.4	47.7
PLA/Integument	61.9	59.7	163.9	91.0	22.1	21.4	33.0	42.8
PLA/Kernel/Integument	61.5	60.1	162.9	86.1	21.4	20.5	32.2	42.0
PLA/BOFE organoclay	59.4	-	160.7	-	21.3	-	43.2	-
PLA/Kernel/Integument/BOFE organoclay	50.5	-	-	-	15.2	-	27.3	-

Note: T_g = glass transition temperature; T_c = crystallization temperature; T_m = melting temperature; ΔH_c = crystallization enthalpy; ΔH_m = melting enthalpy.

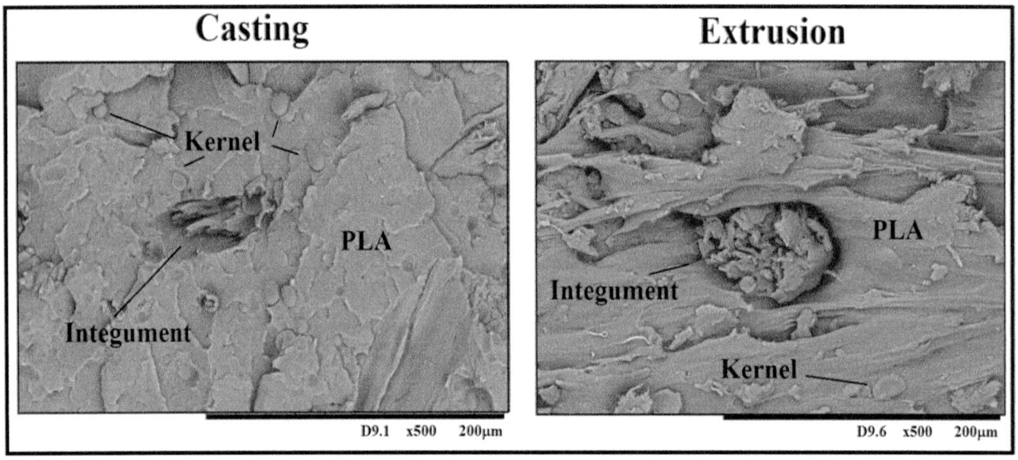

FIGURE 11.13 SEM backscattered electron images of PLA/Kernel/Integument (Analysis conducted by the authors).

organoclay biocomposites displayed a smooth matrix, adhesion to the fibers, and poor attachment to starch granules. However, heterogeneity and roughness were observed in the extruded sheets. The rapid evaporation occurred in extrusion in comparison to casting, results in micropores and cracks due to the expansion of water vapor (Fakhouri et al., 2010), which favored by the high temperature and mechanical shearing, leading to the destructuring of starch granules that could favored the interactions with the plasticizers, improving mechanical and barrier properties (Fitch-Vargas et al., 2016). On the other hand, bubbles/droplets formation in the exit section of the extruder favoring the formation of porous morphology (Horvat et al., 2014). This behavior is explained due to the hydrophobic character of PLA and the hydrophilic character of the fillers and was expected since no compatibilizer was used in these formulations (Wu & Hakkarainen, 2015).

The mechanical behavior during a tensile assay of the obtained materials by casting and extrusion is shown in Table 11.6. It was found that the sample thickness varied considerably, and PLA/Kernel samples presented a higher strength than the others biocomposites in casting process. The samples presented a Young's modulus and strength in the range of PLA blends and the adhesion between

TABLE 11.6
Mechanical Analysis of Some Biocomposites Studied by Casting and Extrusion Processes (Analysis Conducted by the Authors)

	Maximum Tensile Strength (MPa)		Deformation (%)		Elastic Modulus (MPa)	
Sample	Casting	Extrusion – Injection	Casting	Extrusion – Injection	Casting	Extrusion – Injection
PLA	2.6 ± 0.7	57.9 ± 0.7	36 ± 18	5.7 ± 0.9	8 ± 2	2383 ± 157
PLA + Kernel	3.6 ± 0.8	35 ± 2	36 ± 5	1.8 ± 0.2	9.9 ± 0.9	2563 ± 59
PLA + Integument	2.7 ± 0.6	49.5 ± 0.3	23 ± 4	2.6 ± 0.2	11.6 ± 0.4	2882 ± 19
PLA + Kernel + Integument	1.5 ± 0.6	x	15 ± 6	X	10 ± 2	X
PLA + Kernel + Integ + chocolate (clay)	1.4 ± 0.6	x	16 ± 7	X	9.0 ± 0.6	X
PLA + Kernel + Integ + bofe (clay)	1,3 ± 0.2	x	14 ± 1	X	9 ± 2	X

*Extrusion (v = 5 mm/min), **Casting (v = 1 mm/min)

PLA and cellulose fibers and the fiber aspect ratio (which serves as an anchoring surface for PLA and allows stress transfer) increase the sample's Young's modulus (Lima et al., 2019).

For the extrusion process, mechanical properties such as tensile strength, elongation at break, and Young's modulus were studied in a previous work (Lima et al., 2021a) and showed a typical tensile stress–strain curves for the biocomposites observed. The addition of kernel and integument to PLA resulted in an increase in the elastic modulus of all formulations. However, the extent of thermo-oxidative degradation of the filler (kernel) at the processing temperatures (≈220°C in injection molding) may have caused significant disruption of glycosidic bonds of starch, reducing their molecules to sugars (such as pentoses, for example), which would no longer be rigid enough to promote any improvement in mechanical properties, and would be acting only as stress concentrators, therefore the expressive decreases in these properties. Moreover, these degraded fillers could lead to a thermo-oxidative degradation of PLA, breaking its molecule linkage, causing a decrease in its viscosity. When comparing both processes, casting films were proportionally stronger than extruded sheets, possibly because extrusion favors chain alignment in the flow direction, resulting in a more flexible material (Andreuccetti et al., 2012). Therefore, the extrusion process is expected to produce biocomposites with more aligned and organized molecules than casting, which explains the resulting behavior in how water interacts with and permeates the material.

11.4 PHA/MANGO SEED RESIDUE BIOCOMPOSITES

Polyhydroxyalkanoates (PHA) are aliphatic polyesters of R-hydroxyalkanoic acids produced by fermentation (Figure 11.14). Depending on the microorganism, the type of carbon source, and the fermentation parameters, homo- or copolyesters can be obtained with different physical and mechanical properties, varying from thermoplastic to elastomeric characteristics. There are over 125 known PHA (Możejko-Ciesielska & Kiewisz, 2016; Palmeiro-Sánchez & O'Flaherty, 2022; Vicente et al., 2023).

PHA is formed in the cytoplasm of certain microorganisms as stocking material when nutrients are imbalanced (an excess of carbon and a deficit of nitrogen, phosphorous, sulfur, and/or oxygen). The enzymatic intracellular system transfers carbon sources, forming PHA chains. These insoluble chains are used as energy storage for these cells. A single strain of microorganisms can accumulate up to 80% (w/w) of PHA in its cytoplasm. Over 90 bacterial genera can produce PHA (Vicente et al.,

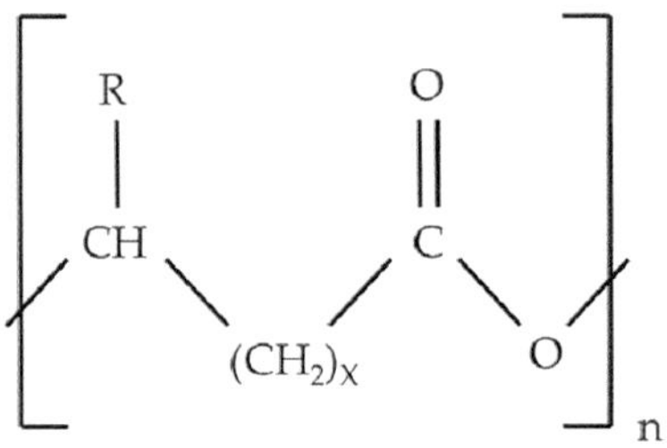

FIGURE 11.14 General structure of PHAs x = the number of methylene groups in mers, ranging from 1 to 8. R = the alkyl side chain (Vicente et al., 2023).

2023). Among the eligible bacteria for PHA production are *Pseudomonas, Bacillus, Alcaligenes, Azotobacter, Burkholderia, Halomonas,* and *Haloferix* (Afreen et al., 2021).

Yeasts can also produce PHA, especially when they are genetically modified to do so. In addition, they use low-price sources of carbon (glucose) to produce PHA. Some yeast modified to obtain PHA are *Y. lipolytica, S. cerevisiae, Rhodotorula minuta RY, Kebeckera spp.*, and *P. pastoris* (Thu et al., 2023). Cyanobacteria/microalgae are also potential organisms for PHA production. During photosynthesis, organic reservoirs are produced when a deficit of N and P occurs in their environment. These reservoirs can be transformed into PHA inside microalgae or used as a food source for bacteria to produce PHA (Afreen et al., 2021).

The microorganisms' nutrient broth can be modified to improve PHA production. The carbon sources used by bacteria to produce PHA are simple molecules characterized into three main groups: monosaccharides (e.g., carbohydrates), hydrocarbons, and triacylglycerol. According to the microorganisms used, each carbon source presents its pathway to produce different PHA (Jiang et al., 2016). To reduce production costs and avoid using substrates for the food industry, some studies proposed using alternative feedstocks for PHA production, such as industrial or municipal wastes (Vicente et al., 2023). Pandley et al. (2024) described a procedure to produce PHA from mango peel with a high PHA yield. The process includes enzymatic hydrolysis of mango peel under mild conditions to convert complex carbohydrates into simple sugars used as raw material for bacterial fermentation.

PHAs are extracted from microorganisms through organic solvents or by adding chemical agents that dissolve their non-PHA parts through cellular lysis. The chlorine solvent route results in low extraction yield and high molecular weight and purity, while the cellular lysis route presents the opposite trend. Since these processes can be considered unfriendly to the environment, some alternatives exist, e.g., ethyl acetate and recyclable surfactants. The multiple strain extraction is more effective through cellular lysis using surfactant (>80% PHA yield) (Pagliano et al., 2021). However, single strains require a sterile environment and sterile processing; multiple strains can overcome the carbon source restrictions and the complex processing, but they are complex to stabilize (Zhou et al., 2023). The PHA extraction process also contributes to the higher price of these polymers than traditional petrochemical polymers, such as polyethylene or polypropylene. After PHA purification, the granules can be processed using the same processing technology used for conventional polymers: extrusion, injection molding, blow molding, electrospinning, etc.

Poly (3-hydroxybutyrate) – PHB and poly (3-hydroxybutyrate-co-3-hydroxyvalerate) – PHBV are the most commercially available members of the PHA family (Corradini et al., 2013; Meereboer et al., 2020). These polymers are biodegradable in both aerobic and anaerobic environments defined by ASTM standards and in soil and marine environments, which makes them excellent candidates for substitute petrochemical polymers in packaging industries, agriculture films, and foams (Meereboer et al., 2020). Due to their biodegradability in body fluid and their biocompatibility to several lines of mammalian cells, they are also applicable to the biomedical field as suture lines, vehicles for drug delivery, scaffolds for tissue engineering, etc. (Kalia et al., 2013).

The homopolymer PHB is the simplest and the most characterized PHA. Considering the general structure of PHAs shown in Figure 11.14, R corresponds to CH_3 and x equals 1. PHB is a hard, brittle, and highly crystalline thermoplastic. It presents values of Young's modulus and tensile strength similar to those of the isotactic polypropylene. However, its elongation at break is significantly lower. Although there are some commercial products made of PHB, its very narrow processing window (degradation temperature very close to the melting temperature) and instability in the melting state make its processing difficult (Garcia-Garcia et al., 2022). Due to the secondary crystallization process, PHB materials usually undergo physical aging after processing, resulting in irregular pores on the surface and limiting flexibility in the final products (Thiré et al., 2011).

The PHBV copolymer is characterized by the insertion of hydroxyvalerate (HV) in PHB polymer chains (R = CH_2CH_3 in Figure 11.14), which leads to lower crystallinity, lesser brittleness, reduced melting and glass transition temperatures, higher thermal stability, and better processability. Moreover, PHBV also presents low cytotoxicity, piezoelectricity, excellent oxygen barrier properties, and ultraviolet radiation resistivity. By controlling the percentage of the HV blocks in PHBV, a broad range of properties can be obtained. PHBV remains noncompetitive in the market due to high production costs and technical limitations, mainly arising from the need for costly substrates and discontinuous production methods (Ibrahim et al., 2021; Meereboer et al., 2020; Atarés et al., 2024). To overcome this issue, biocomposites with low-cost fillers, particularly starch, have been studied.

Starch is attractive filler for PHBV-based composites since it is a renewable, biodegradable, inexpensive, and widely available raw material (Paul et al., 2024). However, the difference in wettability of hydrophobic PHBV and hydrophilic starch leads to low interfacial adhesion between composite phases, resulting, in general, in products with poor mechanical properties. Magalhães and Andrade (2013) proposed the use of organically modified montmorillonite to compatibilizer PHBV/thermoplastic cornstarch (TPS) melt blends. Although the 1:1 PHBV:TPS blends with 10 wt% organoclay showed miscible morphology, a decrease in crystallinity, and Young's modulus and tensile strength at break values higher than those of pure TPS, all the tensile properties were lower than pure PHBV. Another approach to producing PHBV loaded with up to 30 wt% starch is to promote grafting or crosslinking at the interfaces of the polymeric chains (Paul et al., 2024). For this, citric acid, adipic acid, borax, boric acid, or dicumyl peroxide (DCP) can be used to manufacture PHBV/starch bioplastics by extrusion, thermo-compressing, blow molding, etc., with enhanced water resistance and mechanical strength.

Agro-industrial waste containing starch, such as mango seed, has also been investigated as filler for PHBV biocomposites (Aragão et al., 2017; Conceição et al., 2024). Aragão et al. (2017) produced compression molded films of PHBV loaded with 0–10 wt.% mango seed kernel and plasticized with triacetin. The dynamic molecular behavior of these biocomposites was assessed by time domain-nuclear magnetic resonance (TD-NMR) analysis using low-field NMR equipment. The proton spin-lattice relaxation time (T_1H) and the domain curves of the samples are presented in Table 11.7 and Figure 11.15, respectively.

According to Table 11.7, adding different amounts of mango byproduct promoted a decrease in the T_1H values compared to the neat PHBV matrix. This indicates an increase in the system's molecular mobility and, consequently, less brittleness for the biocomposites compared to neat PHBV. This result also suggests a reduction of inter- and intramolecular interaction within the PHBV chains, which is probably related to a good filler dispersion. PHBV composites present domain curves with wider bases than that of neat PHBV, indicating a more significant heterogeneity at the molecular level (Figure 11.15). The higher amount of mango kernel disfavored the homogeneity of the system since a new minor domain appeared in the PHBV10 plot. The decrease in the degree of crystallinity of the materials confirmed the molecular heterogeneity of composite systems. Aragão et al. (2017) also evaluated the PHBV/mango kernel compressed films by migration test under simulated

TABLE 11.7
T_1H Values Obtained for Mango Kernel, PHBV, and their Biocomposites

Samples	T_1H (ms)
Mango kernel	54
Neat PHBV	313
PHBV/5% Kernel	295
PHBV/10% Kernel	285

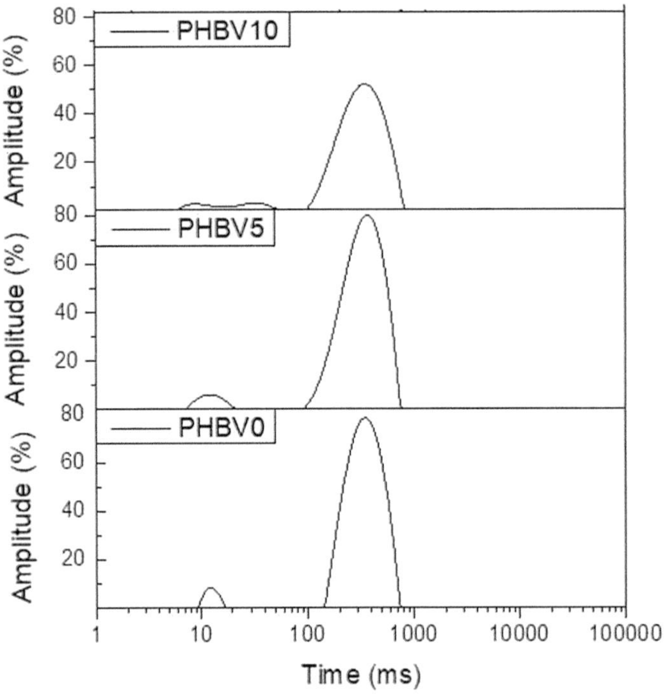

FIGURE 11.15 Distribution domain curves for neat PHBV (PHBV0) and the PHBV/mango kernel biocomposites containing 5 wt % and 10 wt % filler (PHBV5 and PHBV10) (graphics by the authors).

conditions of regular packaging food contact. They verified that these biocomposites have the potential to be used as packaging for nonrefrigerated fatty foods.

PHBV biocomposites containing up to 15 wt% mango seed's integument was produced by injection molding. Adding even a small amount of mango integument into the PHBV matrix accelerates the biodegradation rate of this polymer in soil (Figure 11.16). Initially, all the samples were probably submitted only to a surface erosion process, and the degradation of biocomposites, as determined by mass loss, was slow. However, after 6 weeks, samples began to present significant mass loss. A higher degradation rate was observed for biocomposites with higher filler content (PHBVI15). In this case, the biocomposite degraded three times faster than neat PHBV, presenting only 33% of this initial mass after 9 weeks.

Recently, Conceição et al. (2024) proposed a new application for mango seed kernel: a filler for PHBV filament for fused filament fabrication (FFF) of 3D-printed biodegradable parts. According to the authors, although the cost of PHBV is high, however its high degree of crystallinity could

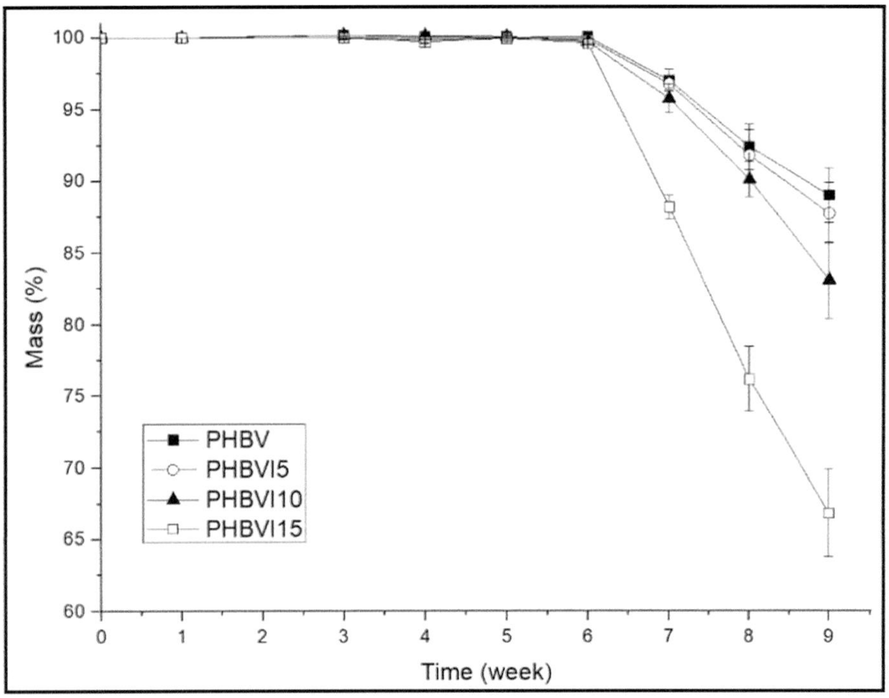

FIGURE 11.16 Biodegradation in terms of mass loss of neat PHBV and PHBV/mango seed´s integument biocomposites containing 5 wt.%, 10 wt.%, and 15 wt.% filler (PHBVI5, PHBVI10, and PHBVI15) (analysis conducted by authors).

impair the adhesion between printed layers and lead to high volumetric contraction during cooling, resulting in parts with poor mechanical properties and dimensional instabilities such as warping. Incorporating mango kernel into the PHBV matrix overcame these issues, and several prototypes were printed with a final quality similar to that of commercial products. This work launches a new alternative for increasing the aggregate value of the mango residues.

11.5 CONCLUSION

In recent years, climate change and the worldwide increase in microplastics have stimulated research toward improving bioplastic properties based on renewable polymers. This aim is to substitute traditional petrochemical plastics for several uses, especially in manufacturing disposable items. The path to sustainable development will be based on innovation and the search for intelligent applications of natural resources. PLA and PHA's biodegradability, low carbon footprint, and renewability are the driving forces for expanding their use in several applications. However, some lower intrinsic properties and higher production costs compared to petrochemical commodity plastics, such as polyethylene and polypropylene, have limited their widespread application. The results presented in this chapter suggest that the manufacture of PLA and PHA biocomposites loaded with mango seed constituents could be a potential strategy to overcome these drawbacks by decreasing the production costs of PLA and PHA-based bioplastics, making these bioplastics more feasible for the market. Moreover, the reuse of residues from the mango agroindustry can add value to this waste and solve environmental problems associated with the mango processing industry while promoting these bioplastics' entry into the large-scale and sizable market, mainly in the packaging and single-use plastics sectors. The biocomposites loaded with mango waste usually present better properties than those loaded with pure starch, probably due to the presence of other components of mango seed, such as oils, proteins, and fibers that should contribute to improving matrix–filler interaction.

Although much work remains to achieve large-scale production of these mango waste-based biocomposites, the presented studies open new horizons for developing materials that significantly contribute to the circular economy concept. Besides, governmental economic incentives and laws to restrict the use of plastic materials for single-use applications are indispensable for increasing biocomposite manufacturing and their widespread commercialization.

REFERENCES

Afreen, R., Tyagi, S., Singh, G. P., & Singh, M. (2021). Challenges and perspectives of polyhydroxyalkanoate production from microalgae/cyanobacteria and bacteria as microbial factories: An assessment of hybrid biological system. *Frontiers in Bioengineering and Biotechnology*, *9*, Article no 624885. https://doi.org/10.3389/fbioe.2021.624885.

Ahlgren, S., Björklund, A., Ekman, A., Karlsson, H., Berlin, J., Börjesson, P., Ekvall, T., Finnveden, G., Janssen, M., & Strid, I. (2013). *LCA of Biorefineries – Identification of Key Issues and Methodological Recommendations*. www.researchgate.net/publication/ 284498552.

Amaral, D. P., Andrade, K. C. S., Costa, C. V. & Carvalho, C. W. P. (2010). *Seminário em Inovação e Tecnologia na Área de Alilmentos*. https://ainfo.cnptia.embrapa.br/digital/ bitstream/item/26992/1/2010-192.pdf.

Ambrose, A., Lohumi, S., Lee, W. H. & Cho, B. K. (2016). Comparative nondestructive measurement of corn seed viability using Fourier transform near-infrared (FT-NIR) and Raman spectroscopy. *Sensors and Actuators B: Chemical, 224* (1), 500–506. https://doi.org/10.1016/j.snb.2015.10.082.

Andrade, L. A., Barrozo, M. A. S. & Vieira, L. G. M. (2016). Thermo-chemical behavior and product formation during pyrolysis of mango seed shell. *Industrial Crops and Products*, *85*, 174–180. https://doi.org/10.1016/j.indcrop.2016.03.004.

Andreuccetti, C., Carvalho, R. A., Galicia-García, T., Martinez-Bustos, F., González-Nunez, R. & Grosso, C. R. F. (2012). Functional properties of gelatin-based films containing *Yucca schidigera* extract produced via casting, extrusion and blown extrusion processes: A preliminary study. *Journal of Food Engineering*, *113* (1), 33–40. https://doi.org/10.1016/j.jfoodeng.2012.05.031.

Aouada, F. A., Mattoso, L. H. C. & Longo, E. (2012). *Workshop da Rede de Nanotecnologia Aplicada ao Agronegócio, Fortaleza*. www.alice.cnptia.embrapa.br/handle/doc/934071.

Aragão, D. I., Cardoso, P. H. M., Lima, E. M. B. & Thiré, R. M. S. M. (2017). *Proceedings of Fifth International Conference on Natural Polymers, Biopolymers and Biomaterials: Applications from macro to nanoscale*.

Arora, N., Fatima, T., Mishra, I., Verma, M., Mishra, J. & Mishra, V. (2018). Environmental sustainability: Challenges and viable solutions. *Environmental Sustainability*, *1*, 309–340. https://doi.org/10.1007/s42398-018-00038-w.

Atarés, L., Chiralt, A., González-Martínez, C. & Vargas, M. (2024). Production of polyhydroxyalkanoates for biodegradable food packaging applications using *Haloferax mediterranei* and agrifood wastes. *Foods*, *13*, Article no 950. https://doi.org/10.3390/foods13060950.

Bangar, S. P., Manoj, K. & Whiteside, W. S. (2021). Mango seed starch: A sustainable and eco-friendly alternative to increasing industrial requirements. *International Journal of Biological Macromolecules*, *183*, 1807–1817. https://doi.org/10.1016/j.ijbiomac.2021.05.157.

Carrot, S. P. J. M. & Carrot, M. M. L. R. (2007). Lignin – from natural adsorbent to activated carbon: A review. *Bioresource Technology*, *98* (12), 2301–2312. https://doi.org/10.1016/j.biortech.2006.08.008.

Cataneo, C. B., Callari, V., Gonzaga, L. V., Kuskoski, E. M. & Fett, R. (2008). Atividade antioxidante e conteúdo fenólico do resíduo agroindustrial da produção de vinho. *Ciências Agrárias*. *29* (1), 93–102. https://doi.org/10.5433/1679-0359.2008v29n1p93.

Chang, Y., Guo, K., Guo, L., Liu, X., Chen, G., Liu, H., Yang, H. & Chen, J. (2016). Ploy (lactic acid)/organo-modified montmorillonite nanocomposites for improved eletret properties. *Journal of Electrostatics*, *80*, 17–21. https://doi.org/10.1016/j.elstat.2016.01.001.

Choudhary, P., Devi, T. B., Tushir, S., Kasana, R. C., Popatrao, D. S. & Narsaiah, K. (2023). Mango seed kernel: A bountiful source of nutritional and bioactive compounds. *Food and Bioprocess Technology*, *16*, 289–312. https://doi.org/10.1007/s11947-022-02889-y.

Ciolacu, D., Ciolacu, F. & Popa, V. I. (2011). Amorphous cellulose – Structure and characterization. *Cellulose Chemistry and Technology*, *45* (1), 13–21. www.researchgate.net/publication/279897864.

Conceição, M. D. N. D., Anaya-Mancipe, J. M., Coelho, A. W. F., Cardoso, P. H. M. & Thiré, R. M. S. M. (2024). Application of starch-rich mango by-product as filler for the development of an additive manufacturing filament compound. *International Journal of Biological Macromolecules*, *260*, Article no 129519. https://doi.org/10.1016/j.ijbiomac.2024.129519.

Cordeiro, E. M. S., Nunes, Y. L., Mattos, A. L. A., Rosa, M. F., Sousa Filho, M. S. M. & Ito, E. N. (2014). Polymer biocomposites and nanobiocomposites obtained from mango seeds. *Macromolecular Symposia*, *344* (1), 39–54. https://doi.org/10.1002/masy.201300217.

Corradini, E., Ferreira, F. C., Agnelli, J. A. M., Marconcini, J. M., Mattoso, L. H.C. & Rosa, M. F. (2013). Absorção de água, solubilidade em água, propriedades mecânicas e morfológicas de compósitos de glúten de milho e poli (hidroxibutirato-co-valerato) (PHBV) reforçados com fibras de coco verde. *Polímeros*, *23* (6), 807–813. https://doi.org/10.4322/polimeros.2014.001.

Cruz, R. A., Bassinello, P. Z. & Ascheri, D. P. R. (2010) *Seminário Jovens Talentos, Santo Antônio de Goiás: Embrapa Arroz e Feijão*. www.alice.cnptia.embrapa.br/alice/handle/doc/883291.

Fakhouri, F. M., Costa, D. L. M., Carvalho, C. W. P., Cardoso, R., Yamashita, F., Innocentine-Mei, L. H. & Queiroz, F. P. C. (2010). *International Symposium on Food Extrusion*.

Fitch-Vargas, P. R., Aguilar-Palazuelos, E., de Jesús Zazueta-Morales, J., Vega-García, M. O., Valdez-Morales, J. E., Martínez-Bustos, F., & Jacobo-Valenzuela, N. J. (2016). Physicochemical and microstructural characterization of corn starch edible films obtained by a combination of extrusion technology and casting technique. *Journal of Food Science*, *81* (9), 2224–2232. https://doi.org/10.1111/1750-3841.13416.

Florêncio, I. M., Souza, M. S. R., Gondim, S. R., Cavalcanti, M. T. & Florentino, E. R. (2012). *Encontro Nacional de Educação, Ciência e Tecnologial*. www.editorarealize.com.br/editora/anais/enect/2012/Poster_811.pdf.

França, F. A. (2014). *Caracterização nutricional e avaliação do potencial antioxidante de farinhas obtidas de resíduos de frutas*. www2.uesb.br/ppg/ppgecal/wp-content/uploads/2017/04/FL%C3%81VIA-FRAN%C3%87A.pdf.

García, A. M., García, A. I., Cabezas, M. A. L. & Reche, A. S. (2015). Study of the influence of the almond variety in the properties of injected parts with biodegradable almond shell based masterbatches. *Waste and Biomass Valorization*, *6*, 363–370. https://doi.org/10.1007/s12649-015-9351-x.

Garcia-Garcia, D., Quiles-Carrillo, L., Balart, R., Torres-Giner, S. & Arrieta, M. P. (2022). Innovative solutions and challenges to increase the use of Poly (3-hydroxybutyrate) in food packaging and disposables. *European Polymer Journal, 178,* Article no 111505. https://doi.org/10.1016/j.eurpolymj.2022.111505.

Genkina, N. K., Kozlov, S. S., Martirosyan, V. V. & Kiseleva, V. I. (2014). Thermal behavior of maize starches with different amylose/amylopectin ratio studied by DSC analysis. *Starch*, *66* (7-8), 700–706. https://doi.org/10.1002/star.201300220.

Gil-Castell, O., Badia, J. D., Kittikorn, T., Strömberg, E., Ek, M., Karlsson, S. & Ribes-Greus, A. (2016). Impact of hydrothermal ageing on the thermal stability, morphology and viscoelastic performance of PLA/sisal biocomposites. *Polymer Degradation and Stability*, *132*, 87–96. https://doi.org/10.1016/j.polymdegradstab.2016.03.038.

Gonzáles-Seligra, P., Guz, L., Ochoa-Yepes, O. & Goyanes, S. F. (2017). Influence of extrusion process conditions on starch film morphology. *LWT – Food Science and Technology*, *84*, 520–528. https://doi.org/10.1016/j.lwt.2017.06.027.

González, M. P. E. & Montoya, V. H. (2007). Characterization of mango pit as raw material in the preparation of activated carbon for wastewater treatment. *Biochemical Engineering Journal*, *36* (3), 230–238. https://doi.org/10.1016/j.bej.2007.02.025.

Gorrasi, G., Milone, C., Piperopoulos, E., Lanza, M. & Sorrentino, A. (2013). Hybrid clay mineral-carbon nanotube-PLA nanocomposite films. Preparation and photodegradation effect on their mechanical, thermal and electrical properties. *Applied Clay Science, 71*, 49–54. https://doi.org/10.1016/j.clay.2012.11.004.

Henrique, M. A., Silvério, H. A., Neto, W. P. F. & Pasquini, D. (2013). Valorization of an agro-industrial waste, mango seed, by the extraction and characterization of its cellulose nanocrystals. *Journal of Environmental Management*, *121*, 202–209. https://doi.org/10.1016/j.jenvman.2013.02.054.

Hizukuri, S. (1985). Relationship between the distribution of the chain length of amylopectin and the crystalline structure of starch granules. *Carbohydrate Research, 141* (2), 295–306. https://doi.org/10.1016/s0008-6215(00)90461-0.

Horvat, M., Ladiges, D. & Schuchmann, H. P. (2014). Extrusion cooking of corn starch by a novel nucleation die. *Food and Bioprocess Technology, 7*, 654–660. https://doi.org/10.1007/s11947-013-1109-5.

Huang, J., Zhao, L., Huai, H., Li. E., Zhang, F. & Wei, C. (2015). Structural and functional properties of starches from wild *Trapa quadrispinosa*, japonica, mammillifera and incise. *Food Hydrocolloids, 48*, 117–126. https://doi.org/10.1016/j.foodhyd.2015.02.003.

Ibrahim, M. I., Alsafadi, D., Alamry, K. A. & Hussein, M. A. (2021). Properties and applications of poly (3-hydroxybutyrate-co-3-hydroxyvalerate) biocomposites. *Journal of Polymers and the Environment, 29*, 1010–1030. https://doi.org/10.1007/s10924-020-01946-x.

Iqbal, M., Saeed, A. & Zafar, S. I. (2009). FTIR spectrophotometry, kinetics and adsorption isotherms modeling, ion exchange, and EDX analysis for understanding the mechanism of Cd^{2+} and Pb^{2+} removal by mango peel waste. *Journal of Hazardous Materials, 164* (1), 161–171. https://doi.org/10.1016/j.jhazmat.2008.07.141.

Ismail, S., Mansor, N. & Man, Z. (2017). A study on thermal behaviour of thermoplastic starch plasticized by [Emim] Ac and by [Emim] Cl. *Procedia Engineering, 184*, 567–572. https://doi.org/10.1016/j.proeng.2017.04.138.

Jaszkiewicz, A., Meljon, A., Bledzki, A. K. & Radwanski, M. (2016). Gaining knowledge on the processability of PLA-based short-fibre compounds – A comprehensive comparison with their PP counterparts. *Composites Part A: Applied Science and Manufacturing, 83*, 140–151. https://doi.org/10.1016/j.compositesa.2015.09.022.

Jiang, G., Hill, D., Kowalczuk, M., Johnston, B., Adamus, G., Irorere, V. & Radecka, I. (2016). Carbon sources for polyhydroxyalkanoates and an integrated biorefinery. *International Journal of Molecular Sciences, 17*, Article no 1157. https://doi.org/10.3390/ijms17071157.

Kalia, V. C., Patel, S. K. S. & Lee, J. K. (2013). Exploiting polyhydroxyalkanoates for biomedical applications. *Polymers, 15,* Article no 1937. https://doi.org/10.3390/polym15081937.

Khoo, R. Z., Ismail, H. & Chow, W. S. (2016). Thermal and morphological properties of poly (lactic acid)/nanocellulose nanocomposites. *Procedia Chemistry, 19*, 788–794. https://doi.org/10.1016/j.proche.2016.03.086.

Kist, B. B. (2023). *Anuário Brasileiro de Horti&fruti.* www.scribd.com/document/689790135/Anuario-Brasileiro-de-Horti-Fruti-23

La Fuente, C. I. A., Val Siqueira, L., Augusto, P. E. D. & Tadini, C. C. (2022). Casting and extrusion processes to produce bio-based plastics using cassava starch modified by the dry heat treatment (DHT). *Innovative Food Science & Emerging Technologies, 75*, Article no 102906. https://doi.org/10.1016/j.ifset.2021.102906.

Lazzari, E., Schena, T., Primaz, C. T., Maciel, G. P. S., Machado, M. E., Cardoso, C. A. L., Jacques, R. A. & Caramão, E. B. (2016). Production and chromatographic characterization of bio-oil from the pyrolysis of mango seed waste. *Industrial Crops and Products, 83*, 529–536. https://doi.org/10.1016/j.indcrop.2015.12.073.

Lima, E. M. B., Lima, A. M., Minguita, A. P. S., Rojas dos Santos, N. R., Pereira, I. C. S., Neves, T. T. M., Costa, G. L. F., Moreira, A. P. D., Middea, A., Neumann, R., Tavares, M. I. B. & Oliveira, R. N. (2019). Poly (lactic acid) biocomposites with mango waste and organo-montmorillonite for packaging. *Journal of Applied Polymer Science, 136*, Article no 47512. https://doi.org/10.1002/app.47512.

Lima, E. M. B., Middea, A., Marconcini, J. M., Corrêa, A. C., Pereira, J. F., Guimaraes, A. V., de Lima, J. F., dos Anjos, M. R., de Castro, I. M., Oliveira, R. N., Moreira, C. R., de Paiva, M. M., Rangel, F. L. C. & Neumann, R. (2021a). Biodegradable PLA based nanocomposites for packaging applications: The effects of organo-modified bentonite concentration. *Journal of Applied Polymer Science, 138*, Article no e50907. https://doi.org/10.1002/app.50907.

Lima, E. M. B., Middea, A., Neumann, R., Thiré, R. M. S. M., Pereira, J. F., de Freitas, S. C., Penteado, M. S., Lima, A. M., Minguita, A. P. S., Mattos, M. C., Teixeira, A. S., Pereira, I. C. S., dos Santos, N. R. R., Marconcini, J. M., Oliveira, R. N. & Corrêa, A. C. (2021). Biocomposites of PLA and mango seed waste: Potential material for food packaging and a technological alternative to reduce environmental impact. *Starch, 73*, Article no 2000118. https://doi.org/10.1002/star.202000118.

Lima, E. M. B., Middea, A., Reis, F. S., Mateus, D. N., de Amorim, R. G., Pereira, I. C. S., dos Santos, N. R. R., Mattos M. C., Minguita, A. P. S., Anjos, M. R., Neumann, R., Oliveira, R. N. & Tavares, M. I. B. (2024). Influence of the microstructure in the biodegradability process of eco-friendly materials based on polylactic acid and mango seed for food packaging to minimize microplastic generation. *J Journal of Applied Polymer Science*, *141*, Article no e55449. https://doi.org/10.1002/app.55449.

Lima, E. M. B., Oliveira, R. N., Middea, A., de Castro, I. M., Mattos, M. C., Neves, T. T. M., Costa, G. L. F., Neumann, R. & Tavares, M. I. B. (2020). Degradation of PLA biocomposites containing mango seed and organo montmorillonite minerals. *Journal of Natural Fibers*, *19* (5), 1783–1791. https://doi.org/10.1080/15440478.2020.1788488.

Lima, E. M. B.; Middea, A.; Pereira, J. F.; Pereira, I. C. S.; Santos, N. R. R.; Oliveira, R. N. & Neumann, R. (2017). *IX Workshop de Nanotecnologia Aplicada ao Agronegócio; Embrapa Instrumentação.*

Lobo, A. R. & Silva, G. M. L. (2003). Resistant starch and its physicochemical properties. *Revista de Nutrição*, 16 (2), 219–226. https://doi.org/10.1590/S1415-52732003000200009.

Loof, D., Hiller, M., Oschkinat, H. & Koschek, K. (2016). Quantitative and qualitative analysis of surface modified cellulose utilizing TGA-MS. *Materials*, *9*, Article no 415. https://doi.org/10.3390/ma9060415.

Lorevice, M. V, Moura, M. R., Zucolotto, V. & Mattoso, L. H. C. (2009). *Jornada Científica-Embrapa São Carlos.*

Magalhães, N. F. & Andrade, C. T. (2013). Properties of melt-processed poly (hydroxybutyrate-cohydroxyvalerate)/starch 1:1 blend nanocomposites. *Polímeros*, *23* (3), 366–372. http://dx.doi.org/10.4322/polimeros.2013.034.

Malacrida, C. R., Angelo, P. M., Andreo, D. & Jorge, N. (2007). Composição química e potencial antioxidante de extratos de sementes de melão amarelo em óleo de soja. *Revista Ciência Agronômica*, *38* (4), 372–376. www.redalyc.org/pdf/1953/195317465007.pdf.

Maldonado-Celis, M. E., Yahia, E. M., Bedoya, R., Landázuri, P., Loango, N., Aguillón, J., Restrepo, B. & Ospina, J.C. G. (2019). Chemical composition of mango (*Mangifera indica L.*) fruit: Nutritional and phytochemical compounds. *Frontier in Plant Science*, *10*, 1–21. https://doi.org/10.3389/fpls.2019.01073.

Manhongo, T. T., Chimphango, A., Thornley, P. & Röder, M. (2021). An economic viability and environmental impact assessment of mango processing waste-based biorefineries for co-producing bioenergy and bioactive compounds. *Renewable and Sustainable Energy Reviews, 148*, Article no 111216. https://doi.org/10.1016/j.rser.2021.111216.

Meereboer, K. W., Misra, M. & Mohanty, A. K. (2020). Review of recent advances in the biodegradability of polyhydroxyalkanoate (PHA) bioplastics and their composites. *Green Chemistry*, *22*, 5519–5558. https://doi.org/10.1039/D0GC01647K.

Meireles, C. S., Filho, G. R., Ferreira Jr, M. F., Cerqueira, D. A., Assunção, R. M. N., Ribeiro, E. A. M., Poletto, P. & Zeni, M. (2020). Characterization of asymmetric membranes of cellulose acetate from biomass: Newspaper and mango seed. *Carbohydrate Polymers*, *80* (3), 954–961. https://doi.org/10.1016/j.carbpol.2010.01.012.

Mestres, C., Matencio, F., Pons, B., Yajid, M. & Fliedel, G. (1996). A rapid method for the determination of amylose content by using differential-scanning calorimetry. *Starch*, *48* (1), 2–6. https://doi.org/10.1002/star.19960480103.

Możejko-Ciesielska, J. & Kiewisz, R. (2016). Bacterial polyhydroxyalkanoates: Still fabulous? *Microbiological Research*, *192*, 271–282. https://doi.org/10.1016/j.micres.2016.07.010.

Nascimento Filho, W. B. & Franco, C. R. (2015). Avaliação do potencial dos resíduos produzidos através do processamento agroindustrial no Brasil. *Revista Virtual de Química*, *7* (6), 1968–1987. https://doi.org/10.5935/1984-6835.20150116.

Nascimento, T. A., Carvalho, C. W. P., Andrade, K. C. S., Freitas, S. P. & Calado, V. M. A. (2010). *Congresso Brasileiro de Ciência e Tecnologia de Alimentos.*

Neto, W. P. F., Silvério, H. A., Dantas, N. O. & Pasquini, D. (2013). Extraction and characterization of cellulose nanocrystals from agro-industrial residue – Soy hulls. *Industrial Crops and Products*, *42*, 480–488. https://doi.org/10.1016/j.indcrop.2012.06.041.

Odom, I. E. (1984). Smectite clay minerals: properties and uses. *Philosophical Transactions of the Royal Society A*, *311*, 391–409. https://doi.org/10.1098/rsta.1984.0036.

Oun, A. A. & Rhim, J. W. (2016). Characterization of nanocelluloses isolated from Ushar (*Calotropis procera*) seed fiber: Effect of isolation method. *Materials Letters*, *168*, 146–150. https://doi.org/10.1016/j.matlet.2016.01.052.

Pagliano, G., Galletti, P., Samorì, C., Zaghini, A. & Torri, C. (2021). Recovery of polyhydroxyalkanoates from single and mixed microbial cultures: A review. *Frontiers in Bioengineering and Biotechnology*, *9*, 1–28. https://doi.org/10.3389/fbioe.2021.624021.

Palmeiro-Sánchez, T. & O'Flaherty, V. (2022). Polyhydroxyalkanoate bio-production and its rise as biomaterial of the future. *Journal of Biotechnology, 348*, 10–25. https://doi.org/10.1016/j.jbiotec.2022.03.001.

Pandey, V. K., Shafi, Z., Tripathi, A., Singh, G., Singh, R. & Rustagi, S. (2024). Production of biodegradable food packaging from mango peel via enzymatic hydrolysis and polyhydroxyalkanoates synthesis: A review on microbial intervention. *Current Research in Microbial Sciences*, *7*, Article no 100292. https://doi.org/10.1016/j.crmicr.2024.100292.

Pardauil, J. J. R., Souza, L. K. C., Molfetta, F. A., Zamian, J. R., Rocha Filho, G. N. & da Costa, C. E. F. (2011). Determination of the oxidative stability by DSC of vegetable oils from the Amazonian area. *Bioresource Technology*, *102* (10), 5873–5877. https://doi.org/10.1016/j.biortech.2011.02.022.

Paul, M. A., Alexandre, M., Degée, P., Henrist, C., Rulmont, A. & Dubois, P. (2003). New nanocomposite materials based on plasticized poly (l-lactide) and organo-modified montmorillonites: thermal and morphological study. *Polymer*, *44* (2), 443–450. https://doi.org/10.1016/S0032-3861(02)00778-4.

Paul, V., Rai, S., Vikranta, U., Nandan, A., Agarwal, A. & Tripathi, A. D. (2024). Starch-PHA blend-based biopolymers with potential food applications. *Starch*, *76*, Article no 2300131. https://doi.org/10.1002/star.202300131.

Perotti, G. F., Tronto, J., Bizeto, M. A., Izumi, C. M. S., Temperini, M. L. A., Lugao, A., Parra, D. & Constantino, V. R. L. (2013). Biopolymer-clay nanocomposites: Cassava starch and synthetic clay cast films. *Journal of the Brazilian Chemical Society*, *25* (2), 320–330. http://dx.doi.org/10.5935/0103-5053.20130300.

Rodrigues Filho, J. G. V., da Cruz, S. F., de Assunção, R. M. N. & Motta, L. A. C. (2012). Synthesis and characterization of methylcellulose from cellulose extracted from mango seeds for use as a mortar additive. *Polimeros*, *22* (1), 80–87. https://doi.org/10.1590/S0104-14282012005000011.

Santos Jr., L. A., Thiré, R. M. S. M., Lima, E. M. B., Racca,L. M. & da Silva, A. L. N. (2018). Mechanical and thermal properties of environment friendly composite based on mango's seed shell and high-density polyethylene. *Macromolecular Symposia*, *381*, Article no 1800125. https://doi.org/10.1002/masy.201800125.

Santos, A. E. O., Assis, J. S., Berbert, P. A., Santos, O. O., Batista, P. F. & Gravina, G. A. (2011). Influência de biofilmes de fécula de mandioca e amido de milho na qualidade pós-colheita de mangas 'Tommy Atkins'. *Revista Brasileira de Ciências Agrárias*, *6* (3), 508–513. https://doi.org/10.5039/agraria.v6i3a755.

Shankar, S. & Rhim, J. W. (2016). Preparation of nanocellulose from micro-crystalline cellulose: The effect on the performance and properties of agar-based composite films. *Carbohydrate Polymers*, *135*, 18–26. https://doi.org/10.1016/j.carbpol.2015.08.082.

Silva, A. K. P., Figueiredo, M. C. B. & Casimiro Filho, F. (2021). *Organizando espaços: a busca por um desenvolvimento sustentável.* www.alice.cnptia.embrapa.br/alice/handle/doc/1140401.

Solis-Fuentes, J. A., Hernandez-Mede, M. R. & Duran-de-Bazua, M. C. (2007). Isothermal crystallization kinetics of mango (*Mangifera indica*) almond seed fat. *Journal of Applied Sciences*, *7* (11), 1538–1542. https://doi.org/10.3923/jas.2007.1538.1542.

Souza Junior, F. G., Silva, A. M., Oliveira, G. E., Costa, R. M., Fernandes, E. R. & Pereira, E. D. (2015). Conducting and magnetic mango fibers. *Industrial Crops and Products*, *68*, 97–104. https://doi.org/10.1016/j.indcrop.2014.09.032.

Thiré, R. M. da S. M., Arruda, L. C. & Barreto, L. S. (2011). Morphology and thermal properties of poly (3-hydroxybutyrate-co-3-hydroxyvalerate)/attapulgite nanocomposites. *Materials Research*, *14* (3), 340–344. https://doi.org/10.1590/S1516-14392011005000046.

Thu, N. T. T., Hoang, L. H., Cuong, P. K., Viet-Linh, N., Nga, T. T. H., Kim, D. D., Leong, Y. K. & Nhi-Cong, L. T. (2023). Evaluation of polyhydroxyalkanoate (PHA) synthesis by Pichia sp. TSLS24 yeast isolated in Vietnam. *Scientific Reports*, *13*, Article no 3137. https://doi.org/10.1038/s41598-023-28220-z.

Vicente, D., Proença, D. N. & Morais, P. V. (2023). The role of bacterial polyhydroalkanoate (PHA) in a sustainable future: A review on the biological diversity. *International Journal of Environment Research and Public Health*, *20*, Article no 2959. https://doi.org/10.3390/ijerph20042959.

Vieira, E. L, Pereira, M. E. C., Santos, D. B. & Lima, M. A. C. (2009). Aplicação de biofilmes na qualidade da manga 'Tommy atkins'. *Magistra, Cruz das Almas-BA*, *21* (3), 165–170. https://ainfo.cnptia.embrapa.br/digital/bitstream/item/18187/1/Auxiliadora_2009.pdf.

Warren, F. J., Gidley, M. J. & Flanagan, B. M. (2016). Infrared spectroscopy as a tool to characterise starch ordered structure—a joint FTIR–ATR, NMR, XRD and DSC study. *Carbohydrate Polymers*, *139*, 35–42. https://doi.org/10.1016/j.carbpol.2015.11.066.

Wu, D. & Hakkarainen, M. (2015). Recycling PLA to multifunctional oligomeric compatibilizers for PLA/starch composites. *European Polymer Journal*, *64*, 126–137. https://doi.org/10.1016/j.eurpolymj.2015.01.004.

Xavier, I. F., Leite, G. A., Medeiros, E. V. & Moraes, P. L. D. (2009). Qualidade pós-colheita da manga 'Tommy Atkins' comercializada em diferentes estabelecimentos comerciais no município de Mossoró-RN. *Revista Caatinga*, *22* (4), 7–13. https://periodicos.ufersa.edu.br/caatinga/article/view/568/pdf.

Yamoneka, J., Malumba, P., Blecker, C., Gindo, M., Richard, G., Fauconnier, M. L., Lognay, G. & Danthine, S. (2015). Physicochemical properties and thermal behaviour of African wild mango (*Irvingia gabonensis*) seed fat. *LWT – Food Science and Technology*, *64* (2), 989–996. https://doi.org/10.1016/j.lwt.2015.06.035.

Yatnatti, S., Vijayalakshmi, D. & Chandru, R. (2014). Processing and nutritive value of mango seed kernel flour. *Current Research in Nutrition and Food Sci*ence, *2* (3), 170–175, https://doi.org/10.12944/CRNFSJ.2.3.10.

Zein, R. E., EL-Bagoury, A. A. & Kassab, H. E. (2005). Chemical and nutritional studies on mango seed kernels. *Journal of Agriculture Science*, *30*, 3285–3299. https://doi.org/10.21608/jfds.2005.237761.

Zeleznak, K. J. & Hoseney, R. C. (1987). The glass transition starch. *Cereal Chemistry*, *64* (2), 121–124. www.cerealsgrains.org/publications/cc/backissues/1987/Documents/64_121.pdf

Zhou, W., Bergsma, S., Colpa, D. I., Euverink, G. J. W. & Krooneman, J. (2023). Polyhydroxyalkanoates (PHAs) synthesis and degradation by microbes and applications towards a circular economy. *Journal of Environmental Management*, *341* (118033), 1–13. https://doi.org/10.1016/j.jenvman.2023.118033.

Zobel, H. F. (1988). Molecules to granules: A comprehensive starch review. *Starch*, *40* (2), 44–50. https://doi.org/10.1002/star.19880400203.

12 Unveiling the Potential of Litchi Seed Starch

E. I. O. Ajayi, A. Nkumah, H. O. Faleke, and Pinki Saini

12.1 INTRODUCTION

Litchi (*Litchi chinensis*) is a member of the soapberry family (Sapindaceae) and is cultivated in tropical and warm subtropical regions worldwide. It holds significant commercial value due to its delicious and nutritious fruit (Li, 2021). The fruit is believed to have originated in Southern China and Northern Vietnam before spreading to at least 20 countries (Singh et al., 2023). Major litchi-producing countries include China, Taiwan, Thailand, India, South Africa, Madagascar, Mauritius, and Australia (FAO). While litchi is primarily consumed domestically in producing countries, it is also exported to key global markets such as France, Germany, the United Kingdom, Hong Kong, Singapore, the Philippines, Japan, Thailand, the United States, and Canada. Additionally, the litchi market has been expanding into Middle Eastern countries. Most litchi produced worldwide is consumed fresh, with a smaller portion processed into frozen, canned, or dried products, as well as beverages and jellies (FAO). The fruit consists of a sweet, juicy aril as its edible part, while the peel and seed, which together make up approximately 30–40% of the fruit's weight, are considered nonedible. These byproducts are often discarded during processing or fresh consumption, contributing to waste management challenges and environmental pollution.

Global litchi production is estimated to be approximately 2.7 million tons, generating around 0.54 million tons of seeds annually (Cronje et al., 2023). Litchi seeds measure 1–3.3 cm in length and 0.6–1.2 cm in width, with an oblong to concave shape and a dark brown to chocolate-like color (Bangar & Kumar, 2021). Their composition includes starch polysaccharides (40.7%), proteins (4.93%), crude fiber (24.5%), oil content (3.2%), and essential minerals such as magnesium (0.28%), calcium (0.21%), and phosphorus (0.11%) (Kaur et al., 2024). Developing efficient utilization strategies for litchi seeds could enhance the economic feasibility of the litchi processing industry. A comprehensive review has been published on the phytochemical profile, biological activities, and applications of litchi seeds in food (Punia & Kumar, 2021). Due to their high starch content, litchi seeds present a promising nonconventional starch source with diverse applications. Valorizing this underutilized starch source could reduce dependence on conventional starches while addressing environmental concerns associated with seed waste. For starch to be suitable for various applications, it must possess certain desirable characteristics. Litchi seed starch (LSS) granules are round to oval, with a diameter ranging from 3 to 10 μm. Research indicates that starch isolated using citric acid retains more moisture than that extracted with alkali, as the presence of hydrophilic hydroxyl groups contributes to higher amylose content. Additionally, LSS exhibits a high water-holding capacity (WHC; 462–587, w/w) and significant thermal stability, with weight loss due to thermal degradation occurring at 360 °C, making it suitable for various food industry applications (Jaiswal & Kumar, 2015).

DOI: 10.1201/9781003534914-12

A thorough understanding of the composition, physicochemical, and functional properties of LSS is essential for exploring its potential in both food and nonfood applications. However, limited research has been conducted in this area. This review compiles existing studies on the isolation techniques, composition, structure, and techno-functional properties of LSS. Furthermore, it provides recommendations for its potential applications in the food, pharmaceutical, and medical industries. To illustrate the key components discussed, an overview of the review is presented in Figure 12.1.

12.2 EXTRACTION AND OPTIMIZATION OF LSS

Litchi seeds, which account for approximately 10–20% of the total fruit weight, represent an underutilized resource with significant potential for starch extraction and industrial applications. These seeds contain about 40.7% starchy polysaccharides, making them a promising nonconventional starch source (Kaur et al., 2024). However, the efficiency of starch recovery from litchi seeds is influenced by several factors, including seed maturity, extraction techniques, and processing conditions. Various methods have been employed to extract starch from litchi seeds, with alkaline, acidic, or neutral steeping being the most common approaches (Kumari et al., 1993). Some examples of the starch extraction methods for litchi seeds are provided in Table 12.1. The choice of extraction method plays a crucial role in determining both the yield and purity of the extracted starch.

12.2.1 Extraction Techniques and Processing Conditions

The standard procedure for LSS isolation typically involves wet milling techniques combined with either acidic or alkaline steeping to break down cellular components and facilitate the release of starch granules. The process begins with washing, soaking, and homogenizing the seeds in various steeping solutions, including water, acidic solutions (such as citric acid), or alkaline solutions (such as sodium hydroxide) (Kaur et al., 2024). Once the seeds are properly homogenized, the resulting slurry undergoes a series of sieving steps to separate starch from fibers, proteins, and other nonstarch components. The purified starch suspension is then subjected to centrifugation, where the heavier starch granules settle as sediment while the supernatant, containing soluble impurities, is discarded. This washing and centrifugation process is repeated multiple times to improve starch purity, ultimately yielding a relatively refined starch product.

Despite these established techniques, the starch recovery efficiency from litchi seeds remains comparatively lower than that of other starch sources such as corn, potatoes, and mung bean. This is primarily due to the tight cellular matrix within litchi seeds, which makes starch granules less accessible and more resistant to extraction (Punia & Kumar, 2021). Researchers have been exploring optimized methods to improve starch yield, purity, and functional properties to enhance the feasibility of using LSS as a commercial alternative.

12.2.2 Effect of Seed Maturity on Starch Content and Yield

The starch content in litchi seeds varies significantly depending on seed maturity at the time of harvest. Kumari et al. evaluated two prominent litchi cultivars, 'Kasba' and 'Deshi', and found that starch levels fluctuate during different stages of seed development. Their study revealed that seeds harvested between 57 and 74 days after flowering contained the highest starch concentrations (82–94 µg/mg of fresh tissue), while overly matured seeds exhibited a substantial decline in starch content (Kumari et al., 1993). This suggests that optimal starch recovery from litchi seeds requires harvesting before full maturity, as prolonged maturation may lead to the conversion of starch into other biochemical constituents, reducing overall yield. Additionally, the structural composition of the seed at different maturity stages influences the ease of starch extraction (Zhang et al., 2024a).

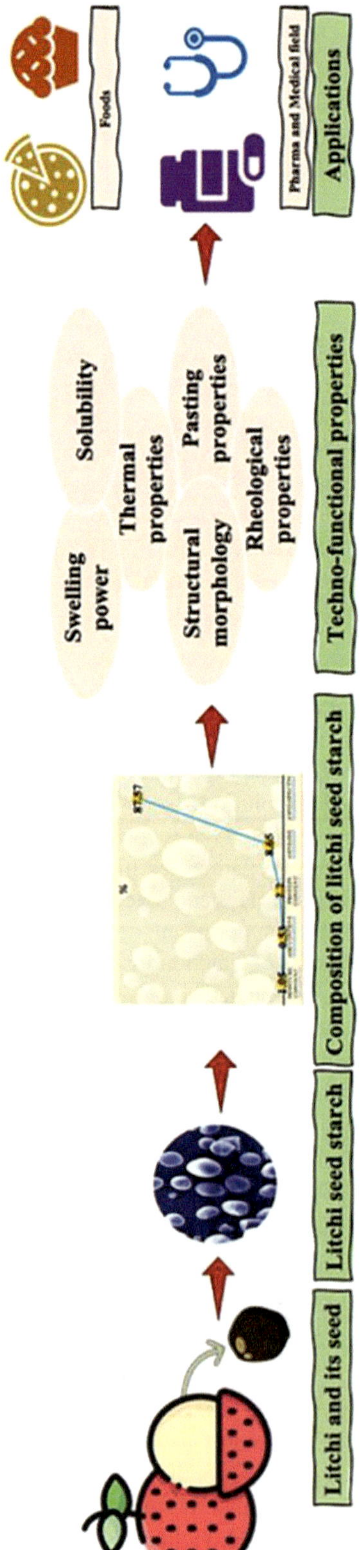

FIGURE 12.1 Overview of the litchi seed starch review (Jaiswal & Kumar, 2015).

TABLE 12.1
Methods of Starch Isolation from Litchi Seeds

Source	Method	Steeping Condition	Yield	Inference	Reference
Litchi seeds (Cultivar: Shahi, Muzzafarpur, Bihar, India)	Wet milling	Alkali: 0.5% NaOH	12.6%	Low yield	(Sahni et al., 2020)
Litchi seeds (Cultivar: Shahi, Muzzafarpur, Bihar, India)	Wet milling	Acidic: 0.3% citric acid	11%	Low yield	(Sahni et al., 2020)
Litchi seeds, local market Sirsa, India	-	Distilled water with 0.16% sodium hydrogen sulphite for 12 h at 50°C	14.17%	Relatively higher yield	(Thakur et al., 2021)

Younger seeds may contain more loosely packed starch granules, allowing for better extraction efficiency, whereas matured seeds may have denser starch–protein interactions, making separation more challenging. Understanding these physiological changes is essential for optimizing litchi seed processing and maximizing starch recovery.

12.3 COMPARATIVE EXTRACTION YIELDS AND EFFICIENCY

Several studies have compared the efficiency of different chemical steeping methods for starch extraction from litchi seeds. An acidic extraction method using citric acid (0.3%, w/w) yielded an 11% starch recovery, while an alkaline extraction method using sodium hydroxide (0.5%, w/w) achieved a slightly higher yield of 12.6% (Kumari et al., 1993). These results indicate that alkaline steeping marginally enhances starch yield, likely due to its effectiveness in breaking down protein-starch interactions. However, both methods remain inefficient, leaving nearly two-thirds of the total starch content unextracted.

Compared to other starch sources, LSS recovery rates are significantly lower (Punia & Kumar, 2021). For example, conventional starch extraction methods for maize kernels and potatoes using sodium bisulfite have yielded recoveries ranging from 43% to 64% and 61% to 67%, respectively. A study investigated the impact of soaking mung beans in a 0.2% sodium bisulfite ($NaHSO_3$) solution at 30°C for 48 hours, resulting in an increase in starch extraction ratio from 83.7% to between 89.7% and 96.2%. Similarly, soaking in a 0.04% sodium hydroxide (NaOH) solution under the same conditions enhanced the extraction ratio to between 93.4% and 98.3% (Wang et al., 2005). These findings highlight the need for improved processing strategies to enhance the efficiency and scalability of LSS extraction.

12.4 STRATEGIES FOR ENHANCING STARCH RECOVERY

Given the low extraction efficiency of current methods, researchers are actively investigating ways to modify and optimize starch isolation protocols to improve yield and quality. Some potential strategies include:

- Enzymatic Pretreatment: Enzymes such as amylases, cellulases, and proteases can partially degrade the seed matrix, facilitating starch release while minimizing structural damage to the granules (Joshi, 2018).
- Optimized Steeping Conditions: Adjusting factors such as pH, steeping time, and temperature can enhance starch solubility and reduce losses during extraction.
- Dual-Stage Extraction: Combining acidic and alkaline treatments sequentially may help break down different seed components, increasing starch recovery.

- Ultrasonic-Assisted Extraction (UAE): The application of ultrasound waves has been shown to disrupt plant cell walls and improve starch extraction efficiency in other botanical sources (Shen et al., 2023).
- Modified Wet-Milling Techniques: Incorporating continuous washing cycles and fine-tuning centrifugation parameters could yield higher starch purity.

By integrating these strategies, it may be possible to improve starch extraction efficiency substantially, making LSS a more competitive alternative in the commercial starch industry. Future advancements in extraction optimization, including enzymatic hydrolysis, ultrasonic-assisted extraction, and improved milling techniques, could significantly enhance starch yield and quality. As the demand for sustainable, functional, and plant-based starch sources continues to grow, LSS presents an exciting opportunity for further research and industrial application. By addressing current extraction inefficiencies and refining processing protocols, LSS could emerge as a viable alternative to conventional starches, unlocking new possibilities in both food and nonfood industries.

12.5 STRUCTURAL CHARACTERISTICS OF LSS

The size and morphology of starch granules vary significantly across plant species (Lindeboomet al., 2004). Various instruments, including scanning electron microscopy (SEM), transmission electron microscopy (TEM), light microscopy, normal light microscopy (NLM), and polarized light microscopy (PLM), are used to examine starch structure (Ai & Jane, 2018). Among these, SEM is the most commonly used technique for analyzing starch granule morphology. LSS granules exhibit a bimodal size distribution, consisting of large (A-type) granules with diameters ranging from 7.88 to 10.18 μm and small (B-type) granules measuring 3.01 to 4.22 μm (Bangar & Kumar, 2021). These SEM findings align with observations of conventional starch sources, such as corn (1.1–62.2 μm), potato (1.9–124.5 μm), and mung bean (1.6–74.0 μm) starch, which also exhibit bimodal distributions (Sahni et al., 2020). SEM images comparing LSS with commercial starches are presented in Figure 12.2. SEM analysis of LSS revealed round to oval granules with smooth surfaces, lacking clefts or irregularities (Thory & Sandhu, 2017). In comparison, corn starch granules have been reported to be less smooth and more angular than those from litchi, potato, and mung bean sources (Sahni et al., 2020). Due to the presence of both A-type and B-type granules, LSS offers diverse application potential-type starch granules demonstrate a higher gelatinization temperature and lower gelatinization enthalpy (ΔH) than A-type granules (Zhang et al., 2024b). Starches predominantly composed of B-type granules have been suggested as fat replacers, whereas those with a higher proportion of A-type granules may be suitable for manufacturing biodegradable plastic films and noncarbon copy paper (Zhang et al., 2022).

12.6 THERMAL CHARACTERISTICS

Heating starch in the presence of excess water leads to hydration and swelling of the starch granules, ultimately forming a paste. This process involves the melting of crystallites, breaking of hydrogen bonds, unwinding of double helices, and uncoiling of amylopectin chains, resulting in the collapse of the granule structure. This phenomenon, known as starch gelatinization, is characterized by the loss of birefringence in the granules. Extensive research has been conducted to measure and define starch gelatinization, focusing on the phase changes involved and their impact on starch digestibility. During this process, the disaggregated starch chains tend to retrograde into partially ordered structures that differ from those found in native granules. Reducing retrogradation is a key area of interest, as it negatively affects the quality of starch-rich products. The susceptibility of starch to enzymatic degradation is largely influenced by the extent of structural disruption during gelatinization and the nature of aggregates formed during retrogradation (Wang et al., 2015). Understanding

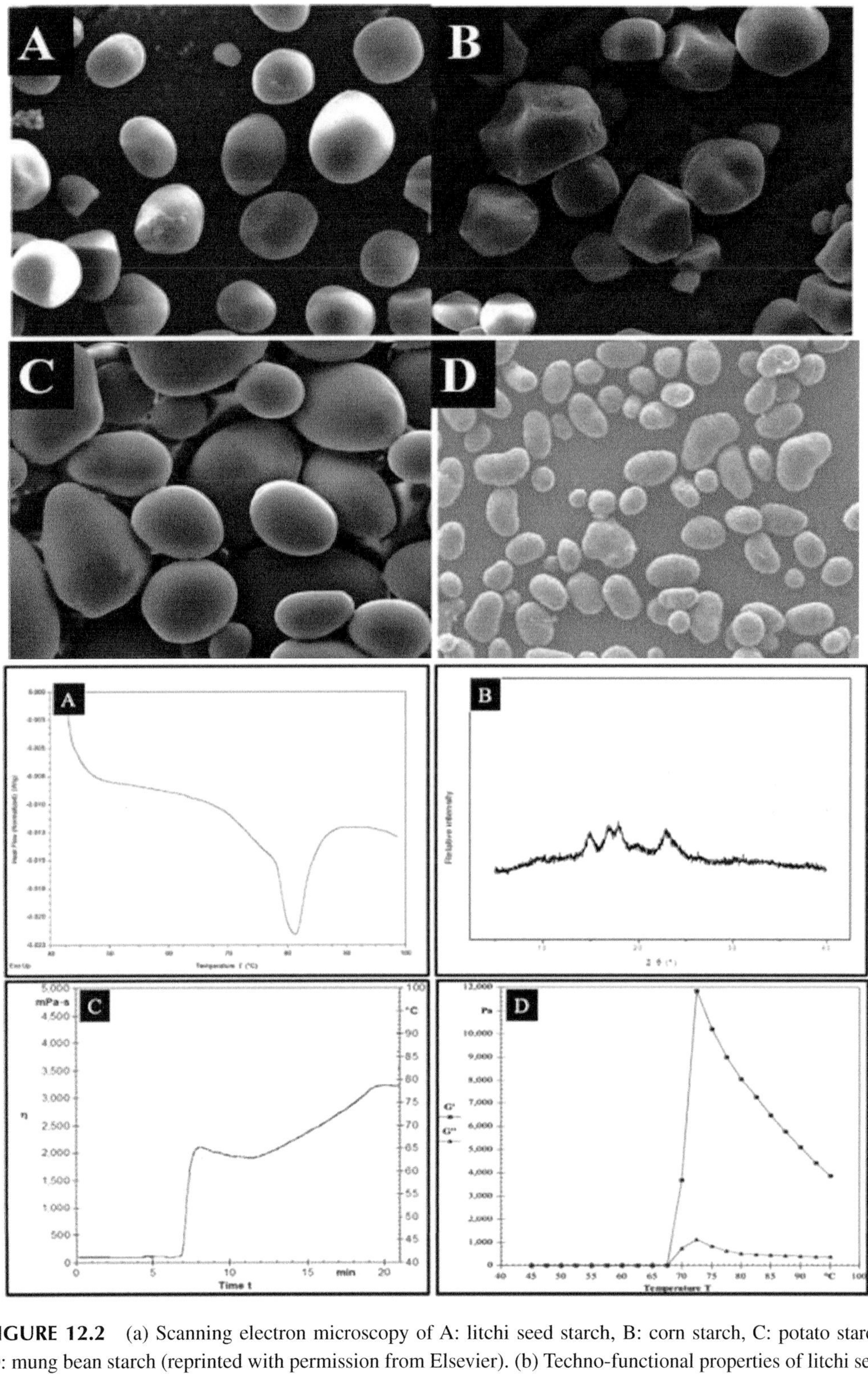

FIGURE 12.2 (a) Scanning electron microscopy of A: litchi seed starch, B: corn starch, C: potato starch, D: mung bean starch (reprinted with permission from Elsevier). (b) Techno-functional properties of litchi seed starch. A) Thermal characteristics, B) X-ray diffraction pattern, C) pasting property, and D) storage and loss modulus of litchi seed starch under varying temperature.

the thermal characteristics of starch and controlling its gelatinization behavior are essential for optimizing its functionality in various food and industrial applications. Extensive research has been conducted to measure and define starch gelatinization, focusing on characterizing the phase changes involved and their impact on starch digestibility (Zhang et al., 2024a). During this process, disaggregated starch chains often retrograde into partially ordered structures, differing from those present in native granules. Reducing retrogradation is a key research area, as it negatively affects the quality of starch-rich products. The susceptibility of starch to enzymatic degradation is primarily influenced by the extent of structural disruption during gelatinization and the nature of aggregates formed during retrogradation.

The gelatinization parameters of LSS include onset temperature (T_o) at 73.3°C, peak temperature (T_p) at 77.7°C, conclusion temperature (T_c) at 82.6°C, with a gelatinization range of 9.3°C and an enthalpy change (ΔH) of 11.6 J/g. Thermogravimetric analysis of LSS reveals a three-phase weight loss between 100 and 600°C. The first phase (0–150 °C) results in a weight loss of 10.57–14.22%, attributed to the evaporation of bound water. The second phase (260–350°C) causes a major weight loss of 64–67%, while the third phase (360–400°C) corresponds to starch degradation (Punia & Kumar, 2021; Jaiwal & Kumar, 2015). The thermal characteristics of LSS are illustrated in Figure 12.2b (A). Compared to starches from cereals, pulses, and tubers, corn starch exhibits two distinct endotherms, whereas a single endotherm characterizes potato and mung bean starches. The first endothermic peak indicates the melting and gelatinization of starch crystallites, while the second peak corresponds to the dissociation of amylose–lipid complexes, which are common in cereal starches but either absent or present in minimal amounts in pulse and tuber starches (Xing et al., 2017).

Among these, LSS exhibits the highest peak temperature (T_p), followed by corn, potato, and mung bean starch. The enthalpy change (ΔH), which measures the energy required to disrupt the molecular double-helical order, is highest in potato starch (13.15 J/g), followed by corn (11.78 J/g), litchi (11.6 J/g), and mung bean (9.61 J/g) starch. Higher gelatinization transition temperatures are associated with a greater degree of crystallinity, which enhances structural stability and makes starch granules more resistant to gelatinization. The superior thermal stability of LSS suggests its potential application as a viscosifier and texturizer in soups, sauces, bakery, and dairy products (Zhang et al., 2022).

12.7 NUTRITIONAL BENEFIT AND APPLICATION OF LSS

Research on various nonconventional starch sources has increased since the last decade, whereas much of the scientific evaluation on functional properties (SP, solubility power, water-holding capacity, size) of LSS has been carried out in the last 5 years (Esua *et al.*, 2024; Carvalho *et al.*, 2024). The interpretation of these properties is helpful in annotating specific applications to LSS. Various suggested applications of the LSS are presented in Figure 12.3. Functional ingredients in foods LSS showed higher water holding capacity ranging from 462 to 587 (%, w/w). Variations in the water holding capacity were due to variable pH used to extract starch. This property shows that LSS has a three-dimensional structure, which does not allow rapid water release. High water holding capacity acts as a critical player in the formation of food texture, especially in the case of baked dough and comminuted meat products (Warner, 2023). The thermal profiling of the LSS showed higher stability during thermal processing, which started degrading when the temperature reached from 360 to 400 °C as reflected by thermal profiling using a thermo gravimetric analyzer. LSS can therefore be used in food items (canned foods) that undergo higher processing temperatures. While solubility was 2 and 1 (%, w/w) in alkali and acid-extracted starch, respectively, at room temperature, SP of LSS was found to be between 3.5 and 4.7 g/g (Jaiswal & Kumar, 2015). The solubility of alkali-extracted starch increased to 8.6 g/g and 6.7% at 90°C, which makes it a useful ingredient for food applications such as thickening and gelling agents (Punia & Kumar, 2021). They can be used as a

FIGURE 12.3 Possible uses for the starch from litchi seeds. (a) Litchi pulp is processed for juice and drinks once it is harvested (b)–(k) use of litchi seed starch as a functional food component in meals (Punia & Kumar, 2021).

thickening agent in soups, sauces, gravies, toppings, and salad dressings and as a gelling agent in jam, jelly, and marmalade (Figure 12.3).

LSS is gaining recognition for its unique nutritional properties, particularly in terms of digestibility, glycemic index, and bioactive potential. The composition of LSS, including its amylose-to-amylopectin ratio, influences its digestibility and subsequent metabolic effects. Research has indicated that LSS has a relatively higher amylose content than traditional starch sources, contributing to a slower digestion rate and a lower glycemic index (Zhang et al., 2021). This characteristic makes it a promising dietary component for individuals managing blood glucose levels, such as those with diabetes, with its attending oxidative stress and subtle inflammation (Izu et al., 2024). Additionally, LSS contains resistant starch, which resists enzymatic digestion and reaches the colon, where it undergoes fermentation by gut microbiota (Thory & Sandhu, 2017; Morales et al., 2022). This process produces short-chain fatty acids (SCFAs), such as butyrate, which have been linked to improved insulin sensitivity and reduced inflammation (Higgins, 2004). Thus, incorporating LSS into the diet may support metabolic health and glycemic control.

Studies have demonstrated that LSS extracts inhibit lipid peroxidation and enhance the activity of antioxidant enzymes such as superoxide dismutase (SOD), offering protective effects against oxidative damage (Izu et al., 2024; Upadhyaya & Upadhyaya, 2017). Furthermore, dietary polyphenols found in LSS have been associated with reduced risks of hypertension and atherosclerosis by modulating inflammatory pathways and improving endothelial function (Zhang et al., 2021). These findings suggest that LSS can serve as a functional ingredient with potential therapeutic benefits beyond essential nutrition.

The function of LSS in gut health and microbiota modification is one of its noteworthy health advantages. By promoting the growth of advantageous gut bacteria like *Bifidobacterium* and *Lactobacillus*, the resistant starch portion of LSS functions as a prebiotic (Chen et al., 2024). SCFAs are produced when the gut bacteria ferment the starch of litchi seeds, and they help to reduce inflammation and preserve the integrity of the gut barrier. Furthermore, it has been demonstrated that butyrate, one of the main SCFAs generated, has anticarcinogenic qualities, which may reduce the incidence of inflammatory bowel disorders and colon cancer (Topping & Clifton, 2001). These gut health advantages further highlight LSS's potential as a dietary component that promotes immunological and digestive health.

LSS also plays a role in weight management and obesity prevention due to its slow digestion rate and impact on satiety. Starches with higher resistant starch content have been associated with prolonged gastric emptying and increased secretion of gut hormones such as peptide YY (PYY) and glucagon-like peptide-1 (GLP-1), which regulate appetite and food intake (Feinle & Horowitz, 2021). Experimental studies have shown that rodents consuming LSS-enriched diets exhibited lower weight gain and improved lipid metabolism than those consuming conventional starches (Guo et al., 2021). Furthermore, the bioactive compounds in LSS contribute to cardiovascular health by reducing oxidative stress and improving lipid profiles. Studies have reported that LSS consumption decreases levels of low-density lipoprotein (LDL) cholesterol while increasing high-density lipoprotein (HDL) cholesterol, reducing the risk of atherosclerosis and heart disease (Queiroz et al., 2017). These metabolic and cardiovascular benefits highlight the potential of LSS as a valuable dietary component in promoting long-term health.

Given its diverse health-promoting properties, LSS has significant potential for applications in functional food development. The incorporation of LSS into low-glycemic index food products, prebiotic formulations, and antioxidant-enriched supplements can provide consumers with additional health benefits (Figure 12.3). Food industries are exploring plant-derived starches with bioactive properties as functional ingredients in bakery products, dairy alternatives, and nutraceutical formulations (Kadyan et al., 2023). As research continues to explore the mechanisms underlying the health benefits of LSS, further studies are needed to optimize its extraction, improve bioavailability, and expand its applications in health-promoting food products. Overall, LSS presents an exciting opportunity for dietary interventions aimed at enhancing metabolic health, gut microbiota balance, and disease prevention.

12.7.1 Pharmaceutical Application

LSS has attracted much attention due to its possible uses as a pharmaceutical excipient and in drug delivery, especially from the "Shahi" cultivar (Jaiswal & Kumar, 2015). Recent studies have evaluated its micrometric properties, which are crucial in determining its suitability for tablet formulation. The assessment of Carr's index, Hausner's ratio, and angle of repose revealed that LSS possesses an excellent packing arrangement, as indicated by its low Carr's index and minimal void volume. These properties contribute to superior powder flowability, an essential characteristic for efficient tablet manufacturing. Specifically, the starch exhibited a Hausner's ratio of <1.11, a Carr's index of <10%, and an angle of repose of <25°, all suggesting smooth powder flow and ease of compression (Lachman et al., 2003). In contrast, higher values in these micrometric parameters typically indicate poor flowability, leading to processing challenges such as inconsistent tablet weight, segregation, and inefficient die filling (Loo et al., 2023).

Further evaluation of tablets formulated with LSS confirmed that they met the physical quality standards outlined by the European Pharmacopoeia (Chakravorty et al., 2016; European Pharmacopoeia, 2018). These parameters include average weight, hardness, friability, thickness, and disintegration time, which are critical for ensuring pharmaceutical tablets' stability, durability, and efficacy. The ability of LSS to provide adequate hardness and low friability without compromising rapid disintegration suggests that it can function effectively as a binder and disintegrant in solid dosage forms. A well-balanced disintegration time is essential for ensuring proper drug release, and LSS demonstrates a favor in this regard. Additionally, its biodegradable and nontoxic nature enhances its potential as a sustainable alternative to synthetic excipients, aligning with the growing demand for eco-friendly pharmaceutical ingredients.

The promising pharmaceutical attributes of LSS open avenues for further research into its potential modifications and applications in controlled drug release systems. Future studies could explore chemical or enzymatic modifications to enhance its functionality, such as improving its swelling index, water absorption capacity, or compatibility with other active pharmaceutical ingredients (APIs). Additionally, given its film-forming ability, litchi starch could also be investigated as a coating agent for encapsulating drugs, protecting them from degradation while enabling targeted release. Its high stability and superior flowability make it a strong candidate for large-scale pharmaceutical production, offering a natural, efficient, and cost-effective alternative to conventional excipients.

The excellent micrometric properties, compliance with pharmacopeia standards, and potential for further functionalization position LSS as a viable excipient in the pharmaceutical industry. As the demand for plant-based and sustainable pharmaceutical ingredients continues to grow, LSS stands out as a valuable and versatile excipient with significant potential for drug formulation and delivery applications.

12.7.2 Medicinal Applications

Due to its therapeutic benefits in treating chronic illnesses, including rheumatoid arthritis, nocturnal asthma, and angina, targeting APIs in the colon has attracted a lot of study interest (Udaipuria & Bhattacharya, 2025). Colonic drug delivery enables localized treatment, improved bioavailability, and reduced systemic side effects, making it an essential strategy for addressing gastrointestinal and systemic diseases. Various natural polysaccharides, including pectins, gums, inulin, chitosan, and starch, have been widely explored for their role in colon-targeted drug delivery due to their biodegradability, biocompatibility, and ability to resist upper gastrointestinal digestion (Gong et al., 2024; Madhav & Yadav, 2013). With its superior swelling, gelling, and controlled-release qualities for specific pharmaceutical applications, starch stands out as an environmentally benign, economically viable, and sustainable excipient.

With its distinct physicochemical characteristics, LSS has shown great promise as an excipient in formulations for controlled release tablets. Using LSS that has been acid- and alkali-extracted, *in vitro* drug release tests have demonstrated that only 9.9% to 26.3% of the drug is released at an acidic pH of 1.2, indicating low release in the stomach, a critical property for colonic drug targeting (Madhav & Yadav, 2013). The medication's slow-release profile and efficacy in delayed drug delivery to the colon are confirmed by the progressive rise in drug release at physiological pH, which reaches 67% after 5 hours. These findings demonstrate how LSS can prevent the early deterioration of active pharmaceutical substances, guaranteeing that a sizable amount of the medication reaches the colon for the best possible therapeutic effect.

The high amylopectin and low amylose content of LSS plays a pivotal role in its application as a colon-targeting excipient. Amylopectin's branched structure enhances water absorption, forming a gel-like network that slows drug diffusion and prolongs drug release. The starch's low paste clarity, swelling power (SP), peak viscosity, and breakdown viscosity, combined with high solubility and setback viscosity, contribute to its excellent drug retention properties (Thory & Sandhu, 2017). These characteristics are particularly beneficial for colon-targeted formulations for inflammatory bowel disease (IBD) and colon cancer (Chen et al., 2024).The ability of LSS to modulate drug release based on physiological conditions makes it an ideal excipient for delayed and sustained-release formulations, enhancing patient compliance and therapeutic efficacy.

Given its biodegradability, cost-effectiveness, and functional versatility, LSS presents a promising alternative to synthetic excipients in pharmaceutical formulations. Future studies aimed at modifying its structural properties through cross-linking, enzymatic treatment, or nanoparticle-based delivery systems could further optimize its performance for targeted drug delivery. LSS has the potential to revolutionize colon-specific drug delivery systems by providing sustainable, patient-friendly, and therapeutically effective options for controlling chronic illnesses due to its distinct physicochemical profile and controlled-release capabilities.

12.7.3 Application in Food Preservation

LSS presents a promising opportunity to develop innovative starch-based films and coatings, offering a cost-effective, biodegradable, and environmentally friendly alternative to synthetic materials (Punia & Kumar, 2021). Its potential extends beyond basic film-forming abilities due to bioactive compounds, including phenolic compounds (particularly catechin-rich compounds) and bioactive polysaccharides such as LSP-W-4. These bioactive constituents contribute to the antioxidant properties of LSS and enhance its functional capabilities for producing edible coatings and films with added health benefits (Kumar & Khan, 2024). Furthermore, LSS films can be blended with other polysaccharides to improve mechanical strength, flexibility, and barrier properties, making them more effective as food packaging materials. This modification enhances the film's resistance to moisture transfer and extends the shelf life of perishable foods by providing a protective layer against oxidation and microbial contamination.

The antibacterial activity of films made from LSS is one of their most convincing benefits, making them extremely helpful for food safety and preservation. Several foodborne pathogens, such as *Escherichia coli, Staphylococcus aureus, Pseudomonas aeruginosa, Bacillus cereus, Enterobacter spp., and Trichoderma fungi*, have been demonstrated to be susceptible to the broad-spectrum antimicrobial effects of the different phenolic compounds found in litchi seeds (Bhat & Al-daihan, 2014). These antimicrobial properties significantly reduce microbial spoilage in food products, offering additional protection without synthetic preservatives. Research has further validated the antimicrobial efficacy of litchi seed extracts, demonstrating notable inhibitory activity against common pathogens such as *Streptococcus pyogenes*, *Bacillus subtilis, Staphylococcus aureus, Pseudomonas aeruginosa,* and *E. coli* (Mir & Perveen, 2022; Bhat & Al-daihan, 2014; Suriyaprom et al., 2022). These findings strongly support the application of LSS-based films as natural antimicrobial packaging materials, which could significantly enhance food safety while reducing reliance on synthetic additives.

Incorporating LSS into biodegradable food packaging solutions aligns with the growing demand for sustainable, eco-friendly alternatives to plastic-based materials. The combination of antioxidant, antimicrobial, and mechanical properties makes it an excellent candidate for active food packaging, which serves as a physical barrier and actively contributes to preserving food quality. As the food industry shifts towards more sustainable practices, LSS could emerge as a key player in next-generation biopolymer films, providing functional, biodegradable, and health-promoting packaging solutions. Future research should focus on optimizing film formulations, enhancing water resistance, and exploring synergistic interactions with other biopolymers to improve their industrial viability further. Given its abundance, natural bioactivity, and impressive functional properties, LSS-based films hold great promise as a sustainable innovation for the food packaging industry.

12.7.4 Application of LSS in Bioplastics

Starch is made up of amylose and amylopectin, serving as the primary energy reserve in plants. Due to its nontoxic nature, affordability, renewability, and biodegradability, it has gained significant attention for making eco-friendly films. Among various sources, LSS shows promise as a second-generation feedstock for creating sustainable packaging materials especially due to its high starch content, biodegradability, and renewable nature. Very few researches have explored the potential of LSS as a sustainable material for bioplastic production. Studies have focused on its extraction, modification, and application in biodegradable films and coatings. Biodegradable films made from both native and sodium trimetaphosphate-crosslinked litchi kernel starch has been developed. Starch extracted from litchi seeds was cross-linked with varying concentrations (1%, 3%, and 5%) of sodium trimetaphosphate (STMP). These modified starches were then evaluated for their physico-chemical, pasting, rheological, and thermal properties. In addition, bio-films were prepared from the cross-linked starches and tested for key barrier and mechanical characteristics. The findings revealed that increasing the concentration of the cross-linking agent led to a notable reduction in amylose content, swelling power, and solubility. Moreover, both peak viscosity and breakdown viscosity were lower in the cross-linked starches compared to the native starch, indicating changes in their structural and functional behavior. These films demonstrated better resistance to water with a tensile strength of 7.70 MPa (Sharma et al., 2021).

Kumar et al. (2024) applied techniques like ultrasonication and the addition of clove oil to further enhance the properties of LSS-based films. The study revealed that incorporating clove essential oil (EO) and applying ultrasonication had a significant impact on the functional and antimicrobial properties of the edible coatings and films. Among the formulations, the edible coating made from litchi kernel starch treated with both clove EO and ultrasonication showed the smallest particle size (273.9 nm) and the highest zeta potential (–41.4 mV), indicating improved stability. This was followed by coatings prepared using mango kernel seed starch and corn starch, which also exhibited favorable characteristics but to a lesser extent. Laha and Chatterjee (2024) explored blending LSS with other unconventional starches, such as churkha tuber starch, and various plasticizers like glycerol, sorbitol, mannitol, propylene glycol, polyethylene glycol to produce biodegradable films. High amylose containing LSS forms firm and more flexible films in comparison to low amylose containing churkha tuber starch. The studies aimed to address environmental concerns and valorize underutilized plant sources.

Starch (LSS) extracted from litchi seeds and xyloglucan (XG) derived from tamarind kernels were utilized to develop composite films. The addition of XG helped reinforce the weak polymer network of LSS, leading to improvements in rheological, molecular, morphological, mechanical, and water vapor barrier properties. Further enhancement was achieved by incorporating lignin nanoparticles (LNPs) into the LSS-XG matrix, which increased tensile strength to 14.83 MPa and elastic modulus to 0.41 GPa, while reducing surface wettability (80.07°) and water vapor permeability ($5.63 \pm 0.38 \times 10^{-7}$ g $m^{-1}s^{-1}Pa^{-1}$). The phenolic hydroxyl groups present in LNPs also provided the films with strong UV-blocking and antioxidant (free radical scavenging) capabilities. These combined properties effectively helped preserve the quality of coated bananas, minimizing both weight loss and color degradation during storage (Santhosh et al., 2025).

12.8 FUTURE PERSPECTIVES

To fully utilize LSS, the biochemical diversity of litchi species for desirable starch characteristics must be explored. Variations in composition, morphology, structure, and functional properties of native and modified litchi starch provide a foundation for expanding its applications. The use of nonconventional starch sources, such as LSS, offers opportunities to develop value-added food products by achieving tailored functional properties. Native litchi starch has limited functionalities and varies in stability under different pH and temperature conditions. Physical, chemical, and enzymatic modifications are necessary to overcome these limitations to enhance its functional attributes. For example, native starch takes longer to gelatinize, which can be improved by producing pregelatinized starch. This modification reduces gelatinization time, making it suitable for applications in the meat industry, ready-to-use bread mixtures, and fruit pie fillings (Egharevba, 2020).

Modified starches exhibit a diverse range of functional properties that make them valuable ingredients in the food industry, enhancing texture, stability, and shelf life across various applications. For example, oxidized starch is characterized by its lower viscosity, thermolabile nature, and high clarity, making it particularly suitable for confectionery applications, where it serves as a coating for candies and sweets, improving their gloss and texture. Cross-linked starch, on the other hand, is known for its exceptional resistance to high temperatures, shear stress, and acidic conditions, making it an ideal thickening and stabilizing agent in heat-processed foods such as sauces, soups, dips, gravies, fruit pie fillings, puddings, bakery, and dairy products. Acid-modified starch has a significantly lower paste viscosity, which enhances its ability to form smooth, stable emulsions, making it particularly useful in products like mayonnaise and salad dressings, where texture and consistency are critical. Acetylated starch offers multiple functional benefits, including increased viscosity, improved food quality, reduced susceptibility to rancidity, and enhanced performance as a bulking agent, thickener, stabilizer, flavor encapsulant, and clouding agent, all of which contribute to its widespread use in processed foods (Abbas et al., 2010). Similarly, succinylated starch enhances viscosity and freeze–thaw stability, making it an essential ingredient in soups, frozen foods, and various thickened formulations, where it prevents syneresis and ensures product consistency under fluctuating storage conditions (Majeed et al., 2024; Bensaad et al., 2022). The versatility of these modified starches highlights their crucial role in improving food quality, extending shelf life, and enhancing sensory attributes, making them indispensable in modern food formulation and production.

Although modern enzymatic and molecular techniques for starch modification exist, they are still limited and often uneconomical. Therefore, physical and chemical modifications remain the most practical approaches. Since LSS is a newly explored starch source from the litchi fruit processing industry, its modifications remain largely unstudied. Investigating these modification techniques could significantly enhance the industrial applications of LSS.

12.9 CONCLUSION

Litchi seeds represent a valuable nontraditional source of starch, offering notable industrial uses, especially in the food, pharmaceutical, and functional food industries. A preponderance of A-type crystalline structure displays a CA-type starch structure, which is a mixture of A- and B-type allomorphs. Its physicochemical characteristics, such as its moderate water-binding capacity, high gelatinization temperature, and thermal stability, are influenced by its special crystalline structure. LSS's granule shape has a bimodal size distribution, containing both big and tiny granules, increasing its adaptability to a range of uses. The presence of both granule sizes allows for different textural and functional contributions, making it suitable for food processing applications that require a range of viscosities and gel-forming capacities. Compared to conventional starches derived from corn, potato, and mungbean, LSS demonstrates comparable, if not superior, performance in specific applications, particularly regarding heat stability and film-forming abilities.

LSS's strong heat durability is a primary functional characteristic, making it an excellent option as a texturizer and viscosifier in high-temperature culinary applications. Its effectiveness in soups, sauces, bread formulations, and dairy products where stability during extended heating is essential is guaranteed by its capacity to tolerate heat without suffering appreciable deterioration. It's comparatively low viscosity also makes it ideal for use as a texturizer in gravies, thickened food items, and coatings for confections. Since many commercial starches become excessively thick for applications requiring fluidity or losing stability at high temperatures, the balance between thermal stability and viscosity offers a distinct benefit. Litchi starch's retrogradation qualities also point to possible freeze-thaw resilience, which makes it a good choice for frozen food formulations that need to reduce syneresis.

Beyond its traditional food applications, LSS also attracts attention in nonfood sectors, particularly in pharmaceutical and functional food formulations. Due to its film-forming capability, it is being explored as an edible coating material for fruits, vegetables, and other perishable food products to extend shelf life and reduce postharvest losses. Its biodegradable and nontoxic properties make it a sustainable alternative to synthetic coatings. In pharmaceuticals, LSS is being researched as a binder and disintegrant in tablet formulations, where its physicochemical characteristics can influence drug release and absorption. Its potential role as a functional food ingredient is also noteworthy, as starch-derived resistant fractions could contribute to gut health by serving as a prebiotic and promoting beneficial gut microbiota.

The competitive advantage of LSS lies in its diverse functional properties, which enable it to compete with well-established commercial starches such as corn, potato, and mung bean. With ongoing research focusing on optimizing its extraction, modification, and application methods, LSS has the potential to become a viable alternative in various industries. Future studies aimed at improving its gelation properties, enzymatic resistance, and structural modifications could further enhance its utility. As global industries continue to seek sustainable, functional, and versatile sources of starch, LSS presents an exciting and valuable opportunity for both food and nonfood applications, ensuring its place in the growing market of specialty starches.

REFERENCES

Abbas, K., Khalil, S. K., & Hussin, A. S. M. (2010). Modified starches and their usages in selected food products: A review study. *Journal of Agricultural Science*, *2*(2), 90.

Ai, Y., & Jane, J.-l. (2018). Understanding starch structure and functionality. In M. Sjöö & L. Nilsson (Eds.), *Starch in Food* (2nd ed., pp. 151–178). Woodhead Publishing.

Bangar, S. P., Kumar, M., Whiteside, W. S., Tomar, M., & Kennedy, J. F. (2021). Litchi (*Litchi chinensis*) seed starch: Structure, properties, and applications—A review. *Carbohydrate Polymer Technologies and Applications*, *2*, 100080.

Bensaad, D. E., Saleh, M., Ismail, K., Lee, Y., &Ondier, G. (2022). Chemical modifications of starch: A perspective for sweet potato starch. *Jordan Journal of Agricultural Sciences*, *18*(4), 293–308.

Bhat, R. S., & Al-Daihan, S. (2014). Antimicrobial activity of *Litchi chinensis* and *Nephelium lappaceum* aqueous seed extracts against some pathogenic bacterial strains. *Journal of King Saud University–Science*, *26*(1), 79–82.

Carvalho, H. J. M., Barcia, M. T., & Schmiele, M. (2024). Non-conventional starches: Properties and potential applications in food and non-food products. *Macromol*, *4*(4), 886–909.

Chakravorty, A., Chakraborty, M., & Sa, B. (2016). Factors influencing delayed release followed by rapid pulse release of drugs from compression coated tablets for colon targeting. *International Journal of Pharmacy and Pharmaceutical Sciences*, *8*(8), 330–336.

Chen, Z., Liang, N., Zhang, H., Li, H., Guo, J., Zhang, Y., et al. (2024). Resistant starch and the gut microbiome: Exploring beneficial interactions and dietary impacts. *Food Chemistry: X*, *21*, 101118.

Cronje, R. B., Chen, H., Wang, H., Zhou, B., Wu, Z., Yao, L., et al. (2023). Litchi. In *Tropical and Subtropical Fruit Crops* (pp. 419–558). Apple Academic Press.

Egharevba, H. O. (2020). Chemical properties of starch and its application in the food. In *Chemical Properties of Starch* (p. 63).

Esua, O. J., Okonkwo, C. E., Ajani, C. K., Bassey, E. J., Ali, M., Hussain, N., et al. (2024). Litchi seed starch—Isolation, modification, and characterization. In J. M. Lorenzo & S. P. Bangar (Eds.), *Non-Conventional Starch Sources* (pp. 21–59). Academic Press.

European Pharmacopoeia. (2018). *European Pharmacopoeia 9.0* (Vol. Supplement). www.edqm.eu/en/european-pharmacopoeia-ph-eur-9th-edition

Feinle-Bisset, C., & Horowitz, M. (2021). Appetite and satiety control—Contribution of gut mechanisms. *Nutrients*, *13*(10), 3635.

Gong, T., Liu, X., Wang, X., Lu, Y., & Wang, X. (2024). Applications of polysaccharides in enzyme-triggered oral colon-specific drug delivery systems: A review. *International Journal of Biological Macromolecules*, 275, 133623.

Guo, J., Tan, L., & Kong, L. (2021). Impact of dietary intake of resistant starch on obesity and associated metabolic profiles in humans: A systematic review of the literature. *Critical Reviews in Food Science and Nutrition*, *61*(6), 889–905.

Higgins, J. A. (2004). Resistant starch: Metabolic effects and potential health benefits. *Journal of AOAC International*, *87*(3), 761–768.

Izu, G. O., Mapasa, N. P., Nambooze, J., Chukwuma, M. S., Njoya, E. M., Tabakam, G. T., et al. (2024). Epicatechin isolated from *Litchi chinensis* Sonn. (Litchi) fruit peel ethyl acetate extract modulated glucose uptake in Chang cells and suppressed ROS production in RAW 264.7 macrophages. *Antioxidants*, *13*, 1233.

Jaiswal, P., & Kumar, K. J. (2015). Physicochemical properties and release characteristics of starches from seeds of Indian Shahi litchi. *International Journal of Biological Macromolecules*, *79*, 256–261.

Joshi, R. (2018). Role of enzymes in seed germination. *International Journal of Creative Research Thoughts*, *6*, 1481–1485.

Kadyan, S., Park, G., Singh, P., Arjmandi, B., & Nagpal, R. (2023). Prebiotic mechanisms of resistant starches from dietary beans and pulses on gut microbiome and metabolic health in a humanized murine model of aging. *Frontiers in Nutrition*, *10*, 1106463.

Kaur, J., Borah, A., Chutia, H., & Gupta, P. (2024). Extraction, modification, and characterization of native litchi seed (*Litchi chinesis* Sonn.) starch. *Journal of the Science of Food and Agriculture*, *104*(1), 215–224.

Kumar, N., Khan, A. A., Pyngrope, D., Alanazi, A. M., Upadhyay, A., & Shukla, S. (2024). Development and characterization of novel starch (mango kernel and litchi seed) based active edible coatings and films using ultrasonication: Effects on postharvest shelf life of Khasi mandarins. *Sustainable Chemistry and Pharmacy*, *39*, 101610.

Kumari, A., Jha, A., Prasad, J., Singh, A. K., & Prasad, U. (1993). Changes in carbohydrate composition of seeds of *Litchi chinensis* sonn during maturation. *Journal of the Science of Food and Agriculture*, *61*(2), 205–210.

Lachman, L., Lieberman, H., &Kanig, J. (2003). *The theory and practice of industrial pharmacy* (3rd ed., pp. 293–294). Varghese Publishing House.

Laha, S., & Chatterjee, A. (2024). Fabrication and characterization of non-conventional starch based biofilms employing different plasticizers. *International Journal of Applied Sciences and Biotechnology*, *12*(1), 29–38.

Li, H. (2021). Longan fruit tree physiology and its flowering induction. In *Handbook of Plant and Crop Physiology* (p. 77).

Lindeboom, N., Chang, P. R., & Tyler, R. T. (2004). Analytical, biochemical and physicochemical aspects of starch granule size, with emphasis on small granule starches: A review. *Starch-Stärke*, *56*(3-4), 89–99.

Loo, S. J., Seah, X. Y., Heng, P. W. S., & Chan, L. W. (2023). Study of diminutive granules as feed powders for manufacturability of high drug load minitablets. *International Journal of Pharmaceutics*, *638*, 122922.

Madhav, N. S., & Yadav, A. P. (2013). A novel translabial platform utilizing bioexcipients from *Litchi chinensis* for the delivery of rosiglitazone maleate. *Acta Pharmaceutica Sinica B*, *3*(6), 408–415.

Majeed, U., Majeed, H., Liu, X., Shafi, A., Liu, T., Ye, J., et al. (2024). Succinylated starch emulsified eugenol and *Carvacrol nanoemulsions* improved digestive stability, bio-accessibility and *Salmonella typhimurium* inhibition. *International Journal of Biological Macromolecules*, *259*, 129230.

Mir, H., & Perveen, N. (2022). Phytochemistry and therapeutic potential of litchi (*Litchi chinensis* Sonn.): An age-old ingredient in traditional medicine. *Indian Journal of Traditional Knowledge*, *21*(1), 29–39.

Morales-Trejo, F., Trujillo-Ramírez, D., Aguirre-Mandujano, E., Lobato-Calleros, C., Vernon-Carter, E. J., & Alvarez-Ramirez, J. (2022). Ultrasound-assisted extraction of lychee (*Litchi chinensis* Sonn.) seed starch: Physicochemical and functional properties. *Starch-Stärke*, *74*(1–2), 2100092.

Punia, S., & Kumar, M. (2021). Litchi (*Litchi chinensis*) seed: Nutritional profile, bioactivities, and its industrial applications. *Trends in Food Science & Technology, 108*, 58–70.

Queiroz, E. R., Abreu, C. M. D., Rocha, D. A., Sousa, R. V. D., Fraguas, R. M., Braga, M. A., et al. (2017). Lychee (*Litchi chinensis* Sonn.) peel flour: Effects on hepatoprotection and dyslipidemia induced by a hypercholesterolemic diet. *Anais da Academia Brasileira de Ciências*, *90*(1), 267–281.

Sahni, R. K., Kumari, S., Kumar, M., Kumar, M., & Kumar, A. (2020). Status of litchi cultivation in India. *International Journal of Current Microbiology and Applied Sciences*, *9*(04), 1827–1840.

Santhosh, R., Thakur, R., Sarkar, P., & Janaswamy, S. (2025). Active bio-nanocomposites from litchi seed starch, tamarind kernel xyloglucan, and lignin nanoparticles to improve the shelf-life of banana (Musa acuminata). *Food Chemistry*, *463*, 141327.

Sharma, V., Kaur, M., Sandhu, K. S., Nain, V., & Janghu, S. (2021). Physicochemical and rheological properties of cross-linked litchi kernel starch and its application in development of bio-films. Starch-Stärke, *73*(7-8), 2100049.

Shen, L., Pang, S., Zhong, M., Sun, Y., Qayum, A., Liu, Y., et al. (2023). A comprehensive review of ultrasonic assisted extraction (UAE) for bioactive components: Principles, advantages, equipment, and combined technologies. *Ultrasonics Sonochemistry*, *101*, 106646.

Singh, S. K., Marboh, E., & Nath, V. (2023). Litchi. In *Fruit and Nut Crops* (pp. 1–28).

Suriyaprom, S., Mosoni, P., Leroy, S., Kaewkod, T., Desvaux, M., &Tragoolpua, Y. (2022). Antioxidants of fruit extracts as antimicrobial agents against pathogenic bacteria. *Antioxidants*, *11*(3), 602.

Thakur, Y., Thory, R., Sandhu, K. S., Kaur, M., Sinhmar, A., &Pathera, A. K. (2021). Effect of selected physical and chemical modifications on physicochemical, pasting, and morphological properties of underutilized starch from rice bean (*Vigna umbellata*). *Journal of Food Science and Technology*, *58*(12), 4785–4794.

Thory, R., & Sandhu, K. S. (2017). A comparison of mango kernel starch with a novel starch from litchi (*Litchi chinensis*) kernel: Physicochemical, morphological, pasting, and rheological properties. *International Journal of Food Properties, 20*(4), 911–921.

Topping, D. L., & Clifton, P. M. (2001). Short-chain fatty acids and human colonic function: Roles of resistant starch and nonstarch polysaccharides. *Physiological Reviews*, 81(3), 1031–1064.

Udaipuria, N., & Bhattacharya, S. (2025). Novel carbohydrate polymer-based systems for precise drug delivery in colon cancer: Improving treatment effectiveness with intelligent biodegradable materials. *Biopolymers*, *116*(1), e23632.

Upadhyaya, D. C., & Upadhyaya, C. P. (2017). Bioactive compounds and medicinal importance of *Litchi chinensis*. In *The Lychee Biotechnology* (pp. 333–361).

Wang, D., Bai, W., & Li, Z. (2005). Soaking technology for raising starch extraction ratio from mung bean during production of thin noodles from mung bean starch. *Transactions of the Chinese Society of Agricultural Engineering*, *21*, 151–156.

Wang, S., Li, C., Copeland, L., Niu, Q., & Wang, S. (2015). Starch retrogradation: A comprehensive review. *Comprehensive Reviews in Food Science and Food Safety*, *14*(5), 568–585.

Warner, R. D. (2023). The eating quality of meat: IV—Water holding capacity and juiciness. In *Lawrie's Meat Science* (pp. 457–508). Elsevier.

Xing, J.-j., Li, D., Wang, L.-j., & Adhikari, B. (2017). Relationship between biphasic endotherms and multi-stage gelatinization of corn starch in excess water. *LWT-Food Science and Technology*, *81*, 335–342.

Zhang, R.-Y., Chen, P.-X., Liu, A.-B., Zhu, W.-X., Jiang, M.-M., Wang, X.-D., et al. (2024a). Effects of different isolation methods on the structure and functional properties of starch from tiger nut (*Cyperus esculentus* L.) meal. *LWT*, *196*, 115853.

Zhang, S., Wang, Z., Wang, L., Tian, H., Wang, H., Du, C., et al. (2024b). A- and B-type wheat starch granules: The multiscale structural evolution during digestion and the distinct digestion mechanisms. *International Journal of Biological Macromolecules*, *278*, 135033.

Zhang, W.-q., Li, X.-I., Xue, W., Guo, H.-w., & Chang, M. (2021). The anti-proliferation effect of total flavonoids of litchi seed combined with paclitaxel on prostate cancer paclitaxel-resistant cells. *Chinese Pharmacological Bulletin*, *37*(12), 528–534.

Zhang, X., Zhao, L., Zhou, W., Liu, X., Hu, Z., & Wang, K. (2022). Variations in the multilevel structure, gelatinization and digestibility of Litchi seed starches from different varieties. *Foods*, *11*(18), 2821.

Zhang, Y., Jin, D., An, X., Duan, L., Duan, Y., & Lian, F. (2021). Lychee seed as a potential hypoglycemic agent, and exploration of its underlying mechanisms. *Frontiers in Pharmacology*, *12*, 737803.

13 Millet Starch

A Comprehensive Study on its Structure, Functionality, and Industrial Potential

Surabhi Pandey, Anurag Singh, Shivangi Srivastava, and Poornima Singh

13.1 INTRODUCTION

Nutritional content is a key factor that consumers consider when evaluating food products for overall physical well-being and a healthy lifestyle. The development of new food products, therefore, emphasizes ingredients that are not only accessible and affordable but also functional and nutritious. Nutritional quality serves as a primary constraint in supporting a healthy lifestyle among consumers (Abah et al., 2020). Current ecological and socioeconomic challenges negatively impact the yields of staple crops such as corn, rice, and wheat. These issues have constrained grain production and slowed agricultural growth, contributing to rising food prices and severe global food security concerns (Adebiyi et al., 2018). Additionally, under such unstable conditions, local farmers often face resource shortages. Addressing on-field production limitations through innovative technologies is essential to identifying alternative cereal crops that could relieve the burden on major staples. Cereals are crucial to human diets due to their unique flavors, extended shelf life, and ease of preparation. The grass subfamily *Panicoideae* includes millets, nutrient-dense cereal grains rich in dietary fiber, positioning them as a valuable alternative to major cereal crops (Wani et al., 2021; Almaski et al., 2017).

Proteins, carbs, dietary fiber, vitamins, and minerals are all abundant in millets (Pandey and Singh, 2024). They have somewhat less protein than other cereals, but they have a similar carbohydrate amount. Free sugars (2–3%), nonstarchy polysaccharides (15–20%), and starch (60–75%) make up millet grains (Table 13.1). Glucose, fructose, and sucrose are the main free sugars found in millets. The dietary fiber content is influenced by the aleurone layer and other components of the kernel cell wall; approximately 90% of the total dietary fiber is soluble fiber. Millet starch is produced extensively in both southern and northern India and is recognized as a valuable dietary component (Srichuwong et al., 2017). Similar to other cereals, millet starch has an amylose-to-amylopectin ratio of about 25:75. Starch makes up 51–79% of millets, which include a variety of protein fractions, including albumin, globulin, glutelin-like proteins, and cross-linked prolamin. About 70% of the grain's overall composition is made up of millet starch, which is typically composed of 70–80% amylopectin and 20–30% amylose (Mahajan et al., 2021). Granules of millet starch can be tiny or big and spherical, oval, or polygonal (Sandhu et al., 2021). Millet starch has a slower digestion rate for protein and starch compared to other cereals. Certain millet species, such as foxtail millet (*Setaria italica*), finger millet (*Eleusine coracana*), pearl millet (*Pennisetum glaucum*), and sorghum (*Sorghum bicolor*), exhibit promising health benefits (Punia et al., 2021). The applications of millet starch in various food products are influenced by its structural, physicochemical, and

DOI: 10.1201/9781003534914-13

TABLE 13.1
Composition of Different Millet Varieties

Millet	Protein (g)	Fat (g)	Carbohydrate (g)	Fiber (g)	Starch Content (%)	Minerals
Barnyard	11.6	5.8	65.5	12.6	48–60	4.7
Finger	7.3	1.3	66.8	11.1	53–68	2.7
Pearl	11.6	5	61.7	11.4	55–65	2.3
Foxtail	12.3	4.3	60.0	6.7	57–73	3.3
Kodo	8.3	1.4	66.1	6.3	46–61	2.6
Little	8.7	5.3	65.5	6.3	43–58	1.7
Proso	12.5	1.1	70.0	8.5	58–78	1.9

Source: Adapted from IIMR (2017); Das et al. (2019); Kaimal et al. (2021).

functional properties, which allow it to function as a binder, texture modifier, structural provider, and viscosity regulator (Kumar et al., 2023). Millet starch's unique properties, including its hypoglycemic effects compared to other cereal starches, make it valuable in the food and pharmaceutical industries as a swelling agent, ingredient, thickener, emulsifier, gelling agent, and binder (Bangar et al., 2021). This review discusses the structure of millet starches, extraction methods, and their diverse industrial applications with associated health benefits.

13.2 METHOD OF STARCH ISOLATION

Various laboratory techniques are employed to extract starch from different sources, with the method chosen depending on the type of plant tissue providing the starch. This generally entails milling into flour and subsequently soaking straight without milling or soaking in a mild alkaline solution. The gentle alkaline environment facilitates the elimination of proteins and nonstarch carbs. Starch, existing in granule form, can be isolated for various food and non-food applications using enzymatic, physical, and chemical methods, including filtration, centrifugation, and gravity sedimentation (Jane, 2009). Each starch possesses unique physicochemical and structural properties, influenced by its type, variety, and origin (Delcour et al., 2010). In their native form, starches may lack the physical and chemical properties needed for specific applications; however, modification processes enhance these attributes, making them more suitable for both food and nonfood industries, as well as improving their functional properties to produce resistant starch for functional foods (Egharevba, 2019; Dupuis et al., 2014). For some extractions, water is used as the soaking solution, typically at room temperature overnight or longer, while higher temperatures of around 50 °C may be applied for shorter durations. After soaking, softened material is collected via centrifugation or sedimentation techniques after passing through a screen. The extracted starch is then dried, either in an oven at approximately 40°C or by solar drying. Wet milling is another widely used technique, especially for cereal grains, and involves steps such as steeping, coarse milling, degermination, fine milling, screening or centrifugal separation, and drying. Tuber starch extraction often relies on centrifugation, drying, disintegration, and soaking. A general schematic of the starch isolation process is illustrated in Figure 13.1.

13.3 STRUCTURE AND COMPOSITION OF MILLET STARCHES

There is significant variation in the production and chemical makeup of millet starch. With amylose content ranging from 6% to 38.6%, lipid content from 0.16% to 2.9%, protein content from 0.2% to 4.3%, and ash percentages from 0.02% to 1.4%, the starch yield varies from 52% to 68.2% (Bangar

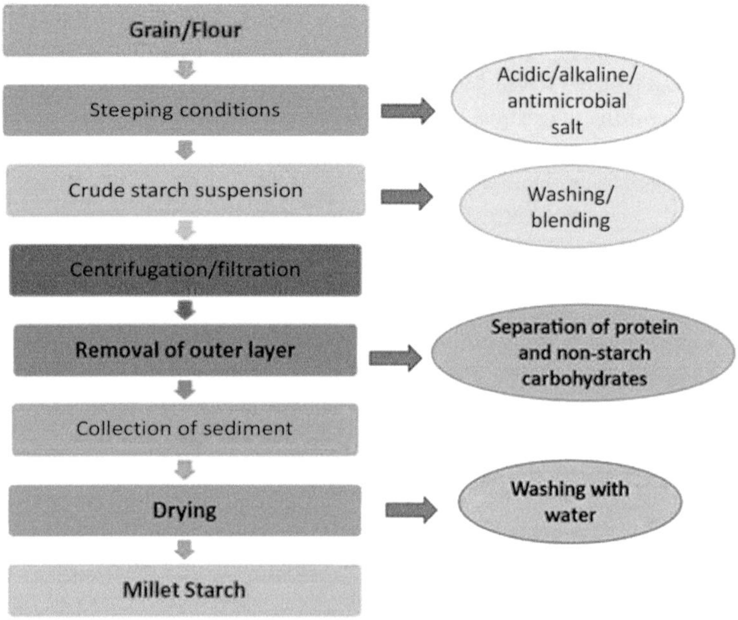

FIGURE 13.1 General process of starch isolation.

et al., 2021). In pearl millet, neutral lipids (NL) and phospholipids (PL) were the main lipid classes in the free and bound lipid extracts, respectively, whereas glycine, glutamine, and aspartic acid were the main amino acids in the starch granules of finger millet (Raungrusmee, 2023). The NL fraction had the highest concentration of linolenic acid, whereas the PL and glycolipid fractions had the highest concentrations of palmitic acid. Significant differences in the lipid content between millet species and genotypes within the same species were observed (Sahaka et al., 2020). This variability in starch composition is attributed not only to the genetic and environmental characteristics of millets but also to the extraction methods used in the studies (Bangar et al., 2021).

The amylose content is crucial in determining the characteristics and possible applications of starch; however, the technique of measurement can significantly influence the reported amylose levels. Defatting starch significantly elevates amylose content, as lipids can create complexes with amylose in a helical configuration (Li et al., 2021). Moreover, iodine may engage with the elongated chain segments of amylopectin, thereby disrupting the color formation in the amylose–iodine complex. Consequently, it is imperative to compare amylose content data from various research within the precise methodological framework employed (He & Wei, 2020). In controlled experimental conditions, genetic variation in amylose content (from zero to 11–27%) has been recorded, providing a significant genetic resource for enhancing starch composition. The advancement of genotypes with diverse amylose concentrations, specifically waxy and high-amylose varieties (surpassing 40% amylose), may enhance the utility of millet (Zhang et al., 2021).

Finger millet has complex granules; however, the majority of the starch granules are simple, having spherical and polygonal shapes (Kumar et al., 2023). Granules are mostly spherical in the mealy kernel of both proso and foxtail millet, while they are frequently angled and polygonal in the flinty endosperm. According to comparative analysis, finger millet has smaller spherical starch granules than pearl, proso, and foxtail millets (Zhu, 2014). Pearl millet grains exhibit pores on their surfaces, with tiny apertures irregularly distributed and often clustered across the granular surface. These pores, along with channels and cavities within the granules, are considered active sites that facilitate the diffusion of chemicals, thus enabling their molecular formation to improve starch reactivity for targeted applications (Singh et al., 2024). Further research is needed on the starch granules with holes from various millet species. Pearl and foxtail millets have a peak granule size of 7.6 μm, whereas proso and finger millets tend to have smaller granules, peaking at 5.7 μm (Bean et al., 2019). Granule sizes range from 0.8 to 24 μm. Granule size variations between species might not be easily distinguished, most likely as a result of genetic and environmental variations (Li et al., 2023).

The outer regions of amylopectin chains primarily crystallize as double helices, forming two configurations that result in A- and B-type polymorphs in granular starch. These polymorphic structures are typically identified through wide-angle X-ray diffraction analysis, with the C-type representing a blend of A- and B-types. In millet starch samples, all studied genotypes displayed the A-type polymorph, which is common among cereal starches. Across four species and four millet genotypes, the percentage crystallinity of starch varied between 27 and 30% (Singh et al., 2024). A comprehensive study employing a variety of methods, such as small-angle X-ray diffractometry would provide more understanding of the granular crystallinity of millet starch across genotypes.

13.3.1 Molecular Structure of Starch (Amylose and Amylopectin)

Amylose and amylopectin are the two main molecular components of starch. Depending on the fractionation technique used, there may also be an intermediate structural element present. Amylopectin has a significantly higher degree of branching than amylose, which has a mostly linear structure with sporadic branches (α- (1 → 6) links) along its linear backbone (α- (1 → 4) connections) (Alzaabi, 2023). The production, physical characteristics, functioning, and possible uses of amylose and amylopectin are all correlated with their molecular structures. With iodine in solution, amylose and amylopectin chains can form helical complexes that result in distinctive red and blue hues, respectively (Ai & Jane, 2024). These branches act as a steric barrier, and iodine and starch's natural lipids may compete to create helices. The blue value and iodine-binding affinity (%) indicate iodine's capacity to build a helix with amylose/amylopectin. In pearl millet, researchers found that the blue values for amylose and amylopectin were 1.3–1.41 and 0.18, respectively. In finger and foxtail millets, Wankhede et al. (1979) determined that the iodine affinity for amylose was 18.2–18.5% and 18.7%, respectively, whereas the values for amylopectin were 0.67–0.72% and 0.64%, respectively. With molecular weights of 105 to 106 for amylose and around 107 for amylopectin, the degree of polymerization (DP) by number for amylose and amylopectin in pearl millet ranges from 1060 to 1250 and 9000 to 9100, respectively. Different measuring methods, including high-performance size-exclusion chromatography and low-pressure gel permeation chromatography, can yield different results for the same chemicals (Zhu, 2014). Regular studies using standardized apparatus setups are crucial for producing comparable results in order to reduce discrepancies in global starch research (Shen *et al.*, 2024). Limit dextrins (LDs), which stand for the internal segments of starch components, are produced when starch molecules are exposed to exo-acting enzymes like phosphorylase and β-amylase, which hydrolyze amylose and amylopectin from the nonreducing ends until branch points are encountered (Adegbaju, 2022). The enzymatically eliminated exterior fraction of amylose or amylopectin is shown by the limit value (β- or φ, β-LV) (%). For amylose, the β-LV (%) varied between 78.4% and 93.7%, whereas for amylopectin, it varied between 52.1% and 58.7%. Amylopectins from four millet species with φ, β-LV (%) ranging from 57.7% to 59.8% (Zhu, 2014).

A-chains and B-chains are the two types of starch unit chains. B-chains have at least one branch, whereas A-chains are linear. One crucial structural characteristic is the ratio of the A-to-B chain. Following debranching, the maltose content of φ, β-limit dextrins (LDs), which are produced by the successive hydrolysis of amylopectin by phosphorylase α and β-amylase, allows for the accurate measurement of A-chains (Bertoft, 2017). Amylopectins from four millet species have A-chain contents ranging from 49.4% to 50.7%. A balanced distribution of Staudinger and Haworth branching patterns inside the structure is indicated by an A-to-B-chain ratio that is close to 1 (Annor,

2013). In order to define the molecular structure of the components that make up starch, the chain length (CL) is crucial. With about four chains per molecule, amylose showed an average chain length of 260–270 glucosyl units. With an external chain length (ECL) of 12–14 units and an internal chain length (ICL) of 4.8–6.3 units, amylopectin demonstrated a chain length (CL) of 18–21 glucosyl units (Sun et al., 2021). Because the methods used to measure and estimate chain length differed within categories and might affect reported values, variations within studies could not accurately reflect intrinsic sample variability. According to research findings, there was far less structural variation across four genotypes in a single investigation than there was over several trials (Collins et al., 2020). Using gel permeation chromatography, amylose may be divided into two categories: long-chain amylose (LCAM) and short-chain amylose (SCAM). For the four millet genotypes, the weight-based ratio of LCAM to SCAM varied between 0.6 and 1.3. Iodine-binding profiles and gel-permeation chromatography of β-limit dextrins (β-LDs) derived from whole starch demonstrated structural variations in amylose branching patterns (Shen et al., 2024). Finger and pearl millets were found to have larger internal chain segments between amylose branches than proso and foxtail millets. To support this finding, chromatographic and enzymatic approaches must be improved in order to accurately measure the distribution of amylose chain lengths. Further research into amylose branching features is crucial since a reliable method for separating mostly linear amylose from its branched variations is required (Taylor & Duodu, 2018). Amylopectin unit chains are believed to assemble within starch granules, where the external structure promotes crystalline formation within the granules and the internal structure, made up of tiny branching units known as "building blocks," constitutes the amorphous areas (Tetlow & Bertoft, 2020). Potential differences in cluster structure are indicated by differences in the internal molecular architecture of amylopectin across four genotypes from other animals. However, more research is still needed on the cluster and structural component organization inside millet amylopectin (Bean et al., 2019).

13.4 MILLET STARCH FUNCTIONALITY

13.4.1 Swelling and Solubilization

Upon heating starch in excess water, the granules absorb the liquid and expand, resulting in the leaching and dissolution of starch components, predominantly as amylose. As the temperature rises, the expanded granules ultimately disintegrate and decompose (Majzoobi & Farahnaky, 2021). The molecular makeup of amylose and amylopectin, the physical interactions between the chemical components inside the granules, the distribution of granule sizes, and the presence of phosphorus groups and lipid–amylose complexes are some of the variables that influence this process (Vamadevan & Bertoft, 2020). While solubility (%) or amylose leaching (AML) (%) indicate the solubilization of granules, swelling factor (SF), or swelling power (SP) are commonly used to quantify the expansion of granules in water upon heating. Whereas SP only indicates the intragranular water, SF refers to both the inter- and intra-granular water components. Starches made from pearl, foxtail, proso, and fonio millets typically have gelatinization temperatures between 50 and 90 °C (Mahajan et al., 2021). In general, millet starch exhibited reduced swelling and solubilization values relative to potato starch at elevated temperatures (e.g., 90 °C). Genetic diversity in swelling properties was noted even among individuals of the same species (Kheto et al., 2022). No distinct threshold in swelling power and solubility was identified to differentiate between several millet species or other grains. Consequently, further genotypes from various millet species must be examined to derive potential statistical inferences.

13.4.2 Rheological Properties

For the rheological evaluation of millet starch, the Brabender Visco-Amylograph (BVA) and Rapid Visco-Analyzer (RVA) are often used. In these procedures, starch is subjected to heating with a significant quantity of water while undergoing continuous shear, and viscosity alterations are observed during a specified temperature cycle (Zhu, 2014). The process termed pasting is affected by multiple parameters, including as water content, shear rate, temperature profile, and the starch structure. The starch concentration throughout many studies varied between 6% and 10%. Consequently, direct comparison data from several researches may yield inaccuracies unless conducted inside the same study under same experimental settings and technical specifications (Karakelle et al., 2020).

Pasting characteristics vary greatly amongst millet species and even between genotypes of the same species. Breakdown (BD) ranged from 183 to 237 RVU, peak viscosity (PV) ranged from 345 to 425 Rapid Visco Units (RVU), setback (SB) ranged from 142 to 188 RVU, and pasting temperature (PT) ranged from 72.0 to 78.5 °C in a study of ten genotypes of foxtail millet. According to Mahajan et al. (2021), five starches that were made from foxtail, barnyard, proso, kodo, and small millets had PV values that ranged from 375 to 520 Brabender Units (BU), BD that ranged from 20 to 50 BU, SB that ranged from 330 to 790 BU, and PT (75.8 to 84.9°C). The restricted variety of genotypes examined for specific species, such as kodo and micro millets, complicates the establishment of a significant classification of these species according to their starch pasting behavior. Besides pasting analysis, dynamic rheological testing can offer additional insights into the physicochemical features of millet starch. The impact of the molecular structure of starch constituents on these rheological properties necessitates further investigation (Kaushik et al., 2024).

13.4.3 Freeze–Thaw Stability (FTS)

Starchy foods go through freeze–thaw cycles in particular food processing and preservation settings. By measuring the amount of water that is released from the starch gel during these cycles—a process known as syneresis—the FTS of millet starch is determined. A partly dehydrated starch gel quickly reabsorbed water to return to its original structure, indicating that proso millet starch was less stable than maize starch. In contrast, another study found that the starch from pearl millet was more stable than the starches from corn and wheat (Vamadevan & Bertoft, 2020). The molecular structure, moisture content, amylose and amylopectin proportions, and temperature history all have an impact on the freeze–thaw resilience of starch. A reduced amylose concentration and an increased ratio of shorter amylopectin unit chains may enhance the freeze–thaw resilience of native starch (Mahajan et al., 2021). Consequently, it is essential to correlate the molecular architecture of starch constituents with the FTS of millet starch, since this knowledge might inform genotype selection and necessitates additional investigation (Kaushik et al., 2024).

13.4.4 Gelatinization

Granular starch, when subjected to heat in the presence of sufficient water over varying temperatures, experiences a phase change from order to disorder termed gelatinization. This process is defined by three essential characteristics: the uptake of water by the amorphous regions, the radial expansion of granules, the degradation of crystalline regions as double helices disintegrate, and the leaching of starch molecules. Gelatinization can be quantified by several methodologies, with the kinetic heat storage method (KHSM) and differential scanning calorimetry (DSC) being the predominant procedures for evaluating millet starch. In these assessments, data like conclusion (Tc), onset (To), peak (Tp) temperatures, and enthalpy change (ΔH) are generally documented (Kheto et al., 2022). The KHSM and DSC procedures exhibit substantial differences in their experimental and instrumental configurations, resulting in varying gelatinization temperature outcomes for the identical starch sample. Recently, DSC has emerged as the favored technique for assessing starch gelatinization, as it accurately quantifies ΔH and the To, Tp, and Tc temperatures, whereas KHSM solely captures temperature data (Kaushik et al., 2024). The ratio of starch to water and the temperature scanning rate significantly influences the gelatinization characteristics assessed by DSC. Significant variations in gelatinization properties were noted for finger millet starch at starch-to-water ratios of 1:2 and 1:3. Consequently, it is essential to account for these experimental settings when juxtaposing results from other investigations. Significant genetic heterogeneity exists in gelatinization capabilities among several species and within genotypes of the same species. Starches from 51 genotypes of foxtail millet demonstrated a transition temperature (Tp) between 62.4 and 75 °C, with a change in enthalpy (ΔH) ranging from 8.2 to 13.5 J/g. In a study of 19 genotypes of proso millet, Tp ranged from 65.8 to 80.2 °C, whereas ΔH varied from 6.4 to 11.4 J/g (Vamadevan & Bertoft, 2020; Hermiati et al., 2023). The correlation between the starch structure and the gelatinization characteristics of millet starch is mostly unexamined. Lipid and amylose can combine to create an inclusion complex within starch granules, resulting in the production of a V-type polymorph, as evidenced by wide-angle X-ray diffraction patterns. This complex is frequently present in grains and significantly affects the physicochemical and nutritional characteristics of starch. DSC was used to analyze the melting properties of the amylose–lipid complex in starches from two genotypes of pearl millet. The onset (To) was 91.7–94.2°C, the peak (Tp) was 99.4–102.8°C, the conclusion (Tc) was 108–110°C, and the enthalpy change (ΔH) was 1.1–1.5 J/g, respectively (Sultana et al., 2024). The structure and makeup of the lipid and amylose contents are responsible for variations in these parameters. The characteristics of lipid and amylose complexes in different millet species, especially in cultivars with high lipid levels, need further research (Hermiati et al., 2023).

13.4.5 Retrogradation

A more ordered structure results from the interaction of amylose and amylopectin molecules with water and one another during the chilling process, which causes gelatinized starch to recrystallize. Moisture content, storage conditions and length, amylose concentration, amylopectin molecular structure, and the presence of nonstarch elements like lipids are some of the parameters that affect retrogradation (Ai & Jane, 2024). Retrogradation can be evaluated using many approaches, including DSC and rheometry for millet starch. Retrograded starch from pearl millet exhibited diminished melting temperatures and an expanded melting temperature range, alongside a decreased enthalpy change (ΔH) relative to its gelatinization properties, possibly attributable to the less stable nature of the crystals generated during retrogradation. Retrograded finger millet starch showed a higher enthalpy change (ΔH) and a lower onset temperature (T_0) than the starches made from foxtail, pearl, and proso millets, according to a comparative DSC analysis (Sultana et al., 2024). However, this might not fully capture species diversity, as other genotypes need to be investigated. Furthermore, a separate investigation using rheological testing showed that while xanthan gum may worsen retrogradation, tapioca starch reduced it when finger millet starch was refrigerated. There are other aspects of millet starch retrogradation that require further research, including kinetics and morphological changes (Hermiati et al., 2023).

13.5 STARCH MODIFICATION

To enhance its functional qualities for a range of uses, native starch can be altered by enzymatic, chemical, or physical means. Modifications of millet starch, utilizing chemical, physical, and malting processes, have been documented, as illustrated in Figure 13.2.

13.5.1 Chemical Modification

The structure and characteristics of native starch can be altered by acids that break the glycosidic linkages in α-glucan chains. Granules' amorphous portions are noticeably more vulnerable to acid hydrolysis than their crystalline counterparts. According to a research, 88.2% of the starches from three millet genotypes were hydrolyzed by acid hydrolysis (HCl, 2.2 M) at 25 °C for 20 days (Punia

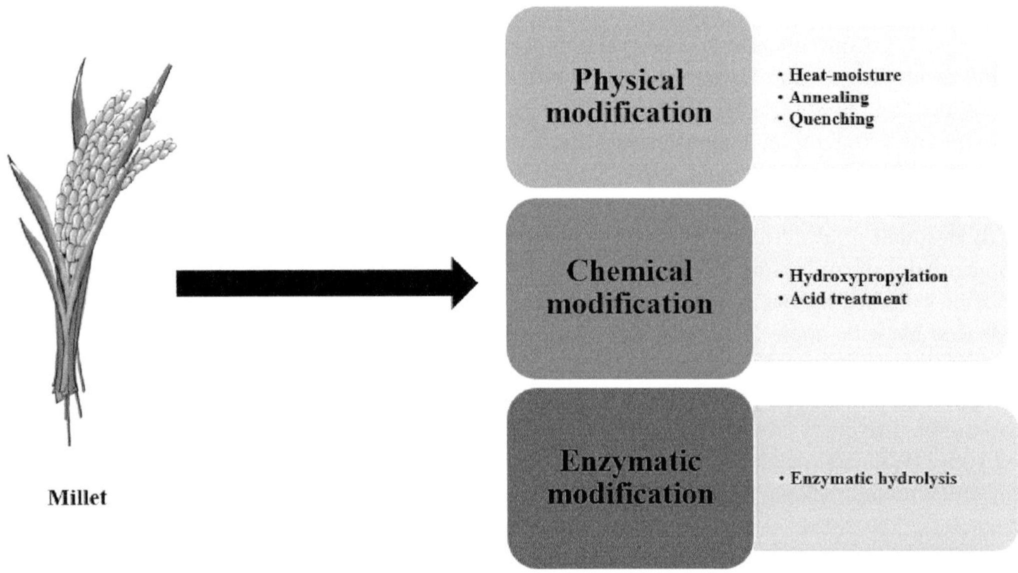

FIGURE 13.2 Methods of modification of millets.

& Kumar, 2022). Variations in granule shape and the physical interactions between amylose and amylopectin are reflected in the genotypes' varying amounts of hydrolysis (Friero et al., 2024). One area that has lately drawn scientific attention is the production of starch nanocrystals using acid treatment. Millet starch with finer granules might be a useful nanofabrication substrate. Other functional groups, such as the hydroxypropyl group, can take the place of the hydroxyl groups in starch and participate in processes. According to Kaur et al. (2023), hydroxypropylation of finger millet starch mostly took place at the carbon-6 position, as demonstrated by carbon-13 nuclear magnetic resonance (NMR) spectroscopy. This change increased peak viscosity while pasting while decreasing turbidity, pasting temperature, setback value, syneresis, and retrogradation. Previous studies have shown that the type of starch used determines how hydroxypropylation affects pasting viscosity. Therefore, it is expected that modifying starch from different millet genotypes or species will provide a wide range of characteristics. Other chemical changes to take into account include cross-linking, oxidation, alkaline treatment, and substitution processes such as acetylation, succinylation, octenyl succinylation, and carboxymethylation in addition to hydroxypropylation and acid treatments (Singh et al., 2024).

13.5.2 Physical Modification

For finger millet starch, a variety of hydrothermal processes have been used, including annealing, heat–moisture, and quenching. With a moisture content limited to less than 35%, heat–moisture treatment (HMT) typically occurs at temperatures between 100 and 130° C. Heating starch with a moisture content of around 40–65% at temperatures below the gelatinization threshold is known as annealing (ANN). The process of quickly cooling overheated starch slurry to much lower temperatures (such as −180°C) is known as quenching (QUE) (Kaur et al., 2023). While ANN and HMT can amplify structural flaws in the crystalline regions, these treatments help the amylose and amylopectin chains within the granules make contact and reassociation. Both HMT and ANN elevated the gelatinization temperatures and ΔH of finger millet starch, while QUE diminished the gelatinization temperatures. Although the molecular processes of HMT and ANN are comprehensively understood, additional study is required to elucidate the impacts of QUE (Friero et al., 2024). Physical alterations are typically regarded as more environmentally sustainable and consumer-friendly for starch alteration than chemical and enzymatic approaches. In addition to hydrothermal treatments, other physical ways to enhance starch qualities encompass pulsed electric processing, high-pressure treatment, ultrasonication, irradiation, and thermal pre-gelatinization, among others (Singh et al., 2024).

Malting encompasses a series of physicochemical and enzymatic techniques applied to grains. Upon malting finger, pearl, and foxtail millets, the resultant starch exhibits diminished granule size, enhanced solubility and gelatinization temperature, and decreased intrinsic viscosity and swelling power, and comparable enzyme susceptibility relative to native starch (Malleshi et al., 1986). Notwithstanding these alterations, starch modification via malting remains underutilized, and its economic viability for both alimentary and nonalimentary uses requires additional investigation. Alternative enzymatic methods exist to bestow distinctive qualities upon starch; however, they have not yet been utilized on millet starch.

13.5.3 Enzymatic modification

Enzymes such as transferases, phosphorylases, glucosidases, and amylases can be utilized to modify starch by changing its structure, function, and properties. Additionally, certain varieties of foxtail and proso millets contain waxy starch, resulting in sticky textures (Serna-Saldivar & Espinosa-Ramírez, 2019). Millets have long been recognized for their hypoglycemic properties, primarily due to the larger size of their starch granules, which reduces the surface area available for enzyme hydrolysis. This leads to less porous starch surfaces, allowing enzymes to penetrate the interior and slowing down degradation (Tester et al., 2006). Research showed that treating foxtail millet starch with α-amylase reduced its swelling power and increased its solubility (Dey & Sit, 2017). Malting, an enzymatic and physicochemical process alters the starch of finger, pearl, and foxtail millets, resulting in reduced granule size, increased solubility, higher gelatinization temperature, and decreased intrinsic viscosity and swelling power, while maintaining comparable enzyme susceptibility to native starch (Malleshi et al., 1986). The impact of lactic acid bacteria fermentation on the composition and physical characteristics of glutinous (waxy) proso millet starch was examined (Bian et al., 2022). Fermentation led to the breakdown of starch chains in the amorphous areas by microbial enzymes and acids, producing fermented starch characterized by increased amylose concentration and crystallinity, alongside diminished molecular weight, swelling power, and solubility. Enzymatic hydrolysis can produce oligosaccharides with bioactive features, including immunomodulatory and antioxidant effects (Lu et al., 2023). Although malting is not commonly employed for starch modification and its economic viability in food and nonfood industries is still questionable, alternative enzymatic techniques are available to improve starch characteristics, although they have not been utilized for millet starch.

13.6 THERAPEUTIC ROLE OF MILLET STARCH

Recent years have seen significant advancements in millet grain research. Millets have demonstrated biological activity and the capacity to create a variety of bioactive chemicals that may have therapeutic utility. Crude extracts and isolated compounds from different millet grains have attracted a lot of attention from the pharmaceutical industry in recent decades. The antibacterial, antidiabetic, anti-inflammatory, anticancer, and antioxidant properties of several millet varieties all backed by verified claims and scientific evidence have been discussed (Table 13.2).

13.6.1 Anti-inflammatory and Antimicrobial Activity

Numerous secondary metabolites found in millet grains have a variety of biological impacts. Bioactive secondary metabolites, including flavonoid and phenolic compounds, are present in millet

TABLE 13.2
Millets with Bioactive Properties for Disease Prevention

Disease	Disease Prevention Compound	Function or Action	Millet Crop	Reference
Cancer	Polyphenols and tannins	Opposes human melanoma cells	Sorghum	Chang et al. (2023)
Cardiovascular diseases (high cholesterol level)	Lignin	Serves as antioxidants that lower triglycerides in plasma	Proso millet, Finger millet, Barnyard millet	Sharma et al. (2021)
	Lecithin, methionine	Eliminate extra fat from the liver and reduce cholesterol	Finger millet	Gupta et al. (2024)
	Sterol	Minimizes cholesterol absorption and prevents its natural production	Sorghum	Harish et al. (2024)
Obesity	High dietary fiber content	Delays food movement from stomach to intestine, slowing digestion and absorption, reducing hunger and consumption	Sorghum	Mondal et al. (2022)
	Tryptophan	Decreases hunger	Finger millet	Yang et al. (2022)
Detoxification	Antioxidants (quercetin, cucurmin, ellagic acid)	Neutralize free radicals and stop cell disruption by scavenging radical cations	Kodo millet, barnyard millet, foxtail millet, finger millet, little millet, sorghum	Dimri et al. (2022)
Diabetes	Fiber, magnesium, vitamins, tannins	Extended digestion of fiber results in slower release of glucose into the blood	Pearl millet, Sorghum, Finger millet	Sachdev et al. (2021)
Gastrointestinal disorders	Alkaline nature of grains	Regulates stomach pH by balancing acidity	Pearl millet	Sharma et al. (2023)
	Fiber	Relieves cramps, bloating, gas, and constipation	Finger millet	Amarnath et al. (2023)
Celiac disease	Gluten-free grains	Grain content does not change the amount of antitransglutaminase antibodies	Sorghum, Pearl millet	Jhan et al. (2020)

cultivars and have antifungal and antibacterial properties (Mahajan et al., 2021). According to Punia et al. (2021), pearl millet protein extract shown superior antifungal effectiveness in comparison to other millet varieties against phytopathogenic fungi such as *Fusarium oxysporum*, *Macrophomina phaseolina*, and *Rhizoctonia solani. Bacillus cereus, Proteus mirabilis, Staphylococcus aureus, Streptococcus pyogenes, Pseudomonas aeruginosa, Yersinia enterocolitica, Listeria monocytogenes, Escherichia coli, Serratia marcescens, Bacillus cereus*, and *Klebsiella pneumoniae* are just a few of the bacterial pathogens that are inhibited by the phenolic and flavonoid compounds found in finger millets (Bangar et al., 2021).

Phenolic chemicals found in finger millet grains have antibacterial properties that might make them a good pharmacological substitute for treating a variety of bacterial and fungal diseases (Thakur et al., 2023). Kimal et al. (2021) looked into the mechanism behind the reduction in fungal function associated with phenolic extracts from finger millet seed coats. As free radicals, these phenolic extracts oxidize fungal cell components and microbial membranes while also inhibiting the enzymatic activity of fungal proteins. However, there has not been much research on millet grains' pharmacological anti-inflammatory properties. Finger millet has been thoroughly investigated as a notable source of bioactive chemicals, with studies concentrating on their impact on skin wound healing in diabetes-induced rats, highlighting the control of inflammation via oxidative stress (Kumar et al., 2024).

13.6.2 Antidiabetic Activity

The antibacterial characteristics of phenolic chemicals present in finger millet grains may provide a viable pharmaceutical solution for addressing diverse bacterial and fungal illnesses (Thakur et al., 2023). Kimal et al. (2021) investigated the mechanism underlying the decrease of fungal function linked to phenolic extracts from the seed coat of finger millet. These phenolic extracts function as free radicals, oxidizing microbial membranes and cellular constituents of fungi while concurrently suppressing the enzymatic activity of fungal proteins. Nonetheless, research on the anti-inflammatory potential of millet grains as pharmacological agents has been scarce. Finger millet has been thoroughly investigated as a notable source of bioactive chemicals, with studies concentrating on their impact on skin wound healing in diabetes-induced rats, highlighting the control of inflammation via oxidative stress (Kumar et al., 2024).

13.6.3 Antioxidant Activity

Phenolic and flavonoid chemicals found in millet grains function as antioxidants by chelating metal ions, preventing radical production, preventing cellular damage from free radicals, and boosting the body's natural antioxidant system (Dekka et al., 2023). DPPH reduction capacity and ferric lowering antioxidant potential were used to evaluate the antioxidant characteristics of many millet cultivars, including Kodo, Finger, Little, Foxtail, Barnyard, Great millet, and their white counterparts (Kumari et al., 2024a). The antioxidant properties and phenolic composition of millet grains may act as antiaging agents, protecting cells against metabolic syndrome. Kaur et al. (2023) examined the antioxidant properties of finger millet, confirming the free radical scavenging ability of the phenolic compounds present in the seed coat. According to Sachdev et al. (2023), the presence of antioxidant chemicals such flavonoids and phenolics in whole grains may provide protection against a number of chronic illnesses. By measuring total phenolic and flavonoid content, ferrous ion chelation activity, and singlet oxygen scavenging ability, Tomar et al. (2022) assessed the antioxidant capacity of seven millet varieties and came to the conclusion that phenolic extracts from millets demonstrated anticancer properties by scavenging peroxyl and hydroxyl radicals. Furthermore, Abah et al. (2020) investigated how processing affected the antioxidant and nutraceutical properties of tiny millets (*P. sumatrense*), finding that roasting greatly increased the phenolic content and boosted the capacity to scavenge free radicals. Mirzababaee et al. (2022) examined the effects of a millet-enriched diet on hypertension and myocardial damage in animal models, finding that such a diet significantly reduced blood pressure and concluded that the antioxidative properties of millets help protect against cardiac damage induced by high salt intake.

13.6.4 Regulating Lipid Levels

Increased serum lipid levels are associated with various risk factors contributing to hyperlipidemia and obesity. High levels of low-density lipoproteins (LDL) in the serum can lead to arterial blockages, resulting in cardiovascular issues. Other contributing factors include fat accumulation in adipocytes, which can lead to fatty liver and obesity (Jhan et al., 2020). A diet high in fats combined with low energy expenditure is often cited as a significant risk factor. Treatment for lipid-related disorders typically involves cholesterol-lowering medications like statins, including atorvastatin, fluvastatin, pravastatin, and lovastatin (Wang et al., 2021). However, the complications associated with statin use have prompted the search for alternative therapeutic agents to alleviate these side effects. Phytochemicals derived from various plant species have shown potential in regulating lipid levels (Deka et al., 2023). Specifically, phytochemicals from minor millets are recognized for their ability to reduce fat content. Below is a comprehensive overview of millet phytochemicals that can influence lipid levels. The release of free radicals from lipids, which causes cellular damage, is largely dependent on lipid peroxidation (Wang et al., 2021). Thus, preventing lipid peroxidation is essential to managing lipid degradation. Due to their high ferulic acid content, little millet and foxtail millet have the ability to cause lipid peroxidation in a liposome system. Extracts from kodo millet, which are rich in phenolic acids, have been shown to increase HDL while decreasing total cholesterol, serum triglycerides, and very low-density lipoprotein (VLDL). Phytochemicals made from millet extracts may effectively lower cholesterol levels as lowering LDL is a key goal in the treatment of lipid disorders (Purewal et al., 2024). The importance of preventing LDL oxidation is demonstrated by the significant inhibition of the copper-catalyzed oxidation of human LDL cholesterol in vitro by phenolic extracts from kodo millet, which include caffeic acid, p-coumaric acid, and ferulic acid. According to Wang et al. (2023), copper can help unsaturated fatty acids in cholesterol oxidize. Because millet phenolic compounds scavenge free radicals and chelate cupric ions, they may stop cholesterol from oxidizing. By lowering triglyceride and total cholesterol levels, foxtail millet methanol extract has been demonstrated to reduce fat accumulation in HepG2 cells. Genes linked to hepatic lipid metabolism (FAS, SREBP-1C, and HMGCR) were downregulated in rats fed a high-fat diet supplemented with millet (Zhang et al., 2023).

13.7 INDUSTRIAL APPLICATIONS OF MILLET STARCHES

13.7.1 Tuwo Production

Tuwo is a traditional dish from Northern Nigeria, made using millet, maize, or sorghum, depending on the preference of the producer. It is a solid meal shaped into balls or swallows, prepared from flour made from these grains. According to Sharma *et al.* (2022), the process of making tuwo involves sorting, dehulling, winnowing, and milling millet into fine flour, which is then sieved. Water is heated to a boil in a pot, and a portion of the flour is combined with cold water to create slurry. The slurry is incorporated into the boiling water and boiled thoroughly. The sieved millet flour is then gradually added to the pot containing the boiling water and slurry, stirred until the desired consistency is reached. The mixture is heated for another 10–15 minutes, stirred again, and then served with a soup of choice (Gautam et al., 2021).

13.7.2 Millet Ball Production "Fura"

Fura, which is mostly made out of millet or sorghum, is a staple meal for the Fulanis and Hausas. Millet is commonly processed into rich porridges in tropical Africa, which are known by different names across the continent (Ge et al., 2023). West Africa, particularly Nigeria, Ghana, and Burkina Faso, is home to the thick, semi-solid fura dumpling cereal dish. Usually made using wet millet flour and spices, it is shaped into spheres and cooked for half an hour. In a mortar and pestle, the cooked dough is then ground with hot water until it reaches a smooth, somewhat elastic, cohesive consistency (Bhatt et al., 2023). After forming the dough into 25–30 g balls, it is sprinkled with flour. Fura balls are dissolved in either fermented whole milk (kindrimo) or fermented skim milk (nono) to make porridge, and sugar is optional. In Nigeria, this dish is known as "fura da nono" (Yang et al., 2022) and is a popular lunch choice. Fura is manufactured for both business and residential markets. Grain is being ground into flour using modern techniques rather than the traditional mortar and pestle process. Because producers place a high priority on waste minimization and container identification, Fura is typically offered with little packaging. Fura has only one day of shelf life at room temperature, and visible mold growth often appears the next day (Mondal et al., 2022). One major barrier to widespread manufacturing has been the short shelf life. Consequently, enhancing fura's processing, packaging, and shelf-life is essential for scaling up production (Harish et al., 2024).

13.7.3 Gruel Production

Millet is commonly used to make gruel or breakfast dishes, which are later, processed into foods like pap, "ogi," or "akamu" and consumed with various snacks for better nutrition. These gruels are often enriched with additional ingredients such as soybeans or ginger (Gupta et al., 2024). Raw millet is carefully sifted, washed, and steeped for 72 hours before the water is drained to make the gruel. Some producers change the water every 24 hours, while others allow fermentation to continue for the full 72 hours. After fermentation, the grains are cleaned and ground using sanitized machinery (Sharma et al., 2021). Muslin cloths are used to filter the resulting grain slurry, and the filtrate is let to settle before the water is removed. Depending on the producer's preferences, the slurry can then be cooked and transformed into products like pap, ogi, akamu, kwokwo, or other meals (Chang et al., 2023).

13.7.4 Local Alcoholic Beverage Production

In Western Africa, including Nigeria, barley has historically been the only cereal utilized to make alcoholic drinks. Nigeria's economy has not profited from this dependency; rather, it has depleted resources and enriched other nations, hence sustaining Nigeria's reliance (Amarnath et al., 2023). By investigating the use of locally available grains in the production of alcoholic drinks, recent research has attempted to lessen this reliance; millet has played a significant role in these advancements. Alcoholic drinks are generally divided into three main categories: beers, wines, and spirits (Sharma et al., 2023). While spirits are manufactured by fermenting and then distilling, which raises the alcohol level, lower-alcohol beverages, such as wine and beer, are prepared by fermenting plant sources high in sugar or starch. Burukutu, pito, ogogoro, and palm wine are typical native alcoholic beverages in Nigeria. Fermented cereal grains are used to make the traditional beer known as burukutu (Sachdev et al., 2021). Cereals are important sources of food for people all over the world because they include a variety of starches with different physicochemical characteristics and molecular structures. The main component of millet, a significant grain grown in Nigeria, is starch. Burukutu's sour taste, which is ascribed to lactic acid, its pH range of 3.3–3.5, and the turbid appearance caused by yeast and suspended particles are its defining characteristics. It contains around 26.7 g of carbohydrate and 5.9 g of protein per liter, along with vitamins, iron, manganese, magnesium, potassium, and calcium. Distinct areas have distinct names for this beverage like *Dolo* in Burkina Faso, *Pito* in Ghana, *Burukutu* or *Otika* in Nigeria, *Bilibili* in Chad, *Mtama* in Tanzania, *Kigage* in Rwanda, and *Techoukoutou* in Benin and Togo. Malting, steeping, germination, milling, mashing, boiling, fermentation, and maturation are among the manufacturing methods that vary substantially according on the region (Dimri et al., 2022). A list of various millet-based food products has been tabulated in Table 13.3.

13.7.5 Local Nonalcoholic Beverage

A popular nonalcoholic, noncarbonated grain beverage in Northern Nigeria, kunu, also known as kunun-zaki, is becoming more and more well-liked in the South. It fulfils various functions, including a breakfast beverage, appetiser, weaning sustenance, and possesses therapeutic attributes. Kunu, a nutritious beverage, is composed of several cereal grains such as millet, sorghum, maize, and rice

TABLE 13.3
Applications of Modification Techniques on Millet Starch Characteristics and their Industrial Uses

Starch Source	Modification Treatment	Key Findings	Industrial Application	References
Proso	AMT (acid modification)	Decreased amylose concentration and white bran content; enhanced starch paste clarity	Noodles, canned, and frozen foods	Deka et al. (2023)
	HTM (hydrothermal)	Elevated amylose content, WBC; decreased paste clarity		
Kutki	AMT	Enhanced swelling, viscosity, and structural integrity of the gel	Fat replacer	Kumari et al. (2024b)
	NS (native starch)	Increased amylose; enhanced pasting and textural characteristics	Thickener, stabilizer, gelling agent	
Finger	Oxidized and acetylated	Reduced disintegration time and friability; enhanced crushing and tensile strength	Capsule and tablet formation	Kaur et al. (2023)
	NS	Superb solubility, swelling index, and water vapor permeability	Thin and flexible food packaging film	
	UAH (ultrasound-assisted hydration)	Enhanced structural morphology, white blood cell count, solubility	Food products	
Teff	NS	Enhanced radical scavenging activity and mechanical characteristics	Antioxidant packaging material	Sachdev et al. (2023)
	Pregelatinized	Accelerated disintegration, enhanced compressibility	Chloroquine phosphate tablets	
Foxtail	AUS, USA	Enhanced starch stability and gelation	Food products	Tomar et al. (2022)
	NS	Enhanced opacity, consistent three-dimensional cellular and hexagonal configuration	Starch gel	
	NS	Improved packing material (antimicrobial, antioxidant)	Cheese packaging	
Barnyard	Cross-linking	Enhanced tensile strength, barrier characteristics, and mechanical qualities	Starch-based films	Abah et al. (2020)
	NS	Enhanced antioxidant activity; reduced solubility and water vapor permeability (WVP)	Starch film with borage oil	
Pearl	HP (hydroxy propylated)	Enhanced uniformity, texture, flavor, and palatability	White sauces	Bangar et al. (2024)
	SUS, OXS, ACS	Improved cold storage durability	Custards	
	HMT, MT, ST	Elevated amylose, transition temperatures, swelling capacity	Film formation	
	HP, succinylated, OXS, ACS	Enhanced cold storage stability, texture, and sensory attributes; reduced syneresis	Custards	
	Octenyl succinic anhydride	Elevated paste viscosity and water solubility	Biodegradable starch films	
	Cross-linked	Lower moisture, solubility, WVP	Films loaded with fenugreek oil	
Kodo	NS	Reduced WBC, swelling capacity, viscosity, dispensability; increased fat absorption, solubility		Babu et al. (2024)

(Bhatt et al., 2023). This beverage is esteemed throughout all social strata in Northern Nigeria and is utilized either as a nutritional supplement or a means to satisfy thirst. Kunu is readily available and provides a more nutritious alternative to carbonated drinks, which frequently lack health advantages. It comprises an intricate amalgamation of macromolecules, encompassing proteins, carbohydrates, and lipids (Ge et al., 2023). The primary cereals used in kunu production are millet, maize, guinea corn, and rice. Alligator pepper (*Afromonium melegueta*), black pepper (*Piper guineense*), red pepper (*Capsicum species*), ginger (*Zingiber officinalis*), and kakandoru or eru are necessary components for creating kunu, and each one has a distinct function in the cooking process. The primary element, water, serves as the solvent for the remaining constituents, which only include trace quantities of inorganic substances. Kunu's high protein, carbohydrate, and vitamin B content are responsible for its great nutritional value (Gautam et al., 2021). Kunu is recognized as a beverage in certain urban areas, especially in Hausa regions, with its quality largely depending on the ingredients and the production methods employed. The product can be produced within two days and stored in the refrigerator for an additional three days (Sharma et al., 2022). However, storing kunu overnight in warm conditions without refrigeration can lead to quality deterioration and potential spoilage, posing health risks (Zhang et al., 2023). This spoilage is often due to improper handling, ongoing fermentation of carbohydrates, and enzymatic activity on the ingredients (Wang et al., 2023). Therefore, proper formulation and carbonation are essential. Carbonated beverages are popular for their sharp, distinctive flavor and are typically made from carbon dioxide, water, flavoring, and sweet syrups. The addition of CO_2 increases the beverage's acidity, which can inhibit the growth of certain microorganisms, thereby prolonging shelf life (Deka et al., 2023). Carbonated kunu is expected to differ significantly from other carbonated drinks due to its nutritional value and various health benefits, such as aiding digestion and enhancing nutrient absorption while boosting the immune system against microbial invasion. Currently, carbonated beverages in Nigeria are expensive, with a 50Cl bottle averaging N100.00, offering minimal nutritional value due to high sugar content and artificial additives (Wang et al., 2023). In contrast, kunu is highly nutritious, cost-effective, and made from locally sourced grains that are staples in the diet. However, its rapid spoilage due to microbial activity poses challenges (Jhan et al., 2020). To prepare kunu, millet grains are initially washed and then immersed in double their amount of water for 24 hours. Following rinsing, spices including ginger, red pepper, cloves, and black pepper are included. The soaked grains and spices are processed in a machine, ground, and sieved to eliminate husks, while the supernatant is collected from the slurry (Wang et al., 2021). One-half of the slurry is combined with boiling water, agitated for 5 minutes, and then cooled to 35°C prior to amalgamation with the remaining slurry. Water is introduced, combined, and permitted to settle. The mixture is subsequently filtered through muslin cloth, sweetened with granulated sugar, and completely combined to produce the fresh millet beverage, which is then bottled in plastic containers (Sachdev et al., 2021).

13.8 CONCLUSION

The increasing demand for sustainable and health-focused products highlights millet starch as a significant resource in food technology and environmental applications. As consumers look for alternatives that support health and sustainability, the relevance of millet starch becomes more evident. With its unique structural and functional properties, millet starch offers distinct opportunities for both food and industrial use. Its granular structure enables effective water absorption and gelatinization, making it an ideal ingredient for creating gels, thickening sauces, and stabilizing emulsions. This versatility is particularly beneficial in gluten-free formulations, where millet starch acts as a vital binder and texture enhancer, improving the mouthfeel and overall quality of gluten-free products. This highlights the need for more study and development to maximize its industrial potential, improve extraction and processing methods, and strengthen its contribution to environmental and nutritional sustainability.

REFERENCES

Abah, C. R., Ishiwu, C. N., Obiegbuna, J. E., & Oladejo, A. A. (2020). Nutritional composition, functional properties and food applications of millet grains. *Asian Food Science Journal*, *14* (2), 9–19.

Adebiyi, J. A., Obadina, A. O., Adebo, O. A., & Kayitesi, E. (2018). Fermented and malted millet products in Africa: Expedition from traditional/ethnic foods to industrial value-added products. *Critical Reviews in Food Science and Nutrition*, *58* (3), 463–474. https://doi.org/10.1080/10408398.2016.1188056

Adegbaju, M. S. (2022). *Analysis of interactions between glucan, water dikinase and either isoamylase or starch branching enzymes in detrmination of starch structure* (Doctoral dissertation, Stellenbosch: Stellenbosch University).

Ai, Y., & Jane, J. L. (2024). Understanding starch structure and functionality. In *Starch in food* (pp. 55–77). Woodhead Publishing.

Almaski, A., Thondre, S., Lightowler, H., & Coe, S. (2017). Determination of the polyphenol and antioxidant activity of different types and forms of millet. *Proceedings of the Nutrition Society*, *76* (OCE1), E5. https://doi.org/10.1017/S0029665117000052

Alzaabi, A. Z. (2023). *The effect of increasing resistant starch content of food starches through chilling and reheating process on postprandial glycaemia and lipidaemia* (Doctoral dissertation, University of Surrey).

Amarnath, M. S., Muhammed, A., Antony, A. K., Malini, B., & Sunil, C. K. (2023). White finger millet starch: Physical modification (annealing and ultrasound), and its impact on physicochemical, functional, thermal and structural properties. *Food and Humanity*, *1*, 599–606.

Annor, G. A. (2013). *Millet starches: Structural characteristics and glycemic attributes* (Doctoral dissertation, University of Guelph).

Babu, A. S., Olalekan, A. S. A., Hegde, S. V., & Jaganmohan, R. (2024). Foxtail millet starch: structure, functionality, and applications. In *Non-Conventional Starch Sources* (pp. 225–251). Academic Press.

Bangar, S. P., Ashogbon, A. O., Dhull, S. B., Thirumdas, R., Kumar, M., Hasan, M., ... & Pathem, S. (2021). Proso-millet starch: Properties, functionality, and applications. *International Journal of Biological Macromolecules*, *190*, 960–968. https://doi.org/10.1016/j.ijbiomac.2021.09.064

Bangar, S. P., Balakrishnan, G., Navaf, M., & Sunooj, K. V. (2024). Recent Advancements on barnyard millet starch: A sustainable alternative to conventional starch. *Starch-Stärke*, *76* (9–10), 2300232.

Bean, S. R., Zhu, L., Smith, B. M., Wilson, J. D., Ioerger, B. P., & Tilley, M. (2019). Starch and protein chemistry and functional properties. *Sorghum and Millets*, 131–170. https://doi.org/10.1016/B978-0-12-811527-5.00006-X

Bertoft, E. (2017). Understanding starch structure: Recent progress. *Agronomy*, *7* (3), 56.

Bhatt, P., Kumar, V., Rastogi, H., Malik, M. K., Dixit, R., Garg, S., ... & Singh, S. (2023). Functional and tableting properties of alkali-isolated and phosphorylated barnyard millet (*Echinochloa esculenta*) starch. *ACS Omega*, *8* (33), 30294–30305.

Bian, X., Chen, J. R., Yang, Y., Yu, D. H., Ma, Z. Q., Ren, L. K., ... & Zhang, N. (2022). Effects of fermentation on the structure and physical properties of glutinous proso millet starch. *Food Hydrocolloids*, *123*, 107144. https://doi.org/10.1016/j.foodhyd.2021.107144

Chang, L., Zhao, N., Jiang, F., Ji, X., Feng, B., Liang, J., ... & Du, S. K. (2023). Structure, physicochemical, functional and in vitro digestibility properties of non-waxy and waxy proso millet starches. *International Journal of Biological Macromolecules*, *224*, 594–603.

Collins, R. L., Brand, H., Karczewski, K. J., Zhao, X., Alföldi, J., Francioli, L. C., ... & Talkowski, M. E. (2020). A structural variation reference for medical and population genetics. *Nature*, *581* (7809), 444–451.

Das, S., Khound, R., Santra, M., & Santra, D. K. (2019). Beyond bird feed: Proso millet for human health and environment. *Agriculture*, *9* (3), 64.

Deka, G., Dutta, H., & Sharma, P. (2023). A review: Hydrothermal processing of millets and effect on techno-functional and nutritional properties. *Journal of Food Process Engineering*, *46* (9), e14408.

Dekka, S., Paul, A., Vidyalakshmi, R., & Mahendran, R. (2023). Potential processing technologies for utilization of millets: An updated comprehensive review. *Journal of Food Process Engineering*, *46* (10), e14279.

Delcour, J. A., Bruneel, C., Derde, L. J., Gomand, S. V., Pareyt, B., Putseys, J. A., ... & Lamberts, L. (2010). Fate of starch in food processing: From raw materials to final food products. *Annual Review of Food Science and Technology*, *1* (1), 87–111.

Dey, A., & Sit, N. (2017). Modification of foxtail millet starch by combining physical, chemical, and enzymatic methods. *International Journal of Biological Macromolecules*, *95*, 314–320.

Dimri, S., & Singh, S. (2022). A brief review on millet starch. *Bhartiya Krishi Anusandhan Patrika*, *37* (2), 126–132.

Dupuis, J. H., Liu, Q., & Yada, R. Y. (2014). Methodologies for increasing the resistant starch content of food starches: A review. *Comprehensive Reviews in Food Science and Food Safety*, *13* (6), 1219–1234.

Friero, I., Martínez-Subirà, M., Romero, M. P., & Moralejo, M. (2024). Improving functional and nutritional profiles of barley flours with diverse starch types through pearling. *Food Chemistry*, *460*, 140611.

Gautam, N., Garg, S., & Yadav, S. (2021). Underutilized finger millet crop for starch extraction, characterization, and utilization in the development of flexible thin film. *Journal of Food Science and Technology*, *58*, 4411–4419.

Ge, Y., Shi, Y., Wu, Y., Wei, C., & Cao, L. (2023). Preparation, structure, and in-vitro hypoglycemic potential of debranched millet starch-fatty acid composite resistant starch. *Food Chemistry: X*, *20*, 100929.

Gupta, R., & Gaur, S. (2024). Investigating the effect of natural fermentation in modifying the physico-functional, structural and thermal characteristics of pearl and finger millet starch. *Journal of the Science of Food and Agriculture*, *104* (4), 2440–2448.

Harish, M. S., Bhuker, A., & Chauhan, B. S. (2024). Millet production, challenges, and opportunities in the Asia-pacific region: a comprehensive review. *Frontiers in Sustainable Food Systems*, *8*, 1386469.

He, W., & Wei, C. (2020). A critical review on structural properties and formation mechanism of heterogeneous starch granules in cereal endosperm lacking starch branching enzyme. *Food Hydrocolloids*, *100*, 105434.

Hermiati, E., Sondari, D., & Sunarti, T. C. (2023). Extraction and classification of starch from different sources: Structure, properties, and characterization. In *Handbook of Natural Polymers, Volume 1* (pp. 19–60). Elsevier.

IIMR. (2017). Nutritional and Health Benefits of Millets; Indian Institute of Millets Research: Hyderabad, India; ICAR: New Delhi, India.

Jane, J. L. (2009). Structural features of starch granules II. In *Starch* (pp. 193–236). Academic Press.

Jhan, F., Shah, A., Gani, A., Ahmad, M., & Noor, N. (2020). Nano-reduction of starch from underutilised millets: Effect on structural, thermal, morphological and nutraceutical properties. *International Journal of Biological Macromolecules*, *159*, 1113–1121.

Kaimal, A. M., Mujumdar, A. S., & Thorat, B. N. (2021). Resistant starch from millets: Recent developments and applications in food industries. *Trends in Food Science & Technology*, *111*, 563–580.

Karakelle, B., Kian-Pour, N., Toker, O. S., & Palabiyik, I. (2020). Effect of process conditions and amylose/amylopectin ratio on the pasting behavior of maize starch: A modeling approach. *Journal of Cereal Science*, *94*, 102998.

Kaur, B., Singh, A., Suri, S., Usman, M., & Dutta, D. (2023). Minor millets: A review on nutritional composition, starch extraction/modification, product formulation, and health benefits. *Journal of the Science of Food and Agriculture*, *103* (10), 4742–4754.

Kaushik, A., Saxena, D. C., & Singh, S. (2024). Exploring the potential of browntop millet (*Urochloa ramosa*) starch: Physicochemical, morphological, thermal and rheological properties across four cultivars. *International Journal of Biological Macromolecules*, *278*, 134923.

Kheto, A., Das, R., Deb, S., Bist, Y., Kumar, Y., Tarafdar, A., & Saxena, D. C. (2022). Advances in isolation, characterization, modification, and application of Chenopodium starch: A comprehensive review. *International Journal of Biological Macromolecules*, *222*, 636–651.

Kumar, N., Singh, R., Upadhyay, A., & Giri, B. S. (2024). Application and functional properties of millet starch: Wet milling extraction process and different modification approaches. *Heliyon 10* (3), e25330. https://doi.org/10.1016/j.heliyon.2024.e25330

Kumar, S. R., Tangsrianugul, N., & Suphantharika, M. (2023). A review on isolation, characterization, modification, and applications of proso millet starch. *Foods*, *12* (12), 2413. https://doi.org/10.3390/foods12122413

Kumari, P., Pareek, V., Kajla, P., & Khurana, S. (2024a). Composition, structure and functionality of starch isolated from kodo millet. In *Non-Conventional Starch Sources* (pp. 253–278). Academic Press.

Kumari, S., Kaur, B. P., & Manonmani, T. (2024b). Ultrasound modified millet starch: Changes in functional, pasting, thermal, structural, in vitro digestibility properties, and potential food applications. *Food Hydrocolloids*, *153*, 110008.

Li, L., Liu, Z., Zhang, W., Xue, B., & Luo, Z. (2021). Production and applications of amylose-lipid complexes as resistant starch: recent approaches. *Starch-Stärke*, *73* (5–6), 2000249.

Li, M., Daygon, V. D., Solah, V., & Dhital, S. (2023). Starch granule size: Does it matter?. *Critical Reviews in Food Science and Nutrition*, *63* (19), 3683–3703.

Lu, C., Zhao, Z., Huang, G., Liu, J., Ye, F., Chen, J., ... & Lei, L. (2023). The contribution of cell wall integrity to gastric emptying and in vitro starch digestibility and fermentation performance of highland barley foods. *Food Research International*, *169*, 112912.

Mahajan, P., Bera, M. B., Panesar, P. S., & Chauhan, A. (2021). Millet starch: A review. *International Journal of Biological Macromolecules*, *180*, 61–79. https://doi.org/10.1016/j.ijbiomac.2021.03.063

Majzoobi, M., & Farahnaky, A. (2021). Granular cold-water swelling starch; properties, preparation and applications, a review. *Food Hydrocolloids*, *111*, 106393.

Malleshi, N. G., Desikachar, H. S. R., & Tharanathan, R. N. (1986). Physico-chemical properties of native and malted finger millet, pearl millet, and millet starches. *Starch*, *38*, 202–205.

Mirzababaee, S. M., Ozmen, D., Hesarinejad, M. A., Toker, O. S., & Yeganehzad, S. (2022). A study on the structural, physicochemical, rheological and thermal properties of high hydrostatic pressurized pearl millet starch. *International Journal of Biological Macromolecules*, *223*, 511–523.

Mondal, D., Awana, M., Aggarwal, S., Das, D., Thomas, B., Singh, S. P., ... & Krishnan, V. (2022). Microstructure, matrix interactions, and molecular structure are the key determinants of inherent glycemic potential in pearl millet (*Pennisetum glaucum*). *Food Hydrocolloids*, *127*, 107481.

Egharevba, O. (2019) Chemical properties of starch and its application in the food industry. In Chemical Properties of Starch, IntechOpen.

Pandey, S., & Singh, A. (2024). Millet milk: A potential cattle milk alternative—extraction and value addition. *Journal of Food Quality*, *2024* (1), 3964937. https://doi.org/10.1155/2024/3964937

Punia, S., & Kumar, M. (Eds.). (2022). *Functionality and Application of Colored Cereals: Nutritional, Bioactive, and Health Aspects*. Elsevier.

Punia, S., Kumar, M., Siroha, A. K., Kennedy, J. F., Dhull, S. B., & Whiteside, W. S. (2021). Pearl millet grain as an emerging source of starch: A review on its structure, physicochemical properties, functionalization, and industrial applications. *Carbohydrate Polymers*, *260*, 117776.

Purewal, S. S., Siroha, A. K., Salar, R. K., Kaur, A., Singh, R. S., Kennedy, J. F., & Bangar, S. P. (2024). Influence of solid-state fermentation on the functional characteristics of pearl millet starch: An innovative and cost-efficient approach to modify starch. *Carbohydrate Polymer Technologies and Applications*, *7*, 100449.

Raungrusmee, S. (2023). Underutilized cereals and pseudocereals' nutritional potential and health implications. In *Pandemics and Innovative Food Systems* (pp. 163–193). CRC Press.

Sachdev, N., Goomer, S., & Singh, L. R. (2021). Foxtail millet: A potential crop to meet future demand scenario for alternative sustainable protein. *Journal of the Science of Food and Agriculture*, *101* (3), 831–842.

Sachdev, N., Goomer, S., Singh, L. R., Pathak, V. M., Aggarwal, D., & Chowhan, R. K. (2023). Current status of millet seed proteins and its applications: A comprehensive review. *Applied Food Research*, *3* (1), 100288.

Sahaka, M., Amara, S., Wattanakul, J., Gedi, M. A., Aldai, N., Parsiegla, G., ... & Carrière, F. (2020). The digestion of galactolipids and its ubiquitous function in nature for the uptake of the essential α-linolenic acid. *Food & Function*, *11* (8), 6710–6744.

Sandhu, K. S., Siroha, A. K., Punia, S., Sangwan, L., Nehra, M., & Purewal, S. S. (2021). Effect of degree of cross linking on physicochemical, rheological and morphological properties of Sorghum starch. *Carbohydrate Polymer Technologies and Applications*, *2*, 100073. https://doi.org/10.1016/j.carpta.2021.100073

Serna-Saldivar, S. O., & Espinosa-Ramírez, J. (2019). Grain structure and grain chemical composition. In *Sorghum and millets* (pp. 85–129). AACC International Press.

Sharma, R., Bhandari, M., Sharma, S., & Bhardwaj, R. (2022). Compositional, structural and functional characteristics of millets as modified by bioprocessing techniques: A review. *Journal of Food Processing and Preservation*, *46* (10), e16885.

Sharma, R., Sharma, S., Dar, B. N., & Singh, B. (2021). Millets as potential nutri-cereals: A review of nutrient composition, phytochemical profile and techno-functionality. *International Journal of Food Science & Technology*, *56* (8), 3703–3718.

Sharma, S., Thakur, K., Sharma, R., & Bobade, H. (2023). Molecular morphology & interactions, functional properties, rheology and in vitro digestibility of ultrasonically modified pearl millet and sorghum starches. *International Journal of Biological Macromolecules, 253*, 127476.

Shen, Q., Wang, S., Wang, H., Liang, J., Zhao, Q., Cheng, K., ... & Mao, Z. (2024). Revolutionizing food science with mass spectrometry imaging: A comprehensive review of applications and challenges. *Comprehensive Reviews in Food Science and Food Safety, 23* (4), e13398.

Singh, P., Singh, R., Bhadauria, V., & Singh, H. (2024). The functionality and extraction of protein from sorghum, finger millet, and kodo millet: A review. *International Journal of Food Science & Technology, 59* (1), 512–521.

Srichuwong, S., Curti, D., Austin, S., King, R., Lamothe, L., & Gloria-Hernandez, H. (2017). Physicochemical properties and starch digestibility of whole grain sorghums, millet, quinoa and amaranth flours, as affected by starch and non-starch constituents. *Food Chemistry, 233*, 1–10. https://doi.org/10.1016/j.foodchem.2017.04.019

Sultana, A., Punia Bangar, S., & Whiteside, W. S. (2024). Effect of stearic acid modification on properties of pearl millet starch. *Biomass Conversion and Biorefinery, 15* (6), 8745–8753.

Sun, J., Ren, J., Hu, X., Hou, Y., & Yang, Y. (2021). Therapeutic effects of Chinese herbal medicines and their extracts on diabetes. *Biomedicine & Pharmacotherapy, 142*, 111977.

Taylor, J. R., & Duodu, K. G. (Eds.). (2018). *Sorghum and millets: chemistry, technology, and nutritional attributes*. Elsevier.

Tester, R. F., Qi, X., & Karkalas J. (2006). Hydrolysis of native starches with amylases. *Animal Feed Science and Technology, 130* (1–2): 39–54

Tetlow, I. J., & Bertoft, E. (2020). A review of starch biosynthesis in relation to the building block-backbone model. *International Journal of Molecular Sciences, 21* (19), 7011.

Thakur, K., Sharma, S., & Sharma, R. (2023). Morphological and functional properties of millet starches as influenced by different modification techniques: A review. *Starch-Stärke, 75* (3–4), 2200184.

Tomar, M., Bhardwaj, R., Verma, R., Singh, S. P., Krishnan, V., Kansal, R., ... & Sachdev, A. (2022). Interactome of millet-based food matrices: A review. *Food Chemistry, 385*, 132636.

Vamadevan, V., & Bertoft, E. (2020). Observations on the impact of amylopectin and amylose structure on the swelling of starch granules. *Food Hydrocolloids, 103*, 105663.

Verma, V. C., Agrawal, S., Tripathi, M. K., & Kumar, A. (2021). Millet starch: Current knowledge and emerging insights of structure, physiology, glycaemic attributes and uses. *Millets and Millet Technology*, 121–142.

Wang, H., Wu, E., Ma, Q., Zhang, H., Feng, Y., Yang, P., ... & Feng, B. (2023). Comparison of the fine structure and physicochemical properties of proso millet (*Panicum miliaceum* L.) starch from different ecological regions. *International Journal of Biological Macromolecules, 249*, 126115.

Wang, Q., Li, L., & Zheng, X. (2021). Recent advances in heat-moisture modified cereal starch: Structure, functionality and its applications in starchy food systems. *Food Chemistry, 344*, 128700.

Wani, I. A., Sultan, N., Qadir, N., & Ahmad, T. (2021). Physico-chemical, functional and antioxidant properties of wild barnyard grass (*Echinochloa crusgalli* L.) seed flour as affected by gamma-irradiation. *Radiation Physics and Chemistry, 183*, 109409. https://doi.org/10.1016/j.radphyschem.2021.109409

Wankhede, D. B., Shehnaj, A., & Raghavendra Rao, M. R. (1979). Carbohydrate composition of finger millet (*Eleusine coracana*) and foxtail millet (*Setaria italica*). *Qualitas plantarum, 28*, 293–303.

Yang, T., Ma, S., Liu, J., Sun, B., & Wang, X. (2022). Influences of four processing methods on main nutritional components of foxtail millet: A review. *Grain & Oil Science and Technology, 5* (3), 156–165.

Zhang, T., Li, K., Ding, X., Sui, Z., Yang, Q. Q., Shah, N. P., ... & Corke, H. (2021). Starch properties of high and low amylose proso millet (*Panicum miliaceum* L.) genotypes are differentially affected by varying salt and pH. *Food Chemistry, 337*, 127784.

Zhang, X., Shen, Q., Yang, Y., Zhang, F., Wang, C., Liu, Z., ... & Cheng, R. (2023). Structural, functional and mechanistic insights uncover the role of starch in foxtail millet cultivars with different congee-making quality. *International Journal of Biological Macromolecules, 242*, 125107.

Zhu, F. (2014). Structure, physicochemical properties, and uses of millet starch. *Food Research International, 64*, 200–211.

14 Utilization of Sago Starch for Biodegradable Plastics

Shahwar Siddiqui, Akansha Srivastava, and Neelam Yadav

14.1 INTRODUCTION

The very first true plastic completely manufactured using manmade materials were discovered in 1907. Its name was given to the discoverer, scientist Leo Hendrik Baekeland. This material was chosen due to its properties as an electrical insulator (Worm et al., 2017). Some of the benefits of plastics include that they are very strong, lightweight, and durable. This makes them favorable for a wide range of applications, and this is why they are an integral part of nearly all business. Plastics have substituted paper and glass in most of packaging (Thakur & Thakur, 2016). The development of plastics made life much easier because it produces so many valuable things. However, the lifetime of plastics is such that they do not break down quickly, and sometimes do not decompose. Today, our environment has become a lot less attractive because it suffers from pollution due to dumped plastics. Removing plastic is really a highly procedural task. Toxic fumes such as dioxins are released when plastic is burnt, These are hazardous substances to the environment and contributors to global warming. This makes the recycling of plastics a very involved process because there are so many different polymers to recycle under different procedures of recycling (Yamori et al., 2001; MacArtain et al., 2007; Teas et al., 2009). Plastics, being synthetic polymers, have several advantages such as strength, flexibility, persistence, and endurance; yet, because they are nonbiodegradable, they are harmful to our environment. The constantly increasing need for plastic is causing the depletion of natural resources such as petroleum, as well as the accumulation of waste plastic in the environment. This has prompted several researchers to produce biodegradable plastics, sometimes known as bioplastics. Bioplastics, which are made from natural organic compounds such as complex polysaccharides, proteins, and lipids, have the potential to be eco-friendly options than harmful synthetic plastics. Bioplastics can be disintegrated by microorganisms in soil without releasing pollutants. Green plastics/bioplastic have several advantages over petroleum-based plastics, including nontoxicity, easy recyclability, less usage of fossil fuels, less energy consumption, renewable in nature, and eco-friendly. Application of bioplastics is not new, its usage started in early 19th century with the wrapping of candies to wider applications by means of new researches (Paxman et al., 2008; Morita & Nakano, 2002; Buck et al., 2006; Shan et al., 1999).

14.2 PETROLEUM-BASED PLASTICS: A THREAT TO THE ENVIRONMENT

Petroleum-based plastics are not biodegradable and harm the environment by Polluting oceans, freshwater bodies, and land (Gasperi et al., 2018; McCormick et al., 2014). Problems are many, and drawbacks are numerous of plastics. Nonbiodegradable material does not dissolve in environmental conditions and remains in the environment for hundreds of years after disposal (Sanyang et al.,

DOI: 10.1201/9781003534914-14

2016). The production of plastics is energy-intensive and environmentally hazardous (Azahari et al., 2011; Avérous & Pollet, 2012; Jones et al., 2013). With the increased growth in the economy, usage of plastic has increased and thus leftover to landfills. Salem et al. (2017) also reported the numerous adverse effects that influence human and animal health, environmental degradation, and sanitation. Impacts on human health by using plastics occur through three pathways: particle toxicity, chemical toxicity, and a carrier for pathogens and parasites (Vethaak & Leslie, 2016). Studies have revealed that plastic particles adversely affect the lungs and intestines. Additionally, small particles can pass through cell membranes, placentas, and blood–brain barriers to cause harm to health (Galloway, 2015). According to McCormick et al. (2014), plastic particles may facilitate the spread of pathogens and parasites. Harmful microbes, such as bacteria, may proliferate as they can grow on plastic particles by encountering wastewater from treatment plants or households. It may proliferate. Infected particles of plastic could then mix with dishwashing or drinking water and allow humans to be exposed to dangerous diseases. Microplastics (MP) have been recognized in edible fish and may enter human body by biomagnification (Alfaro-Núñez et al., 2021; Goswami et al., 2020; James et al., 2020). In species, the impairments due to MP comprised minor disruptions of biological systems to considerable adverse effects that led to morbidity (Mallik et al., 2021). Ma et al. (2020) reported the three categories of harm they identified included the physiological injury from MPs by its accumulation in the digestive system, the interference of the organisms' energy flow by MPs expulsion as pseudoforces, and internal body tissues exposed to MPs through transfer within the body. Besides, there is an entry path for biological pollutants and heavy metals into aquatic environments (Gholizadeh & Patimar, 2018). MPs may change the predatory feeding behavior of fish and induce MP-induced confusion between MPs and natural food items, thus triggering under nutrition and MP concentration in critical organs such as gills, intestine, and stomach (Lu et al., 2018; Güven et al., 2017; Greven et al., 2016). MPs have also been found in fish muscles and meat, which is the most relevant human consumable (Thiele et al., 2021; Barboza et al., 2020a, 2020b; Akhbarizadeh et al., 2018; Abbasi et al., 2018). Accumulation of MPs have been found to cause trigger in behavioral alterations, genotoxicity, hormone dis balance, neural disturbance, growth retardation, and metabolic disturbance (Güven et al., 2017; Choi et al., 2018). Plastic units or their suspended particles appear to be hazardous, leading to occupational illnesses. Inhaled particles can cause bronchial asthma, fibrosis of intestine, and granulomas and inflammatory changes in bronchial and peribronchial tissue, inflammatory, and fibrotic changes in the bronchial and peribronchial tissue, allergic reactions, and interalveolar septa lesions in humans (Beckett, 2000). Microplastic particles having size less than 2.5 μm are absorbed through endocytosis by M cells (specialized epithelial cells of the mucosa-associated lymphoid tissues) in Peyer's patches to reach the gastrointestinal tract (GIT) once ingested. The particles can either be transported from the lumen of the alimentary canal to the mucosal lymphoid tissues or transvers to the paracellular persorption. Persorption is the kneading of solid particles through the gaps placed in the single-layer epithelium at the tips of the villi of human GIT (desquamation zones) and into the circulatory system. Microplastics as well as their distinctive properties of hydrophobicity and chemical nature cause inflammation and supposed to have an accumulative toxic effect that depends on its concentration (Wright & Kelly, 2017). This was further substantiated with presence of microplastics in men at a gastro-intestinal level as well as in stools. Concentration of MP was found to be 20 microfibers, mostly PE and PP (ranging in size between 5 and 500 mm), in 10 g of stool (Harvey & Watts., 2018; Schwabl et al., 2019). Smith et al. (2018) stated that human excretory system is capable of removing up to 90% of ingested micro and nanoplastics.

14.3 BIOPLASTICS: CONCEPT AND USAGE

Bioplastics are produced from natural resources such as polysaccharides, proteins, and lipids. The more common sources of polysaccharides for bioplastic production are starch, chitin, lignin, and cellulose. Natural macromolecules include gelatine, casein, gluten, plant oils, and animal by products (Song et al., 2009). Bioplastics, derived from renewable organic material, hold great promise as an alternative to traditional plastics. They are predicted to reduce the consumption of fossil fuels and plastic wastes (bin Ahmad Shamsuddin et al., 2015). Compared to conventional plastics, bioplastics have a significantly lower carbon footprint since the carbon assimilated by the plants that produce them is only recirculated. Since environmental pollution created by nondecomposable materials is on the rise, formulation of biodegradable products for common usage is highly in demand. Starch-based bioplastics are mostly very attractive owing to their plenty availability, renewability, environment friendly, and biodegradability; however, more study is required to develop bioplastics so that they are both economically and practically viable (Shafqat et al., 2020). Many starch-rich sources have been explored for bioplastics, and considering the abundance availability of sago, its starch can also be considered for use in the development of biodegradable plastic composites.

14.4 SAGO STARCH AND ITS UTILIZATION FOR PRODUCTION OF BIODEGRADABLE THERMOPLASTIC

The biopolymer starch holds a very promising outlook. Starch owing to its abundance availability in plant kingdom, low cost, unique structure, and harmless degradation process can be regarded as the most important raw material for bioplastics. Starch is composed of two polysaccharides: amylose and amylopectin, consisting of α-D-glucopyranosyl units (French, 1984). Amylose is a linear (1-4) linked α-D-glucan, whereas amylopectin is highly branched with a short chain of (1-4) linked α-D-glucose and (1-6) linked α-linked branches (Kennedy et al., 1983). The ratio of amylose to amylopectin in the starch sources is usually 20:80 (Orford et al., 1987). Sago palm is one of the important crops that produces starch while its frequently used species are *M. longispinum, M. sylvestre, M. microcanthum, M. sanguinea,* and *M. rumphii*. Sago starch is made up of oval granules with diameters ranging from 20 to 40 μm and moisture content between 10.6% and 20.0%. The total amylose content (lipid-free starch) in sago starch ranged between 24% and 31%. Gelatinization temperature of sago starch surpasses than corn, peas, and potatoes but lower than sweet potatoes, Tania, and yam (Tian et al., 1991; Veletudie et al., 1995). These resulted in essential features of sago starch, including ease of gelatinization, high viscosity with correct extraction, and easy molding (Takahashi, 1986). As a result, results have demonstrated that sago starch has a higher potential for producing biodegradable polymers. Sago starch can also be treated into thermoplastic using heating and shearing processes. With a low decomposition temperature, granular starch calls for a plasticizer in processing. There are attractive characteristics of starch derived from renewable resources, including low cost and abundant supply, as well as environmental friendliness (Takahashi, 1986).

14.5 SAGO STARCH: COMPOSITION AND STRUCTURE

14.5.1 Composition

The functional properties of starch are determined by its chemical composition (Srichuwong & Jane, 2007). Percentage of amylose in sago starch ranges from 21.4 to 30 (Martínez et al., 2018; Uthumporn et al., 2014), but its concentration could be different based on the application of the quantitative methods (Zhu, 2017). Sago starches isolated from the four different phases of development palms have similar contents of amylose (Uthumporn et al., 2014). The minor constituents in sago starch include small amounts of protein (~0.3%), lipids (~0.1%), fiber (~0.3%), and ash (~0.4%) (Karim et al., 2008a). Phosphorus-containing compounds have been reported at intermediate concentrations, and most of the phosphorus residues are bound to C6 position of glucosyl residues in starch extracted from sago (Takeda et al., 1989).

14.5.2 Granular and Chemical Structure

Shape of sago starch granules has been reported to be spherical or ovals. Most of the particles are usually complete, while others have truncated ends. The form of the granule could be attributed to mechanical activity during the extraction process of starch (Villas-Boas et al., 2019). Sago starch granules are oval in shape, hence, resemble a temple bell image. Their size varies from 8 to 240 μm. The average diameter is 37.59 μm. The size distribution of sago starch granules was slightly narrower than potato starch, indicating more forms and, in particular, sizes of granule are uniform. The shape and dimensions of starch granules regulates the molecular structure of starches which significantly influence their physico-chemical characteristics and enzymatic digestion. The starch crystal structure is divided into types A–C, taking into consideration the ordering of amylopectin side chains and the double helices structure formation (Perez & Bertoft, 2010).

14.5.3 Amylopectin and Amylose Distribution in Sago Starch

Amylopectin consists of chains of 107–108 of molecular having α- (1,4)-linked D-glucosyl units connected by α-(1,6)-linkages, (degree of polymerization: 12,000–40,000) (Hizukuri, 2003). Conversely, the amylose is composed mainly of a linear structure of glucose units linked by α-(1,4) bonds and has 105–106 of molecular weight (degree of polymerization: 600–36,000) which is entangled in the amylopectin. Takeda et al. (1989) reported that a smaller molecule containing slightly more long chains than the high-viscosity amylopectin is the cause for low viscosity of sago starch amylopectin. The amylopectin chain of sago is consisting of four fractions with different chain lengths: Fraction I (22.9 ± 0.8%), Intermediate Fraction (3.9 ± 0.4%), Fraction II (17.2 ± 0.4%), and Fraction III (56.0 ± 1.4%) (Ishii et al., 1990). Takahashi & Hirao (1994) reported that Fraction III of sago amylopectin falls (52.3%) between the amylopectin of sweet potato and mung bean. Srichuwong et al. (2005) also found the average chain length of the amylopectin extracted from 14 kinds of sago starch samples in Fiji, Samoa, and Vanuatu and found to be 41.0%, 53.8%, and 5.2%, respectively.

14.5.4 Development of Biofilm from Sago Starch

14.5.4.1 Extraction from Bulk Sago

This entails extraction of sago starch by weighing the bulk sago, then blending it in a blender with 500 ml of water into smooth confection. Sago has 25% water and thus has to be filtered. Then, water containing sago starch must be cooked in an oven at 100°C for nearly 1 day. Sago starch is weighed in a beaker, then the sago starch and water (50 ml) is combined. Then, heat treatment is given at 70°C. Sago starch solution is stirred, and 5ml of glycerin and 5 ml of acetic acid is added while stirring. Here, glycerin acts as a plasticizer while acetic acid is the coplasticizer used to enhance the tensile strength. Stirring is continued till the solution becomes colorless and springy. Solution is poured into a petri dish and allowed for drying at room temperature for 3–4 days. Zawawi et al. (2017) has determined that this plastic is actually biodegradable by running degradability tests through Fourier Transform Infrared (FTIR) wherein the composition of alkane and alkyne of sago starch-based biodegradable polymers were also tested. After 3 days, the existence of alkane and alkyne were reduced. After 14 days, the alkane has fully depleted in the biodegradable plastic. Alkane and alkyne compositions had declined significantly. At 14th day, similar results were gained with further breakdown of the plastic. Compression test showed compressive strength up to 2.6 MPa.

14.5.4.2 Starch Extraction from Sago Pith Waste

The raw material sago pith waste is applied for the synthesis of biodegradable polymers with enhanced mechanical qualities and rapid degradation. The process requires the weighing of sago pith waste up to 1 kg, and soaking sago pith waste in clean water (2 litres) in such a way that starch solution is formed. Then, starch solutions are sundried for 12 hours and sieved through 100 size mesh (Fauziah *et al.*, 2020). Sago waste starch in different concentration levels of 5 g, 10 g, 15 g, and 20 g was mixed with 3 ml glycerin and 100 ml aqua des. In the second step, solutions were mixed at a speed of 600 rpm for 5 min with a magnetic stirrer. Solutions were heated in a magnetic stirrer at the speed of 600 rpm at 150°C for 25 min. After each mixed solution, the stock is poured into mould and dried in oven with a temperature of 120°C for 12 hours. By using this method, sago pith waste starch bioplastics was obtained. In this case, the plastic-based sago waste can absorb the most amount of water when the composition contains 20 g of starch with the value of 6.075%. The starch 10 g had the highest tensile strength and elongation with values being 2.694 N/mm^2 and 18.015%, respectively. The soil burial method of test proved that the biodegradable plastic can be destructed with more than 80% presentation after 20 days. The incorporation of a high concentration of starch would accelerate the degradation process by reducing moisture absorption. It is noted that sago pith waste starch possesses a potential to produce highly viable biodegradable plastic (Fauziah et al., 2020).

14.5.4.3 Rapid Ultrasound-Assisted Sago Starch Extraction

Tan et al. (2021) extracted sago pith waste starch from ultrasonication before converting the obtained starch into bioplastic film. Ultrasonic assisted extraction produced 71.4% sago starch of particle size <250 μm, solid loading of 10%, ultrasonic amplitude of 70%, and duty cycle of 83% in 5 minutes. The fast ultrasonic technique proved to be better than the conventional extraction (60.9% in 30 min) method. Ultrasound extraction yielded starch with higher purity since lower protein and ash concentrations were obtained compared to control starch. Ultrasound treatment brought about irregularity and disturbance of the surfaces of starch granules and therefore increased the starch swelling power, bringing about preparation for forming the solution. Finally, the extracted sago starch using ultrasound was used in making a bioplastic film with a solution casting technique. The formed bioplastic film was brown in color having tensile strength of 0.9 ± 0.1 MPa, Young's modulus of 22 ± 0.8 MPa, elongation at break of 13.6 ± 2.0%, and water vapor permeability (WVP) of $1.11 \pm 0.1 \times 10^{-8} gm^{-1}s^{-1}Pa^{-1}$. The study found that indeed, sago pith starch can be suitable candidate for bioplastic material.

14.6 PHYSICOCHEMICAL PROPERTIES OF SAGO STARCH

Sago palm fiber has strong physicochemical qualities that make it suitable for composite applications.

14.6.1 Gelatinization

The starch gelatinization process breaks the intermolecular bonding between starch molecules, and the addition of heat and water results in more water–hydrogen bonding sites, hydroxyl hydrogen groups, and oxygen bonds. DSC study showed that sago starch has higher gelatinization temperatures than A-type starches, but equal or lesser temperatures than B- and C-type starches (Vamadevan et al., 2013). Sago starch has more elevated gelatinization temperature and lower peak viscosity as compared to the A-type starches which makes it a good stabilizer (Achudan et al., 2020). Karim et al., (2008a) suggested that the sago starch possesses greater swelling ability and lesser enzyme digestibility than the maize and pea starches (Karim et al., 2008b). In comparison with the starches of corn and potato, gels of the starch possessed higher values of hardness, adhesiveness, and springiness. Pasting study indicated that the peak viscosity of sago starch was < potato starch but > cassava, sweet potato, rice, wheat, and maize starches (Karim et al., 2008b). Length of internal chains of sago amylopectin was higher than the A-type starches. Vamadevan et al. (2013) reported that the longer internal chain of amylopectin assisted in packing double helices in granules, which in turn stabilizes sago starch.

14.6.2 Retrogradation

Vamadevan et al. (2013) further pointed out that retrograded sago starch has a higher peak melting temperature than retrograded A-type starches but lower than retrograded B-type starches, due to the fact that the internal chain length of sago amylopectin is longer than that of A-type starches but shorter than that of B-type starches. The ΔH of retrograded sago starch has been found to be higher than most A-type starches but lower than most B-type starches. Studied have reported thermal properties of retrograded starch by differential scanning calorimetry (DSC). Bioavailability of starch was measured at 37°C by porcine pancreatic α-amylase. Amylose degraded freely at 5°C, whereas recrystallization of amylopectin was aided by successive storage at 5°C followed by 30°C. Higher retrogradation and melting enthalpy values and melting temperature at 37°C conferred lower susceptibility to enzymes in sago starch and its derivatives.

14.7 NUTRITIONAL PROPERTIES OF SAGO STARCH

14.7.1 Digestibility

The enzyme digestibility of sago starch is lower compared with cereal starches. According to Jane et al. (2003), native C-type starches are more resistant to enzymes compared with A-type starches but are more prone than B-type starches. Retrogradation reduces starch digestibility very significantly. Sago starch is also not readily hydrolyzed by enzymes because of its starch granule structure, which protects the amylose and amylopectin chains from enzymatic attacks. As such, native sago starch's resistant starch would likely be an RS2 (Du et al., 2020). Martínez et al. (2018) noted that storage of starch for 7 days resulted in a reduction of 65% in the rate of in vitro digestion of sago starch. Double helices and intermolecular interactions were evident in the structural property analysis based on chains consisting of over 15.5 peak DP of amylopectin units. Martínez et al. (2018) observed that double helices and intermolecular interactions inhibited the retrogradation of starch.

14.7.2 Fermentation

Research shows that the gut fermentation of resistant starch (RS) has positive health effects (Koropatkin et al., 2012). Several researchers have conducted studies on the in vitro fermentation of sago RS (Arshad et al., 2018; Purwani et al., 2012; Tan et al., 2014). Purwani et al. (2012) autoclaved native sago starch, stored it at 4°C for 12–14 hours, followed by hydrolysis with enzyme amylase and/or pullulanase to achieve a fraction of RS. The microorganisms *Eubacterium rectale* DSM 17629 and *Clostridium butyricum* BCC B2571 were used for fermentation. The *C. butyricum* BCC B2571 RS fermentation yields short chain fatty acids (acetate, propionate, and butyrate) at the molar ratio of 1.8:1:1 and 1.7:1:1.2 for E. DSM 17629 rectale. Pre-treatment of starch with pullulanase gave a significantly higher generation of short-chain fatty acid than that obtained by the treatment with amylase. Purwani et al. (2012) found that fermentation of sago RS produced a higher short chain fatty acids than rice RS. Total bacterial count and concentration of RS of retrograded sago and cassava starches were higher than that of native sago starch when fermented for 24 hours (Arshad et al., 2018). Tan et al. (2014) prepared an autoclaving and cooling cycle-based sago RS percentage. The resultant RS, which was treated with HCl solution, was utilized for the preparation of the HCl-RS fraction. HCl-RS and RS showed resistance towards hydrolysis by stomach acid and enzymes. The RS fractions were given to *Lactobacillus, Bifidobacterium, Escherichia coli, Campylobacter coli*, and *Clostridium perfringens*. The RS fractions showed a growth-promoting effect on all the microorganisms. The RSs fed to *Bifidobacterium* sp. FTDC8943 had greatest prebiotic activity.

14.8 APPLICATIONS OF SAGO STARCH

Sago starch has been utilized in a variety of food and nonfood purposes, as discussed in this section.

14.8.1 Rubber Latex-Based Composite Films

The addition of polysaccharides like sago starch to natural rubber (NR) latex composites may further improve the biodegradability of the films. The dispersing agents added to sago starch to enhance dispersion may affect the rate of decomposition in NR latex films. Sodium alginate and carboxymethyl cellulose (CMC) are amphiphilic polymers based on polysaccharides and short-chain molecule surfactants. This is different from the more commonly used dispersing agent, anchoid. Polysaccharide group blended with the dispersing agent of the sago starch dispersion of NR latex sheets tends to increase the rate of degradation, whereas the time to disintegrate is reduced. The hydrophilicity of sodium alginate and CMC may be another reason why it can be regarded as an effective dispersing agent. In addition, sodium alginate and CMC can gelate and enhance the viscosity of NR latex films by forming hydrocolloids. Afiq and Azura (2013) blended natural rubber latex with native sago starch up to 25 phr. The higher percentage of starch lowered tensile strength and elongation at break (%), and improved tear strength. Addition of sago starch also encouraged film degradation in soil. Kiing et al. (2013) carried out the gelatinization of acetylated and hydroxypropylated sago starches followed by blending with natural rubber latex and glycerol to cast films. Acetylation showed superior interaction of starch with rubber latex that resulted in higher adhesion. The surface of casted films was uniform and homogenous in nature. Rubber latex has decreased the adsorption of water and degradability of films in the natural environment. Addition of starch copolymers enhanced colloidal stability, reduced tensile strength, increased elongation at break, and imparted temperature stability to the films made up from latex. Misman et al. (2015) reported that mixing of starch copolymer into latex films enhanced the aging rate.

14.8.2 Low-Density Polyethylene (LDPE)/Polyvinyl Alcohol (PVA) Based Composite Films

The low-cost, easily processable plastics containing polystyrene, polypropylene, polyethylene, polyvinyl chloride, and polyethylene terephthalate are mechanically hard and strong, but unfortunately cannot biodegrade and hence cause adverse effect to the eco system (Kiing et al., 2013). Starch addition in plastic film formulations minimizes environmental degradation (Beg et al., 2016; Ismail & Zaaba, 2014; Mohamed et al., 2018; Sarifuddin et al., 2014). Melt-blended native sago starch at 5-30% by weight with low-density polyethylene (LDPE) was an additive in injection moulding to produce a composite material. The presence of starch decreased the melt flow rate, tensile strength, and impact strength of the composite but increased the flexural strength, tensile modulus, and flexural modulus. Studies have used twin-screw extruder for the manufacture of a PVA/native sago starch composite plasticized with sorbitol and epoxidized soybean oil (ESO). The final composite was exposed to the outside environment for 2 weeks. ESO resulted in improved elongation at break, tensile strength, and modulus of the composite compared to sorbitol both before and after exposure (Mohamed et al., 2018).

14.8.3 Starch-Gelatin-Based Edible Films

Edible films have been prepared by mixing sago starch with gelatine of fish and glycerol for film synthesis (Al-Hassan & Norziah, 2012; Nafchi et al., 2013).

14.8.4 Formation of Nanoparticle/Nanofiber

"Green" starch nanoparticles (SNPs) can be fabricated directly from sago (Metroxylon Sagu) starch granules using high-pressure homogenization. This is eco-friendly, highly feasible, low-cost, chemical-free method of synthesis, and capable of producing bulk amount. Chin et al. (2012) synthesized fluorescent carbon nanodots (50–80 nm) through the carbonization of sago SNPs followed by oxidation of the surface. In addition, native sago starch was employed for silver nanoparticles synthesis as the capping agent that capsuled in proven controlled release capability for antibacterial and biomedical uses. Hunyek et al. (2017) utilized starch as a bonding agent for synthesis through green sol–gel of $CoFe_2O_4$ nanocomposites with different nanoparticle sizes in the range of 30 nm–500 nm based on metal nitrate and starch mass ratio. Tay et al (2012) and Yusof et al. (2019) also found that starch is used to develop nanoparticles for various applications. Yusof et al. (2019) fabricated nanofibers by electrospinning using up to 20% carboxymethyl sago starch in Poly-L-lactic acid (PLLA). Nanofibers with less carboxymethyl starch had a more uniform shape and were free of beads. Composite nanofibers prepared from carboxymethyl starch poly (L-lactic acid) have significant biomedical applications (Yusof et al., 2019).

14.8.5 Pharmaceutical and Medical Applications

Poeloengasih and Anggraeni (2014) cast native sago starch in glycerol for making films. Tensile strength decreased with an increase in the percentage of glycerol; however, elongation at break and water vapor penetrability increased. The amylose content in the starch increased tensile strength but reduced elongation at break. Poeloengasih and Anggraeni (2014) reported water solubility as low for the films, thereby making them suitable for the application of hard capsules. Widodo and Hassan (2015) employed pregelatinized sago starch as part of tablet formulations. For a longer time of pregelatinization, it was observed that the sago starch had higher compressibility in comparison of maize-derived commercial product Spress® B820. Wound dressing applications of biomaterials from sago starch were also prepared by Pushpamalar et al. (2013) and Ramnath et al. (2012). Ramnath et al. (2012) cross-linked sago starch (SG), collagen (C), and soya protein (SY) into composite films by glutaraldehyde for wound dressing applications. It has been seen that the inclusion of SG, SY, and glutaraldehyde found to improve the mechanical strength and texture smoothness of the developed bio film.

14.8.6 Food and Feed Product Formulations

Javanmard et al. (2012) attempted to use hot starch sago for thickening mango jam. Addition of starch at higher concentration increased the viscosity, storage (G′) and loss (G'') moduli, of jam and its hardness. Sago starch was used as a thickening agent in porridge (Syahariza & Yong, 2017). The rheological characteristics of rice porridge prepared using 1–3% sago starch were compared with cassava starch and commercial thickeners. The results revealed that the sago starch yielded thicker and more elastic rice porridge than that of the cassava starch. According to Syahariza and Yong (2017), the sago starch can be a thickener in preparing rice porridge for dysphagic individuals. Sago starch had been also used in preparation of aquafeed (Umar et al., 2013). Extrusion was used to prepare nitrogenous feed with variable starch and moisture percentages from sago to cassava (20:0 to 0:20). The feed with 40% moisture content had the highest floating capacity.

14.8.7 Ceramic Foam

Thermolysis of poly organo borosil zane with native sago starch resulted in Si–B–C–N foams (Figure 14.1) (Kousaalya et al., 2015). The starch present was entirely destroyed at 850°C. Increasing starch concentration declined foam density, ranging from 0.4 to 0.25 g/cm^3 (Kousaalya et al., 2015).

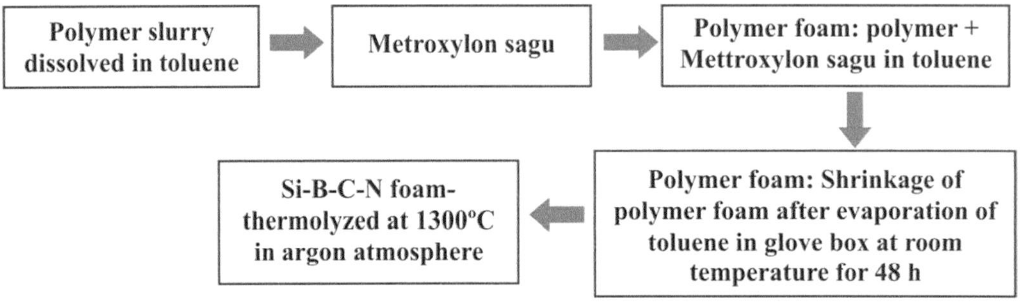

FIGURE 14.1 Flow diagram for formulation of Si–B–C–N foam with sago starch (Kousaalya et al., 2015).

14.9 SAGO STARCH VERSUS OTHER STARCHES

Sago starch is abundant as well as inexpensive (Karim et al., 2008a). Manufacturing cost of sago starch is lower compared to other sources of starch with larger yield. After 8 years of growth cycle, sago can produce up to 25 tons of starch with air dry per hectare. Sago produces more starch than other crops even including maize, rice, and cassava. Native sago palm starch is among the most productive starchy crops worldwide. The intermediate features of its structural and physicochemical characteristics were reportedly the nature of native sago starch compared to A-type and B-type starches. According to Martínez et al. (2018), in terms of susceptibility to enzyme during retrogradation, sago starch outperformed other 12 starches. Some technical applications might be found in the qualities of sago starch. Growing sago palms has a huge environmental advantage. The palm tree's fibrous roots absorb heavy metals, pollutants, and feces while holding the silt. Global warming will eventually expand the region where sago palms thrive. This would make the sago palm more resilient. Sago forests effectively serve as carbon sinks for sequestration. This minimizes the greenhouse gases and solves problem of climate change.

14.10 CONCLUSION

In conclusion, the utilization of sago starch for the production of biodegradable plastics presents a promising and sustainable alternative to conventional petrochemical-based polymers. Sago starch, a renewable and abundant resource, offers several advantages including its biodegradability, reduced environmental impact, and potential for lowering production costs. The incorporation of sago starch into plastic materials aligns with the growing need for eco-friendly solutions that address the environmental challenges posed by plastic waste. Advances in material science and processing technologies continue to enhance the performance and applicability of sago starch-based plastics, making them viable for a varied and diverse range of applications from packaging to agricultural films. However, further research is needed to optimize the properties of sago starch-based plastics, such as their mechanical strength, thermal stability, and shelf life, to ensure they meet the demands of various industries. Additionally, addressing the scalability of production processes and evaluating the long-term environmental impacts are crucial for the widespread adoption of these materials. By continuing to explore and invest in sago starch-based biodegradable plastics, we can contribute to a more sustainable future and mitigate the adverse effects of plastic on the earth.

REFERENCES

Abbasi, S., Soltani, N., Keshavarzi, B., Moore, F., Turner, A., & Hassanaghaei, M. (2018). Microplastics in different tissues of fish and prawn from the Musa Estuary, PersianGulf. *Chemosphere*, *205*, 80–87. https://doi.org/10.1016/j.chemosphere.2018.04.076

Achudan, S. N., Dos Mohamed, A. M., Abd Rashid, R. S., & Mittis, P. (2020). Yield and physicochemical properties of starch at different sago palm stages. *Materials Today: Proceedings*, *31*, 122–126. https://doi.org/10.1016/j.matpr.2020.01.341

Afiq, M. M., & Azura, A. R. (2013). Effect of sago starch loadings on soil decomposition of natural rubber latex (NRL) composite films mechanical properties. *International Biodeterioration & Biodegradation*, *85*, 139–149. https://doi.org/10.1016/j.ibiod.2013.06.016

Akhbarizadeh, R., Moore, F., & Keshavarzi, B. (2018). Investigating a probable relationship between microplastics and potentially toxic elements in fish muscles from northeast of Persian Gulf. *Environmental Pollution*, *232*, 154–163. https://doi.org/10.1016/j.envpol.2017.09.028

Alfaro-Núñez, A., Astorga, D., Cáceres-Farías, L., Bastidas, L., Soto Villegas, C., Macay, K., & Christensen, J. H. (2021). Microplastic pollution in seawater and marine organisms across the Tropical Eastern Pacific and Galápagos. *Scientific Reports*, *11* (1), 6424.

Al-Hassan, A. A., & Norziah, M. H. (2012). Starch–gelatin edible films: Water vapor permeability and mechanical properties as affected by plasticizers. *Food Hydrocolloids*, *26* (1), 108–117. https://doi.org/10.1016/j.foodhyd.2011.04.015

Al-Salem, S. M., Antelava, A., Constantinou, A., Manos, G., & Dutta, A. (2017). A review on thermal and catalytic pyrolysis of plastic solid waste (PSW). *Journal of Environmental Management*, *197*, 177–198. https://doi.org/10.1016/j.jenvman.2017.03.084

Arshad, N. H., Zaman, S. A., Rawi, M. H., & Sarbini, S. R. (2018). Resistant starch evaluation and in vitro fermentation of lemantak (native sago starch), for prebiotic assessment. *International Food Research Journal*, *25* (3), 951–957.

Avérous, L., & Pollet, E. (2012). *Environmental silicate nano-biocomposites* (Vol. 1). London, UK: Springer.

Azahari, N. A., Othman, N., & Ismail, H. (2011). Biodegradation studies of polyvinyl alcohol/corn starch blend films in solid and solution media. *Journal of Physical Science*, *22* (2), 15–31.

Barboza, L. G. A., Cunha, S. C., Monteiro, C., Fernandes, J. O., & Guilhermino, L. (2020a). Bisphenol A and its analogs in muscle and liver of fish from the North East Atlantic Ocean in relation to microplastic contamination. Exposure and risk to human consumers. *Journal of Hazardous Materials*, *393*, 122419. https://doi.org/10.1016/j.jhazmat.2020.122419

Barboza, L. G. A., Lopes, C., Oliveira, P., Bessa, F., Otero, V., Henriques, B., ... & Guilhermino, L. (2020b). Microplastics in wild fish from North East Atlantic Ocean and its potential for causing neurotoxic effects, lipid oxidative damage, and human health risks associated with ingestion exposure. *Science of the Total Environment*, *717*, 134625. https://doi.org/10.1016/j.jhazmat.2020.122419

Beckett, W. S. (2000). Occupational respiratory diseases. *New England Journal of Medicine*, *342* (6), 406–413. doi:10.1056/NEJM200002103420607

Beg, M. D. H., Kormin, S., Bijarimi, M., & Zaman, H. U. (2016). Preparation and characterization of low-density polyethylene/thermoplastic starch composites. *Advances in Polymer Technology*, *35* (1). https://doi.org/10.1002/adv.21521

bin Ahmad Shamsuddin, D. M., Chew, B. C., & Shin, L. H. (2015). Quality function deployment for bio plastics adoption in Malaysian industry. *Journal of Technology Management and Business*, *2* (2). https://publisher.uthm.edu.my/ojs/index.php/jtmb/article/view/1139

Buck, C. B., Thompson, C. D., Roberts, J. N., Müller, M., Lowy, D. R., & Schiller, J. T. (2006). Carrageenan is a potent inhibitor of papillomavirus infection. *PLoS Pathogens*, *2* (7), e69. https://doi.org/10.1371/journal.ppat.0020069

Chin, S. F., Yazid, S. N. A. M., Pang, S. C., & Ng, S. M. (2012). Facile synthesis of fluorescent carbon nanodots from starch nanoparticles. *Materials Letters*, *85*, 50–52. https://doi.org/10.1016/j.matlet.2012.06.082

Choi, J. S., Jung, Y. J., Hong, N. H., Hong, S. H., & Park, J. W. (2018). Toxicological effects of irregularly shaped and spherical microplastics in a marine teleost, the sheepshead minnow (*Cyprinodon variegatus*). *Marine Pollution Bulletin*, *129* (1), 231–240. https://doi.org/10.1016/j.marpolbul.2018.02.039

Du, C., Jiang, F., Jiang, W., Ge, W., & Du, S. K. (2020). Physicochemical and structural properties of sago starch. *International Journal of Biological Macromolecules*, *164*, 1785–1793. https://doi.org/10.1016/j.ijbiomac.2020.07.310

Fauziah, A. R., Lanuru, M., Syahrul, M., Werorilangi, S., & Demmallino, E. B. (2020). Utilization of solid waste from sago flour industry (sago pith waste) as biodegradable plastic. *Advances in Environmental Biology*, *14* (1), 42–48.

French, D. (1984). Organization of starch granules. In *Starch: Chemistry and technology* (pp. 183–247). Academic Press. https://doi.org/10.1016/B978-0-12-746270-7.50013-6

Galloway, T. S. (2015). Micro-and nano-plastics and human health. *Marine Anthropogenic Litter*, 343–366.

Gasperi, J., Dris, R., Rocher, V., & Tassin, B. (2018). Microplastics in the continental area: An emerging challenge. *Current Opinion in Environmental Science & Health*, *1*, 1–5. doi: 10.1016/j.coesh.2017.10.002

Gholizadeh, M., & Patimar, R. (2018). Ecological risk assessment of heavy metals in surface sediments from the Gorgan Bay, Caspian Sea. *Marine Pollution Bulletin*, *137*, 662–667. https://doi.org/10.1016/j.marpolbul.2018.11.009

Goswami, P., Vinithkumar, N. V., & Dharani, G. (2020). First evidence of microplastics bioaccumulation by marine organisms in the Port Blair Bay, Andaman Islands. *Marine Pollution Bulletin*, *155*, 111163. https://doi.org/10.1016/j.marpolbul.2020.111163

Greven, A. C., Merk, T., Karagöz, F., Mohr, K., Klapper, M., Jovanović, B., & Palić, D. (2016). Polycarbonate and polystyrene nanoplastic particles act as stressors to the innate immune system of fathead minnow (*Pimephales promelas*). *Environmental Toxicology and Chemistry*, *35* (12), 3093–3100. https://doi.org/10.1002/etc.3501

Güven, O., Gökdağ, K., Jovanović, B., & Kıdeyş, A. E. (2017). Microplastic litter composition of the Turkish territorial waters of the Mediterranean Sea, and its occurrence in the gastrointestinal tract of fish. *Environmental Pollution*, *223*, 286–294. https://doi.org/10.1016/j.envpol.2017.01.025

Harvey, F., & Watts, J. (2018). Microplastics found in human stools for the first time. *The Guardian*, 22. www.theguardian.com/environment/2018/oct/22/microplastics-found-in-human-stools-for-the-first-time

Hizukuri, S. (2003). Molecular structure of starch. *Encyclopedia of starch science. Asakura Shoten, Tokyo*, 11–38.

Hunyek, A., Sirisathitkul, C., Mahaphap, C., Boonyang, U., & Tangwatanakul, W. (2017). Sago starch: Chelating agent in Sol-gel synthesis of cobalt ferrite nanoparticles. *Journal of the Australian Ceramic Society*, *53*, 173–176. https://doi.org/10.1007/s41779-017-0022-1

Industries 2 (pp. 430–431). Oxford: Pergamon Press

Ishii, Y., Nagashima, N., Kawabata, A., & Nakamura, M. (1990). A study of chain length distribution in amylopectins from tropical starches. *Journal of Applied Glycoscience*, *37* (3), 251–257. https://doi.org/10.5458%2Fjag1972.37.251

Ismail, H., & Zaaba, N. F. (2014). Effects of poly (vinyl alcohol) on the performance of sago starch plastic films. *Journal of Vinyl and Additive Technology*, *20* (2), 72–79. https://doi.org/10.1002/vnl.21348

James, K., Vasant, K., Padua, S., Gopinath, V., Abilash, K. S., Jeyabaskaran, R., ... & John, S. (2020). An assessment of microplastics in the ecosystem and selected commercially important fishes off Kochi, south eastern Arabian Sea, India. *Marine Pollution Bulletin*, *154*, 111027. https://doi.org/10.1016/j.marpolbul.2020.111027

Jane, J. L., Ao, Z., Duvick, S. A., Wiklund, M., Yoo, S. H., Wong, K. S., & Gardner, C. (2003). Structures of amylopectin and starch granules: How are they synthesized?. *Journal of Applied Glycoscience*, *50* (2), 167–172. https://doi.org/10.5458/jag.50.167

Javanmard, M., Chin, N. L., Mirhosseini, S. H., & Endan, J. (2012). Characteristics of gelling agent substituted fruit jam: studies on the textural, optical, physicochemical and sensory properties. *International Journal of Food Science and Technology*, 47(9), 1808–1818.

Jones, A., Zeller, M. A., & Sharma, S. (2013). Thermal, mechanical, and moisture absorption properties of egg white protein bioplastics with natural rubber and glycerol. *Progress in Biomaterials*, *2*, 1–13.

Karim, A. A., Nadiha, M. Z., Chen, F. K., Phuah, Y. P., Chui, Y. M., & Fazilah, A. (2008a). Pasting and retrogradation properties of alkali-treated sago (*Metroxylon sagu*) starch. *Food Hydrocolloids*, *22* (6), 1044–1053. https://doi.org/10.1016/j.foodhyd.2007.05.011

Karim, A. A., Tie, A. P. L., Manan, D. M. A., & Zaidul, I. S. M. (2008b). Starch from the sago (*Metroxylon sagu*) palm tree—properties, prospects, and challenges as a new industrial source for food and other

uses. *Comprehensive Reviews in Food Science and Food Safety*, *7* (3), 215–228. https://doi.org/10.1111/j.1541-4337.2008.00042.x
Kennedy, J. F., Griffiths, A. J., & Atkins, D. P. (1983). The application of hydrocolloids: Recent development, future trends. *Gums and Stabilisers for Food Industries*, *2*, 430–431.
Kiing, S. C., Dzulkefly, K., & Yiu, P. H. (2013). Characterization of biodegradable polymer blends of acetylated and hydroxypropylated sago starch and natural rubber. *Journal of Polymers and the Environment*, *21*, 995–1001. https://doi.org/10.1007/s10924-013-0576-0
Koropatkin, N. M., Cameron, E. A., & Martens, E. C. (2012). How glycan metabolism shapes the human gut microbiota. *Nature Reviews Microbiology*, *10* (5), 323–335. https://doi.org/10.1038/nrmicro2746
Kousaalya, A. B., Kumar, R., & Sridhar, B. T. N. (2015). Thermal conductivity of precursor derived Si–B–C–N ceramic foams using Metroxylon sagu as sacrificial template. *Ceramics International*, *41*(1), 1163–1170.
Lu, K., Qiao, R., An, H., & Zhang, Y. (2018). Influence of microplastics on the accumulation and chronic toxic effects of cadmium in zebrafish (*Danio rerio*). *Chemosphere*, *202*, 514–520. https://doi.org/10.1016/j.chemosphere.2018.03.145
Ma, H., Pu, S., Liu, S., Bai, Y., Mandal, S., & Xing, B. (2020). Microplastics in aquatic environments: Toxicity to trigger ecological consequences. *Environmental Pollution*, *261*, 114089. https://doi.org/10.1016/j.envpol.2020.114089
MacArtain, P., Gill, C. I., Brooks, M., Campbell, R., & Rowland, I. R. (2007). Nutritional value of edible seaweeds. *Nutrition Reviews*, *65* (12), 535–543.
Mallik, A., Xavier, K. M., Naidu, B. C., & Nayak, B. B. (2021). Ecotoxicological and physiological risks of microplastics on fish and their possible mitigation measures. *Science of the Total Environment*, *779*, 146433. https://doi.org/10.1016/j.scitotenv.2021.146433
Martinez, M. M., Li, C., Okoniewska, M., Mukherjee, I., Vellucci, D., & Hamaker, B. (2018). Slowly digestible starch in fully gelatinized material is structurally driven by molecular size and A and B1 chain lengths. *Carbohydrate Polymers*, *197*, 531–539. https://doi.org/10.1016/j.carbpol.2018.06.021
Martínez, M. M., Román, L., & Gómez, M. (2018). Implications of hydration depletion in the in vitro starch digestibility of white bread crumb and crust. *Food Chemistry*, *239*, 295–303. https://doi.org/10.1016/j.foodchem.2017.06.122
McCormick, A., Hoellein, T. J., Mason, S. A., Schluep, J., & Kelly, J. J. (2014). Microplastic is an abundant and distinct microbial habitat in an urban river. *Environmental Science & Technology*, *48* (20), 11863–11871. https://doi.org/10.1021/es503610r
www.ilvo.vlaanderen.be/Portals/68/documents/Mediatheek/Mededelingen/177_microplastics.pdf
Misman, M. A., Azura, A. R., & Hamid, Z. A. A. (2015). The physical and degradation properties of starch-graft-acrylonitrile/carboxylated nitrile butadiene rubber latex films. *Carbohydrate Polymers*, *128*, 1–10. https://doi.org/10.1016/j.carbpol.2015.04.004
Mohamed, R., Mutalib, N. W. A., Norizan, M. N., Hirzin, R. S. N., & Isa, S. A. M. (2018, July). Effect of different plasticizers on tensile properties of PVA/sago starch system before and after weathering exposure. In *AIP Conference Proceedings* (Vol. 1985, No. 1). AIP Publishing. https://doi.org/10.1063/1.5047163
Morita, K., & Nakano, T. (2002). Seaweed accelerates the excretion of dioxin stored in rats. *Journal of Agricultural and Food Chemistry*, *50* (4), 910–917. https://doi.org/10.1021/jf0111920
Nafchi, A. M., Nassiri, R., Sheibani, S., Ariffin, F., & Karim, A. A. (2013). Preparation and characterization of bionanocomposite films filled with nanorod-rich zinc oxide. *Carbohydrate polymers*, *96* (1), 233–239. https://doi.org/10.1016/j.carbpol.2013.03.055
Orford, P. D., Ring, S. G., Carroll, V., Miles, M. J., & Morris, V. J. (1987). The effect of concentration and botanical source on the gelation and retrogradation of starch. *Journal of the Science of Food and Agriculture*, *39* (2), 169–177. https://doi.org/10.1002/jsfa.2740390210
Paxman, J. R., Richardson, J. C., Dettmar, P. W., & Corfe, B. M. (2008). Daily ingestion of alginate reduces energy intake in free-living subjects. *Appetite*, *51* (3), 713–719. https://doi.org/10.1016/j.appet.2008.06.013
Pérez, S., & Bertoft, E. (2010). The molecular structures of starch components and their contribution to the architecture of starch granules: A comprehensive review. *Starch-Stärke*, *62* (8), 389–420. https://doi.org/10.1002/star.201000013
Poeloengasih, C. D., & Anggraeni, F. D. (2014). Exploring the characteristics of sago starch films for pharmaceutical application. *Starch-Stärke*, *66* (11–12), 1103–1108. https://doi.org/10.1002/star.201400017

Purwani, E. Y., Purwadaria, T., & Suhartono, M. T. (2012). Fermentation RS3 derived from sago and rice starch with *Clostridium butyricum* BCC B2571 or *Eubacterium rectale* DSM 17629. *Anaerobe*, *18* (1), 55–61.

Pushpamalar, V., Langford, S. J., Ahmad, M., Hashim, K., & Lim, Y. Y. (2013). Preparation of carboxymethyl sago pulp hydrogel from sago waste by electron beam irradiation and swelling behavior in water and various pH media. *Journal of Applied Polymer Science*, *128* (1), 451–459. https://doi.org/10.1002/app.38192

Ramnath, V., Sekar, S., Sankar, S., Sankaranarayanan, C., & Sastry, T. P. (2012). Preparation and evaluation of biocomposites as wound dressing material.

Sanyang, M. L., Sapuan, S. M., Jawaid, M., Ishak, M. R., & Sahari, J. (2016). Effect of plasticizer type and concentration on physical properties of biodegradable films based on sugar palm (*Arenga pinnata*) starch for food packaging. *Journal of Food Science and Technology*, *53*, 326–336.

Sarifuddin, N., Ismail, H., & Ahmad, Z. (2014). Incorporation of kenaf core fibers into low density polyethylene/thermoplastic sago starch blends exposed to natural weathering. *Molecular Crystals and Liquid Crystals*, *603* (1), 180–193. https://doi.org/10.1080/15421406.2014.976516

Schwabl, P., Köppel, S., Königshofer, P., Bucsics, T., Trauner, M., Reiberger, T., & Liebmann, B. (2019). Detection of various microplastics in human stool: A prospective case series. *Annals of Internal Medicine*, *171* (7), 453–457. https://doi.org/10.7326/M19-0618

Shafqat, A., Tahir, A., Mahmood, A., Tabinda, A. B., Yasar, A., & Pugazhendhi, A. (2020). A review on environmental significance carbon foot prints of starch based bio-plastic: A substitute of conventional plastics. *Biocatalysis and Agricultural Biotechnology*, *27*, 101540. https://doi.org/10.1016/j.bcab.2020.101540

Shamsuddin, R. M., Verbeek, C. J. R., & Lay, M. C. (2015). Protein-intercalated bentonite as nano-reinforcement in bloodmeal-based bioplastics. *Journal of Polymer Materials, 32*(4), 435–449.

Shan, B. E., Yoshida, Y., Kuroda, E., & Yamashita, U. (1999). Brief communication immunomodulating activity of seaweed extract on human lymphocytes in vitro. *International Journal of Immunopharmacology*, *21* (1), 59–70. https://doi.org/10.1016/S0192-0561 (98)00063-0

Smith, M., Love, D. C., Rochman, C. M., & Neff, R. A. (2018). Microplastics in seafood and the implications for human health. *Current Environmental Health Reports*, 5, 375–386.

Song, J. H., Murphy, R. J., Narayan, R., & Davies, G. B. H. (2009). Biodegradable and compostable alternatives to conventional plastics. *Philosophical Transactions of the Royal Society B: Biological Sciences*, *364* (1526), 2127–2139.

Srichuwong, S., & Jane, J. I. (2007). Physicochemical properties of starch affected by molecular composition and structures: A review. *Food Science and Biotechnology*, *16* (5), 663–674.

Srichuwong, S., Sunarti, T. C., Mishima, T., Isono, N., & Hisamatsu, M. (2005). Starches from different botanical sources I: Contribution of amylopectin fine structure to thermal properties and enzyme digestibility. *Carbohydrate Polymers*, *60* (4), 529–538.

Syahariza, Z. A., & Yong, H. Y. (2017). Evaluation of rheological and textural properties of texture-modified rice porridge using tapioca and sago starch as thickener. *Journal of Food Measurement and Characterization*, *11*(4), 1586–1591.

Takahashi, S. (1986). Some useful properties of sago starch in cookery science. In *The 3rd International Sago Symposium, Sago-'85, Tokyo, Japan. May 20-23.*

Takahashi, S., & Hirao, K. (1994). Studies on the physical and the chemical properties of sago starch, and its performance when used in the preparation of Japanese sweets. *Bull Facul Home Econ, Kyoritsu Women's University*, *40*, 59–64.

Takeda, Y., Takeda, C., Suzuki, A., & Hizukuri, S. (1989). Structures and properties of sago starches with low and high viscosities on amylography. *Journal of Food Science*, *54* (1), 177–182. https://doi.org/10.1111/j.1365-2621.1989.tb08596.x

Tan, S. X., Andriyana, A., Lim, S., Ong, H. C., Pang, Y. L., & Ngoh, G. C. (2021). Rapid ultrasound-assisted starch extraction from sago pith waste (SPW) for the fabrication of sustainable bioplastic film. *Polymers*, *13* (24), 4398. https://doi.org/10.3390/polym13244398

Tan, X. W., Romainor, A. N. B., Chin, S. F., & Ng, S. M. (2014). Carbon dots production via pyrolysis of sago waste as potential probe for metal ions sensing. *Journal of Analytical and Applied Pyrolysis*, *105*, 157–165. https://doi.org/10.1016/j.jaap.2013.11.001

Tay, S. H., Pang, S. C., & Chin, S. F. (2012). A facile approach for controlled synthesis of hydrophilic starch-based nanoparticles from native sago starch. *Starch-Stärke*, *64* (12), 984–990. https://doi.org/10.1002/star.201200056

Teas, J., Baldeón, M. E., Chiriboga, D. E., Davis, J. R., Sarriés, A. J., & Braverman, L. E. (2009). Could dietary seaweed reverse the metabolic syndrome?. *Asia Pacific Journal of Clinical Nutrition, 18* (2), 145–154.

Thakur, V. K. and Thakur, M. K. (2016). Handbook of Sustainable Polymers. Pan Stanford Publishing, Singapore. pp. 212–245.

Thiele, C. J., Hudson, M. D., Russell, A. E., Saluveer, M., & Sidaoui-Haddad, G. (2021). Microplastics in fish and fishmeal: An emerging environmental challenge?. *Scientific Reports, 11* (1), 2045.

Tian, S. J., Rickard, J. E., & Blanshard, J. M. V. (1991). Physicochemical properties of sweet potato starch. *Journal of the Science of Food and Agriculture, 57* (4), 459–491. https://doi.org/10.1002/jsfa.2740570402

Umar, S., Kamarudin, M. S., & Ramezani-Fard, E. (2013). Physical properties of extruded aquafeed with a combination of sago and tapioca starches at different moisture contents. *Animal Feed Science and Technology, 183* (1-2), 51–55.

Uthumporn, U., Wahidah, N., & Karim, A. A. (2014, August). Physicochemical properties of starch from sago (*Metroxylon sagu*) palm grown in mineral soil at different growth stages. In *IOP Conference Series: Materials Science and Engineering* (Vol. 62, No. 1, p. 012026). IOP Publishing.

Valetudie, J. C., Guadeloupe, L., Colonna, P., Bouchet, B., & Gallant, D. J. (1995). Gelatinization of sweet potato, tania and yam tuber starches. *Starch-Stärke, 47* (8), 298–306. https://doi.org/10.1002/star.19950470804

Vamadevan, V., Bertoft, E., & Seetharaman, K. (2013). On the importance of organization of glucan chains on thermal properties of starch. *Carbohydrate Polymers, 92* (2), 1653–1659. https://doi.org/10.1016/j.carbpol.2012.11.003

Vethaak, A. D., & Leslie, H. A. (2016). Plastic debris is a human health issue. *Journal: Environmental Science & Technology, 50* (13), 6825–6826.

Villas-Boas, F., Yamauti, Y., Moretti, M. M., & Franco, C. M. (2019). Influence of molecular structure on the susceptibility of starch to α-amylase. *Carbohydrate Research, 479*, 23–30. https://doi.org/10.1016/j.carres.2019.05.001

Widodo, R. T., & Hassan, A. (2015). Compression and mechanical properties of directly compressible pregelatinized sago starches. *Powder Technology, 269*, 15–21.

Worm, B., Lotze, H. K., Jubinville, I., Wilcox, C., & Jambeck, J. (2017). Plastic as a persistent marine pollutant. *Annual Review of Environment and Resources, 42* (1), 1–26.

Wright, S. L., & Kelly, F. J. (2017). Plastic and human health: A micro issue?. *Environmental Science & Technology, 51* (12), 6634–6647.

Yamori, Y., Miura, A., & Taira, K. (2001). Implications from and for food cultures for cardiovascular diseases: Japanese food, particularly Okinawan diets. *Asia Pacific Journal of Clinical Nutrition, 10* (2), 14 https://doi.org/10.1111/j.1440-6047.2001.00227.x

Yusof, M. R., Shamsudin, R., Zakaria, S., Abdul Hamid, M. A., Yalcinkaya, F., Abdullah, Y., & Yacob, N. (2019). Fabrication and characterization of carboxymethyl starch/poly (l-lactide) acid/β-tricalcium phosphate composite nanofibers via electrospinning. *Polymers, 11* (9), 1468. https://doi.org/10.3390/polym11091468

Zawawi, Z. A. M., Akom, N. F., Dose, D., Syauwye, A., Ahmad, R. A., & Yusoff, E. (2017). Biodegradable plastic from sago starch. *Journal Mechanical Engineering Department Politeknik Kuching Serawak*, 1 (1), 46–54.

Zhu, F. (2017). Plasma modification of starch. *Food Chemistry, 232*, 476–486.

15 Marine-Derived Biopolymers as Potential Bioplastics in Food Sectors

R. Rajam, R. Sai Nandhini, P. Karthik, and T. Venkateshwari

15.1 INTRODUCTION

With the increase in population worldwide, plastics have become essential in today's culture and gained a stable market due to their strength, lightweight, affordability, durability, and versatility. The synthetic polymer-based plastic packaging sector is flourishing continuously because of its special qualities, which include low production costs and variety in its application and usage. The global plastic packaging market was worth around $384.76 billion in 2023, and it is expected to grow to around $531.27 billion by 2032. Plastics are extensively used in industries because they are inexpensive and easy to extrude and mould. The food packaging sector is the largest consumer of plastics, and it contributes huge plastic waste (Sharma et al., 2023). The global food packaging market alone generated $400.2787 billion in revenue in 2024. This market is expected to grow to $562.3227 billion by 2030. Urbanization and shifts in dietary habits, increased internet usage, and e-commerce preferences by consumers have increased the use of plastic packaging. Modern consumers lifestyles are mostly characterized by convenience, particularly plastic packaging enabled by offering compact, lightweight, and easily accessible solutions. Foods and food products like single-serve snacks, juices, and beverages vastly utilize the benefits of plastic packaging's durability and simplicity (Sharma et al., 2023). In 2023, global plastics production reached 413.8 million metric tonnes, and roughly half of this volume is thrown after a single use. It is anticipated that the quantity able to produce bioplastics worldwide will rise from 2.18 million metric tons in 2023 to 7.43 million metric tons in 2028.

Plastics are produced from by-products of fossil fuels, and it takes hundreds of years to decompose due to poor degradability. Nondegradable nature and improper disposal system of plastic waste damage the terrestrial and marine ecosystems and endanger human and wildlife health. Conventional plastic disposal techniques like incineration emit harmful chemicals and greenhouse gases into the atmosphere contributing to air pollution and affecting the climate (Pathak et al., 2023). Hence, alternative approaches to reducing plastic pollution like recycling, chemical treatment, and landfill disposal also appear to be ineffective (Ayala et al., 2024a). To phase out the nondegradable conventional plastics, massive production of biodegradable plastics known as 'bioplastics' is presently the top priority in the packaging sector. Materials that spontaneously decompose or break down into carbon dioxide and water due to the action of microorganisms in soil or water are referred to be 'biodegradable' (Ghasemlou et al., 2024). Biodegradable polymers are not always biobased, and plastics derived from bio-based sources do not always exhibit biodegradability. Thus, 'Bioplastics' constitute a broad kind of materials that are bio-based or bio-degradable, or both. Bioplastics are new generation plastics materials that are either entirely or partially derived from renewable biomass.

Bioplastics are classified according to their biological origin or degradation capacity (Figure 15.1). The bioplastics such as biopolyethylene terephthalate (bio-PET), biopolypropylene (bio-PP), biopolyvinyl chloride (bio-PVC), biopolyethylene (bio-PE), and biopolyamides (bio-PAs) are

 DOI: 10.1201/9781003534914-15

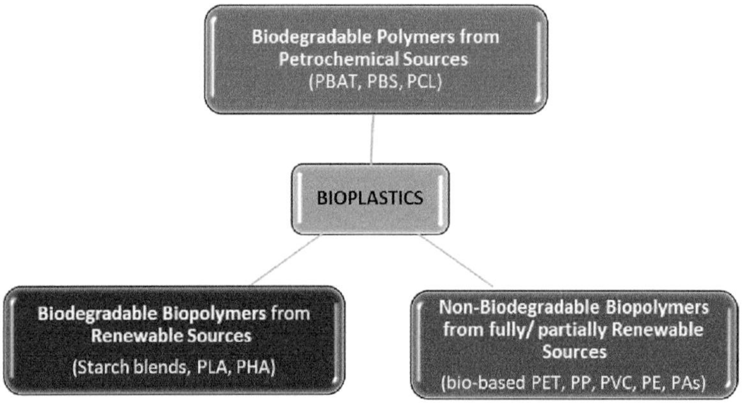

FIGURE 15.1 Types of bioplastics.

produced from natural bioresources; however, these are not having biodegradation characteristics (Rahman & Bhoi, 2021). Biodegradable plastics such as polyhydroxyalkanoates (PHA), polylactic acid (PLA), and certain starch blends from renewable sources exhibited biodegradation properties at certain stages (Fernandes et al., 2020). The degrading properties of PHA derived from agrosolid debris including rice husk, bagasse, straw, potato and mango peels, and wheat bran vary depending on the media and are correlated with the polymer's structure (Coppola et al., 2021). PLA, a bioplastic, is made from processed starchy substances from plants. Polymers are synthesized through bioprocess by various renewable sources such as microorganisms, plants, animals, and algae are known as biopolymers. Marine-derived bioplastics are obtained from marine organisms such as microbes, algae, and marine animals.

The most widely used feedstock sources for market-dominant bio-based bioplastics include sugarcane, maize, sugar beetroot, wheat, cassava, potatoes, and barley. However, due to specific limitations including brittleness, poor strength, and high production costs, plant-based sources account for one percent of the entire each year's output of bioplastics (Tennakoon et al., 2023). Further, plant-based renewable resources availability depends upon the volume of crop production which further relies on climatic conditions and seasons. Also, utilization of edible feedstocks raises market demand and prices while endangering food security. The limitation highlights the importance of a viable alternate biomass supply is needed (Irfan et al., 2025). Biomass resources for the first-generation feedstock compete with food production and second-generation feedstocks are inedible plant-based materials that involve a substantial amount of land usage directly or indirectly. Biomass derived from marine resources is considered a third-generation feedstock for bioplastics, as its development does not require land use and has negligible environmental footprints.

Seaweeds are identified as a sustainable bioresource for bioplastics. The growth rates of seaweeds are approximately ten times higher than those of terrestrial crops (Zhang et al., 2022). Furthermore, the cultivation of seaweeds can limit global warming by consuming CO_2 during photosynthesis releasing O_2 as a byproduct into the atmosphere. Seaweeds found to be suitable for bioplastic production are *Ulva* sp., *Kappaphycus* sp., *Gracilaria* sp., *Gelidium* sp. and *Sargassum* sp. (Lim et al., 2021a). The total polysaccharide content of brown seaweed species varies from 4% to 70% of dry weight and is prevalent in cold water continental coastal locations. Red seaweeds are the largest group of marine macroalgae found deep in the sea and have 40–50% polysaccharides in their dry matter. Fresh water is home to the majority of green seaweeds. Ulva and Enteromorpha are rich sources of acidic polysaccharides and Ulva contains polysaccharides up to 65% of its dry weight (Zhang et al., 2022). Earlier marine algae are collected from natural sources. The growing demand for food and medicinal use of seaweed-based products promoted seaweed farming. Controlled culture methods are designed for large-scale cultivation of specific seaweed species in saline environments through regulation of nutrients, pH, light, and salinity.

Seaweeds synthesize polysaccharides are part of their cellular structure and as a way to store energy. Because of their excellent film-forming properties, biocompatibility, nontoxicity, and ease of production, macroalgal polysaccharides have been extensively studied for potential application in food packaging and the bioplastic sector for the past two decades. Polysaccharides are extracted by mechanical, chemical, and biological methods. The method employed for the synthesis of bioplastics is depends on various factors, including the specific algae species utilized, the solvents selected, the environmental impact considerations, cost and time requirements, quantity needed, scalability potential, and the desired properties of the final product (Lim et al., 2021a). Pressurized solvent extraction, enzyme-assisted extraction (EAE), supercritical fluid extraction (SFE), ultrasound-assisted extraction, and microwave-assisted extraction (MAE) are some of the latest extraction methods developed and are compared with the conventional extraction technologies (Grosso et al., 2015). Seaweed bioplastic films frequently exhibit poor mechanical properties and water barrier properties when utilized alone. To overcome the limitations, plasticizing agents and additives are combined with seaweed polysaccharides to strengthen the intermolecular bonds which in turn improve the strength, flexibility, and durability of biopolymers derived from algae (Devadas et al., 2021; Lim et al., 2021a). Glycerol, sorbitol, and polyols are the most common plasticizers used in the synthesis of bioplastics from microalgae and carboxy methyl cellulose is used to improve the mechanical properties (Cao et al., 2025). Water-resistant additives may be the focus of future research, but this may have an impact on the films' biodegradability (Lim et al., 2021b).

The physical properties and potential manufacturing methods of seaweed-derived plastics were reported, with some studies utilizing experimental data for further comparisons. However, the environmental effects of the production and end-of-life of such seaweed-based plastics are not quantitatively evaluated. Ayala et al. (2023) conducted a life cycle assessment (LCA) at an experimental pilot scale to evaluate the environmental impacts associated with the production of a bioplastic film, incorporating the carbon uptake from seaweed growth. More research is needed to scale up and commercialize seaweed-based polymers, which are still in the early phases of development. Marine algae are a very promising option because of their short harvesting cycles, high reproductive capability and capacity to thrive on non-arable soil. They are a new, innovative, and possibly beneficial biomass source for the production of bioplastics as it is able to grow on non-arable terrain and are harvested quickly. According to recent studies, microalgae hold great promise for additional investigation and advancement in the realm of carbon capture technologies. In addition to using carbon monoxide (CO), carbon dioxide (CO_2), and nitrogen oxides (NO), such as nitric oxide (NO) and nitrogen dioxide (NO_2) as dietary substances, recent research has demonstrated that 1 kg of algal may consume 1.83 kg of CO_2 (Kiseleva et al., 2024). To reduce CO_2 emissions and replace finite fossil fuels, microalgae are seen as environmentally viable substitutes for the continuous manufacture of bioplastics. Figure 15.2 depicts the ecosystem service of seaweeds.

Microalgae are found to synthesize PHAs, an important ingredient in the production of bioplastics. Characteristics of PHAs derived from microalgae are found to be similar to the polymers obtained from fossil fuels (Chong et al., 2022). Compared to traditional plastic manufacturing, microalgae biofuel–bioplastic refineries reduced the greenhouse gas emissions ranging from 67 to 116%, which shows the comprehensive way to capture CO_2 emission from the industry. Due to high protein content, the microalgae chlorella (51–58%) and spirulina (46–63%) were found to be excellent sources of bioplastics (Irfan et al., 2025). Bioplastics have the potential to replace fossil fuels, improve plastic quality, and reduce the ecological damage caused by the extensive use of polymers derived from natural resources. Compared to traditional plastics, these environmentally friendly substitutes provide several advantages, such as a lower carbon footprint, better quality of life, and the capability to reduce plastic pollution. The environmental benefits of bioplastics are (i) reduction in the utilization of fossil fuel, (ii) lowered emission of greenhouse gases, (iii) diminished

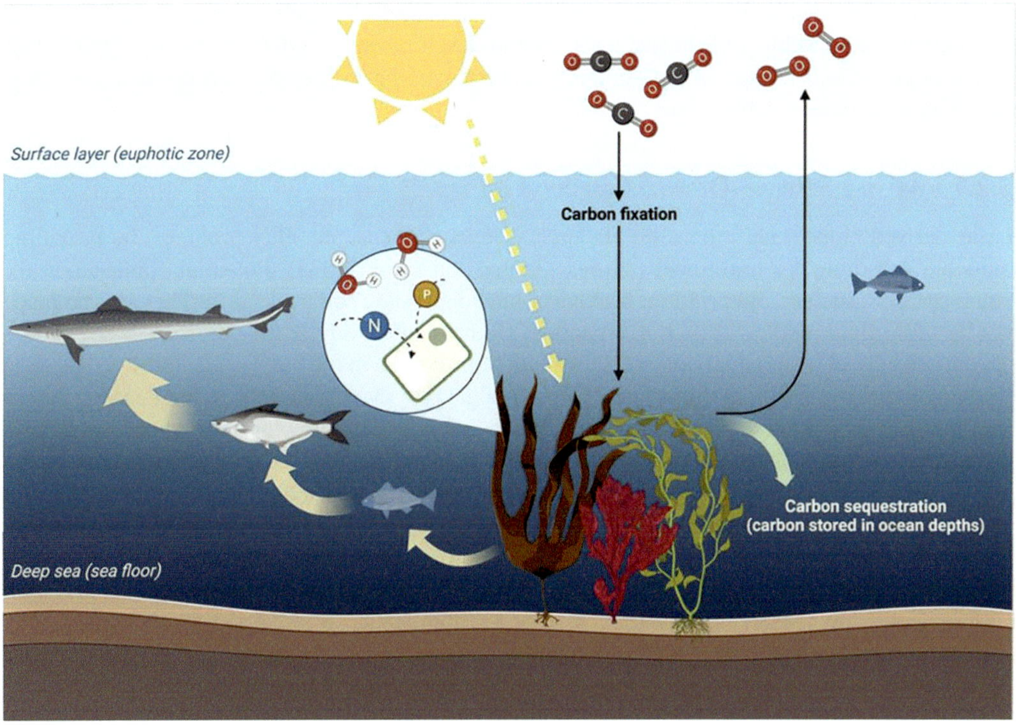

FIGURE 15.2 Eco system service of seaweeds (Xie et al., 2024).

accumulation of plastic wastes, (iv) environment friendly disposal management, and (v) conservation of marine environments (Kumar et al., 2024).

Poly(3-hydroxybutyrate) (PHB), one of the PHA, is a completely biodegradable biopolymer that is produced by cyanobacteria. Hydrophobicity, thermoplastic processability, biodegradability, and biocompatibility are among the qualities of pure PHB that make it a great raw material for biodegradable plastics (Muthulakshmi et al., 2023). Certain marine algae and bacteria diminish the need for freshwater and land footprints, which means there is less competition for agricultural areas and water for the production of biopolymers. Thus, this chapter focusses on the marine biopolymers based-bioplastic, their sources, methods of production, applications in food, challenges in developing bioplastics, regulations, and safety concerns.

15.2 SOURCES OF MARINE BIOPOLYMERS

Marine sources are becoming more and more well-known as cutting-edge resources for creating biological polymers including proteins and polysaccharides (Mahmud et al., 2021). Polysaccharides are usually synthesized by seaweed as parts of their cellular structures and as a way to store energy. The chemical composition of each of these polysaccharides varies depending on the species, life cycle, harvest season, habitat, and extraction technique. These polysaccharides are highly suitable for a variety of applications due to their biocompatibility, nontoxicity, and ability to form hydrogels in response to specific pH levels. Seaweeds have been utilized in regular culinary practices in many countries like China and Japan since ancient times. Significant progress has been made in the development of marine biopolymers, as marine resources like fish, crabs, and mollusks are being more widely used in industrial processes to produce commercial products (Ruocco et al., 2016). Marine ecosystems encompassing over 70% of the Earth's surface harbor a remarkable diversity of organisms. This marine biodiversity, surpassing that observed in both freshwater and terrestrial environments, is the subject of intense research and development, yielding innovative applications (Duarte et al., 2022). Hence, marine biopolymers are gaining importance due to their abundant, renewable, and biodegradable nature.

15.2.1 Marine Seaweed-Derived Biopolymers

Marine-derived biopolymers focus on resource scarcity and sustainability problems by providing a possible substitute for traditional plastics (Tennakoon et al., 2023). Potential applications for these bioplastics include 3D printing, biological medicine, and food packaging. They can be made from bacteria, algae, and other marine organisms (Ranganadhareddy & Chandrasekhar, 2023). The inherent capacity of macroalgal polymers to create films is used in the bioplastics industry, and food packaging heavily relies on biofilms made of macroalgal polysaccharides due to their ease of manufacturing and compatibility. Marine macroalgae are crucial to the growing blue economy because they generate a range of beneficial polysaccharides with potential applications. Their perspective applications in medications, nutritional products, and cosmetics allow them to be utilized as useful components for human health (Benjamin et al., 2022). Seaweed-derived marine biopolymers are present in several phyla, including phaeophytes, or brown algae, rhodophytes, or red algae, and chlorophytes, or green algae (Yang et al., 2021). Figure 15.3 illustrates the major types of macroalga.

More people are considering macroalgae as a potential source of bioactive compounds as they learn more about the health advantages of eating certain foods (Sudhakar et al., 2019). As a renewable resource of biopolymers with exceptional properties for a range of applications, seaweed offers a lot of potential. Seaweed has shown promising outcomes when used as filler in polymer composites, enhancing the mechanical, thermal, and physical characteristics of synthetic polymer matrices. A material made of polymers can be categorized as bioplastic if it is considered bio-based or made of renewable resources that can break down naturally (Mekonnen et al., 2013). Seaweed polysaccharides are extensively used in food, cosmetics, pharmaceuticals, fertilizer, biofuel, waste water treatment etc., and 5.34% of total production is used for the production of bioplastics, which is illustrates in Figure 15.4 (Xie et al., 2024).

Table 15.1 illustrates the characteristics and applications of marine-derived biopolymers in bioplastic production. Bioplastics recognize types of polymeric substances made from bioresources which are considered to be a viable and cost-effective substitute for conventional petrochemical plastics. Numerous environmental problems related to plastics are mitigated by biopolymers and bio-based fibers derived from biodiversity and sustainable wastes (Bátori et al., 2018). Bio-based plastics are a remarkable invention that encourages a sustainable economy and reduces reliance on fossil fuels in favor of bioproducts. The primary sources of marine macroalgal polysaccharides such as laminarin, fucoidan, ulvan, carrageenan, alginate, and agar are especially well-suited for the development of biofilm and bioplastic because of their propensity to form films and their ability to decompose (Nagarajan et al., 2024). Figure 15.5 summarizes the polysaccharides from seaweeds suitable for the production of bioplastics.

15.2.1.1 Ulvan

Ulvan, a sulfated polymer often discovered in Ulva species, is made up of subunits such as rhamnose, xylose, glucuronic acid, and iduronic acid and is extracted from the cellular walls of green seaweeds. The plants included in the study included *Sargassum vulgare, Ulva compressa, Ulva ohnoi, Cladophora pellucida, Nemalion helminthoides, Galaxaura rugosa, Gracilaria sp., Padina pavonica*, and others commonly found around the Eastern Mediterranean beaches. Employing a water and solvent precipitation process, the green seaweed *Ulva fasciata* produced a 28.86% higher yield of ulvan polysaccharides (Li et al., 2023). Moreover, the change in ulvan temperature and their surface charge is influenced by extraction conditions instead of their entire chemical makeup. To

FIGURE 15.3 Major types of macroalgae (Nesic et al., 2024).

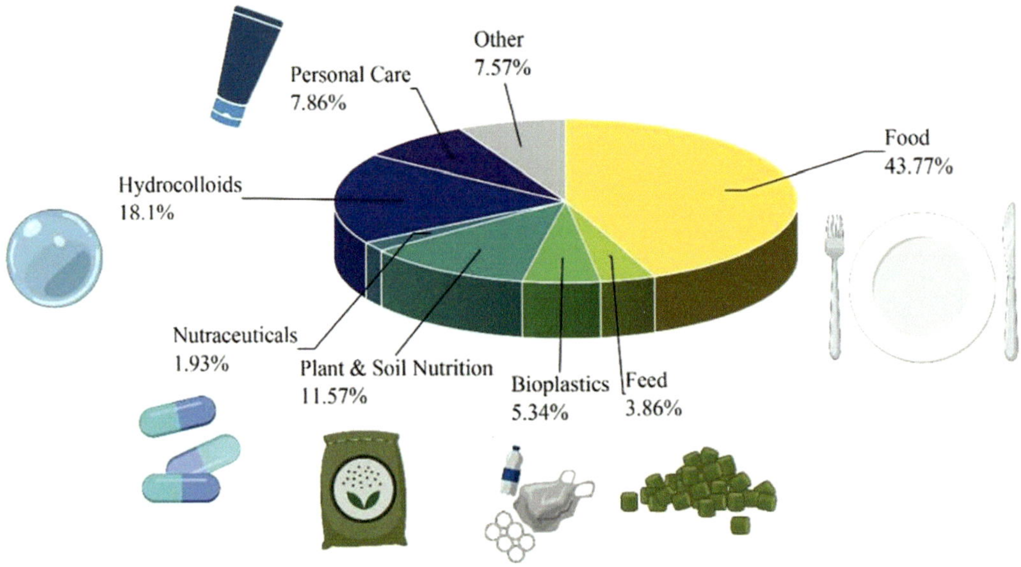

FIGURE 15.4 Current application directions of seaweeds (Xie et al., 2024).

TABLE 15.1
Characteristics and Applications of Marine-Derived Biopolymers in Bioplastic Production

Biopolymer	Source	Processing Method	Applications	Biodegradability	References
Agar	Red algae	Extraction through boiling and filtration	Edible films, food packaging	Biodegradable under natural conditions	Yousefi et al. (2013)
Chitosan	Exoskeletons of crustaceans	Deacetylation of chitin extracted from shells	Antimicrobial films, drug delivery systems	Biodegradable; rate varies with environmental conditions	Hisham et al. (2024)
Laminarin	Brown algae	Water extraction and purification	Biomedical applications, drug delivery	Biodegradable; specific rates under investigation	Pramanik et al. (2024)
Alginate	Brown algae	Alkaline extraction and precipitation	Wound dressings, packaging films	Biodegradable; degradation influenced by environmental factors	Bojorges et al. (2023)
Fucoidan	Brown algae	Extraction using acidic solutions	Biomedical applications, Food Industry	Biodegradable; degradation rate depends on molecular structure	Anisha et al. (2022)
Starch	Green algae and other plants	Extraction and purification	Packaging materials, edible films	Biodegradable; degradation rate varies with environmental conditions	Gupta et al. (2022b)
Carrageenan	Red algae	Extraction with alkaline solutions	Food packaging, edible films	Biodegradable under natural conditions	Rupert et al. (2022)
Ulvan	Green algae	Extraction using water or dilute acid	Edible films, packaging materials	Biodegradable; degradation influenced by environmental factors	Wang et al. (2023)

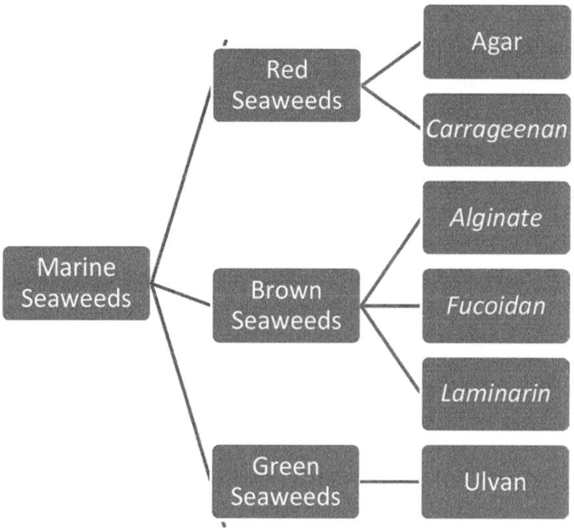

FIGURE 15.5 Polysaccharides from seaweeds suitable for the production of bioplastics.

make Ulvan films, polymers like sorbitol or glycerol are typically used after dry extracts have been dispersed into distilled water.

15.2.1.2 Agar

Red algal membrane components are used to extract agar, a polysaccharide made up of disaccharide repeating units of 3-linked b-D-galactose and 4-linked 3,6-anhydro-a-L-galactose (3,6-AG) residues (Dang et al., 2022). It is composed of two fractions: agarose and agaropectin. The thickening qualities are caused by the nongelling portion of agaropectin, while the gelling properties are caused by the gelling component of agarose. Physical characteristics like melting temperature and gel strength are directly related to its chemical structure. Because a high sulfate content prevents the creation of gel networks, low levels of anionic sulfates encourage the formation of strong gels (Asif et al., 2021). Agar agar, or simply agar, was the first phycocolloid used as a component of food in our society and extends back over three centuries to the far eastern region. Phycocolloids are seaweed-derived gel-forming compounds that have a variety of uses (Pereira, 2016). Agar-based biopolymer-film coatings are synthesized, processed, and characterized on paper materials in compliance with specified guidelines set by certifying organizations. It entails investigating the mechanical properties of agar-coated biopolymer films using profilometry, nanoindentation techniques, and surface properties such as the hydrophobic nature of the layer and the material capability as a gas barrier (Matheus et al., 2023).

15.2.1.3 Alginate

Alginate is a polymer comprising β-D mannuronic acid (M) and α-L guluronic acid (G) residues that is a stiff copolymer found in brown algal cell walls. Alginates that contain additional G blocks have stronger gelling properties, but viscosity is a characteristic. Alginate is a copolymer of mannuronic acid and guluronic acid. A common polymer for encapsulating is alginate, which creates a gelled matrix that is reliable, renewable, and adaptable (Stanisci et al., 2020). The gelled matrix will improve stability by shielding the active ingredient from environmental factors including moisture and heat. A lot of researchers are also interested in alginate due to its moderate gelation qualities, low toxicity, ease of handling, taste and odor that are pleasant, affordability, and biological compatibility (Venugopal & Sasidharan, 2021; Gupta et al., 2022a). Due to their structural and acetylated differences, alginates derived from the external polysaccharides of bacteria like *Pseudomonas aeruginosa* and *Azotobacter vinelandii* are not frequently utilized in commercial environments (Gheorghita Puscaselu et al., 2020). For example, water-soluble powder items such as coffee, coffee-based snacks, powdered milk, and rapid teas are frequently packaged in alginate-based films, which are both edible and fully biodegradable (Robal et al., 2017). In the food industry, alginate is referred to as a food additive (E400).

15.2.1.4 Starch

Marine algae mostly store carbohydrates as starch. The amylose-to-amylopectin proportions and starch content in algae tend to be species, environment, and season-dependent. The ratio of amylose to amylopectin is crucial for starch gelation, which is facilitated by the temperature change process and gelatinization and retrogradation (Prabhu et al., 2019). When grown offshore for 13 months, the marine macroalgae *U. ohnoi* yielded an average of 3.43 tons of starch per hectare annually, demonstrating its potential as a substitute for agricultural crop-based starch, which needs freshwater and arable land. Conventional corn starch and starch derived from the marine microalgae *Klebsormidium flaccidum* exhibit a lot of the same properties (Ramli et al., 2020).

15.2.1.5 Carrageenan

Red algae cell walls include carrageenan, a water-soluble linear polymer, which relies on the repeating disaccharide sequence of β-D-galactose bound at position 3 and α-D-galactose bonded at position 4. The number, location, and level of sulfation within the sulfate ester chains on galactose determine the isomers of carrageenan, k- (kappa), l- (lambda), and i- (iota). Because of its adaptability, carrageenan has a higher potential for usage as a gel-forming substance. L-carrageenan, for instance, produces viscous solutions that do not gel, while k-carrageenan produces brittle, rigid, sturdy gels, and i-carrageenan produces soft, elastic, and weak gels (Dang et al., 2022). To produce semi-refined carrageenans, the seaweed's carrageenans are elevated to roughly 75°C using an alkaline potassium hydroxide solution (Skryplonek et al., 2019). Hydroxide reacts with the sulfate esters to create and carrageenan, increasing the final product's gel strength even if potassium binds to it and causes a gel by preventing the hydrocolloid chains from dissolving (Martiny et al., 2020). The semi-refined carrageenan extraction technique breaks down the gel matrix by heating alkali-treated seaweed fronds to 95–110°C, which produces refined carrageenan. The cost of extracting refined carrageenan is significantly higher than that for preparing semi-refined carrageenan (Zia et al., 2017). Among other applications, carrageenan has been investigated as a therapeutic component in medicinal products for usage as a carrier/stabilizer in micro/nanoparticle combinations, a framework of polymers in oral extended-release pills, and a novel extrusion aid for granular production (Necas & Bartosikova, 2013). For many years, carrageenan was widely used as a food ingredient, and several studies have confirmed its effectiveness. By linking up protein and absorbing water to encourage the development of gels, carrageenan is frequently utilized in food production to enhance taste, thickness, and stability (Gopi et al., 2022).

15.2.1.6 Fucoidans

Sulfated polysaccharides, such as fucoidan, belong to a number of algal polysaccharides that have gained the interest of research. Their numerous pharmacological effects include anticoagulant, antithrombotic, immunomodulatory, antiobesogenic, and anticancer properties (Zhao et al., 2018). The polysaccharide fucoidan is produced by brown macroalgae. Fucoidan levels in brown seaweed are high (Hentati et al., 2020). The main component of fucoidan ($C_6H_{12}O_5$) is the deoxyhexose sugar fucose (Wang et al., 2019). Fucoidan effectiveness as a treatment for cardiovascular conditions is being further investigated. In recent years, an increasing amount of research has shown the significance of fucoidan in the cardiovascular system (Wu et al., 2016). Fucoidan has been used as a nutritious supplement in the food business since Kylin first isolated it from brown algae in 1913 (Luthuli et al., 2019). Fucoidan remarkable anticancer impact has led to the most detailed research in the field of cancer treatment. Although it protects against issues with the liver and urinary system, fucoidan has garnered interest (Chen et al., 2019). Fucoidan-containing brown seaweeds have emerged as an essential component of many Asian diets, especially in China, South Korea, and Japan. Fucoidan is useful for medicinal uses since it is nontoxic, rarely causes adverse reactions, and possesses a number of biological roles.

15.2.1.7 Laminarins

Laminaria digitata was utilized for the production of the polysaccharide laminarin, which was subsequently hydrolyzed at 0 °C with chlorosulfonic acid in pyridine as well as liquid sulfur dioxide. Furthermore, laminarin displays certain biological activities that are appealing for further investigation and might lead to industrial implications. To get the greatest amount of the algal biomass, laminarin might be extracted from brown algae at the same time as the different polysaccharides that are present in the algae cellular walls. Laminarin is found in several algal species, including *Ecklonia kurome, Laminaria japonica, Laminaria digitata, and Eisenia bicyclis*. When laminarin, a glucan, is broken down by microbes into simple sugars like glucose and others, ethanol is produced as a byproduct (Karuppusamy et al., 2022).

15.2.2 Marine Animal-Derived Biopolymers

15.2.2.1 Chitin

Chitin is a primary structural polysaccharide found in the exoskeletons of crustaceans like crabs, shrimp, lobsters, and molluscs. It consists of repeating units of N-acetyl-D-glucosamine (Barikani et al., 2014). The biosynthesis of chitin has been examined in diverse organisms, with the chitin synthase (CS) enzyme being essential to this process. In the yeast *Saccharomyces cerevisiae*, three distinct isoforms of chitin synthase CS I, CS II, and CS III mediate the chitin synthesis (Singh et al., 2017). Chitosan, a derivative of chitin, is formed by deacetylation and is a copolymer comprised of glucosamine (the deacetylated monomer) and N-acetyl-D-glucosamine (the acetylated monomer) (Ali et al., 2024). Due to its abundant availability and low cost, chitin and its derivatives has emerged as promising biopolymers. The chitin exhibits potential properties, including antibacterial and anti-inflammatory effects, biocompatibility, and biodegradability, making them suitable for a variety of applications. The edible and biodegradable films derived from chitosan have demonstrated antimicrobial activity against both gram-positive and gram-negative bacteria (Claverie et al., 2020). Furthermore, the inclusion of nanosized chitosan particles has been shown to influence the mechanical properties of these films, specifically their strength and stiffness (Song et al., 2018). Annually, over 13 million tons of crustaceans are harvested from marine environments globally, resulting in large amounts of food waste (Harkin et al., 2019). The source of chitin significantly influences the properties of chitosan. The food sector uses marine-based chitosan due to its antimicrobial and antioxidant benefits along with as an adhesive to improve preservation (Roy et al., 2017).

15.2.2.2 Marine Collagen

Collagen, a polypeptide, is the most abundant protein found in both animals and humans. Marine collagen, typically type I, can be extracted from the connective tissues of various marine organisms, including sea cucumbers, sea urchins, jellyfish, and starfish. Marine collagens, encompassing both fibrillar and nonfibrillar forms, are characterized by higher viscosities compared to bovine collagen. However, they exhibit lower gelling and melting temperatures than mammalian collagens (Monteiro et al., 2023). Specifically, fish collagen possesses more labile cross-links, rendering it better susceptible to thermal degradation than mammalian collagen. Collagen is employed as a food additive to confirm the presence of adequate animal-derived nutritional fibers in processed meats such as sausages and frankfurters. Furthermore, it serves to enhance the rheological properties of these food products.

15.2.2.3 Heparin

Heparins, part of the glycosaminoglycan (GAG) family, are polysaccharides found in the tissues of both terrestrial and marine vertebrates and invertebrates (Milovanovic & Hayes, 2018). Heparin was the first anticoagulant used clinically, having been discovered by McLean in 1916 during research aimed at identifying a thromboplastic agent. While almost all cell types can synthesis heparan sulphates, mast cells are the main source of the most significant sulphated heparins (Rathod et al., 2023).

15.2.2.4 Gelatin

Gelatin, obtained from collagen through hydrolytic degradation and denaturation, is rich in glycine, proline, and 4-hydroxyproline residues. Marine gelatin-based bioactive films are increasingly being investigated as sustainable alternatives to synthetic materials in food packaging. Despite the significant rise in global demand for marine-derived food products, particularly fish, most commercially available collagen is sourced from mammals. While fish gelatin offers a promising alternative, its application is hindered by challenges such as odor, color, functional properties, and variability in amino acid composition (Milovanovic & Hayes, 2018).

15.2.3 Marine Microorganisms-Derived Biopolymers

There are several researches focused on exploring marine environments to discover novel microorganisms capable of producing PHAs cost-effectively under nonsterile, low-temperature conditions (Milovanovic & Hayes, 2018). Marine microorganisms offer the advantage of reduced contamination risks due to their halotolerance and psychrotolerance (Arun et al., 2009). Several marine-derived microorganisms have been identified as potential PHA producers, including *Alcaligenes latus*, *Bacillus licheniformis*, *Vibrio sp.*, *Ralstonia eutropha*, *Shewanella marisflavi*, and *Pseudomonas putida*, as well as the yeast *Pichia kudriavzevii* isolated from marine seaweed (Bhattacharya et al., 2016). These organisms have shown the capacity to accumulate substantial PHA content with minimal nutrient requirements. Halophiles, which are extremophiles thriving in high-salt environments, provide an additional benefit by inhibiting the growth of contaminating microorganisms without the need for strict sterilization (Możejko-Ciesielska et al., 2023). Therefore, halophilic bacteria like *Halomonas hydrothermalis* and *Halomonas venusta* hold promise as candidates for low-cost PHA production through open, unsterile, and continuous fermentation, bypassing the high costs associated with sterile batch or fed-batch processes (Chmelová et al., 2021).

Studies have identified several marine bacteria as promising low-cost PHA producers. These include *Cobetia*, *Sulfitobacter*, and *Pseudoalteromonas* strains isolated from the seaweed *Ulva* sp. and cultivated using various sugars and acid hydrolysates as carbon sources (Stanley et al., 2018). Additionally, *Bacillus cereus* and *Pseudomonas pseudoalcaligenes* strains, isolated from other seaweeds, have demonstrated PHA production using industrial waste as a substrate (Lee et al., 2024).

15.3 BIOPLASTICS PRODUCTION FROM MARINE-DERIVED BIOPOLYMERS

Bioplastics are bio-based polymers used to make plastic and are also known as biodegradable (Sudhakar et al., 2024). The class of materials known as bioplastics has a wide range of uses in many industries and diverse features. If a material is sustainable, recyclable, or possesses both qualities, it is referred to as a bioplastic. The manufacture of plastics emits fewer greenhouse gases when petroleum-based feedstock is replaced with organic ingredients like starch and different natural fibers. The simplest method to lessen the detrimental effects of petroleum-based plastics is to replace them with recycled, recovered, or renewable resources. These substances can improve product physical and visual qualities while lowering the amount of nonrenewable feedstocks. Algae grow by inorganic substances like C, N, P, trace elements, etc. (Abdel-Raouf et al., 2012). They also generate organic materials like proteins, lipids, and carbohydrates that can be used in a variety of industrial processes, including the synthesis of polysaccharides for bioplastics. To create Ulvan films, dried extracts are usually dissolved in distilled water before plasticizers like sorbitol or glycerol are added (Guidara et al., 2019). Ulvan polysaccharides distinct gelling methods make them appropriate for use in filmmaking, where the type and concentration of plasticizer have a major impact on the final ulvan-derived product characteristics. Regardless of whether the plasticizer is sorbitol or glycerol, it was found that increasing the plasticizer content has increased the thickness of films by up to 40%. The circumstances and methods used for extraction affect the ulvan films optical, thermal, structural, and antioxidant qualities. However, ulvans can provide film compositions independent of the extraction process (Guidara et al., 2020). *U. fasciata*, green seaweed, has a 14.7% amount of cellulose on a dry weight basis. The extracted material was then used to create carboxymethyl cellulose and thin films using a solvent evaporation process. Furthermore, a different study has documented the use of basic alkaline extraction to remove less pure cellulose portions derived from the brown seaweed species *Ascophyllum nodosum*, *Saccharina latissima*, and *Alaria esculenta* to create edible packaging films. Superior mechanical qualities and great cellulose purity were displayed by the resulting films.

Agar was extracted from the red seaweed *Gracilaria corticata* using the hot water extraction method, yielding a high 18.5%. To create environmentally acceptable bioplastics, the film-casting technique was used to combine extracted agar with plasticizing agents such sorbitol and glycerol. Sorbitol increases a film's tensile strength, but glycerol helps to maximize its elasticity. The highest levels of absorption and biological degradation were seen in bioplastic samples made with the least agar content mixed with glycerol. The results showed that the tensile strength and thickness of the film were directly impacted by the volume of agar added to the mixture (Asif et al., 2021). In the process of producing bioplastics, agar low hygroscopicity is beneficial. Since the imperfections in agar have a plasticizing effect, unpurified agar-based lysates are utilized for environmentally friendly and economical filmmaking because no other plasticizers are needed. Carrageenan was extracted from dried red seaweed (*Solieria robusta*) using the hot water extraction technique, and smooth, flexible films with a smooth texture were made using the film casting method. Including sorbitol made the films more resilient to deformation, dissolution, and biological degradation than glycerol blends, and the films thickness, elasticity, and tensile properties improved as the biopolymer concentration increased. To make edible films, semi-refined κ-carrageenan, which is less expensive than refined carrageenan, can be utilized (Farhan & Hani, 2017). The elongation at breakage and tear strength of films made from carrageenan are greatly increased by the incorporation of plasticizers. It has been observed that chitosan-derived films that are edible and biodegradable exhibit antibacterial action against bacteria of all kinds. Additionally, the addition of nanoscale chitosan alters their stiffness and mechanical strength (Gomes et al., 2018).

According to these findings, chitin and chitosan can be combined to create useful bioplastics. Both the chitosan pathway and the chitinolytic pathway are responsible for chitin biodegradation in aquatic environments. Chitinase hydrolyzes chitin to produce oligosaccharides in the chitinolytic pathway, and exochitinase then converts chitin to chitobiose. Up until 2019, PLA and starch mixtures accounted for over 45% of the manufacture of bioplastics. Further study was recently investigated on various bioplastics, such as PHAs and their derivatives, like polyhydroxyalkanoates (PHB) (Kavitha et al., 2018). The output of PHA and PHB is predicted to rise significantly over the next 5 years, despite not exceeding 2% in 2019. PLA and Mater-bi®, which are bioplastics made from starch, vegetable oils, or a mix of the two, have become two of the most significant bioplastics available on the market. Because of their quick growth rates, ease of cultivation, and lack of interference or competition with food supplies, marine sources such as marine algae, marine animal waste, and marine-derived microorganisms represent an attractive alternative to the current bioplastics.

15.4 APPLICATIONS OF MARINE-DERIVED BIOPLASTICS IN FOOD

In food packaging applications, marine-derived biopolymers are becoming more and more popular as environmentally friendly substitutes for traditional plastics. Seaweed, marine animals, and microorganisms are the sources of these biopolymers, which have nontoxicity, biodegradable properties and biocompatibility (Bose et al., 2023). Common marine biopolymers that can be utilized in a variety of food systems, including meat, dairy, fruits, and vegetables, include agar, alginate, chitin/chitosan, and carrageenan (Mahmud et al., 2021). The capability of food packaging to endure the challenges of storage, transit, and packaging depends heavily on its mechanical qualities. Features related to light transmission are essential for marketing, UV protection, and product visibility. Additionally, thermal and moisture stability are important for maintaining the quality and hygiene of packaged goods throughout transit and storage. Good mechanical, permeability, thermal, and optical qualities are displayed by marine biopolymer films; gas permeability is particularly important for prolonging the shelf life of food (Xue et al., 2023). Nevertheless, adjustments such as mixing, cross-linking, and nanomaterial reinforcement can be used to overcome these quality limits.

Packaging materials made from marine biopolymers are prepared by applying various solutions and solvents using different methods, including brushing, coating, dipping, spraying, wrapping, and panning (Vázquez et al., 2019). The materials are frequently discarded after their final use in food packaging and are allowed to break down due to their biodegradable properties. The most crucial and possibly most anticipated aspect of these packaging components is their environmental decomposition potential. Water barrier ability, oxygen barrier ability, edibility, transparency, thickness, and elasticity are a few of the specific innovative qualities (Lionetto & Esposito Corcione, 2021). For instance, polyethylene terephthalate and polyvinylidene chloride are ideal for packaging fresh meat, ham, and sausage because of their remarkable oxygen barrier properties. On the other hand, since polypropylene and polyethylene have high oxygen permeability (OP), they can be used to package fruits and vegetables as they allow post-harvest respiration (Xue et al., 2023). The main applications for bioplastics include flexible food packaging, trash bags, and shopping bags for groceries. Nowadays, a number of major bioplastic corporations, including CocaCola, PepsiCo, and Heinz, have begun using bio-PET bottles for drinks and other food products. Agricultural mulch films, zipper bags, and cling wrap films for nutritious foods are being made using poly butylene-adipate-co-terephthalate based bioplastics (Ghasemlou et al., 2024). These characteristics assist in preserving the food in packaging from oxidation and microbiological contamination.

15.5 CHALLENGES IN DEVELOPING BIOPLASTICS FROM MARINE BIOPOLYMERS

Bioplastics have a lower carbon footprint, and the natural breakdown of biodegradable plastics under specific conditions promotes green disposal and contributes a significant reduction in the quantity of plastic waste. The composition of bioplastic, size, and thickness influence the rate of degradation. The degree of biodegradability of biodegradable bioplastics obtained from different polymers is challenging due to the wide variations in the methodologies adopted to validate the degradability (Lavagnolo et al., 2024). Other challenges in the development of bioplastics from biopolymers include high production costs, limited biodegradability, lack of standardization, technological limitations, and variations in end-of-life management (Kumar et al., 2024). Bioplastics undergo three phases of degradation such as (i) biodeterioration process, or change in the chemical, mechanical, and physicochemical characteristics of the polymer due to the growth of microorganisms on its surface and release of the enzymes' (ii) biofragmentation process, in which the microbes transform the polymers into oligomers and monomers, (iii) bioassimilation process, in which microbes obtain the carbon, energy, and nutrients from the broken-down polymers and transform the carbon into CO_2, biomass, and water (Amobonye et al., 2023). Taxes on nonrecyclable plastic products, regardless of their origin, may hinder the market's adoption of biodegradable plastics, which are designed to decompose naturally through processes like composting. Research to explore new bioplastics is still high, and at the same time, continuous modifications affect the recyclability, compostability, and another form of end-of life treatments. Greenhouse gas emissions and growing public awareness of microplastic pollution in the land and water are opening up new markets for sustainable biodegradable bioplastics. Currently, corn and sugarcane are the primary feedstock for bioplastics whose land utilization requirements directly conflict with the food production.

Industrial scale cultivation of seaweed biomass offers new ways to diminish the dependence on fossil feedstocks, perhaps aiding in climate mitigation and alleviating pressure on terrestrial resources. Even though the findings of algae-based bioplastics on a laboratory scale showed satisfactory results, commercialization and large-scale manufacturing are still facing challenges. The composition of seaweed biomass varies among the species and the identification of appropriate seaweed suitable for the synthesize of bioplastics with desired properties needs to be explored (Krishnan et al., 2024). In addition to seaweed experts, it would be beneficial to involve a larger spectrum of stakeholders and industry professionals in the bioplastics sector. Engaging with specialists from many domains can provide valuable perspectives on the potential and challenges associated with upscaling, leading to a more comprehensive understanding of the process and its consequences (Ayala et al., 2024b). The production cost of bioplastics highly depends on feedstock costs. Currently, bioplastics are mainly produced from corn starch and sugarcane. If the prices of corn and sugarcane are not stable, the fluctuation in prices reduces the demand for bioplastics in the market. Transforming biological resources into bioplastics requires a considerable amount of energy which would affect the overall production cost. When compared with traditional plastics and their well-established waste management systems, bioplastics become less cost-competitive since clear regulations and frameworks for their disposal are lacking.

The seaweed, cyanobacteria, and microalgae are found to be the potential raw material to make third-generation bioplastics; however, it received less attention due to poor economic feasibility (Sadvakasova et al., 2023). Polysaccharides such as alginate, carrageenan, ulvan, fucoidan, and laminarin are suitable candidates for bioplastic components owing to their compatibility with diverse matrices and their film-forming capabilities. Microalgae and cyanobacteria have been identified as promising sources for bioplastics; these can be commercially cultivated in open waters or bioreactors. PHAs synthesized from these organisms are biodegradable and potential alternatives for bioplastic production. Currently, the market for microalgal biorefineries is limited by complex process management, high costs associated with the collection and process of biomass, inconsistent quality of biomass, energy needs, and technological hurdles. To meet the industrial demand, microalgae cultivation must be scaled up and to increase the process efficiency, further research is required to boost the biomass yield through optimized cultivation methods. Shortly, policies and regulations need to support sustainable alternatives to fossil source-based plastics that could affect the market penetration of bioplastics. Though bioplastics reduce the environmental impact, their lifecycle and end-of-life management need to be addressed (Krishnan et al., 2024). Algal farming and breeding have increased substantially in recent years, but the main obstacles facing the algal businesses are a lack of improved varieties and regulatory restrictions, a lack of consistency in the methodologies used in the LCAs, and no specific support and exemptions on bioplastic categories.

Biodegradable bioplastics are becoming more and more popular as an eco-friendly alternative and have drawn the attention of researchers. Biodegradable refers to the breakdown of substances into (i) carbon dioxide during aerobic degradation of methane and carbon dioxide during anaerobic circumstances and (ii) inorganic chemicals and new cellular biomass by naturally occurring microorganisms (Harrison et al., 2018). Bioplastic made from seaweed-based polymers emerging as a promising biobased and biodegradable alternative derived from renewable biomass; additionally, seaweed production offers several environmental advantages. The multidisciplinary approach used in the LCA of seaweed bioplastic from cultivation to biomass processing at the laboratory and pilot scale indicates the potential of seaweed for reducing the environmental impact. Recycling is increasingly becoming a global effort to minimize the amount of plastic waste landfilled. However, health care products are highly contaminated, and recycling is challenging or impracticable in some instances. Recycling resulted in inferior physical properties and raised the products carbon footprint. In one environment, certain bioplastics may biodegrade rapidly, whereas in another, they may take years or never biodegrade at all. Therefore, when describing and defining the biodegradation of biodegradable bioplastic, it is crucial to include the duration and environmental. The biodegradation rate of PLA and other biodegradable bioplastics is only feasible in certain environments, such as industrial composting settings with thermophilic conditions; degradation rate will be low if the environmental conditions are not optimal. High production costs are the main barrier to their market establishment, even though they have the potential to be highly degradable at the end of their life. LCA is the tool to evaluate the environmental impacts. It is necessary to carefully consider the toxicological and physiochemical effects on the environment when evaluating the end-of-life assessment of marine-based bioplastics. The creation of marine-based bioplastics, which are thought to decompose more readily than petroleum-based plastics, is a research need and is therefore expected to rise. Microplastics are fragments of plastic size ranging from 1 um to 5000 um that pose a potential threat to ecology. Government policies and regulations are required for the production, usage, and disposal of biodegradable plastics to be a feasible substitute for conventional plastics (Ayala et al., 2024b). Seaweeds capturing a significant amount of CO_2 can mitigate (i) erosion of coastal lines and (ii) ocean deoxygenation and acidification, both of which are important factors in climate change.

Microalgae are unicellular organisms found in both fresh and marine water environments and are distinguished by their capacity to synthesize both protein and lipids. The photoautotrophic cultivation technique is the most widely adopted method for microalgae cultivation. Macroalgae, also known as seaweeds, are mainly found in marine areas, especially on the seabed as a dense strand. They include green macroalgae (*Enteromoma* spp.), red macroalgae (*Porphyra* spp.), and brown macroalgae (*Laminaria* spp.). The indirect method of obtaining macro algae biomass obtained from natural stocks yields a mixture of species with poor commercial value. Cultivation of targeted microalgal species under a controlled environment is the direct approach. Although algal biomass can have a higher growth rate, the demand for plastics is still high; again sustainability is questioned as seaweeds are found to be an alternate source for other applications as well. Thus, more research is to be focused on utilization of seaweed waste to produce bioplastics, which helps to minimize the waste. Cultivation of seaweed biomass is largely influenced by light intensity, temperature, nutrients, and pollution in water resources which eventually affect the biofilm characteristics. Compared to traditional methods, seaweed films produced using green production techniques are fast, affordable, and environmentally beneficial (Lim et al., 2021a). Green synthesis techniques and seaweed biorefineries can help to overcome this by lowering production costs and, ultimately, the quality of the final product. Poor mechanical and water barrier properties of the biofilms can be improved by blending the raw seaweeds of various species and more research is needed for green production methods of bioplastic production.

15.6 REGULATIONS AND SAFETY CONCERNS RELATED TO MARINE BIOPOLYMERS-BASED BIOPLASTICS

The production of bioplastics from marine sources is a more environmentally friendly process that achieves zero carbon emission targets and has lower greenhouse gas emissions. This process is the top priority set by many national and international laws and regulations. Plastics made of biopolymers are mainly used in food packaging, pharmaceutical, electronics, automotive, cosmetic, and agricultural industries. Recycling, incineration, composting, and anaerobic digestion are the end-of-life solutions for bioplastics (Jayakumar et al., 2023). Bioplastics are the alternate solution for conventional plastics, and the government must educate the consumer and provide efficient infrastructure for collecting, recycling, and composting. Biodegradable bioplastics are to be disposed of in compliance with the regulations set forth by the various regulatory bodies. When biodegradation occurs on a large scale, the byproducts can be released back into the environment without causing the detrimental effect. To eliminate the issues related to plastics, technological advancements, policies and governing regulations should be combined with individual self-driven solutions.

For the chemical or mechanical route of recycling, the bioplastics must be identified and sorted before cleaning. Under controlled conditions, majority of the biodegradable bioplastics degrade completely, thus avoiding waste accumulation. Algae and cyanobacteria are great feedstocks for the synthesis of plastics and polymers because they can grow well in a variety of environments and help reduce carbon emissions (Agarwal et al., 2022). To extract the seaweed hydrocolloids conventional techniques such as cleaning, washing, acid/alkali treatment, filtration and neutralization, precipitation, drying, and milling are employed. However, these methods are not viable since the amount of water utilized, the production process, the reactant, and the chemicals and their appropriate waste treatment all contribute to the high cost. Alternatively, green production methods namely EAE, MAE, photo-bleaching extraction, reactive extrusion, pressurized-solvent extraction, SFE, and ultrasound assisted extraction showed great potential and viability (Lim et al., 2021a).

15.7 CONCLUSION

Conventional plastics, which are mostly made of fossil fuels, are durable and nonbiodegradable, which is constant threat to ecosystems. Petrochemical-based plastics have complex compositions that make them hard to recycle or reuse, and inappropriate disposal system pollutes the land and water, possibly killing millions of animals and decreasing soil fertility. Biobased biodegradable plastics are alternative substitutes for traditional plastics. The synthesis of bioplastics and the carbon-capturing properties of microalgae can be combined to create a closed-loop system that actively lowers plastic waste and carbon emissions at the same time. Production of bioplastic from seaweed can effectively reduce the environmental impacts per kilogram of bioplastic generated, and from an LCA perspective, the overall impacts are less than those derived from fossil fuels. Additionally, bioplastics derived from marine sources are fulfilled with the principles of the circular economy. Seaweeds play a significant role in wastewater treatment by eliminating excess nutrients and heavy metals, thereby providing dual environmental advantages. Rising demand for eco-friendly products has paved the way to explore biopolymers obtained from marine sources such as seaweed, microalgae, aquatic animals, and bacteria for the production of biodegradable bioplastics.

ACKNOWLEDGMENTS

The author (PK) wishes to thank the Faculty of Engineering, Karpagam Academy of Higher Education for their support and help.

REFERENCES

Abdel-Raouf, N., Al-Homaidan, A. A., & Ibraheem, I. B. M. (2012). Microalgae and wastewater treatment. *Saudi Journal of Biological Sciences*, *19*(3), 257–275.

Agarwal, P., Soni, R., Kaur, P., Madan, A., Mishra, R., Pandey, J., Singh, S., & Singh, G. (2022). Cyanobacteria as a promising alternative for sustainable environment: Synthesis of biofuel and biodegradable plastics. *Frontiers in Microbiology*, 13, 939347.

Ali, G., Sharma, M., Salama, E. S., Ling, Z., & Li, X. (2024). Applications of chitin and chitosan as natural biopolymer: Potential sources, pretreatments, and degradation pathways. *Biomass Conversion and Biorefinery*, *14*(4), 4567–4581.

Amobonye, A., Aruwa, C. E., & Pillai, S. (2023). Microbial bioresource for plastic-degrading enzymes. In Sudhir P. Singh and Santosh Kumar Upadhyay (Eds.), *Microbial Bioreactors for Industrial Molecules*, Wiley, pp. 421–441.

Anisha, G. S., Padmakumari, S., Patel, A. K., Pandey, A., & Singhania, R. R. (2022). Fucoidan from marine macroalgae: Biological actions and applications in regenerative medicine, drug delivery systems and food industry. *Bioengineering*, *9*(9), 472.

Arun, A., Arthi, R., Shanmugabalaji, V., & Eyini, M. (2009). Microbial production of poly-β-hydroxybutyrate by marine microbes isolated from various marine environments. *Bioresource Technology*, 100(7), 2320–2323.

Asif, M., Tabassum, A., Tahira, M., & Rahman, A. (2021). Preparation and estimation of physio-mechanical properties of eco-friendly bioplastics of *Gracilaria corticata* from Karachi coast. *Pakistan Journal of Botany*, *53*(4), 1525–1529.

Ayala, M., Arlov, Ø., Nøkling-Eide, K., Sœther, M., Dore, C., Vidal, J., ... & Pizzol, M. (2024a). A supply-chain perspective on producing and upscaling bioplastic from cultivated brown seaweed. *Journal of Cleaner Production*, *444*, 141248.

Ayala, M., Goosen, N., Michalak, L., Thomsen, M., & Pizzol, M. (2024b). Prospective LCA of brown seaweed-based bioplastic: Upscaling from pilot to industrial scale. *Sustainable Production and Consumption*, *52*, 416–426.

Ayala, M., Thomsen, M., & Pizzol, M. (2023). Life cycle assessment of pilot scale production of seaweed-based bioplastic. *Algal Research*, *71*, 103036.

Barikani, M., Oliaei, E., Seddiqi, H., & Honarkar, H. (2014). Preparation and application of chitin and its derivatives: A review. *Iranian Polymer Journal* (English Edition), *23*(4), 307–326.

Bátori, V., Åkesson, D., Zamani, A., Taherzadeh, M. J., & Sárvári Horváth, I. (2018). Anaerobic degradation of bioplastics: A review. *Waste Management*, *80*, 406–413.

Benjamin, T.-A., Ahmad, I., & Sadiq, M. B. (2022). Applications of marine biochemical pathways to develop bioactive and functional products. In *Marine Biochemistry*. CRC Press

Bhattacharya, S., Dubey, S., Singh, P., Shrivastava, A., & Mishra, S. (2016). Biodegradable polymeric substances produced by a marine bacterium from a surplus stream of the biodiesel industry. *Bioengineering*, *3*(4), 34.

Bojorges, H., López-Rubio, A., Martínez-Abad, A., & Fabra, M. J. (2023). Overview of alginate extraction processes: Impact on alginate molecular structure and techno-functional properties. *Trends in Food Science & Technology*, *140*, 104142.

Bose, I., Nousheen, Roy, S., Yaduvanshi, P., Sharma, S., Chandel, V., & Biswas, D. (2023). Unveiling the potential of marine biopolymers: Sources, classification, and diverse food applications. *Materials*, *16*(13), 4840.

Cao, Y., Otsuka, A., Teah, H. Y., Kanematsu, Y., & Kikuchi, Y. (2025). Consumers' knowledge, perceptions, and preferences regarding bioplastics: Evidence from a survey in Japan. *Sustainable Production and Consumption*, *55*, 1–10.

Chen, Q., Liu, M., Zhang, P., Fan, S., Huang, J., Yu, S., Zhang, C., & Li, H. (2019). Fucoidan and galactooligosaccharides ameliorate high-fat diet-induced dyslipidemia in rats by modulating the gut microbiota and bile acid metabolism. *Nutrition (Burbank, Los Angeles County, Calif.)*, *65*, 50–59.

Chmelová, D., Legerská, B., Ondrejovič, M., & Miertuš, S. (2021). Optimization of propagation medium for enhanced polyhydroxyalkanoate production by *Pseudomonas oleovorans*. *Fermentation*, *8*(1), 16.

Chong, J. W. R., Tan, X., Khoo, K. S., Ng, H. S., Jonglertjunya, W., Yew, G. Y., & Show, P. L. (2022). Microalgae-based bioplastics: Future solution towards mitigation of plastic wastes. *Environmental research*, *206*, 112620.

Claverie, M., McReynolds, C., Petitpas, A., Thomas, M., & Fernandes, S. C. (2020). Marine-derived polymeric materials and biomimetics: An overview. *Polymers*, *12*(5), 1002.

Coppola, G., Gaudio, M. T., Lopresto, C. G., Calabro, V., Curcio, S., & Chakraborty, S. (2021). Bioplastic from renewable biomass: a facile solution for a greener environment. *Earth Systems and Environment*, *5*, 231–251.

Dang, B.-T., Bui, X.-T., Tran, D. P. H., Hao Ngo, H., Nghiem, L. D., Hoang, T.-K.-D., Nguyen, P.-T., Nguyen, H. H., Vo, T.-K.-Q., Lin, C., Yi Andrew Lin, K., & Varjani, S. (2022). Current application of algae derivatives for bioplastic production: A review. *Bioresource Technology*, *347*, 126698.

Devadas, V. V., Khoo, K. S., Chia, W. Y., Chew, K. W., Munawaroh, H. S. H., Lam, M. K., Lim, J. W., Ho, Y. C., Lee, K. T., & Show, P. L. (2021). Algae biopolymer towards sustainable circular economy. *Bioresource Technology*, 325, 124702O.

Duarte, C. M., Bruhn, A., & Krause-Jensen, D. (2022). A seaweed aquaculture imperative to meet global sustainability targets. *Nature Sustainability*, *5*(3), 185–193.

Farhan, A., & Hani, N. M. (2017). Characterization of edible packaging films based on semi-refined kappa-carrageenan plasticized with glycerol and sorbitol. *Food Hydrocolloids*, *64*, 48–58.

Fernandes, M., Salvador, A., Alves, M. M., & Vicente, A. A. (2020). Factors affecting polyhydroxyalkanoates biodegradation in soil. *Polymer Degradation and Stability*, *182*, 109408.

Ghasemlou, M., Barrow, C. J., & Adhikari, B. (2024). The future of bioplastics in food packaging: An industrial perspective. *Food Packaging and Shelf Life*, *43*, 101279.

Gheorghita Puscaselu, R., Lobiuc, A., Dimian, M., & Covasa, M. (2020). Alginate: From food industry to biomedical applications and management of metabolic disorders. *Polymers*, *12*(10), 2417.

Gomes, L. P., Souza, H. K., Campiña, J. M., Andrade, C. T., Silva, A. F., Gonçalves, M. P., & Paschoalin, V. M. F. (2018). Edible chitosan films and their nanosized counterparts exhibit antimicrobial activity and enhanced mechanical and barrier properties. *Molecules*, *24*(1), 127.

Gopi, S., Balakrishnan, P., & Bračič, M. (Eds.). (2022). *Biopolymers in nutraceuticals and functional foods* (Vol. 36). Royal Society of Chemistry.

Grosso, C., Valentão, P., Ferreres, F., & Andrade, P. B. (2015). Alternative and efficient extraction methods for marine-derived compounds. *Marine Drugs*, *13*(5), 3182.

Guidara, M., Yaich, H., Benelhadj, S., Adjouman, Y. D., Richel, A., Blecker, C., Sindic, M., Boufi, S., Attia, H., & Garna, H. (2020). Smart ulvan films responsive to stimuli of plasticizer and extraction condition in physico-chemical, optical, barrier and mechanical properties. *International Journal of Biological Macromolecules*, *150*, 714–726.

Guidara, M., Yaich, H., Richel, A., Blecker, C., Boufi, S., Attia, H., & Garna, H. (2019). Effects of extraction procedures and plasticizer concentration on the optical, thermal, structural and antioxidant properties of novel ulvan films. *International Journal of Biological Macromolecules*, *135*, 647–658.

Gupta, R. K., Guha, P., & Srivastav, P. P. (2022a). Natural polymers in bio-degradable/edible film: A review on environmental concerns, cold plasma technology and nanotechnology application on food packaging-A recent trends. *Food Chemistry Advances*, *1*, 100135.

Gupta, V., Biswas, D., & Roy, S. (2022b). A comprehensive review of biodegradable polymer-based films and coatings and their food packaging applications. *Materials*, *15*(17), 5899.

Harkin, C., Mehlmer, N., Woortman, D. V., Brück, T. B., & Brück, W. M. (2019). Nutritional and additive uses of chitin and chitosan in the food industry. *Sustainable Agriculture Reviews 36: Chitin and Chitosan: Applications in Food, Agriculture, Pharmacy, Medicine and Wastewater Treatment*, 1–43.

Harrison, J. P., Boardman, C., O'Callaghan, K., Delort, A. M., & Song, J. (2018). Biodegradability standards for carrier bags and plastic films in aquatic environments: A critical review. *Royal Society Open Science*, *5*(5), 171792.

Hentati, F., Tounsi, L., Djomdi, D., Pierre, G., Delattre, C., Ursu, A. V., Fendri, I., Abdelkafi, S., & Michaud, P. (2020). Bioactive polysaccharides from seaweeds. *Molecules*, *25*(14), 3152.

Hisham, F., Maziati Akmal, M. H., Ahmad, F., Ahmad, K., & Samat, N. (2024). Biopolymer chitosan: Potential sources, extraction methods, and emerging applications. *Ain Shams Engineering Journal*, *15*(2), 102424.

Irfan, F., Tasnim, N., Razzak, S. A., & Uddin, S. (2025). Advances in capturing carbon and producing bioplastics through the microalgal approach towards a sustainable future: A review. *Journal of Cleaner Production*, *486*, 144576.

Jayakumar, A., Radoor, S., Siengchin, S., Shin, G. H., & Kim, J. T. (2023). Recent progress of bioplastics in their properties, standards, certifications and regulations: A review. *Science of The Total Environment*, *878*, 163156.

Karuppusamy, S., Rajauria, G., Fitzpatrick, S., Lyons, H., McMahon, H., Curtin, J., ... & O'Donnell, C. (2022). Biological properties and health-promoting functions of laminarin: A comprehensive review of preclinical and clinical studies. *Marine Drugs*, *20*(12), 772.

Kavitha, G., Rengasamy, R., & Inbakandan, D. (2018). Polyhydroxybutyrate production from marine source and its application. *International Journal of Biological Macromolecules*, *111*, 102–108.

Kiseleva, S. V., Chernova, N. I., Vlaskin, M. S., Grigorenko, A. V., Chunzhuk, E. A., Malaniy, S. Y., ... & Rositskaya, T. V. (2024). Carbon dioxide absorption by microalgae: Analysis of technologies and energy costs. *Thermal Engineering*, *71*(12), 1038–1048.

Krishnan, L., Ravi, N., Kumar Mondal, A., Akter, F., Kumar, M., Ralph, P., & Kuzhiumparambil, U. (2024). Seaweed-based polysaccharides – Review of extraction, characterization, and bioplastic application. *Green Chemistry*, *26*(10), 5790–5823.

Kumar, R., Lalnundiki, V., Shelare, S. D., Abhishek, G. J., Sharma, S., Sharma, D., Kumar, A., & Abbas, M. (2024). An investigation of the environmental implications of bioplastics: Recent advancements on the development of environmentally friendly bioplastics solutions. *Environmental Research, 244,* 117707.

Lavagnolo, M. C., Poli, V., Zampini, A. M., & Grossule, V. (2024). Biodegradability of bioplastics in different aquatic environments: A systematic review. *Journal of Environmental Sciences, 142,* 169–181.

Lee, T., Wearne, S., Wakefield-Rann, R., & Scardifield, K. (2024). Enabling factors and critical extensions for the algal-based bioplastics industry. *Journal of Cleaner Production*, *457*, 142365.

Li, C., Tang, T., Du, Y., Jiang, L., Yao, Z., Ning, L., & Zhu, B. (2023). Ulvan and Ulva oligosaccharides: A systematic review of structure, preparation, biological activities and applications. *Bioresources and Bioprocessing*, *10*(1), 66.

Lim, C., Yusoff, S., Ng, C. G., Lim, P. E., & Ching, Y. C. (2021a). Bioplastic made from seaweed polysaccharides with green production methods. *Journal of Environmental Chemical Engineering*, *9*(5), 105895.

Lionetto, F., & Esposito Corcione, C. (2021). Recent applications of biopolymers derived from fish industry waste in food packaging. *Polymers*, *13*(14), Article 14.

Luthuli, S., Wu, S., Cheng, Y., Zheng, X., Wu, M., & Tong, H. (2019). Therapeutic effects of fucoidan: A review on recent studies. *Marine Drugs*, *17*(9), 487.

Mahmud, N., Islam, J., & Tahergorabi, R. (2021). Marine biopolymers: Applications in food packaging. *Processes*, *9*(12), 2245.

Martiny, T. R., Pacheco, B. S., Pereira, C. M. P., Mansilla, A., Astorga–España, M. S., Dotto, G. L., Moraes, C. C., & Rosa, G. S. (2020). A novel biodegradable film based on κ-carrageenan activated with olive leaves extract. *Food Science & Nutrition*, *8*(7), 3147–3156.

Matheus, J. R. V., Dalsasso, R. R., Rebelatto, E. A., Andrade, K. S., Andrade, L. M. D., Andrade, C. J. D., ... & Fai, A. E. C. (2023). Biopolymers as green-based food packaging materials: A focus on modified and unmodified starch-based films. *Comprehensive Reviews in Food Science and Food Safety*, *22*(2), 1148–1183.

Mekonnen, T., Mussone, P., Khalil, H., & Bressler, D. (2013). Progress in bio-based plastics and plasticizing modifications. *Journal of Materials Chemistry A*, *1*(43), 13379–13398.

Milovanovic, I., & Hayes, M. (2018). Marine gelatine from rest raw materials. *Applied Sciences*, *8*(12), 2407.

Monteiro, L. P., Borges, J., Rodrigues, J. M., & Mano, J. F. (2023). Unveiling the assembly of neutral marine polysaccharides into electrostatic-driven layer-by-layer bioassemblies by chemical functionalization. *Marine Drugs*, *21*(2), 92.

Możejko-Ciesielska, J., Ray, S., & Sankhyan, S. (2023). Recent challenges and trends of polyhydroxyalkanoate production by extremophilic bacteria using renewable feedstocks. *Polymers*, *15* (22), 4385.

Muthulakshmi, L., Rajam, R., Mohan, S., & Karthik, P. (2023). Recent advances in the development of phb (polyhydroxybutyrate)-based packaging materials. In Sneh Punia Bangar and Anil Kimar Siroha (Eds.), *Biopolymer-Based Films and Coatings*. Taylor & Francis, pp. 265–289.

Nagarajan, D., Senthilkumar, G., Chen, C.-W., Karmegam, N., Praburaman, L., Kim, W., & Dong, C.-D. (2024). Sustainable bioplastics from seaweed polysaccharides: A comprehensive review. *Polymers for Advanced Technologies*, *35*(8), e6536.

Necas, J., & Bartosikova, L. (2013). Carrageenan: A review. *Veterinární Medicína*, *58*(4), 187–205.

Pathak, G., Nichter, M., Hardon, A., Moyer, E., Latkar, A., Simbaya, J., Pakasi, D., Taqueban, E., & Love, J. (2023). Plastic pollution and the open burning of plastic wastes. *Global Environmental Change*, *80*, 102648.

Pereira, L. (2016). *Edible Seaweeds of the World.* CRC Press. https://doi.org/10.1201/b19970

Prabhu, M., Chemodanov, A., Gottlieb, R., Kazir, M., Nahor, O., Gozin, M., Israel, A., Livney, Y. D., & Golberg, A. (2019). Starch from the sea: The green macroalga *Ulva ohnoi* as a potential source for sustainable starch production in the marine biorefinery. *Algal Research*, *37*, 215–227.

Pramanik, S., Singh, A., Abualsoud, B. M., Deepak, A., Nainwal, P., Sargsyan, A. S., & Bellucci, S. (2024). From algae to advancements: Laminarin in Biomedicine. *Rsc Advances*, *14*(5), 3209–3231.

Ramli, R. N., Lee, C. K., & Kassim, M. A. (2020). Extraction and characterization of starch from microalgae and comparison with commercial corn starch. *IOP Conference Series: Materials Science and Engineering*, *716*(1), 012012.

Ranganadhareddy, A., & Chandrasekhar, C. (2023). Biopolymer from marine waste biomass and its applications-A review. *Journal of Biochemical Technology*, *14*(2), 87–93.

Rathod, N. B., Bangar, S. P., Šimat, V., & Ozogul, F. (2023). Chitosan and gelatine biopolymer-based active/biodegradable packaging for the preservation of fish and fishery products. *International Journal of Food Science & Technology, 58(2)*, 854–861.

Robal, M., Brenner, T., Matsukawa, S., Ogawa, H., Truus, K., Rudolph, B., & Tuvikene, R. (2017). Monocationic salts of carrageenans: Preparation and physico-chemical properties. *Food Hydrocolloids*, *63*, 656–667.

Roy, J. C., Salaün, F., Giraud, S., Ferri, A., Chen, G., & Guan, J. (2017). Solubility of chitin: Solvents, solution behaviors and their related mechanisms. In *Solubility of Polysaccharides*. IntechOpen, p. 109.

Ruocco, N., Costantini, S., Guariniello, S., & Costantini, M. (2016). Polysaccharides from the marine environment with pharmacological, cosmeceutical and nutraceutical potential. *Molecules*, *21*(5), 551.

Rupert, R., Rodrigues, K. F., Thien, V. Y., & Yong, W. T. L. (2022). Carrageenan From *Kappaphycus alvarezii* (Rhodophyta, Solieriaceae): Metabolism, structure, production, and application. *Frontiers in Plant Science*, *13*, 859635.

Sadvakasova, A. K., Kossalbayev, B. D., Bauenova, M. O., Balouch, H., Leong, Y. K., Zayadan, B. K., Huang, Z., Alharby, H. F., Tomo, T., Chang, J. S., & Allakhverdiev, S. I. (2023). Microalgae as a key tool in achieving carbon neutrality for bioproduct production. *Algal Research*, *72*, 103096.

Sharma, S., Sharma, V., & Chatterjee, S. (2023). Contribution of plastic and microplastic to global climate change and their conjoining impacts on the environment – A review. *Science of The Total Environment*, *875*, 162627.

Singh, R., Shitiz, K., & Singh, A. (2017). Chitin and chitosan: Biopolymers for wound management. *International Wound Journal*, *14*(6), 1276–1289.

Skryplonek, K., Henriques, M., Gomes, D., Viegas, J., Fonseca, C., Pereira, C., Dmytrów, I., & Mituniewicz-Małek, A. (2019). Characteristics of lactose-free frozen yogurt with κ-carrageenan and corn starch as stabilizers. *Journal of Dairy Science*, *102*(9), 7838–7848.

Song, Z., Li, G., Guan, F., & Liu, W. (2018). Application of chitin/chitosan and their derivatives in the paper-making industry. *Polymers*, *10*(4), 389.

Stanisci, A., Tøndervik, A., Gaardløs, M., Lervik, A., Skjåk-Bræk, G., Sletta, H., & Aachmann, F. L. (2020). Identification of a pivotal residue for determining the block structure-forming properties of alginate C-5 epimerases. *ACS Omega*, *5*(8), 4352–4361.

Stanley, A., Punil Kumar, H. N., Mutturi, S., & Vijayendra, S. V. N. (2018). Fed-batch strategies for production of PHA using a native isolate of *Halomonas venusta* KT832796 strain. *Applied Biochemistry and Biotechnology*, *184*(3), 935–952.

Sudhakar, M. P., Kumar, B. R., Mathimani, T., & Arunkumar, K. (2019). A review on bioenergy and bioactive compounds from microalgae and macroalgae-sustainable energy perspective. *Journal of Cleaner Production*, *228*, 1320–1333.

Sudhakar, M. P., Maurya, R., Mehariya, S., Karthikeyan, O. P., Dharani, G., Arunkumar, K., Pereda, S. V., Hernández-González, M. C., Buschmann, A. H., & Pugazhendhi, A. (2024). Feasibility of bioplastic production using micro- and macroalgae- A review. *Environmental Research*, *240*, 117465.

Tennakoon, P., Chandika, P., Yi, M., & Jung, W. K. (2023). Marine-derived biopolymers as potential bioplastics, an eco-friendly alternative. *Iscience*, *26*(4), 106404.

Vázquez, J. A., Meduíña, A., Durán, A. I., Nogueira, M., Fernández-Compás, A., Pérez-Martín, R. I., & Rodríguez-Amado, I. (2019). Production of valuable compounds and bioactive metabolites from by-products of fish discards using chemical processing, enzymatic hydrolysis, and bacterial fermentation. *Marine Drugs*, *17*(3), Article no 3.

Venugopal, V., & Sasidharan, A. (2021). Seafood industry effluents: Environmental hazards, treatment and resource recovery. *Journal of Environmental Chemical Engineering*, *9*(2), 104758.

Wang, H., Cao, Z., Yao, L., Feng, T., Song, S., & Sun, M. (2023). Insights into the edible and biodegradable ulvan-based films and coatings for food packaging. *Foods*, *12*(8), 1622.

Wang, Y., Xing, M., Cao, Q., Ji, A., Liang, H., & Song, S. (2019). Biological activities of fucoidan and the factors mediating its therapeutic effects: A review of recent studies. *Marine Drugs*, *17*(3), 183.

Wu, L., Sun, J., Su, X., Yu, Q., Yu, Q., & Zhang, P. (2016). A review about the development of fucoidan in antitumor activity: Progress and challenges. *Carbohydrate Polymers*, *154*, 96–111.

Xie, C., Lee, Z. J., Ye, S., Barrow, C. J., Dunshea, F. R., & Suleria, H. A. R. (2024). A review on seaweeds and seaweed-derived polysaccharides: Nutrition, chemistry, bioactivities, and applications. *Food Reviews International, 40(5),* 1312–1347.

Xue, W., Zhu, J., Sun, P., Yang, F., Wu, H., & Li, W. (2023). Permeability of biodegradable film comprising biopolymers derived from marine origin for food packaging application: A review. *Trends in Food Science & Technology*, *136*, 295–307.

Yang, Y., Min, Z., Alalawy, A., Almutairi, F., Al Duais, M., Wang, J., & Salama, E.-S. (2021). Identification and characterization of marine seaweeds for biocompounds production. *Environmental Technology & Innovation*, *24*, 101848.

Yousefi, M., Rajabi Islami, H., & Filizadeh, Y. (2013). Effect of extraction process on agar properties of *Gracilaria corticata* (Rhodophyta) collected from the Persian Gulf. *Phycologia*, *52*, 481–487.

Zhang, L., Liao, W., Huang, Y., Wen, Y., Chu, Y., & Zhao, C. (2022). Global seaweed farming and processing in the past 20 years. *Food Production, Processing and Nutrition*, *4*(1), 23.

Zhao, D., Xu, J., & Xu, X. (2018). Bioactivity of fucoidan extracted from *Laminaria japonica* using a novel procedure with high yield. *Food Chemistry*, *245*, 911–918.

Zia, K. M., Tabasum, S., Nasif, M., Sultan, N., Aslam, N., Noreen, A., & Zuber, M. (2017). A review on synthesis, properties and applications of natural polymer based carrageenan blends and composites. *International Journal of Biological Macromolecules*, *96*, 282–301.

16 Life Cycle Assessment and Applications of Starch-Based Biopolymers Towards Net-Zero Carbon Bioeconomy

Gitanjali Jothiprakash and Subburamu Karthikeyan

16.1 INTRODUCTION

The shift of a fossil fuel-based economy to a sustainable bioeconomy has gained significant consideration over recent decades, particularly with the growing environmental concerns linked with petro-based plastics. The global plastic production is estimated to be 400.3 MT in 2022, of which 90% is derived from petrochemical sources (Plastics Europe, 2023). The global $CO_{2eq.}$ emissions are around 54.6 BT, of which 3.3% are due to plastic annually, aggravating pollution and climate change issues (Hannah, 2023).

Following, biopolymers have appeared as probable alternatives, characterised by their renewable source, biodegradability, and a smaller carbon footprint. Starch-based biopolymers may be considered for their abundance, ease of processing, low cost, and favourable physicochemical properties, viz., film-forming ability, hydrophilicity, and compatibility compared to other biopolymers. Starch is estimated to exceed 100 MT annually, mainly from potato (5%), wheat (8%), cassava (12%), and corn (75%) (Forfora et al., 2024). Around 10–15% of global starch production is focused on non-food applications, such as packaging, agriculture, medicine, textiles, pharmaceuticals, etc (Amobonye et al., 2025)

As the world falls towards a net-zero carbon bioeconomy in future, starch-based biopolymers present a promising path for decreasing resource dependence and emissions. In the product life cycle, replacing one tonne of petroleum-based polyethylene with starch–polymer composites resulted in a reduction of 1.5 tonnes of CO_2 emissions (Islam et al., 2025). Considering the complete life cycle of biopolymers from feedstock cultivation to end-of-life disposal, it is crucial to optimise their sustainable deployment and assess their factual environmental impact across various industrial sectors.

16.2 STARCH-BASED BIOPOLYMER RESOURCES

Starch is a naturally available polysaccharide that aids as the main carbohydrate storage molecule in various plants. It is broadly utilised as a renewable source for biopolymer production due to its biodegradability, abundance, and favourable film-forming properties. The corn is the dominant starch source, contributing to around 60% of total starch production globally. The maize kernels yield about 720 kg of starch per tonne of grain, making it a highly effective source (Ranum et al., 2014). The next source is cassava roots, which contribute 250–300 kg of starch per tonne of fresh cassava roots. Cassava yields range from 15 to 20 tonnes per hectare annually (Howeler, 2004). Furthermore, cassava's adaptability to marginal and nutrient-poor soils makes it an efficient choice for sustainable starch production in areas where conventional crops find it difficult to grow.

DOI: 10.1201/9781003534914-16

TABLE 16.1
Properties of Various Starch Types Related to Biopolymer Production

Starch Source	Extraction Yield (kg/T)	Typical Crop Yield (T/ha/year)	Biopolymer Properties	Application	Ref.
Corn	~720	5–10	Moderate barrier, moderate tensile strength	Packaging films	Sobeih et al. (2025)
Cassava	250–300	15–20	Low tensile strength, high flexibility	Biodegradable packaging films	Florencia et al. (2020)
Wheat	450–500	3–6	Moderate moisture barrier	Food packaging, edible films	Wikifarmer, (2020)
Rice	400–450	4–8	Moderate water vapour permeability, Flexible films,	Edible coatings, packaging films	Qun et al. (2023)
Potato	500–600	20–40	Moderate barrier, high transparency	Packaging films, agricultural films	Rashwan et al. (2024)
Sweet Potato	350–400	10–25	Moderate mechanical strength	Biodegradable packaging films	Ye et al. (2020)
Barley	300–350	3–6	Moderate tensile strength	Bioplastics	Zhu (2017)
Banana Peel	100–200	-	Good barrier property	Bioplastics	Seth et al. (2025)
Mango Seed	150–250	-	High moisture resistance, High tensile strength	Bioplastics	Nawab et al. (2022)

Other starch sources are rice, wheat, barley, potato, sweet potato, and emergent sources like tubers (yam), fruit waste (mango seed, banana peel), and underutilised crops (millet). The extraction yield varies significantly among sources, affecting the mechanical properties and barrier properties of the resultant biopolymer. The high-amylose starch (< 40%) yields biopolymer films with good tensile strength and less water vapour permeability, appropriate for food packaging applications (Fatima et al., 2024).

Table 16.1 summarises the starch sources, extraction yields, and properties critical for the performance of the biopolymer. The starch feedstock selection is important in adapting the final product for specific applications, ranging from packaging films to agricultural films, matching durability and biodegradability.

The factor influencing the characteristic performance of starch-based biopolymers is the starch composition, mainly the amylose to amylopectin ratio. This ratio fluctuates significantly among different plant sources and affects the flexibility, mechanical strength, and barrier properties of the biopolymer films. The high-amylose starch is characterised by amylose content <40% and is known to form films with improved tensile strength and lowered water vapour permeability. The high-amylose starch is especially suitable for applications necessitating moisture-resistant and durable packaging materials (Varghese et al., 2022). On the contrary, starch with low amylose content tends to form biopolymer films that are flexible but low resistant to transferring moisture, which may limit certain food packaging situations.

16.3 APPLICATIONS AND MARKET TRENDS

Starch-based biopolymers are gradually being adopted across various sectors due to their functional adaptability and eco-friendly application. These polymers exhibit desirable films, thermoplastic and water-resistant, making them appropriate for varied industrial applications (Flórez et al., 2023). Figure 16.1 illustrates the various applications of starch-based biopolymers.

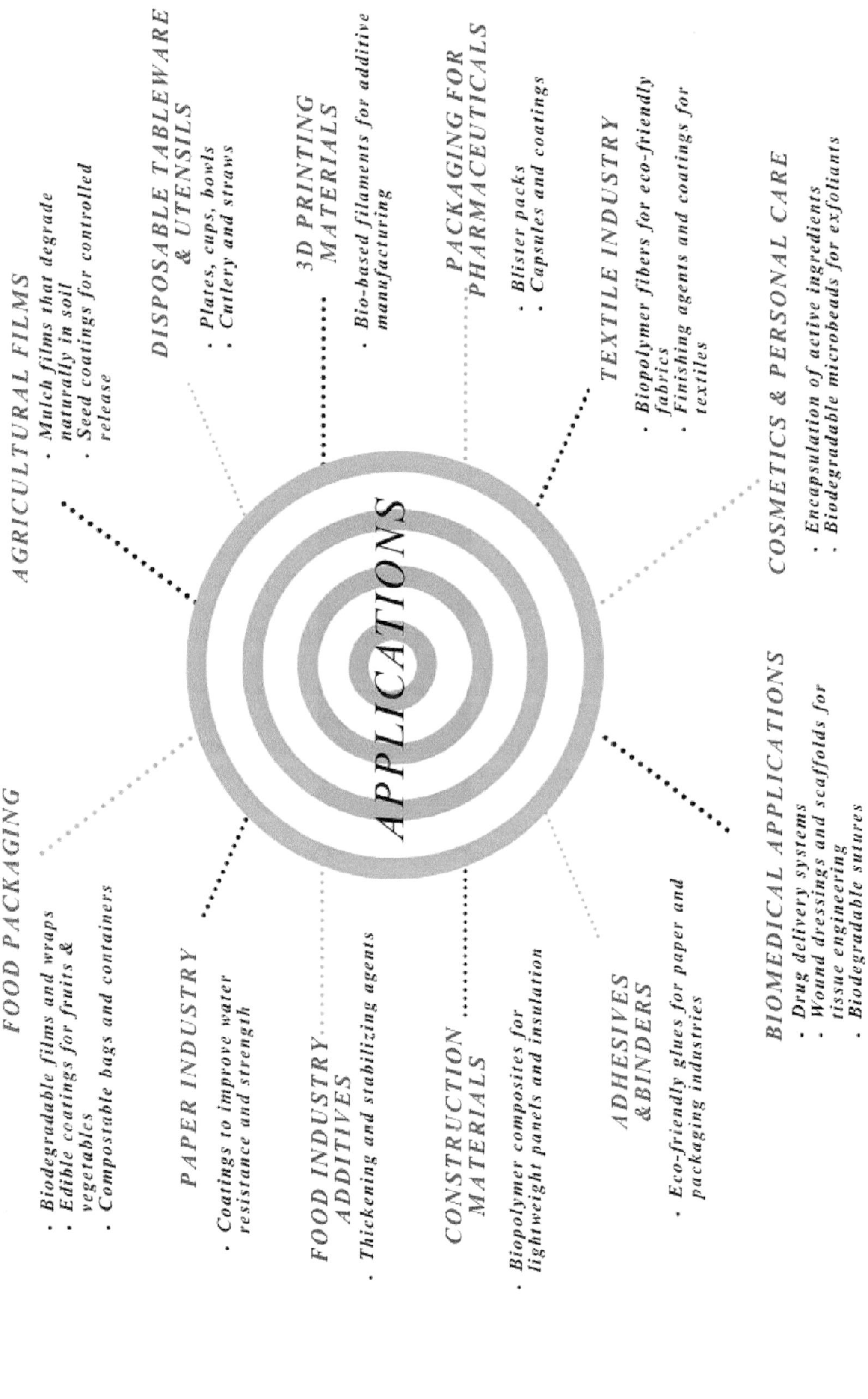

FIGURE 16.1 Applications of starch-based biopolymers.

In the packaging industry, starch-based biopolymer blends presently account for about 15% of the biodegradable plastics market globally, with an estimated value of USD 7.97 billion in 2024. This market is predicted to increase at a CAGR of 6.6% through 2030, driven by rigorous regulations on single-use plastics and intensifying consumer awareness (Grand View Research, 2023). Starch is normally combined with other biodegradable polymers, namely polylactic acid, polycaprolactone, and polybutylene adipate terephthalate to improve flexibility, mechanical strength, and moisture resistance (Agostinho et al., 2023). These blends are broadly utilised in disposable packaging products, such as grocery bags, stretch films, clamshell containers, and food wrapping materials, providing a compostable alternative to conventional plastic packaging material (Dybka-Stępień et al., 2021).

In the agricultural sector, it has found an important role in replacing non-degradable mulch films (polyethylene films). The biodegradable mulch films were valued at around USD 64 million in 2023 in the global market and are anticipated to attain USD 126.8 million by 2032, at a CAGR of 8.1% with starch derivatives including a major share due to their soil compatibility and biodegradability (Fortune Business Insights, 2025). These films conserve soil moisture, suppress weed growth, and decompose naturally. It eliminates the labour-intensive and ecologically damaging disposal processes associated with conventional films. The utilisation of these films supports sustainable agricultural practices by minimising microplastic contamination in soil environments.

In the pharmaceutical and biomedical industries, starch-based materials have proved better biocompatibility, non-toxicity, and chemical modification capacity, making them suitable for innovative therapeutic applications. It is employed in the fabrication of nanocarriers, hydrogels, microparticles, and porous frameworks used for wound dressing, tissue engineering, and drug delivery. The starch-based hydrogel is capable of encapsulating able to 70–80% of active pharmaceutical ingredients as starch composition, enabling measured release profiles in the range of a few hours to several days (Lee & Hwang, 2023). It also supports cellular adhesion and proliferation, making them suitable for regenerative medicine and bioactive wound care.

In the paper and textile industries, modified starches act as functional additives for coating, sizing, and surface modification. In the European starch industry, the annual starch production is estimated at 9.2 million tonnes in 2023. Among this, 7.2 million tonnes of starch and its derivatives are consumed in European Union in the form of native starches (34%), modified starches (18%), and starch sweeteners (48%) (Starch Europe, 2023). The derivatives of starch improve paper stiffness, ink retention, printability, and surface smoothness, contributing to both the visual quality and mechanical performance of resultant products. In the textile industry, starch-based formulations are used in distortion sizing to enhance yarn strength and decrease breakage during weaving (Li et al., 2019).

Emerging and novel applications are being explored in transient electronics and additive manufacturing. The starch-based filaments can be engineered with flexible thermal properties and mechanical properties, ideal for low-impact designs and biodegradable prototypes in 3D printing. Moreover, in transient electronics, the devices that disintegrate after a functional lifespan are leveraging starch biodegradability and solubility for the manufacturing of temporary sensors and circuit substrates (Xiang et al., 2022). Such novelties align with circular economy values for e-waste reduction, offering sustainable solutions. Generally, the multifunctional effectiveness of starch-based biopolymers, coupled with their compatibility and biodegradability with other bio-based materials, acts as a basis of green material innovation across emerging and traditional industrial sectors.

16.4 LIFE CYCLE ASSESSMENT OF STARCH-BASED BIOPOLYMERS

Life cycle assessment (LCA) is a quantitative tool designed to assess the environmental impacts connected with all life stages of a product, from the raw material extraction to final recycling or disposal. In starch-based biopolymers, cradle-to-grave LCA analyses characteristically assess the

TABLE 16.2
Life Cycle Assessment of Starch-Based Biopolymers

Feedstock	Additives	Global Warming Potential kg $CO_{2\,eq}$/kg	Acidification Potential, kg $SO_{2\,eq}$ /kg	Eutrophication Potential kg $PO_{4\,eq}$/kg	Ref.
Corn starch	-	1.88	3.76 E–03	3.20 E–04	Moy et al. (2021)
Cassava starch	-	2	1.66 E–02	9.23 E–03	Papong et al. (2014)
Potato starch	Glycerol	2.54	1.97E–02	9.74E–04	Garcia Rengifo et al. (2022)
Sweet potato starch	Glycerol	3.22	9.16 E–03	3.95E–04	
Mango kernel starch	Glycerol and starch nanocrystals	4.5	1.8E–02	1.19E–03	Muller Carneiro et al. (2023)
Wheat flour starch	PVOH	1.61	1.4E–2	3.52E–03	Guo (2013)

impact categories, namely energy consumption, greenhouse gas emissions expressed as CO_2-equivalent, land occupation, water usage, and acidification potential (Broeren et al., 2017). The various LCA of starch-based biopolymers of different feedstock is given in Table 16.2.

The production of 1 kg of starch-based biopolymer typically results in environmental load of 3.2–5.8 kg of CO_{2eq}, compared to 1 kg of petroleum-based plastics production with 2.6–4.8 kg of CO_{2eq} (PET) and 2.8 (LDPE) (Forfora et al., 2024). It highlights the considerable potential of starch-based materials to decrease carbon footprints in the plastics sector. A substantial portion of the total environmental problem in starch-based biopolymer production originates from the agricultural phase, accounting for 40% of total Green House Gas (GHG) emissions. These emissions start from fertiliser application, particularly nitrogen-based fertilisers and land-use changes (deforestation or soil disturbance). Fertiliser use nitrous oxide emissions, an effective GHG, while land adaptation can lead to carbon stock loss and biodiversity (Bishop et al., 2021).

The agricultural impact can be greatly reduced by using starch from food and farm waste instead of main crops. The utilisation of starch extracted from residues (potato peels, corn husks, or cassava waste) not only reduces the requirement for dedicated land cultivation but also averts methane emissions, which arise from anaerobic disintegration of organic waste in landfills (Valle et al., 2024). The incorporation of agro-waste-derived starch into biopolymer production can reduce total GHG emissions of~5.88 kg of CO_{2eq} /kg while concurrently leading to water conservation and fertiliser input reduction (Ali et al., 2023). LCA underscores the essentiality of feedstock selection and agricultural practices in the environmental performance of these biopolymers. The starch-based plastics are a sustainable biomass source with energy-efficient manufacturing and can offer a lower environmental impact than their petroleum-based plastics, supporting the transition toward a low-carbon and circular bioeconomy.

16.5 BENEFITS AND CHALLENGES IN LCA

Starch-based biopolymers are gradually recognised as eco-friendly alternatives to conventional petro-based plastics, due to their renewability, biodegradability, and relatively small production costs. The benefits and challenges of applying LCA to biopolymers are depicted in Figure 16.2. LCA offers carbon footprint quantification, enabling detailed decision information in selecting materials and designing products sustainably. The assessment aids identify process inadequacies, optimise resource use, and support compliance with certification schemes and environmental regulations. Furthermore, LCA can assist product developers in aligning with principles of the circular economy by providing opportunities to improve end-of-life biodegradability or use agro-residues (Hottle et al., 2017; Gowthaman et al., 2021; Arias et al., 2021).

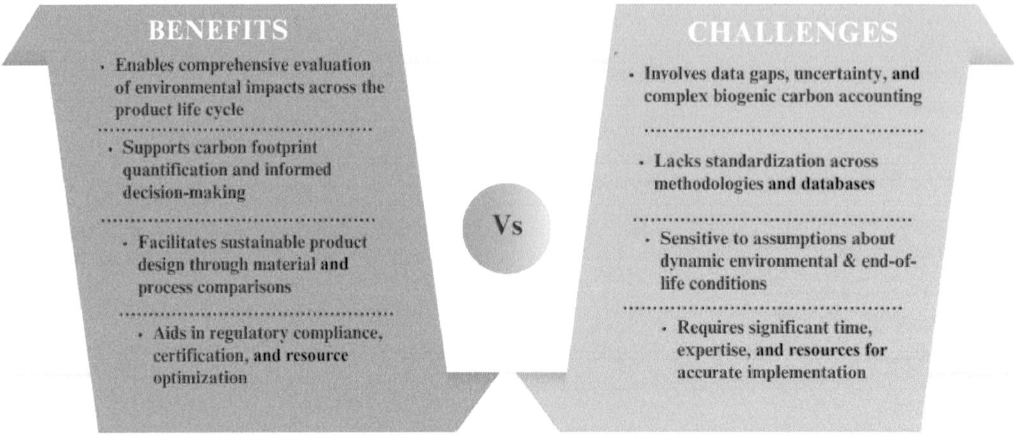

FIGURE 16.2 Benefits and challenges of starch-based biopolymers.

Despite these benefits, LCA implementation for starch-based biopolymers presents numerous challenges. One major challenge is the occurrence of uncertainties and data gaps, especially about biodegradation rates and regional agricultural practices. Also, the difficulty of biogenic carbon accounting, following carbon absorbed during plant growth and released during degradation, can lead to variations. The lack of standardisation in LCA practices and databases further obscures comparisons between products. Besides, LCA results are complex due to assumptions about environmental conditions, disposal situations, and land use changes, which can alter the outcomes. Lastly, conducting LCA demands software tools, specialised information, and considerable resources and time, making it less manageable for small-scale producers (Ita-Nagy et al., 2020; Mtibe et al., 2021; Ragitha et al., 2022; Hernandez et al., 2025). Addressing these challenges is vital to fully realise the environmental benefits of starch-based biopolymers.

16.6 BIODEGRADABILITY AND ENVIRONMENTAL IMPACT OF STARCH-BASED BIOPOLYMERS

Starch-based biopolymers are recognised for inherent biodegradability, particularly under controlled industrial composting conditions. As per International Standards namely, ASTM D6400 and ISO 17088, these materials can achieve more than 90% biodegradation within 90 days when subjected to 58°C at industrial composting facilities. This rapid breakdown outcomes in the polymer conversion into substances, with carbon dioxide, water, and microbial biomass. When exposed to natural soil environments at ambient temperatures, biodegradation of starch-based bioplastics normally occurs over a longer timescale from 6 to 12 months. The exact rate is inclined by environmental factors such as temperature fluctuations, soil moisture levels, and diversity of microbial communities capable of enzymatically degrading starch polymers (Muniyasamy et al., 2013, Fredi & Dorigato, 2021).

The environmental impact of these biopolymers ranges beyond the kinetics of biodegradation. One of their most substantial advantages over conventional plastics is the minimal development of microplastics. Then starch-based materials endure near-complete mineralisation and do not leave determined microplastic fragments that can accumulate in ecosystems. This is a crucial factor in long-term pollution mitigation, particularly in marine environments where microplastics pose considerable risks to aquatic life (Elkaliny et al., 2024). Toxicological assessments highlight the environmental benefits of starch biopolymers. This suggests a reduced ecological risk, which is critical given the extensive apprehensions over plastic pollution in freshwater and marine systems. Furthermore, starch-based bioplastics do not harvest persistent soil residues, which aids sustain soil structure and microbial health, thus supportive sustainable land use and agricultural productivity (Appelqvist & Debet, 1997; Aung et al., 2018).

The environmental benefits of starch-based biopolymers depend profoundly on proper end-of-life management. Improper disposal, such as mixing with non-biodegradable plastics in recycling system or landfilling without acceptable composting conditions, can suggestively obstruct complete biodegradation. Such contamination can compromise recycling efforts and reduce the environmental benefits of these materials. Thus, effective waste segregation approaches, strong product labelling, and public education are vital to maximise the ecological potential of starch-based biopolymers. Executing these measures ensures that the materials achieve their promise of plastic pollution reduction and environmental sustainability promotion.

16.7 NET-ZERO CARBON BIOECONOMY APPROACH

Attaining a net-zero carbon bioeconomy in the starch-based biopolymer sector needs a comprehensive approach that incorporates LCA, renewable energy integration, and sustainable feedstock utilisation. This approach is to understand the material lifetime, including the phase length, extraction methods, and end-of-life management. Starch-based biopolymers are suitable for short-term to medium-term applications, with product lifetimes ranging from 3 to 24 months, depending on the type of application and functional reinforcement. It is used in agricultural films, food packaging, and disposable cutlery (Mousavi-Avval et al., 2023; Dutta & Sit, 2024; Sheldon, 2024). The different approaches for achieving net-zero carbon bioeconomy by using starch-based biopolymers are given in Table 16.3.

TABLE 16.3
Approach to Achieve a Net-Zero Carbon Bioeconomy with Starch-Based Biopolymers

Approach	Description	Remarks	Ref.
Feedstock utilization	Use of second-generation biomass feedstock	Up to 25% reduction in GHG emissions	Gonzalez-Garcia et al. (2021)
Renewable energy inputs	Manufacturing is powered by renewable energy to minimise fossil fuel use	100% renewable energy use	Zhao et al. (2023)
Green extraction methods	Adoption of enzymatic hydrolysis and solvent-free process to reduce energy consumption and chemical use	Reducing energy by 30% and water 50% compared to conventional methods	Rashwan et al. (2024)
Carbon capture integration	Carbon capture technology implementation during polymerisation or fermentation to reduce carbon emissions	Captures 0.1–0.3 kg CO_2 per kg of produced biopolymer	Irfan et al. (2024)
Life cycle emissions	Total GHG emissions associated with starch-based biopolymer production.	1.8–3.7 kg CO_2-eq/kg biopolymer	Ali et al. (2023)
Biodegradability and end-of-life	Complete biodegradation under industrial composting condition	> 90% biodegradation in 90 days at 58°C	Avérous and Pollet (2012)
Circular economy integration	Recycling design, composting, and reducing waste, via closed-loop material cycles	Potential to decrease plastic waste by up to 50% in the packaging sector	Ghasemlou et al. (2024)
Carbon budget alignment	Use of LCA and carbon budget	The GHG reduction target with limiting warming to 1.5°C requires a 45% emissions reduction in the plastics sector by 2030	Glew et al. (2012)

To reduce the environmental problem related to production, green extraction methods, viz., enzymatic hydrolysis, ultrasound-assisted processes, and supercritical CO_2 extraction, are being adopted. Enzymatic hydrolysis can reduce energy consumption by 40–60% compared to the conventional acid hydrolysis method, by maintaining high product yield and purity. The energy-efficient method also improves the energy return on investment with second-generation feedstock material like forest residues and agro-wastes (Pinaeva and Noskov, 2024; Clauser et al. 2025). End-of-life management of starch-based biopolymers plays a critical role in the carbon footprint. The industrially composted biopolymer, when subjected to controlled conditions, can achieve over 90% biodegradation within 90 days, converting into carbon dioxide, water, and microbial biomass. In natural conditions, the soil biodegradation process is slow, occurring over 6–12 months, depending on moisture content, temperature, and microbial source. Notably, these biopolymers do not leave persistent residues or microplastic fragments, thereby mitigating long-term ecological risks and preserving soil health (Folino et al., 2023; Ahsan et al., 2023; Tamoor et al., 2023).

Biogenic carbon accounting supports the starch-based biopolymers. Subsequently, starch is a renewable material derived from plant biomass; it characteristically absorbs CO_2 during cultivation. This absorbed carbon from the atmosphere, upon degradation, is released back into the atmosphere, forming a closed-loop cycle that does not contribute to an increase in net atmospheric carbon (Hosha et al., 2020; Wang et al., 2021). The second-generation feedstock utilisation, such as lignocellulosic biomass and non-food agricultural residues, is especially beneficial as it evicts competition with food crops and can decrease life cycle GHG emissions by 80–90 % compared to petro-based plastics (Morais et al., 2022). Added emission reductions can be attained by integrating carbon capture techniques into the polymer production process. During polymerisation or fermentation stages, CO_2 released by microbial breakdown can be captured and either repurposed or stored, offsetting 15–20% of residual emissions depending on the capture efficiency. It cost around $60–675 per metric ton CO_2 removed across the various fermentation pathways (Clauser et al., 2025). The incorporation of carbon capture systems enhances the production cycle and moves toward carbon neutrality.

By quantifying environmental impacts and aligning with carbon budgeting frameworks with the Intergovernmental Panel on Climate Change (IPCC) to meet climate target, LCA supports the plastics industry in reducing its emissions by over 90 % by 2050. The target is to reduce emissions from 1.8 to 3.5 kg CO_2-eq. per kg of petro-based polymer to a range of 0.3–0.4 kg CO_2-equivalent per kg of biopolymer (Morais et al., 2022, Zheng & Suh, 2019). In the starch-based biopolymer industry, the net-zero carbon emissions are viable through an integrated approach that combines sustainable resource utilisation, biogenic carbon accounting, low-energy processing techniques, and environmental impact assessment methodologies (Olawumi et al., 2024). This strategy not only aligns with global climate commitments but also paves the route for a circular, resilient, and low-impact bioeconomy.

16.8 CONCLUSIONS AND FUTURE TECHNOLOGICAL ADVANCEMENTS

Starch-based biopolymers are a sustainable alternative to conventional petro-based plastics, offering several benefits that align with the net-zero carbon bioeconomy goals. The life-cycle advantages range from the use of renewable resources to their inherent biodegradability, positioning them as key players in the shift toward circular economy models. The raw materials not only decrease fossil resources dependency but also reduce environmental pollution through improved end-of-life degradation. Despite these advantages, numerous challenges persist that limit wider adoption, mainly related to the barrier performance, mechanical strength, and scalability of production processes.

The future technological advancements are vital to unlock the complete potential of starch-based biopolymers. One critical area of development involves improving their mechanical and barrier properties to make them more competitive with traditional plastics. Recent research highlights the incorporation of nanocomposite materials, such as starch combined with nano-clays, which can enhance film strength, flexibility, and resistance to moisture and gases. These nanocomposites demonstrate promising improvements, sometimes increasing tensile strength by up to 50% and reducing permeability by 30–40%, thereby broadening the range of potential applications in packaging and beyond. Furthermore, the scale-up of enzymatic and solvent-free processing methods represents a significant advancement towards greener production pathways. These methods decrease energy consumption and avoid harmful chemical solvents, addressing health and environmental concerns associated with conventional extraction methods. Industrial-scale implementation of such processes could lead to a decrease in overall production costs while sustaining product quality, finally supporting the commercial feasibility of starch-based biopolymers.

Another emergent frontier is the incorporation of artificial intelligence into LCA tools. AI-driven environmental monitoring systems can allow sustainability metrics and real-time tracking of carbon footprints during the biopolymer manufacturing process. This dynamic feedback loop enables quick optimisation, confirming that production improves transparency for stakeholders and remains within targeted environmental thresholds. The AI convergence, big data analytics, and sustainability science hold promise for transforming how environmental impacts are managed and measured in the biopolymer industry. By promoting interdisciplinary collaboration across biotechnology, material science, and environmental engineering, starch-based biopolymers have the potential in the global push for sustainability. Sustained innovation in refining eco-friendly manufacturing techniques, enhancing material performance, and harnessing digital tools for environmental oversight will be influential in scaling these bio-based resources to meet growing demands. Finally, it will contribute not only to decreasing the carbon footprint of plastics but also to producing circular economic systems, resilient that protect environmental health for future generations.

REFERENCES

Agostinho, B., Silvestre, A.J., Coutinho, J.A. and Sousa, A.F., 2023. Synthetic (bio) degradable polymers–when does recycling fail? *Green Chemistry*, 25(1), p. 13–31. https://doi.org/10.1039/D2GC02726G

Ahsan, W.A., Hussain, A., Lin, C. and Nguyen, M.K., 2023. Biodegradation of different types of bioplastics through composting—A recent trend in green recycling. *Catalysts*, 13(2), p. 294. https://doi.org/10.3390/catal13020294

Ali, S.S., Abdelkarim, E.A., Elsamahy, T., Al-Tohamy, R., Li, F., Kornaros, M., Zuorro, A., Zhu, D. and Sun, J., 2023. Bioplastic production in terms of life cycle assessment: A state-of-the-art review. *Environmental Science and Ecotechnology*, 15, p. 100254. https://doi.org/10.1016/j.ese.2023.100254

Amobonye, A., Bendoraitiene, J., Peciulyte, L. and Rutkaite, R., 2025. Review of recent advancements in starch modification: Improving the functionality of starch-based films. *International Journal of Biological Macromolecules*, 315, p. 144354. https://doi.org/10.1016/j.ijbiomac.2025.144354

Appelqvist, I.A. and Debet, M.R., 1997. Starch-biopolymer interactions—a review. *Food Reviews International*, 13(2), p. 163–224. https://doi.org/10.1080/87559129709541105

Arias, A., Feijoo, G. and Moreira, M.T., 2021. Evaluation of starch as an environmental-friendly bioresource for the development of wood bioadhesives. *Molecules*, 26(15), p. 4526. https://doi.org/10.3390/molecules26154526

Aung, S.P.S., Shein, H.H.H., Aye, K.N. and Nwe, N., 2018. Environment-friendly biopolymers for food packaging: Starch, protein, and poly-lactic acid (PLA). In *Bio-based materials for food packaging: Green and sustainable advanced packaging materials*, Singapore: Springer. p. 173–195. https://doi.org/10.1007/978-981-13-1909-9_8

Avérous, L., and Pollet, E. (2012). Biodegradable polymers. In *Environmental silicate nano-biocomposites*, London: Springer. p. 13–39.

Bishop, G., Styles, D. and Lens, P.N., 2021. Environmental performance of bioplastic packaging on fresh food produce: A consequential life cycle assessment. *Journal of Cleaner Production*, 317, p. 128377. https://doi.org/10.1016/j.jclepro.2021.128377

Broeren, M.L., Kuling, L., Worrell, E. and Shen, L., 2017. Environmental impact assessment of six starch plastics focusing on wastewater-derived starch and additives. *Resources, Conservation and Recycling*, 127, p. 246–255. https://doi.org/10.1016/j.resconrec.2017.09.001

Clauser, N., Scown, C.D., Pett-Ridge, J. and Sagues, W.J., 2025. A techno-economic assessment of carbon dioxide removal pathways via biochemical conversion of lignocellulose to biofuels and bioplastics. *Renewable and Sustainable Energy Reviews*, 216, p. 115714. https://doi.org/10.1016/j.rser.2025.115714

Dutta, D. and Sit, N., 2024. A comprehensive review on types and properties of biopolymers as sustainable bio-based alternatives for packaging. *Food Biomacromolecules*, 1(2), p. 58–87. https://doi.org/10.1002/fob2.12019

Dybka-Stępień, K., Antolak, H., Kmiotek, M., Piechota, D. and Koziróg, A., 2021. Disposable food packaging and serving materials—Trends and biodegradability. *Polymers*, 13(20), p. 3606. https://doi.org/10.3390/polym13203606

Elkaliny, N.E., Alzamel, N.M., Moussa, S.H., Elodamy, N.I., Madkor, E.A., Ibrahim, E.M., Elshobary, M.E. and Ismail, G.A., 2024. Macroalgae bioplastics: a sustainable shift to mitigate the ecological impact of petroleum-based plastics. *Polymers*, 16(9), p. 1246. https://doi.org/10.3390/polym16091246

Fatima, S., Khan, M.R., Ahmad, I. and Sadiq, M.B., 2024. Recent advances in modified starch based biodegradable food packaging: A review. *Heliyon*. 10(6), p. e27453. https://doi.org/10.1016/j.heliyon.2024.e27453

Florencia, V., López, O.V. and García, M.A., 2020. Exploitation of by-products from cassava and ahipa starch extraction as filler of thermoplastic corn starch. *Composites Part B: Engineering*, 182, p. 107653. https://doi.org/10.1016/j.compositesb.2019.107653

Flórez, M., Cazón, P. and Vázquez, M., 2023. Selected biopolymers' processing and their applications: A review. *Polymers*, 15(3), p. 641. https://doi.org/10.3390/polym15030641

Folino, A., Pangallo, D. and Calabrò, P.S., 2023. Assessing bioplastics biodegradability by standard and research methods: Current trends and open issues. *Journal of Environmental Chemical Engineering*, 11(2), p. 109424. https://doi.org/10.1016/j.jece.2023.109424

Forfora, N., Azuaje, I., Kanipe, T., Gonzalez, J.A., Lendewig, M., Urdaneta, I., Venditti, R., Gonzalez, R. and Argyropoulos, D., 2024. Are starch-based materials more eco-friendly than fossil-based? A critical assessment. *Cleaner Environmental Systems*, 13, p. 100177. https://doi.org/10.1016/j.cesys.2024.100177

Fortune Business Insights, 2025. Biodegradable Mulch Film Market Size, Share & Industry Analysis, By Polymer (Thermoplastic Starch (TPS), Starch Blended with Polylactic Acid (PLA), Starch Blended with Polyhydroxyalkanoate (PHA), Aliphatic Aromatic Copolyesters (AAC), and Others), By Application (Fruits & Vegetables, Grains & Oilseeds, and Flowers & Plants), and Regional Forecast, 2024-2032. www.fortunebusinessinsights.com/biodegradable-mulch-film-market-102759

Fredi, G. and Dorigato, A., 2021. Recycling of bioplastic waste: A review. *Advanced Industrial and Engineering Polymer Research*, 4(3), p. 159–177. https://doi.org/10.1016/j.aiepr.2021.06.006

García-Rengifo, A.R., Rojas-Bringas, P.M., De-la-Torre, G.E. and Torres, F.G., 2022. Environmental impact of peanut skin-reinforced native starch foams modified by acetylation. *Environmental Quality Management*, 31, p. 89–99. https://doi.org/10.1002/tqem.21754

Ghasemlou, M., Barrow, C.J. and Adhikari, B., 2024. The future of bioplastics in food packaging: An industrial perspective. *Food Packaging and Shelf Life*, 43, p. 101279.https://doi.org/10.1016/j.fpsl.2024.101279

Glew, D., Stringer, L.C., Acquaye, A.A. and McQueen-Mason, S., 2012. How do end of life scenarios influence the environmental impact of product supply chains? Comparing biomaterial and petrochemical products. *Journal of Cleaner Production*, 29, p. 122–131. https://doi.org/10.1016/j.jclepro.2012.02.005

Gowthaman, N.S.K., Lim, H.N., Sreeraj, T.R., Amalraj, A. and Gopi, S., 2021. Advantages of biopolymers over synthetic polymers: Social, economic, and environmental aspects. In *Biopolymers and their industrial applications*, Elsevier. p. 351–371. https://doi.org/10.1016/B978-0-12-819240-5.00015-8

Grand View Research. 2023. Starch-based packaging market size, share & trends analysis report by product (bags & pouches, films & wraps, bottles), by end-use (food & beverages, e-commerce), by region (North America, Europe, Asia Pacific), and segment forecasts, 2025 – 2030. www.grandviewresearch.com/industry-analysis/starch-based-packaging-market-report#:~:text=Starch%2Dbased%20Packaging%20Market%20Trends,pollution%20and%20promote%20biodegradable%20alternatives

Guo, M., 2012. *Life cycle assessment (LCA) of light-weight eco-composites*, Singapore: Springer Science.

Guo, M., Stuckey, D.C. and Murphy, R.J., 2013. End-of-life of starch–polyvinyl alcohol biopolymers. *Bioresource Technology*, 127, p. 256–266. https://doi.org/10.1016/j.biortech.2012.09.093

Hannah Ritchie, 2023. "How much of global greenhouse gas emissions come from plastics?" Published online at OurWorldinData.org. https://ourworldindata.org/ghg-emissions-plastics

Hernández-Centeno, F., Hernández-González, M., López-De La Peña, H.Y., Rodríguez-González, C.A., Rios-Velasco, C., Zamudio-Flores, P.B. and Tirado-Gallegos, J.M., 2025. Functionalized starch biomaterials as a possible alternative in the future of food packaging: Advantages, disadvantages, and perspectives. In *Food security, safety, and sustainability*, Apple Academic Press. p. 225–260. https://doi.org/10.1201/9781003518440

Hottle, T.A., Bilec, M.M. and Landis, A.E., 2017. Biopolymer production and end of life comparisons using life cycle assessment. *Resources, Conservation and Recycling*, 122, p. 295–306. https://doi.org/10.1016/j.resconrec.2017.03.002

Howeler, R.H., 2004. Cassava in Asia: Present situation and its future potential in agro-industry. *Journal of Root Crops*, 30, p. 81–92. http://ciat-library.ciat.cgiar.org/articulos_ciat/CASSAVA_IN_ASIA___PRESENT_S.pdf

Islam, M.M., Chaudry, S., Thornton, A.W., Haque, N., Lau, D., Bhuiyan, M. and Pramanik, B.K., 2025. Environmental footprint of polylactic acid production utilizing cane-sugar and microalgal biomass: An LCA case study. *Journal of Cleaner Production*, 496, p. 145132. https://doi.org/10.1016/j.jclepro.2025.145132

Ita-Nagy, D., Vázquez-Rowe, I., Kahhat, R., Chinga-Carrasco, G. and Quispe, I., 2020. Reviewing environmental life cycle impacts of biobased polymers: Current trends and methodological challenges. *International Journal of Life Cycle Assessment*, 25, p. 2169–2189. https://doi.org/10.1007/s11367-020-01829-2

Irfan, F., Tasnim, N., Razzak, S.A. and Uddin, S., 2024. Advances in capturing carbon and producing bioplastics through the microalgal approach towards a sustainable future: A review. *Journal of Cleaner Production*, 486, p. 144576. https://doi.org/10.1016/j.jclepro.2024.144576

Lee, C.-S., and Hwang, H. S., 2023. Starch-based hydrogels as a drug delivery system in biomedical applications. *Gels*, 9(12), p. 951. https://doi.org/10.3390/gels9120951

Li, H., Qi, Y., Zhao, Y., Chi, J. and Cheng, S., 2019. Starch and its derivatives for paper coatings: A review. *Progress in Organic Coatings*, 135, p. 213–227. https://doi.org/10.1016/j.porgcoat.2019.05.015

Morais, A.M., Morais, R.M., Drew, D., Mustakhimov, I. and Lackner, M., 2022. Biodegradable bio-based plastics toward climate change mitigation. In *Handbook of Climate Change Mitigation and Adaptation (1987-2029)*. Cham: Springer International Publishing. https://link.springer.com/content/pdf/10.1007/978-3-030-72579-2_91.pdf

Mousavi-Avval, S.H., Sahoo, K., Nepal, P., Runge, T. and Bergman, R., 2023. Environmental impacts and techno-economic assessments of biobased products: A review. *Renewable and Sustainable Energy Reviews*, 180, p. 113302. https://doi.org/10.1016/j.rser.2023.113302

Moy, C.H., Tan, L.S., Shoparwe, N.F., Shariff, A.M. and Tan, J., 2021. Comparative study of a life cycle assessment for bio-plastic straws and paper straws: Malaysia's perspective. *Processes*, 9, p. 1007. https://doi.org/10.3390/pr9061007

Mtibe, A., Motloung, M.P., Bandyopadhyay, J. and Ray, S.S., 2021. Synthetic biopolymers and their composites: Advantages and limitations—An overview. *Macromolecular Rapid Communications*, 42(15), p. 2100130. https://doi.org/10.1002/marc.202100130

Müller-Carneiro, J., Figueir^edo, M.C.B. de, Rodrigues, C., Azeredo, H.M.C. and de Freire, F., 2023. Ex-ante life cycle assessment framework and application to a nano-reinforced biopolymer film based on mango kernel. *Resources, Conservation and Recycling*, 188 https://doi.org/10.1016/j.resconrec.2022.106637

Muniyasamy, S., Anstey, A., Reddy, M.M., Misra, M. and Mohanty, A., 2013. Biodegradability and compostability of lignocellulosic based composite materials. *Journal of Renewable Materials*, 1(4), p. 253–272. https://doi.org/10.7569/JRM.2013.634117

Nawab, A., Alam, F., Hadi, A. and Lutfi, Z., 2022. Development and characterization of edible film made from mango kernel starch. *Journal of Packaging Technology and Research*, 6(1), p. 63–72. https://doi.org/10.1111/1541-4337.12265

Olawumi, M.A., Omigbodun, F.T. and Oladapo, B.I., 2024. Integration of sustainable and net-zero concepts in shape-memory polymer composites to enhance environmental performance. *Biomimetics*, 9(9), p. 530. https://doi.org/10.3390/biomimetics9090530

Papong, S., Malakul, P., Trungkavashirakun, R., Wenunun, P., Chom-In, T., Nithitanakul, M. and Sarobol, E., 2014. Comparative assessment of the environmental profile of PLA and PET drinking water bottles from a life cycle perspective. *Journal of Cleaner Production*, 65, p. 539–550. https://doi.org/10.1016/j.jclepro.2013.09.030

Pinaeva, L.G. and Noskov, A.S., 2024. Biodegradable biopolymers: Real impact to environment pollution. *Science of The Total Environment*, 947, p. 174445. https://doi.org/10.1016/j.scitotenv.2024.174445
Plastics Europe, 2023. Plastics – the fast Facts 2023. https://plasticseurope.org/knowledge-hub/plastics-the-fast-facts-2023/
Qun, Z.H.O.U., Rui, Y.U.A.N., Zhang, W.Y., Gu, J.F., Liu, L.J., Zhang, H., Wang, Z.Q. and Yang, J.C., 2023. Grain yield, nitrogen use efficiency and physiological performance of Indica/Japonica hybrid rice in response to various nitrogen rates. *Journal of Integrative Agriculture*, 22(1), p. 63–79. https://doi.org/10.1016/j.jia.2022.08.076
Ragitha V. M. and Edison, L.K., 2022. Safety issues, environmental impacts, and health effects of biopolymers. In *Handbook of biopolymers*, Singapore: Springer Nature Singapore. pp. 1–27. https://doi.org/10.1007/978-981-16-6603-2_54-1
Ranum, P., Peña-Rosas, J.P. and Garcia-Casal, M.N., 2014. Global maize production, utilization, and consumption. *Annals of the New York Academy of Sciences*, 1312(1), p. 105–112. https://doi.org/10.1111/nyas.12396
Rashwan, A.K., Younis, H.A., Abdelshafy, A.M., Osman, A.I., Eletmany, M.R., Hafouda, M.A. and Chen, W., 2024. Plant starch extraction, modification, and green applications: a review. *Environmental Chemistry Letters*, 22(5), p. 2483–2530. https://doi.org/10.1007/s10311-024-01753-z
Seth, A. K., Vellaparambil Hariharan, S. and Jaideva, R. (2025). Harnessing banana peel waste: development of starch-based bioplastics for green packaging. *Starch-Stärke*, 77, e202400178. https://doi.org/10.1002/star.202400178
Sheldon, R.A., 2024. Waste valorization in a sustainable bio-based economy: The road to carbon neutrality. *Chemistry–A European Journal*, 30(54), p. e202402207. https://doi.org/10.1002/chem.202402207
Sobeih, M. O., Sawalha, S., Hamed, R., Ali, F. and Kim, M. P. (2025). Starch-derived bioplastics: Pioneering sustainable solutions for industrial use. *Materials*, 18(8), p. 1762. https://doi.org/10.3390/ma18081762
Tamoor, M., Samak, N.A. and Xing, J., 2023. Life cycle assessment and policy for the improvement of net-zero emissions in China. *Cleaner Engineering and Technology*, 15, p. 100663. https://doi.org/10.1016/j.clet.2023.100663
Valle, C., Voss, M., Calcio Gaudino, E., Forte, C., Cravotto, G. and Tabasso, S. (2024). Harnessing agri-food waste as a source of biopolymers for agriculture. *Applied Sciences*, 14(10), p. 4089. https://doi.org/10.3390/app14104089
Varghese, S., Awana, M., Mondal, D., Rubiya, M.H., Melethil, K., Singh, A., Krishnan, V. and Thomas, B., 2022. Amylose–amylopectin ratio: Comprehensive understanding of structure, physicochemical attributes, and applications of starch. In *Handbook of biopolymers*, Singapore: Springer Nature Singapore. p. 1–30. https://link.springer.com/content/pdf/10.1007/978-981-16-6603-2_48-1.pdf
Wang, F., Harindintwali, J.D., Yuan, Z., Wang, M., Wang, F., Li, S., Yin, Z., Huang, L., Fu, Y., Li, L. and Chang, S.X., 2021. Technologies and perspectives for achieving carbon neutrality. *The innovation*, 2(4). https://doi.org/10.1016/j.xinn.2021.100180
Wikifarmer. 2020. *Yield, Harvest, Storage of Wheat*, European Union. https://wikifarmer.com/library/en/article/yield-harvest-storage-of-wheat
Xiang, H., Li, Z., Liu, H., Chen, T., Zhou, H. and Huang, W., 2022. Green flexible electronics based on starch. *NPJ Flexible Electronics*, 6(1), p. 15. https://doi.org/10.1038/s41528-022-00147-x
Ye, F., Li, J. and Zhao, G., 2020. Physicochemical properties of different-sized fractions of sweet potato starch and their contributions to the quality of sweet potato starch. *Food Hydrocolloids*, 108, p. 106023. https://doi.org/10.1016/j.foodhyd.2020.106023
Zhao, X., Wang, Y., Chen, X., Yu, X., Li, W., Zhang, S., Meng, X., Zhao, Z.M., Dong, T., Anderson, A. and Aiyedun, A., 2023. Sustainable bioplastics derived from renewable natural resources for food packaging. *Matter*, 6(1), p. 97–127. https://doi.org/10.1016/j.matt.2022.11.006
Zheng, J. and Suh, S., 2019. Strategies to reduce the global carbon footprint of plastics. *Nature Climate Change*, 9(5), p. 374–378. https://doi.org/10.1038/s41558-019-0459-z
Zhu, F., 2017. Barley starch: Composition, structure, properties, and modifications. *Comprehensive Reviews in Food Science and Food Safety*, 16(4), p. 558–579. https://doi.org/10.1111/1541-4337.12265

Index

M

N